The Working Mind

The Working Mind

Meaning and Mental Attention in Human Development

Juan Pascual-Leone and Janice M. Johnson

The MIT Press
Cambridge, Massachusetts
London, England

The open access edition of this book was made possible by generous funding from Arcadia – a charitable fund of Lisbet Rausing and Peter Baldwin.

ARCADIA
A charitable fund of Lisbet Rausing and Peter Baldwin

This book was set in Stone Serif and Stone Sans by Westchester Publishing Services. Printed and bound in the United States of America.

Library of Congress Cataloging-in-Publication Data

Names: Pascual-Leone, Juan, author. | Johnson, Janice M., author.
Title: The working mind : meaning and mental attention in human development / Juan Pascual-Leone and Janice M. Johnson.
Description: Cambridge : The MIT Press, 2021. | Includes bibliographical references and index.
Identifiers: LCCN 2020027072 | ISBN 9780262045551 (hardcover)
Subjects: LCSH: Developmental psychology. | Cognition. | Task analysis.
Classification: LCC BF713 .P378 2021 | DDC 155.4/13--dc23
LC record available at https://lccn.loc.gov/2020027072

10 9 8 7 6 5 4 3 2 1

Contents

Preface vii

Acknowledgments xiii

I Foundations of a Causal Constructivist Theory: Semiotic Processes

1 Dialectical Constructivism: The Working Mind Underlying Working Memory 3

2 Problems of Cognitive Developmental Theory 33

3 Emergence of Mental Attention in Infancy: From Sensorimotor to Symbolic Processing 61

4 Meaning, Mental Attention, and the Symbolic Function 119

II Theory of Constructive Operators (TCO)

5 Schemes/Schemas and Their Causal Constructivist Learning 141

6 Automatic Attention: Effortless, Perceptual, and Personal 181

7 Mental Attention, Intelligence, and Consciousness 205

III TCO, Task Analysis, and Neuroscience

8 Process Analysis and Mental Task Analysis: Foundations 257

9 Process and Mental Task Analysis: Methods and Examples 289

10 The Working Mind Inside a Working Brain: A Neuropsychological Introduction 317

11 The Working Mind Inside a Working Brain: Functional Dimensions in Brain Semiotics 359

12 The Working Mind Model across Human Domains 397

Appendix: On Metasubjective Task Analysis (MTA) 413
Glossary 421
Notes 427
References 435
Index 471

Preface

This book highlights a general process model of the psychological organism "from within" (i.e., from a subject's own processing perspective). It takes the form of a scientific essay with explanations, model definitions, and illustrations of relevant empirical data across periods of child development. It does not review systematically relevant literature. We believe this theory of the "psychological organism" (Theory of Constructive Operators—TCO), and its methods of process/task analysis, can advance understanding of human intelligence, mental attention, problem solving, stages of development, cognitive complexity measurement, brain semantics, and the constructivist *causal-organismic* tradition within human science (which recognizes psychological processes as rooted or embodied in the organism, and having multiple organismic causes that interact and combine, see chapter 1). The theory also can help to explain motivation and the intertwining of cognitive and affective processes. Our method of process/task analysis "from within" the person's processing (*metasubjective* analysis) can expand qualitative, subjective analysis (e.g., phenomenology, hermeneutics) into causal-organismic models of processing in mental performance or tasks, often yielding quantitative, interval-scale estimates of process or task complexity. We use precise terminology to express our ideas, and the reader may find some of it unusual. To facilitate familiarization with terminology, we include a glossary at the end of the book.

By *psychological organism* (or psychological brain) we mean a general model of sensorimotor or mental processes analyzed "from within," to explain complexities of human performance. Such a general model could provide operative tools for real-life behavior or mental analysis, study of thinking, and scientific investigation. Epistemologically, the TCO model is mainly based on European (particularly French) constructivist epistemology, that is, an empirically grounded and critical rationalism. Our work has an epistemological influence from A. R. Luria (1973, 1979), which is why we adopted the title *The Working Mind*, to tacitly suggest Luria's famous book, *The Working Brain*. We present the TCO model informally, but in an explicit manner, open to expansion. *Rationalist epistemology* (meta-theory of knowing) primordially considers knowledge a product of the working mind, whereas *empiricist epistemology* interprets knowledge mostly as modeled by constraints of outer Reality.[1] Constructivism is a dialectical middle way for these two epistemologies.

The two complementary types of epistemology/metatheory in human science are empiricism (or more precisely *meta-empiricism*—empirical-method empiricism with much theoretical sophistication) and *constructivism* (empirical rationalism). Empiricist or meta-empiricist theories tend to be local and descriptive, whereas constructivist ones are general and causal-organismic. The TCO is a constructivist general-organismic causal model. The two types (in method and theory) are incommensurable: we cannot present one by using presentational methods of the other. For instance, one does not clarify constructivist theories by critically reviewing empiricist research and judging whether constructivism meets empiricist constraints. Like the theory of evolution, constructivist theories must be argued using metatheoretical biological arguments and by interpreting concrete and diverse examples that use the same constructivist organismic-causal model—something we do repeatedly in this book.

From this constructivist middle position, knowledge seems to emerge with the mind's repeated syntheses and compromises between Reality and the psychological organism's own organizing principles and operators (brain resources). Notice that Reality (with capital R to signal uninterpreted actual reality) can be regarded as packages of interrelated Resistances to goal-directed activity, as we discuss in detail in chapter 5. An important developmental construct is mental-attentional capacity. This is a maturational causal factor of working memory, constituted by several resource processes that we call hidden operators (see chapter 7). The capacity of mental attention grows with age in childhood, allowing children to cope with increasing complexity, particularly within misleading situations (in which some aspects strongly induce unsuitable task responses or habits).

Seven Founding Principles

Our dialectical constructivism (dialectical because it has dynamic competitive processes in it) gives special importance to developmental (including learning) and evolutionary processes, to create models and knowledge about the human biological and psychological organism. We highlight seven founding principles.[2]

1 Reflective Abstraction

Reflective abstraction, as Piaget called it (Vygotsky's internalization), brings into the conscious and unconscious mind recurrent aspects (i.e., functional regularities, probabilistic "invariants") from active experience or goal-directed activity in the world, and from directed mentation (the mind's work). Learning takes place by adapting innate or acquired schemes to actual situations of agency/praxis, and current thought. Such mental construction is a dialectical enterprise. This adaptation will incorporate, within a functional invariant, task-relevant *resistances* or restrictions conjointly imposed by the Real to the actions and cognitions of agents. Piaget called *accommodation* this adaptive change to Reality. Reality resistances, when facilitating for a task, are called affordances; when misleading or distracting, they are called obstacles or encumbrances. Notice that Reality prior to knowing can be construed, for any given individual, as constituted by many distinct coordinated packages of mutually entwined resistances, elicited during agency/praxis; these packages are species-specific.

2 Individual Constructivism

Individual constructivism is recursive and provokes emergence of distinct levels of processing (some sensorimotor, other symbolic) that interact to cause functional structures for knowing (i.e., schemes, schemas). Schemes adapt to increase their effective complexity as individual development progresses. These processes are qualitatively recursive (in the mathematical sense) because they apply on their own products—schemes of schemes of schemes, and so forth. These complex schemes (often called schemas) are organized in the person at functionally distinct semantic-pragmatic levels, to constitute the substratum for valid information processing. Piaget, Case, and many others call this diverse substratum *functional structures*.

Complexity in constructivism corresponds to what physicist Murray Gell-Mann (1994) called *effective complexity*, that is, the nonrandom aspects of a system "roughly characterized as the length of a concise description of the regularities of that system or string" (Gell-Mann, 1994, p. 50). He concluded: "Effective complexity is then related to the description of regularities of a system by a complex adaptive system [such as a person] that is observing it." Gell-Mann understood by "regularities," as Piaget did, the relevant probabilistic *functional invariants* (Ullmo, 1967) that emerge during repeated situations, generating relational meaning. Tolman (1959) called these task-relevant interrelations "means-end readinesses." Random aspects vary, but functional invariants are preserved over repetitions.

Concepts, repeatable patterns of action, stable representations, and repeatable relations are examples of invariants. These schemes often carry and produce expectancies.

The effective complexity of schemes/schemas relates to the number of invariants ("regularities") essential to characterize the given experience (e.g., person, object, thought, performance, situation).

To maintain effective reasoning, abstraction within misleading situations always demands mental-attentional capacity. Mental capacity is a maturational causal factor of working memory, constituted by several hidden operators (see chapter 7). The theory predicts that power of mental attention generally increases by one unit (one additional scheme that can be held in mind) every two years during normal childhood (Pascual-Leone, 1970). We present data to validate this prediction. We also outline (and give numerous examples of) a method of task analysis to quantify mental-attentional demand of tasks. Effective complexity of schemes involved in a task is both objective and *metasubjective*—it should be analyzed objectively, but "from within" the intended agency (e.g., person, robot) coping with the task.

3 Constructivist Complexity from Combination of Different Schemes

Growth of effective complexity during development is action-driven and holistic (schemes coordinate toward *functional totalities* of perceiving, thinking, or doing). However, inside the brain these functional totalities appear deconstructed into units (circuits or networks that are information carriers of schemes) or discrete but interrelated functional systems; psychologically these units are schemes. Schemes must be self-consistent (noncontradictory) in their internal functioning. Once formed, they are recursive, constituting flexible hierarchies or families of schemes.

4 Schemes Are Self-Propelling

Schemes are self-propelling (Piaget's assimilation function), and consequently, each behavioral, perceptual, or cognitive act is overdetermined by the strongest cluster of mutually compatible schemes activated in the current situation. *Overdetermined* means multidetermined by schemes and hidden operators that interact in nonlinear ways to produce performance. This is a multidetermination in which a dynamic synthesis produces truly novel results that are beyond a simple combination of constituent factors (see chapters 1 and 6). The dominant set of compatible schemes in a situation overdetermines overt or mental behavior.

5 Functionality of Schemes

Cognitive schemes are functional systems that carry with them a *truth value*. They tacitly make inferences or expectancies anticipating results/outcomes of their application (i.e., a plausible or probabilistic truth, the match of expected effects with encountered

Reality). Schemes are formed by extracting repeatable functional components from the given sort of agency (or thinking) from which they were abstracted, and to which they later can be applied. This functional component carries with it (in sufficiently complex schemes) the inference or expectancy of ensuing results. *Affective schemes* (pure feelings) do not carry a truth value; they carry a *vital value* (i.e., a life-significant organismic appraisal or feeling evaluation).

6 Emergence of Conflicts among Schemes

Schemes are self-propelling, autonomous, and internally consistent, but they are often in contradiction with other schemes. When several mutually contradictory schemes are activated and compete for application, conflict or misleadingness emerges. These problem-situations are where maturational levels or stages of child development can be found. Stages are largely caused by complexity of tasks beyond the mental-attentional capacity of age groups of children.

7 Hidden Organismic Operators and Principles Regulate Schemes

Dynamic coordination and synthesis of schemes, and their activation in tasks or learning processes, are regulated by hidden operators and principles—brain infrastructural resources. These operators and principles relate to constructs from the psychology and neuroscience literatures.

Our book discusses schemes/schemas, hidden operators, and principles in detail, as well as how together they organismically overdetermine performances. We illustrate this "from within" (metasubjective) theory with numerous task analyses of distinct age-group performances. The theory is comprehensive in that it incorporates ideas and findings from many diverse branches or schools of psychology: Piagetian theory, learning theory, Gestalt theory, psychoanalysis, cognitive psychology, and neuroscience. We do not aim to reject, replace, or dismiss other approaches, but rather recognize their value and attempt to integrate their findings into a more general, causal-organismic theory of development.

Acknowledgments

Although very personal, this book reflects a collective enterprise: We owe much to mentors, scholars, students, and friends. This collective helped and influenced us to make the TCO model and book possible. Early mentors for Juan Pascual-Leone were his professors in the Faculty of Medicine (Valencia, Spain) and Casa Salud Valdecilla (University of Valladolid, Spain); Jean Piaget and Barbel Inhelder (University of Geneva, Switzerland); and Herman Witkin (SUNY Downstate Medical Center, New York). Others who most directly contributed their empirical or theoretical research, advice, or personal influence include Robbie Case, Anik deRibaupierre, Sergio Morra, Nancy Benson-Hamstra, Ronald Miller, Glenys Parkinson, Doba Goodman, Jud Burtis, Leslie Greenberg, John Todor, Sandra Cunning, Nancie Im-Bolter, Alba Agostino, Enzo Verrilli, Marie Arsalidou, Mariela Giuliano, Cheryl Lee, Antonio Pascual-Leone, Steven J. Howard, Michael Shayer, Andreas Demetriou, and many others.

This list includes M.A. and Ph.D. psychology students, most of them from York University, whose work with us (often unpublished) has been important to advance our theorizing. A small group of senior undergraduates from York deserves mention and gratitude: the volunteer members of the "Book Club" seminar at our Developmental Processes Lab. They assisted, discussed, and advised on the chapters; drafted figures and tables; and helped with references. The Book Club members are Dana Burlac, ZiYi Chua, Xuan Feng, Yehuda Gabler, Vladyslav Glezin, Adriana Milani, and Rodrigo de Urioste. They have made a major contribution to this book. We also benefited from advice of scholars and researchers who read and commented on chosen chapters (Marie Arsalidou, Ronald Miller, Alvaro Pascual-Leone Sr., Antonio Pascual-Leone, M. Angeles Cerezo, Alvaro Pascual-Leone Jr., and finally, four insightful anonymous reviewers who read and commented on the book). We are grateful for their generous donation of time that led to many book improvements.

Our research was funded for many years by grants from the Social Sciences and Humanities Research Council of Canada, Natural Sciences and Engineering Research Council of Canada, and internal grants from York University.

Two people personally inspired our many years of working and writing on this book: Dr. Alvaro Pascual-Leone Sr., JPL's late twin brother, and our son/stepson Dr. Antonio Pascual-Leone. This book is dedicated to Alvaro and to our sons/stepsons (Antonio, Rafael, & Carlos).

I Foundations of a Causal Constructivist Theory: Semiotic Processes

1 Dialectical Constructivism: The Working Mind Underlying Working Memory

A brief description of the working mind, formulating epistemological notions relevant to model it organismically, is presented. We discuss the need to adopt in constructivist science a perspective "from within" the subject matter (or the subject). Causal mechanisms of organismic dynamics are introduced, among them resistances of reality to human action and the overdetermination process that synthesizes human action. Constructivist-learning (without mental attention) versus constructivist maturational-attention theories are discussed. We examine organismic-causal processes that explain emergence of working memory and task analyze the Raven Matrices Test to illustrate how constructivist causal processes can explain working memory and the emergence of fluid intelligence.

> One demonstrates the real, one cannot show it.... In fact, because the object appears as a complex of relations, one must capture it by multiple methods.
>
> —Bachelard, 1934/1987, p. 16, translated by JPL

> There is indeed a problem: On which factors depend initial disequilibrations? This was a new problem for us because heretofore we had mistakenly considered initial disequilibrations as not needing explanation or resulting from difficulties of synthesis.
>
> —Piaget, 1974, p. 155, translated by JPL

> For Hegel... the power of the spirit lies in synthesis as the mediation of all contradictions.
>
> —Gadamer, 1971, p. 105

The *working mind* is the active unconscious or conscious processes (affective and cognitive) that synthesize meaningful experiences and produce problem solving and learning. These processes involve mental attention directed to the person's own thoughts, whether addressed to the external situation or internal states. They also involve associative knowledge and perceptual attention driven by expectancies, which stem from inferentially constructed knowledge schemes. This working mind is the product of a *working brain* (Luria, 1973). To model a working mind, one must adopt a perspective from within the subject's processes. Such description should include internal and external processes, interpreting external situations and tasks in terms of the mind/brain working processes involved. Explicitly functional neuropsychological theorizing (modeled from within the subject's mind) is necessary. This sort of macro theorizing can be interpreted within the working brain as neural circuits and brain regulations, as we attempt to do in chapters 10 and 11.

The preface provides an overview of the theory. As mentioned there, we use the term *metasubjective* for our approach to modeling processes *from within* (Pascual-Leone, 2013; Pascual-Leone, Pascual-Leone, & Arsalidou, 2015; Pascual-Leone & Johnson, 2017). This approach aims to capture in explicit process models the evolving *functional totality* of a subject's inner processes. Here, *totality* stands for all intercoordinated processes that in the subject produce and control behavior. We call it *functional*, because this totality of processes is psychologically described by the structure of the process's own functioning. Our approach contrasts with often-adopted modeling from an *observer's perspective*, "from outside" the subject.

The Copernican Revolution can be used to illustrate the constructivist or metasubjective versus the observer's meta-empiricist approaches collaborating to advance knowledge in science (Pascual-Leone, 2012b, 2013; Pascual-Leone & Johnson, 2017). Our interpretation of Feyerabend (1978) suggests that the insight that brought Copernicus to his heliocentric viewpoint was a metasubjective turn (i.e., toward dialectical constructivist thinking). Ptolemy's intuitions about movement of the sun and stars were more congruent with the everyday experience of people who look at the sky and watch the sun and moon move about with them. Copernicus, however, rejected the observer's perspective and placed himself intuitively within the cosmos itself: he experienced the dynamic object-interaction and activity of this cosmos from within it. He anticipated rationally, against sensorial appearances, that taking the sun as rotation axis in our planetary system is a more congruent solution for this dynamic functional totality—a simpler solution for the empirically found repeatable (inter-) relations (Ullmo, 1967). At the time the Ptolemaic system could indeed make more precise, albeit empiricist, predictions than Copernicus's own (Feyerabend, 1978). Copernicus's metasubjective turn promoted a new *constructivist* form of empirical research from within the cosmos itself (not just an external observer's analysis as in Ptolemy). This new way of making science provides a causal process account: placing emphasis on internal constraints and external resistances of nature as a functional totality, not on the local empirical appearances (Bachelard, 1981).

Similarly, in metasubjective psychology what are relevant are the internal constraints that the organism brings to a task, as well as the reality resistances opposed to the organism and its activities. *Resistances* are obstacles to goal-directed action, caused by situations[1] or our organism. Metasubjective theories help to clarify such resistances. Talking about science, Bachelard (1980) said, "The first specific instance of the notion of matter is the *resistance*" (p. 10; translated by JPL). Metasubjective analysis can help us to model the internal constraints and resistances of the organism, as a dynamic totality within the task.

This book offers, in the form of a scientific essay, a metasubjective general theory/model of neuropsychological processes. This theory is dialectical-constructivist in its epistemology and neo-Piagetian in its origin. The focus of the theory is developmental: explaining how and why the working mind's ability to cope with truly novel situations (often called *problem solving*) increases with age, demarcating stages. It does so by using constructs (schemes and organismic regulations—hidden constructive operators) that are constant functional manifestations of the working brain. We consider the theory to be organismic and constructivist-causal.

The term *organismic* (Pascual-Leone, 1984) was introduced by Goldstein (1934/2000; Werner & Kaplan, 1984), who spoke of "equalization" processes (in the sense of Piaget's later "equilibration") of the "organism as a whole." *Organismic* here refers to the organism as a very active, organized functional totality with its own essential nature, which is purposeful and dialectically driven toward maintaining balance/equilibration of the working mind (Pascual-Leone, 2014). Working-mind processes are rooted in the brain and can be modeled from within as discrete macroprocesses that codetermine task performances. Such modeling is *metasubjective analysis* (Arsalidou, Pascual-Leone, & Johnson, 2010; Pascual-Leone, 1995, 2013; Pascual-Leone et al., 2015).

A theory or model is *descriptive* when it offers ways to express encountered phenomena and structural findings. In contrast, a model is *causal* when its constructs are distinct from descriptive constructs to be explained and can be independently anchored on experience via experimentation. These causal constructs account for change that descriptive constructs (data) undergo with experience, maturation, and organismic transformations. Note that by "causal" we mean organismic causal overdetermination, as discussed below. Descriptive and causal theories or models can be *local* versus *general* (Pascual-Leone, 1978). These distinct sorts of theories/models are all jointly needed. Combined, they yield two dimensions of variation (i.e., local vs. general and descriptive vs. causal) that can be crossed. Thus, simplifying, there are local descriptive, general descriptive, local causal (Pepper's, 1942, "mechanistic"), and general causal

(Pepper's "organismic") theories or models. This is important because the more general a causal theory is, the more distinctly differentiated it will be from the descriptive structural theories it aims to explain and coordinate. The more local, the less differentiated causal theories will tend to be from their descriptive theories, eventually leading the distinction to collapse (Pascual-Leone & Johnson, 2005). Note that a causal theory/model (unlike purely descriptive ones) must have an explicit sequential account of how change-as-process occurs: showing how consequent conditions emerge via causal overdetermination from context and antecedent organismic conditions. So defined, organismic-causal general models are much needed but rare in cognitive development or in psychometrics (Gottfredson, 2016).

It is important for a causal theory to address problem solving. From an organismic-causal perspective, problem-solving processes are those that dynamically synthesize *truly novel* (external or mental) *performances*. These are novel performances that are not directly learned nor maturationally acquired, nor are they automatic results of learned coordination. They result instead from "creative" *dynamic syntheses*, often generating truly novel complex schemes that can help to solve intended problems and remain in the person's repertoire (long-term memory) as potential components of future solution alternatives (see Shipstead, Lindsey, Marshall, & Engle, 2014). As Gestalt psychologists (Koffka, 1935/1963) and others have assumed, dynamic syntheses in misleading situations (typical of problem solving) result from various *organismic resource factors* (general-purpose brain operators) whose interactions cause representational and operative syntheses, making intelligence, the symbolic function, and learning possible. Our theory has four distinct characteristics: it is founded epistemologically in dialectical constructivism; it is metasubjective; it yields a powerful, truly novel method of metasubjective task analysis; and its constructs are interpretable in the working brain.

Why Dialectical Constructivism?

A theory of development and learning is *constructivist* when it minimizes (albeit recognizing) both content-bound maturational predeterminations and passive empiricist (i.e., simple associative) learning, emphasizing instead the learning/internalization from experience of *functional invariants* (i.e., recurrent functional patterns) that express or embody Resistances that Reality presents to the person's agency or praxis (a view pioneered by Cassirer, 1923/1953, 1938, 1929/1957; Gonseth, 1936/1974; Ullmo, 1967). Such a theory creates suitable, situated internal models that emphasize innate biogenetic determinants enabling adaptation via constructivist learning. These models serve as dynamic functional structures for *agency* or *praxis* (i.e., conscious or unconscious goal-directed activity addressed to the environment), structures that cope with types of situations. Cassirer called them symbolic forms (Cassirer, 1929/1957, 1944/1966), and Piaget either *schemes* or *functional structures/schemas* (more recently called chunks by empiricists)—all of them stable or invariant coordinations of schemes. These internal models are the result of neural plasticity, which internalizes suitably adapted, relevant, probabilistically invariant patterns of coping with experience, thus increasing life adaptation and survival.

The organism encounters anomalies in its (implicit or explicit) expectations about reality. Such anomalies occur whenever situations offer unexpected Resistances to the subject's agency/praxis (*Resistances* in capitals, to signify Reality as such). Piaget and others called these Resistances *perturbations*. Perturbations elicit in the organism an endogenous process of functional arousal, with evaluations and change (often unconscious) toward re-equilibration. This is part of what Piaget called *optimizing equilibration* (successful adaptive rebalancing of the working mind).

As both Piaget and Tolman emphasized, meaning-bearing processes in the organism (i.e., Piaget's schemes and structures/schemas) carry within them *expectancies* about what should happen next, conditional to what has happened before. In developmental constructivism the meaning-bearing processes are unitized into schemes/schemas that carry local expectancies about what leads to what in the current situation. They are packaged into unitized schemes, that is, functional dynamic totalities, local mini-systems. Because schemes are distinct and relatively segregated, they are able to functionally differentiate between distinct and segregated Resistances of the situation, thus allowing emergence of selective adaptive expectancies. This is so whether these expectancies bear on external reality or on results from one's actions (examples are Tolman's [1959, 1961] means-end expectancies and Piaget's operations).

When expectancies are violated by experience (errors of anticipation), a loss of organismic dynamic balance occurs. Piaget (1975/1985) called it *disequilibration*. It produces organismic arousal that mobilizes both automatic-perceptual and mental attention to (perhaps automatically) explore and search for concrete differences between the new situation (in which violation of expectancy has happened) and other "normal" situations that conform to the expectancy. This is the organism's initial attempt to resolve anomalies and restore equilibrium. Such an organism is constructivist: it seeks to increase adaptation by reworking or reorganizing meaning-bearing (cognitive or affective) inner processes (schemes) to resolve anomalies, eliminate resistances, and simplify processing.

Notice the epistemological affinities of this developmental constructivism with current dynamic field theories, dynamic system theories, or computational rationality

(Gershman, Horvitz, & Tenenbaum, 2015; Schöner, 2014; Spencer, Perone, & Buss, 2011). They are all dynamic process theories that produce performances by synthesizing in situ multiple distinct processes (in our case, schemes) that, in their coordination, make the (internal or external) performance emerge. The big difference of developmental/dialectical constructivism, as we present it in this book, is the qualitative-process schemes or schemas (described above and in chapter 5), together with explicit organismic hidden operators and principles that provide a causal account for dynamic descriptive formulations offered by other field or system probabilistic-process approaches.

Piaget's constructivism was a biologically grounded *empirical rationalism*, which envisioned human organisms as seeking to adapt by developing internal intelligent models that cause re-equilibrations, which in turn expand the scope of situations in which the organism can be well adapted (Pascual-Leone, Escobar, & Johnson, 2012). For clarity, we should distinguish two kinds of constructivism. The first, *categorical constructivism*, is clearly found in Piaget's early writings (up to the late 1960s), as it is found in Kant's and (often tacitly) in that of many current cognitive scientists. This is a constructivism in which higher-order organizing principles or categories are posited top-down by dictum and used as organizers without sufficient developmental, learning, or evolutionary justification. In Kant (1929/1965) the transcendental intuitions of space and time and transcendental categories such as quality, relation, substance, quantity, causality, and so forth arguably play this role, making experience possible.

In Piaget's writings there are stage-defining descriptive *psycho-logic models* meant to express metasubjective operative processes. They are inferred from functional-relational patterns in data across subjects, whose temporal change Piaget characterized into developmental stages: sensorimotor, preoperational, concrete operational, and formal operational. These models were formulated in interaction with dynamic/causal functional-system categories, such as schemes/structures, assimilation versus accommodation, physical/empirical abstractions versus reflective abstractions, equilibration, disturbances, and so on. Nonetheless these dynamic/causal categories are not well coordinated with Piaget's descriptive psycho-logical stage models. His causal process for stage changes is not formulated explicitly, leaving the theory causally inadequate. A more explicit *dialectical constructivism* is needed. This is the second kind of constructivism.

Dialectics (Mihalits & Valsiner, 2020; Pascual-Leone, 2014) is a mode of reasoning that emphasizes change within, and interactions among, process constituents (dynamic constructs) to bring about balance or coordination. This form of thinking may help to explain conflicts and competition among constituents and therefore the emergence of truly novel outcomes. To clarify dialectical constructivist epistemology, we critically analyze Piaget's approach. We do so because these criticisms also apply to many empiricist cognitive theories. Note that constructivist theories are empirical although not empiricist.

The later Piaget[2] explicitly became a dialectical constructivist (Inhelder, Garcia, & Voneche, 1977; Piaget, 1980, 1985) and made some attempt to coordinate his psychological models with dynamic process constructs; however, he never achieved a detailed integration (e.g., Inhelder et al., 1977; Piaget, 1974, 1975/1985; Piaget & Garcia, 1987). This problem is illustrated by his failure to formulate explicit and separate causal mechanisms to account for conflict (dialectical contradictions and negations) and explain equilibration processes (e.g., the third epigraph). Such omissions are fundamental, because negations/conflicts, re-equilibrations, and disequilibrations express central causal mechanisms in Piaget's and other dynamic theories (Inhelder et al., 1977; Piaget, 1974, 1975/1985). Indeed, as Piaget explicitly stated (in Inhelder et al., 1977), "What the notion of equilibration adds is, in contrast, the causal dimension; it is the work of the subject himself… to speak of progressive equilibration is to give the causal process (psychologically formulated) that generates structures, constructing them step by step" (p. 114, translation by JPL).

Sixteen lines later he added: "Thus what is causal in the theory of equilibration is the attempt, whether succeeded or failed, to explain what mathematicians *do* and not the nature of logico-mathematical structures" (p. 114, translation by JPL). However, attempts to explain *how* children (or mathematicians!) come to solve Piagetian tasks were not successful within Piagetian theory, as his collaborators and others have recognized (Inhelder et al., 1977; Valsiner, 2006). In his later work Piaget (1980) seriously sought a dynamic/dialectical reformulation of his theory, stating that "dialectics constitutes the inferential mechanism of equilibration" (p. 223, translation by JPL).

To understand dynamic theories of metasubjective change, including Piagetian theory, we should functionally explain dialectics and dialectical systems. Dialectics (a qualitative precursor of dynamic-systems analysis) studies competition among processes. Dialectics means the dynamic interaction that occurs between two or more sources (or distinct causal factors/determinants) of process, which often are in competition or mutual contradiction, even though they may functionally complement one another to codetermine outcomes. Leonardo da Vinci, in his analysis of the architectural arch, provides us with an example illustrating essential characteristics of dialectical systems. Reflecting on this milestone of architectural ingenuity, used since Roman times, da Vinci said, "What is an arch? An arch is nothing else than a strength caused by two weaknesses… as one withstands the downfall of the other the two weaknesses are converted into a single strength" (da Vinci, 1959, p. 210).

Each segment of an arch, due to gravity, thus exemplifies a dynamic constituent of this functional totality whose normal output is to fall. However, the two segments oppose each other (are in contradiction), and each in fact regulates or adaptively modifies the effect of the other. Together they cause a new outcome or invariant, the cancellation of the effects of gravity resulting in stability of the arch. Thus described, an arch illustrates a minimal dialectical system constituted by two subsystems, the arch segments, in dynamic interaction. We will say that two or more dynamic systems constitute an overall functional totality or *dialectical system* (**DS**) whenever:

(**DS1**) each (sub) system is *contradictory* with the others in its functional effects

(**DS2**) effects of each of them *regulate*, compensate, or adaptively modify effects of the others

(**DS3**) all of them are jointly needed to generate an *emergent* functional *invariant*, which often is a truly novel performance

Contradictory means that the two terms have effects vis-à-vis the total system that lead to different dynamic outcomes, which may at least in part cancel one another. For instance, each segment of an arch falls literally against the other segment. *Regulate* means that each of the subsystems can in fact influence the dynamic consequences of the other(s), setting dynamic limits to its/their manifestations; the falling of each arch segment is curtailed by the falling of the other segment. *Emergent* is a property or entity that suddenly appears: that is, a truly novel happening that marks the existence of a replicable *invariant,* probabilistically invariant because it reoccurs as a truly novel event (performance pattern) whenever opposing dialectical subsystems interact in appropriate circumstances. Notice that dialectics also involves a focus on processes and on the changing nature of experience and reality (as dynamic field/system theories currently do). As Lenin writes, quoting Engels: "Dialectics is the science of the general laws of motion, both of the external world and of human thought" (Marx, Engels, & Lenin, 1977, p. 374).

An example of dialectical emergence in cognitive development is babies learning to walk (W). By about 12 months a baby has learned, by trial and error, to stand on her feet unaided. The product of this intentional learning is a scheme system for *free standing* (call this skill W1). Then, because she wants to walk unaided, she learns to unbalance herself from this position, by raising one leg and hip and immediately letting herself fall momentarily on this leg in a controlled manner (call this skill W2). As it falls, the leg becomes firm to allow the child to land on it. Let us call *controlled falling* the operative scheme system of W2. At this moment, the baby is *standing* on one leg

(W3), and she unbalances herself on the other leg and hip (W2′, similar to W2). Then she makes firm this moving leg to enable her landing on it (W3′, similar to W3). This walking sequence recurs: W1-W2-W3-W2′-W3′-W2-W3-W2′-W3′...

After some persistence in practice, the child can walk by herself if, and only if, she has enough sensorimotor maturational attention to boost together all the relevant schemes in order to produce a suitable dynamic synthesis. The infancy literature usually assumes that no working mind is needed, because the walking scheme is innate. Some conditions or parameters necessary for walking indeed are innate, but not the walking skill itself. This process analysis shows that walking is synthesized by the 12-month-old infant's mind: synthesized as a controlled form of regulated falling, regulated by the alternation of the scheme *falling* (i.e., W2 and W2′) with the scheme *standing* (W3 and W3′). These two contradictory scheme systems (like the two arms of da Vinci's arch) regulate each other with their controlled alternation, permitting dynamic emergence of this truly novel repeatable outcome (new performance invariant), the skill of walking.

Emergence of functional invariants generated by dialectical/dynamic systems enables humans and other animals, during development or learning (with brain maturation or neuroplasticity), to achieve truly new characteristic, behavioral or mental, performances of all sorts. These performances are not innate nor, strictly speaking, learned, but dynamically synthesized. Thus, different truly novel perceptual-motor or mental acts appear, again and again, under suitable circumstances.

Dialectical systems are continuously acquired, dynamically synthesized, throughout development. Age-typical performances characteristic of developmental stages provide examples, but there are many others: creativity, problem-solving strategies, self-acquired techniques, human social interactions, life adaptations, and so on. When one looks closely, most innovations result from dialectical/dynamic syntheses that often are serendipitous.

In his later work Piaget studied this sort of constructivist dialectics using two functional categories first introduced by Hegel (the founder of modern dialectics): affirmations and negations, two complementary forms of performance process (Inhelder et al., 1977; Piaget, 1974, 1975/1985). *Affirmations* are processes (of any level of complexity) that are more or less congruent with the intended current cognitive-structural processing (by stage-related developmental schemes/structures) and congruent with the intended agency/praxis. *Negations* are often *distracting processes*, incongruent or dialectically contradictory with the current affirmations, or they are *misleading processes* that induce performances that lower probability of the intended praxis. However, in a positive sense, negations can also be inverse operations in a well-equilibrated operational system, or even "perturbations" that favor development by disequilibrating inadequate schemes or structures of the cognitive state (e.g., in Piaget's terms, when a preoperational structure is negated by newly experienced contradictory and perturbing feedback).

Piaget used affirmations and negations categorically, without making sufficiently clear that they are relative to the praxis at hand. Nonetheless, he raised an important point: to enhance performance by differentiation of responses, one must regulate (i.e., compensate by means of suitable affirmations or negations) all negation or affirmation processes that may emerge. For instance, the person who in ancient Rome invented the architectural arch discovered how to place the two arms in perfect opposition, each of them compensating the other (as regulated negation/affirmation). Only then could a truly new object, the stable arch, emerge. This case also illustrates, by using a minimal number of dialectical opposites, Piaget's claim that in *well-equilibrated* systems each affirmation is compensated or balanced by its own negation, and vice versa (Piaget, 1974, 1975/1985). As Piaget emphasized, affirmations tend to be observed and appear first developmentally, because negations (or inverse operations) are harder to notice (i.e., less salient, less facilitated) in ordinary situations. However, Piaget seemed unaware that negations are typically more salient than affirmations within misleading situations, as we shall illustrate. In either case, until a mutual compensation between affirmations and negations occurs, a truly equilibrated (e.g., a Piagetian operational) system is not possible (Piaget, 1975/1985).

In psychological processes, affirmations are schemes (or organismic dispositions) relevant for the intended praxis. Negations are generally either external features of the situation that are inconvenient and must be overcome, or internal schemes (habits, automatized schemes, or organismic tendencies) that are misleading, because they propitiate unwanted performances. In either case, success depends on applying affirmative schemes that can dynamically compensate and neutralize unwanted negations. For instance, in the well-known Stroop task (naming the ink color of words that name other irrelevant colors), the affirmations are name schemes for the actual ink color; the negations are (color) schemes being cued by the word names, which tend to be more salient when reading is automatized. In the Piagetian conservation of substance task that uses Plasticine, negations are either representations of external features suggesting greater amount (e.g., a long sausage) or internal schemes (e.g., a tendency learned in everyday life and facilitated by the field factor, the neo-Gestaltist simplicity principle, is to equate a larger perceptual appearance with a bigger object). As Piaget's descriptive (not causal) analyses imply, success in misleading situations, in which dialectical negations must be controlled, depends on the ability to neutralize all misleading factors with the help of suitably chosen affirmative schemes, which the persons either have in their repertoire or can synthesize dynamically.

Causal Mechanisms for Organismic Dynamics

Piaget and his collaborators (Garcia, 1980; Inhelder et al., 1977; Pascual-Leone, 1987; Piaget, 1975/1985) recognized a main objection to their dynamic theory: It gives "description and provides no explanation" (Piaget, 1975/1985, p. 147). It offers "optimizing/progressive equilibration" as a causal organismic process to explain truly novel performances and stages of cognitive development but gives no coherent analytical account of the organismic factors that may cause equilibrations. Piaget's equilibration construct is the name for a problem. Indeed, this *optimizing/progressive equilibration* (OE) involves three subprinciples that Piaget failed to distinguish (Pascual-Leone & Goodman, 1979), although they metatheoretically express well the synthetic disposition of the working mind. It is an active (self-propelling, endogenous) disposition of the organism to undergo restructuration (i.e., structural change) spontaneously via dynamic syntheses, in order to:

(OE1) maximize the internal consistency among its functional parts

(OE2) maximize the scope of adaptation (its functional payoff) in dealing with the environment: that is, maximize the number of different types of situations to which the organism can adapt without having to learn (i.e., change its internal structures)

(OE3) minimize internal complexity (structural cost) in its organization

These three subprinciples should lead to extraction of new invariants across variable functional aspects in the situations.

To explain analytically how the working mind could satisfy these three subprinciples in a constructivist manner, we must assume the existence of multiple, qualitatively different general-purpose brain resources, or *hidden operators,* that could generate multiple learning mechanisms. Because the organism is very active and self-propelling, its functional schemes and structures (i.e., complex schemes) serve as self-organizing dynamic systems driven by *assimilation* (self-propelling disposition to apply). Schemes also *accommodate* (adapt) when they must, when Resistances from Reality cannot be avoided and cause intolerable anomalies. In chapters 10 and 11 we discuss neuroscientific evidence suggesting that the assimilation and accommodation functions may be expressed in distinct complementary pathways of the cortex. A neglected implication of this sort of theorizing is that any performance typically should be *overdetermined* by all schemes that could apply due to the assimilation tendency. Thus, performances often should result from *dynamic syntheses* (as dialectical thinkers often have claimed; e.g., Gadamer, 1971; Kant, 1929/1965; Luria, 1973; Marx et al., 1977; Vygotsky, 1978; Vygotsky & Luria, 1994). Such syntheses, combining task-applicable schemes that can epistemically reflect Reality, may be related to processes that Peirce called abduction (Buchler, 1955; Johansen, 1993).

For Piaget, dynamic syntheses are carried out by what he called *regulations*—a notion he left unexplicated (Garcia, 1980; Inhelder et al., 1977; Pascual-Leone, 1969, 1970, 1987, 1988; Pascual-Leone & Johnson, 2005; Piaget, 1967, 1975/1985). Regulations cause optimizing equilibration when different dialectical factors are involved and some organismic resources (cause of regulations) intervene, which brings about adaptive syntheses. These resources (for us brain capacities or *operators*) should be information free, general purpose, distinct, and independent from schemes. Such general-purpose capacities (hidden operators and principles in this book) were ignored by Piaget as they often are by current researchers.

Consider again briefly dialectical processes. Luria (1973), Piaget (1980, 1975/1985), Vygotsky (Kozulin, 1990), and others agree with this idea: there are three complementary types of (complex) dialectics, which together promote individual adaptation and development: (1) *purely external dialectics* between the individual and his or her life context; (2) *external intersubjective dialectics,* from each person to other persons, often psychosocial and sociocultural; and (3) *purely internal dialectics* interrelating psychological processing components (cognitive, affective, and personal schemes, with general-purpose brain capacities) within the individual. These three distinct sorts of dialectics lead to complex interactive situations that should be experienced subjectively as well as objectively.

In this respect, we must emphasize the difference between *facilitating* and *misleading situations* (Arsalidou et al., 2010; Pascual-Leone, 1987; Pascual-Leone & Johnson, 2005, 2017; Pascual-Leone & Morra, 1991). This important distinction is not found in Piaget or Vygotsky and often is ignored in current cognitive science and neuroscience. Metasubjectively, a situation is facilitating when it activates only schemes compatible with the task at hand. It is misleading when it elicits schemes that interfere with the task.

Facilitating situations elicit schemes that contribute to (or do not interfere with) the person's task. Such situations traditionally have been used by learning researchers. Development appears continuous when studied using facilitating situations, a linear growth function being its characteristic curve. Misleading situations, in contrast, are typical of problem-solving paradigms. Such situations reveal coping, problem-solving levels in cognitive development. Development appears as discontinuous in misleading situations, exhibiting (at times stepwise) nonlinear growth curves in tasks as a function of chronological age (Arsalidou et al., 2010; Johnson, Fabian, & Pascual-Leone, 1989; Pascual-Leone, 1970; Pascual-Leone & Baillargeon, 1994; Pascual-Leone & Johnson, 2005, 2011; Pascual-Leone & Morra, 1991). Note that this is found more easily in cross-sectional studies, where learning, due to repeated testing (as found in longitudinal studies), does not occur. However, longitudinal studies are necessary to validate cross-sectional results.

Thus, stages of development are found reliably (as relatively culture-fair, age-bound differences) only in chosen misleading situations. In these misleading paradigms, performance tends to conform to what descriptive, nonlinear system theories, like catastrophe theory (Camba, 2014; Molenaar & van der Maas, 1994; van der Maas, 1993) would predict. This has led to claims that development should be explained with nonlinear system theories (e.g., Thelen & Smith, 1994), although the organismic causal factors that produce these "catastrophes" in performance are never clearly investigated.

Stages of development appear in misleading situations because mental attention is needed by the working mind to inhibit and control unsuitable habits (due to overlearned schemes or innate reflexes), cued by situations and by organismic factors that make them salient. These misleading schemes induce competing mistaken strategies, or interfering processes (dialectical contradictions, negations), relative to intended/desirable action or results. Individuals must use problem solving in misleading situations—nonautomatized methods (e.g., invention and creative, synthesized truly novel performances) to cope with demands. Dynamic syntheses that generate the task solution are caused by compatible dominant schemes applying together to overdetermine performance.

Performance and the mind's mentation are always overdetermined by all schemes or schemas activated by the situation and boosted or controlled by active hidden operators of the brain. These dynamic syntheses are achieved automatically by the principle of *Schemes' Overdetermination of Performance* (*SOP*). This *SOP* principle was first formulated by Freud (Rapaport, 1960) and used in human science by sociologists (Callari & Ruccio, 1996) among others. It postulates that performance and mentation always result from unifying adaptive integrations (*SOP*) of the activated processes (this is a position also taken by current dynamic field theories, e.g., Schöner, 2014, although less explicitly than dialectical constructivism). Schemes/schemas, hidden operators, and organismic principles competitively coadapt to "negotiate" the resulting performance or mentation.

To clarify this important *SOP* dialectics, consider an invented "Freudian" example of how *SOP* works: A high-ranking executive is discussing business in the office with an attractive female subordinate executive. As he excitedly argues some business points, he comes closer to her and in his mind notices her lovely face and attractive attire. He is mindfully aware that he should not express his attraction to her, and he continues calmly arguing the business points. However, unconsciously, he stretches his right hand to lean on the wall she is standing near. His arm is now in front of her, as a friendly barrier. She does not react; he is at a reasonable distance. They continue their conversation, holding the mutual posture, with his arm-and-wall now looping around her in a symbolic (perhaps unconscious) "proper" embrace.

Note that his posture emerged from the synthesis of several contradictory schemes (**sch**): (**sch1**) desire, perhaps suppressed from consciousness; (**sch2**) a wish to embrace her, perhaps unconscious; (**sch3**) some ethical schemes of proper conduct related to women colleagues; (**sch4**) interest in the conversation leading him to move closer to her; (**sch5**) some unconscious desire to ease his tired feet, and so on. The "Freudian" interpretation, which the *SOP* principle and our Theory of Constructive Operators (TCO) in fact predict, is that **sch1**, compelling although perhaps unconscious, elicits activation in his working mind of **sch2**, which in turn elicits activation of **sch3**. This personal (affective *and* cognitive, emotional) conflict unconsciously promotes (via *SOP*) a new bodily posture that also is congruent with, and induced by, **sch4** and **sch5**. This is the posture of leaning on the wall: symbolically (but without moral transgression) he is now embracing her (at a distance) with the help of the wall. The example illustrates how *SOP* helps to solve conflict situations (with any sort of schemes, affective or not) by symbolically working out compromises among competing or compatible activated schemes. This is a new way of looking at psychological and brain processes as products of two sorts of operators, schemes (subjective operators) and hidden operators, which codetermine processes in the working mind by automatically producing dynamic syntheses among affirmative and negative schemes.

Overdetermination is a multidetermination in which causal factors interact in nonlinear ways, within the organism and the working mind. It explains dynamic/dialectical syntheses in a manner consistent with neuroscience. (For example, the brain's cortical spreading of activation among connected neurons, regulated by lateral inhibition, is an overdetermination process.) Overdetermination can also be seen as expressing Piaget's assimilation function, which compels schemes to apply together when compatible, or as a generalization of Sherrington's neural principle of a final common path (neuronal spreading of activation and its convergence; Edelman, 1987; Edelman & Tononi, 2000; McFarland & Sibly, 1975; Sherrington, 1906). According to this principle, all schemes active within a situation tend to apply together, if compatible, to codetermine performance. Sherrington seems to have shared this generalized principle of convergence: "Where it is a question of 'mind' the nervous system does not integrate itself by centralization upon one pontifical cell. Rather it elaborates a million-fold democracy whose each unit is a cell" (1940, p. 277). Thus, performance at every moment is synthesized by the currently *dominant* (most activated) *cluster of compatible schemes/processes.* These schemes often compete with interfering schemes. The probability of this performance is proportional to the relative dominance of the current cluster of compatible schemes generating it, relative to other clusters.

Dynamic syntheses result from application of hidden operators and organismic principles (Pascual-Leone & Goodman, 1979), often coordinated by executive processes. Piaget's theory lacked these organismic causal constructs (e.g., executive processes, mental attention, attentional inhibition/interruption, *SOP*; Morra, Gobbo, Marini, & Sheese, 2008; Pascual-Leone, 1987, 1996b; Pascual-Leone & Baillargeon, 1994; Pascual-Leone & Johnson, 2005, 2017). Resolution of this scheme competition obeys three optimizing-equilibration rules (**OE1**, **OE2**, and **OE3**) mentioned above. Rules **OE1** and **OE2** (i.e., maximize internal consistency and maximize scope of adaptation) result from an overdetermination of performance (*SOP*) by action schemes regulated with executive schemes, which set the course (cognitive goals) and direction (in action or thought) to serve current intentions (affective goals or motives). These **OE1** and **OE2** equilibration processes are biased by dialectical intertwining between cognitive and affective processes, the origin of complex motivation. Piaget and many current cognitive scientists lack clear models interfacing cognition with affects (affective schemes) to generate complex motivation (affective goals or motives).

Regarding **OE3** (i.e., minimize internal complexity in the organism), another factor, only informally used and mentioned by Piaget and others, is important. This is the brain resource that Gestalt psychologists (Koffka, 1935/1963) and Piaget called (internal) *field factor* and we call *F*-operator (Pascual-Leone, 1989; Pascual-Leone & Morra, 1991). A neuroscientific interpretation of this factor is lateral inhibition processes in the cortex (e.g., Edelman, 1987; see chapter 10). Psychologically, this factor expresses a performance-closure mechanism akin to the neo-Gestaltist principles of "Pragnanz," "minimum," "simplicity/simplexity," and "S-R compatibility" (Berthoz, 2012; Pascual-Leone & Morra, 1991; Proctor & Reeve, 1990; Rock, 1983).

This organismic *F*-factor, in interaction with *SOP*, produces a sort of *minimax-function* effect that we formulate as follows: The performance produced will tend to be such that it minimizes the number of schemes directly applying to inform performance (including perception or representation), and it does so while maximizing the set of distinct, salient empirical aspects or features (i.e., applied low-level schemes) that, directly or indirectly, inform this experience (Pascual-Leone, 1987, 1995, 1996b; Pascual-Leone & Johnson, 1991, 2005; Pascual-Leone & Morra, 1991).

For example, errors in the picture completion subtest of the Weschler Intelligence Scale, such as failing to see the missing doorknob in the picture of a door, are caused by this *F* minimax mechanism. This *F*-factor prevents application of low-level (local-perceptual) schemes, like the simple doorknob scheme, when higher-order automatized schemes, like the structure of a standard door with its own doorknob, can also be applied: *F* suppresses, via lateral inhibition (see chapter 10), the application of the simpler lower-level doorknob scheme.

Aspect **OE3** of optimizing equilibration can be explained by *F* in conjunction with *SOP*, together with the need to minimize processing. Optimizing equilibration, and other organismic resolution-syntheses of performance, can be explained by assuming brain capacities and principles other than schemes that Piaget's theory clearly lacks, as do most current learning theories. Our theory (the TCO) is not simply an elaboration of Piaget's theory, because Piaget's theory is a *developmental cognitive-learning theory*. For him, development occurs because learning accumulates and (aided by unexplicated maturation) causes development. Piaget (possibly contrary to Binet and Spearman) did not believe that there is another factor such as mental "energy" or endogenous attention that is subject to maturation and can function as specific organismic cause of psychological-developmental stages. When one of us (Juan Pascual-Leone) personally told Piaget that mental attention, or "mental energy," grows with his stages, he was in disbelief. Piaget refused to entertain this hypothesis and told JPL that he would never be able to prove this idea. But we have proved it, and the TCO is a result. Since most psychologists are also cognitive-learning developmental theorists (and not mental-attention developmental theorists), stressing the theoretical difference with Piaget is necessary, even though the TCO explicates Piaget's theory very well.

Piaget did not explicate the maturation concept, nor had he other general-purpose innate "central" constructs like *F*, or maturational mental attention (which we call *M*-operator), or maturational attentional inhibition/interruption (which we call *I*-operator). Current cognitive-science researchers often resemble Piaget in their lack of enough organismic-causal (working-mind) operators to explain performance and developmental growth.

Two Types of Constructivist Developmental (Neo-Piagetian) Theories

Neo-Piagetian theories explain cognitive growth as caused by incrementation in the processing of *effective complexity*, due to acquisition of schemes/schemas that generate performance, facilitated by increase in mental attention, working memory, and other resources. Developmental constructivists have different views on what causes transitions from one stage to the next. Some neo-Piagetians, and most adult working-memory researchers, support what we might call the *constructivist learning* group of theories. They see developmental growth as caused by some form of insightful (since the role of consciousness often is assumed) constructivist learning. Piaget is part of this group and would have equated constructivist learning with psychogenetic/developmental intelligence. For Piaget, developmental intelligence had four distinct main causal factors: *maturation* (innate organismic determinants), *specific learning* (cognitive, affective, and motivational modes related to tasks and particular situations, across domains), *general/social learning* (affective, psychosocial, cultural, historical modes and knowledge—across domains), and *equilibration* (organismic internal balance; developmental, problem-solving, and learning adaptation). Demetriou and colleagues (Demetriou & Spanoudis, 2018; Demetriou, Spanoudis, & Shayer, 2014; Spanoudis, Demetriou, Kazi, Giorgala, & Zenonos, 2015) call *cognizance* a related encompassing construct for which, unlike Piaget, they seem to consider consciousness the primary causal factor (as do some other cognitive scientists, e.g., Dehaene, 2014).

Constructivist learning often is seen as producing durable internalization of recurrent functional patterns of processing and behavior, achieved by way of complex schemes (schemas, functional structures, chunks) at various complexity levels and in different content domains. Constructivist learning theorists interpret the growth in working memory or mental attention (which Piaget called "field of centration" or "field of equilibrium") as a product of cognitive learning. Demetriou's current theory (Demetriou & Spanoudis, 2018; Demetriou et al., 2014; Pascual-Leone, 2019; Spanoudis et al., 2015) may be the version of constructivist learning theories that has been more thoroughly investigated psychometrically. This and other neo-Piagetian theories are discussed by Arsalidou and Pascual-Leone (2016), and we refer readers to this source.

Other neo-Piagetians, who constitute the *maturational-attention group* of theories, agree with the importance of constructivist learning but think that this sort of learning (in contrast to associative learning) is possible only with the maturational growth of a limited resource: that is, mental/executive attention, often misnamed as working memory (e.g., Case, 1998; Halford, Cowan, & Andrews, 2007; Halford, Wilson, Andrews, & Phillips, 2014; Morra & Borella, 2015; Pascual-Leone, 1970; Pascual-Leone & Baillargeon, 1994; Pascual-Leone & Johnson, 2005, 2011; Pascual-Leone, Johnson, & Agostino, 2010; Petersen & Posner, 2012). Maturational attention is a key determinant of working memory, but this is not the only determinant. Working memory, as it is commonly understood, demands other organismic-causal factors, perhaps all the factors important for a working mind. This maturational attention group of theories has shown that mental-attentional mechanisms grow in power as a function of age in normal children; this growth (along with other causes) relates to emergence of developmental stages.

For Robbie Case cognitive abilities improve within substages due to the qualitative gain in proficiency of *central conceptual structures* (mental-operational complex schemes) caused by constructivist learning and mental attention. Case (and many others) confused mental attention with working memory, even though he used learning constructs of Pascual-Leone (Case, 1998; Pascual-Leone & Goodman, 1979). Theoretical relations between Case's and Pascual-Leone's theory can be found in Pascual-Leone et al. (2010), Case (1998), and Morra et al. (2008). Another maturational-attention neo-Piagetian researcher who confounds maturational/endogenous mental attention (an executive-driven sort of attention) with working memory is Halford. Halford and associates (Andrews & Halford, 2011; Halford et al., 2007, 2014) have focused on how to assess the effort or memory load of their relational working-memory construct. They attempt this appraisal by estimating complexity only in terms of number of interrelated terms to be jointly considered in the task at hand. Their Relational Complexity Theory states that meaning occurs when a link is formed via interrelations (e.g., "cat" and "lion" are related because both are felines, and one is smaller than the other—two binary relations—and may also be related because "their particular trainer prevents the lion from attacking the cat"—a ternary relation). Meaning, and usually mental demand, accrue as higher-order relations (unary, binary, ternary, quaternary relations, etc.) are formed. However, Halford's relational complexity is not the only task-relevant sort of effective complexity. The mental demand of tasks can also accrue with other not interrelated but relevant relations, or pieces of knowledge, such as those that must be kept in mind for later use in a task. Nonetheless relational complexity is perhaps the most important aspect of knowledge, as Cassirer (1938, 1923/1953, 1929/1957) early emphasized.

For us, working memory can be best understood as a name for the working mind: a product of multiple causal-organismic factors that include maturational (endogenous) mental attention. Pascual-Leone is the founder of neo-Piagetian approaches and initiator of the maturational attention group of theories (Pascual-Leone, 1970; Pascual-Leone & Baillargeon, 1994). As mentioned, the TCO adopts the viewpoint of subjects' own processes: a metasubjective perspective (Pascual-Leone & Johnson, 2017). Within this theory, behavior is generated by mediation of various general-purpose hidden operators, or information-free brain resources (mental attentional effort, the field factor of simplicity, learning mechanisms, and other resources), which modulate the activation and functioning of self-propelling schemes. All this is governed by the principle of Schemes' Overdetermination of Performance (*SOP*), which dynamically/automatically synthesizes performance out of activated schemes (Pascual-Leone, 1970, 1987, 2014; Pascual-Leone & Johnson, 2005, 2017). As first formulated by Pascual-Leone (1970), mental-attentional capacity (*M-capacity*) increases (in its behavioral measure) every other year after the age of 3, reaching a limit of seven units at 15 to 16 years (this count of seven is obtained only when all necessary schemes, figurative as well as operative and their essential parameters, have been counted, if they are not boosted by some other resources such as affect or learning). This maximal *M*-capacity of seven is also found in adults (Miller, 1956), although adults often use a capacity of four or five, unless highly motivated and challenged (Cowan, 2005, 2016; Cowan, Ricker, Clark, Hinrichs, & Glass, 2015).

Although these issues are discussed in detail in the chapters to follow, table 1.1 and figure 1.1 give an overview of the TCO as coordination of hidden *resource operators* that apply to various kinds of action and executive schemes activated by the situation, to lead them to overdetermine (*SOP*) and cause performance and perhaps to change (adapt, learn) the schemes in question. Table 1.1 (discussed in detail in chapter 7) presents the organismic (hidden) resource operators of the working mind as processes in the brain. They are listed in a plausible order of their evolutionary emergence.

Figure 1.1 sketches diagrammatically how these operators codetermine functioning and adaptation of all types of schemes (executives, operatives, parameters, figuratives) and thus control *SOP* results. Operative schemes stand for processes that produce transformations and operations, mental or behavioral acts. These operatives can be ordinary action schemes (e.g., a grasping action blueprint) or executive schemes, which carry plans and regulate action schemes and hidden resource operators. Figurative schemes stand for the "objects" (psycho-*logical* entities, usually distal objects that in logic may be called "arguments") on which operatives apply (e.g., I see the piece of banana—figurative scheme—and the view elicits in me a desire to grasp and eat it—operative schemes). Parameters are a sort of relational figurative that stipulates the suitable conditions for operatives applying on figuratives (e.g., the banana looks too ripe for eating, so I resist taking it).

Critical Realism of Piaget and Cognitive Science

Constructivism appears as a "middle way" between rationalism and (meta-) empiricism. It produces operative/procedural and figurative/declarative models with the coordination of schemes and schemas active in the working mind. This is a result of organismic construction, based on perception, mentation, actions, and feedback from actions. However, Piaget defined schemes solely in terms of their *operating* (action)

Table 1.1
TCO's hidden operators listed in order of their likely evolutionary emergence

Operator	Description	Main Brain Region
A	Set of **affective** processes that intervene in motivation and attentive arousal.	Brainstem, hypothalamus, extended amygdala, limbic system
C	Both the process of **content** learning and the schemes derived from associative content learning.	Thalamus, Brodmann primary and secondary areas
F (*SOP*)	The **field** operator, which acts as a binding mechanism in the brain and brings closure to mental representations in a neo-Gestaltist manner. It often functions intertwined with the principle of Schemes' Overdetermination of Performance (SOP)	All areas
LC	The process of automatized **logical-structural** learning derived from ***C* learning** through overpractice.	Right hemisphere (RH)
T	**Temporarily** and effortlessly collates **sequences** of figurative schemes, thus facilitating the coordination that constitutes distal objects.	Hippocampal complex, occipito-temporal cortex
S	Effortlessly coordinates relations of coexistence among activated schemes, during operative activity (praxis). It, thereby, facilitates emergence of **spatial** schemes or schemas.	Hippocampal complex, occipito-parietal cortex
LA, B, LB	**Psychosocial and self-schemas** (*B*). **Logical-structural** learning primed by strong **affects** (*LA*), or by the personal **being** preferences—including emotions (*LB*).	Limbic system, orbito- and medial prefrontal, inferotemporal, medial parietal cortex
I	The attentional **interrupt**, which corresponds to the power of central active inhibition of unwanted schemes activated in the situation.	Prefrontal, RH-medial cortex, dorsolateral cortex, basal ganglia, thalamus
M	**Mental** attentional capacity of the individual.	Prefrontal, lateral, and dorsolateral cortex; basal ganglia; thalamus
LM	**Logical-structural learning** caused by the effortful use of **mental** attentional capacity	Left hemisphere tertiary areas, polymodal
E	**Executive** schemes in the person's repertoire, for the task at hand.	Prefrontal, lateral, dorsolateral, and frontopolar areas

Adapted from Pascual-Leone, J. & Johnson, J. (2005). A dialectical constructivist view of developmental intelligence. In O. Wilhelm & R. W. Engle (Eds.), *Handbook of understanding and measuring intelligence* (p. 181). CA: Sage. Copyright 2005 by Sage.

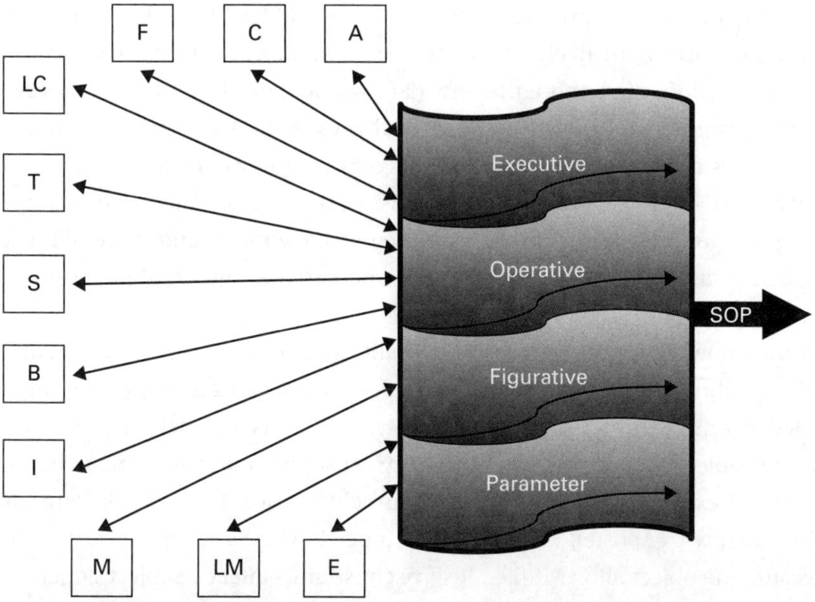

Figure 1.1
Theory of Constructive Operators (TCO) overview, with key organismic hidden-resource operators that can apply on various sorts of schemes, which in turn apply in coordination to synthesize/codetermined (*SOP*) the overt or covert performance. Parameters are figurative (representational) schemes that stipulate relations between operative (procedures) and figurative schemes. (Adapted by M. Arsalidou from Arsalidou, M., Pawliw-Levac, M., Sadeghi, M., & Pascual-Leone, J. [2018]. Brain areas associated with numbers and calculations in children: Meta-analysis of fMRI studies. *Developmental Cognitive Neuroscience, 30,* 246, CC-BY.)

components, often ignoring their *releasing component* (their cuing-features or conditions). For instance, when I see the brown, mushy flesh of a banana, a figurative aspect of the context or local reality, these characteristics automatically induce the working mind to expect the banana to be overripe. Mobilized scheme processes (figurative and at times also operative) serve as cues. B. F. Skinner with his operants (Catania & Harnad, 1988), like Piaget, also focused mostly on action components, segregating figurative aspects or releasing conditions of schemes as "discriminant stimuli"—aspects of the Real situation, as learning/memory theoreticians often do. There is a major constructivist problem when schemes are defined without reference to explicit releasing conditions, however. Unless conditions exist in the scheme that make it context-sensitive (thus producing *situated* knowledge), schemes would be released by any situation and thus should often fail to apply successfully. This would lead the scheme not to appear as a functional invariant likely to succeed when applied. Such unreliability, or functional deterioration, of the potential invariant would make it not (or much less) learnable, having failed its regularity as functional invariant. Organismic schemes emerge in the person's activity as we internalize encountered Resistances from actual Reality (see chapter 5). With the growth of cognitive learning, one's models of reality (coordinated packages of schemes) internalize progressively more, and more refined, resistances, becoming *viable* (pragmatically true) when used in agency or praxis (i.e., willful agency).

Because knowing (epistemology) and Reality (ontology) are so functionally intertwined, causality can be seen (Inhelder et al., 1977) as an attribution of empirically grounded psycho-logical operations to material objects—models that help to constitute the subject's representation. Causality must "be found in some manner as an objective and external reality" (Apostel, in Inhelder et al., 1977, p. 63; translation by JPL). In his reply to Apostel, Piaget agreed: "Indeed, when one arrives to true theories, it is because the object allows it [i.e., Reality constraints affect a subject's operating]; to say it in other words: the object already contained something like my operations [i.e., Piaget's models]" (Piaget, in Inhelder et al., 1977, p. 64; translation by JPL). Piaget, however, never said clearly what this "something like" is. He may not have worked out how causal coordinations (when true) embody relational task-relevant packages of Reality resistances. Learned low-level scheme packages serve as referential domain for more encompassing, higher-level abstractions. Often high-cognitive knowing is abstracting (according to Piaget) not from external reality/Reality itself but from the subjects' own actions or mentations (mental operations). Both ways of abstraction coordinate recurrent probabilistically invariant relations as patterns of activity, which emerge as schemas/structures that take as content abstracted schemes from lower levels, to generate flexible hierarchies. Such is the process of *reflective abstraction*, whose function is to restore adaptive equilibration in the working mind.

Resistances and Two Forms of Accommodation

As emphasized by Peirce, William James, and many others, knowledge has a practical aim: to serve in agency (or praxis—goal-directed activity addressed to the environment). It must embody (usually after progressive approximations, trial and error) the essential constraints (resistances—encumbrances, difficulties, or affordances) that Reality, in each particular type of situation, imposes on our agency. Cassirer (1940/1996) aptly described these resistances as such: "We experience something that stands in opposition to us, which is different from us, and out of this opposition grows our experience of the 'object'" (p. 140).

In fact, from a constructivist perspective, it is useful to think of reality/Reality as a universe of resistances (some sort of patterning that intertwines, e.g., essential characteristics and experiential outcomes) that emerge with the individual's activity in specific situations. Neither Piaget nor cognitive scientists have considered this idea enough. Resistances often have dependency relations among themselves (e.g., when I hit the *correct* pedal the car accelerates, *if* it is running *with* enough gas). Thus reality/Reality is populated by packages of interdependent resistances that may be somewhat different for each biological species. These resistance packages can be interpreted, without falling into empiricist excesses, as indexing recurrent patterns, that is, real complex functional invariants that emerge conditional to specific activities (Gibson, 1979; Nozick, 2001; Ullmo, 1967). The individual can cognize and relate these relational aspects of reality to other invariants, forming valid representative packages that probabilistically maintain invariances of mutual interdependencies of one to another resistance. Examples of such packages are complex *distal objects*, objects that emerge as invariants in the context of life-related praxis, such as cars, people, movie theaters, airports, universities, and so on. When such complex resistances (or packages) are helpful in current praxis, they generate *affordances* (Gibson, 1979). When they are a hindrance, they emerge as *encumbrances* or obstacles. Motivation (the conversion of affective goals into cognitive goals) leads the person to attend to, and internalize, such resistance packages, eventually causing people to have internal models (representations) for what Tolman and Brunswik (1935) called the *causal texture of the environment*.

Piaget used the term *accommodation* for adaptive incorporation of new reality/Reality features into a scheme. He called *assimilation* the application to experience and to reality/Reality (including subjective processes) of schemes already available to the subject. For Piaget, accommodation is always secondary to assimilation: accommodation takes place only when assimilation fails.

This assertion suggests that Piaget did not quite understand the cognitive dialectics of resistances, serving to internalize the causal texture of an animal's environment. Although *Piagetian accommodation* is real and important (a sort of accommodation that is *secondary* because it follows the failure of assimilation), it does not explain all constructivist learning. There are other ways to internalize resistances relevant to optimizing equilibration. There is also a limited *primary accommodation function* (we call it *Gibsonian*, because it was tacitly introduced by J. J. Gibson). It is primary because it occurs without assimilation (in Piaget's sense) and is limited because it is restricted to the concrete here-and-now. Gibson used wine tasting as an intuitive example to illustrate this sort of primary-accommodation learning, which he called *differentiation*. When one begins to taste wine, it is difficult even to distinguish between some red and white wines. Then, with practice, one learns to recognize kind, origin, and grape of many wines and begins to understand the language experts use to describe wine qualities. Because initially there is no discrimination, and this learning occurs with repetition but without feedback, differentiation learning must be the result of primary accommodation, not secondary.

In primary accommodation the lower central nervous system makes discriminations that are not initially passed to higher central nervous system areas (sites of the conscious/preconscious psychological organization). New information comes to the sensorial organs and is received and processed by the brain outside consciousness. As the experience repeats itself, however, the cortex may extract a functional invariant that is progressively moved from lower to higher brain processing areas until it eventually enters the psychological levels, potentially emerging into consciousness. This emergence constitutes a scheme resulting from primary accommodation, because information entered the brain and was abstracted via accommodation as an invariant before it could have been assimilated by psychological schemes.

Without primary accommodations, the refinement of affordances or encumbrances, or many truly novel insights about reality, would be impossible (due to the overlearning of competing schemes that become habits). Optimizing equilibration requires also primary accommodation. This view is consistent with the idea of two sorts of information processing, empirical/physical versus relational/conceptual (Piaget's empirical versus logico-mathematical experience), which often work together and become intertwined. As Langevin, a French physicist, claimed, much of the concrete "is nothing but overused abstract" (Ullmo, 1967, p. 637; Goldstein & Sheerer, 1941).

Organismic-Causal Processes of Working Memory Are Those of Working Mind

Moscovitch (1992) raised the important point that working memory is a process of "working-with-memory," a purely functional capability of memory. It is not a working space (often equated with consciousness, e.g., Dehaene, 2014). Moscovitch made a clear attempt to explain working memory in terms of organismic-causal processes (brain resources) that, in their interaction, produce the functional complex described as working memory. Moscovitch sharply distinguished between specific, more or less automatic (reflex) content- or action-processing *"modules"* (which are low-cognition analyzers or processors, e.g., hippocampal and sensorial functions) *versus "central systems"* (located in prefrontal lobe and related brain areas), often open and accessible to consciousness. The latter organize (i.e., coordinate) related content data obtained from local modules. He also made clear that working-with-memory is a product of procedural-representational central systems, which in the brain often involve prefrontal lobes.

Moscovitch did not, however, present a from-within (metasubjective) organismic-causal model of the working mind that produces this working-with-memory function. To formulate such processes one must differentiate two distinct sorts of functional systems: those serving as information carriers (the information-bearing and context-bound units that we call *schemes*: i.e., schemas, chunks) versus information-free and general-purpose system resources that we call *hidden operators and principles* (such as *SOP* discussed above). Hidden operators express "hardware" constraints, regulations of our brain as an organized functional totality. As already mentioned, this includes automatic/perceptual attention, mental/executive attention, content/associative learning, constructivist/structural/conceptual learning, neo-Gestaltist field factor, and so on. Researchers often conflate these two sorts of organismic-causal processes. As a result, they mistakenly equate mental/endogenous attention with working memory. In our view mental attention is indeed a key maturational constituent of working memory. However, working memory is a broader functional complex constituted by the coordination of mental attention with the repertoire of ordinary schemes and executive schemes; this coordination serves the working mind.

In the remainder of the chapter we briefly illustrate how our constructive-operators model can with advantage explain complex cognitive tasks involving working memory. Our example is the Raven Matrices Test of fluid intelligence (Gf)—a task that could not be explained well with constructs such as "modules" versus "central system" processes, or with more traditional conceptions of working memory. Heitz, Unsworth, and Engle (2005, p. 74) have expressed this lack of clarity: "The most puzzling realization is that we have good reason to implicate attention in Gf, but we are devoid of a suitable explanation for how attention comes into play when performing a task such as the Raven."

Figure 1.2 shows an adapted version of Raven's item C7 from Skuy et al. (2002). In all problems of this sort there is a matrix of figures (upper part of figure 1.2), each of them related to others by distinct relational patterns that run over the figures. One figure from the matrix is missing, and participants must choose from a sample of possible figures (lower part of figure 1.2) the one to best fill the missing slot. Doing so requires analysis of invariant patterns running over figures, to identify the matching one. In the more difficult Raven's items, to identify the missing part, participants must analyze into schemes (figural/relational pattern schemes) the perceptual features and dimensions of variation from every row, column, and at times diagonals, of the item's matrix.

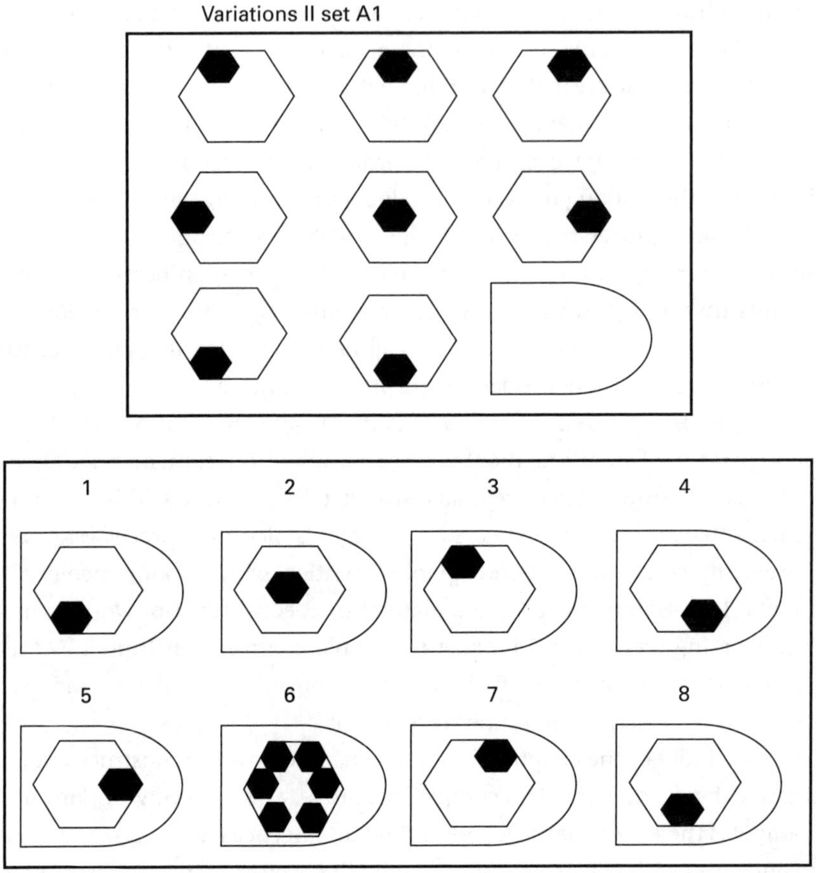

Figure 1.2
Variation on Raven's item C7. (Adapted from *Intelligence, 30*[3], M. Skuy, A. Gewer, Y. Osrin, D. Khunou, P. Fridihin, & J. P. Rushton, Effects of mediated learning experience on Raven's matrices scores of African and non-African university students in South Africa, p. 12. Copyright 2002, with permission from Elsevier.)

Sometimes a second dimension of variation is found on the rows or columns of the matrix, rather than in the diagonals, giving rows or columns two concurrent dimensions instead of one. Participants must identify simple or complex features (f_i, f_j, etc.) that for each dimension constitute a relational figurative invariant characteristic of this dimension, found in the appropriate cells. Such relational invariant makes mutually congruent (a correspondence rule) the various patterned objects within the figures (cells in the matrices) that concretely make up the dimension in question (instantiating its correspondence rule as a concrete token case). This is done via perceptual-cognitive analysis conducted successively for each required dimension[3] and then held in mind. An operative/procedural scheme of general synthesis (OPΣ) then recursively integrates the invariants of all dimensions to produce (via overdetermination) the pattern corresponding to the missing part.

Figure 1.3 outlines generically a possible case for these dimensions of variation. Note that we show OP3 in only one column. This represents a possible need to process a second dimension on the columns (not needed to solve the presented item). Were a second dimension present, as in more difficult items, OP3 would apply on all three columns. The actual process of dynamic problem solving could be roughly expressed by the following formula:

$$\underline{OP\Sigma}\ (\underline{OP4}(\underline{OP3}(\underline{OP2}(\underline{OP1}(\underline{PER\!:\!f_i,\ f_j}\ ))))) \qquad \textbf{(f1)}$$

Here operative processes (OP1, OP2, etc.) for the various relevant dimensions apply, to the right, on the products of previous operations. Notice that "PER:f_i, f_j" is the local perceptual process that Moscovitch may arguably identify as a module; the various OPs in contrast, and their coordination/organization, are the mind's inferential processes that Moscovitch may call central-system's work. But here we are assuming that these are schemes flexibly organized in hierarchies of various levels of processing. The order of operative processes could vary, but to complete the problem, all dimensions must be analyzed to extract or abstract their features' invariant; and these relational features are synthesized by OPΣ into the solution pattern. Against this final synthesis there are factors working: feature saliency differentials (often due to the neo-Gestaltist field or *F*-operator and to familiarity or due to logical-structural associative learning, which we call *LC*-operator). Misleading aspects of these feature-cues could make it harder to identify relevant dimensions of variation. The misleading factors are controlled by means of mental-attentional activation (our *M*-operator) and mental attentional inhibition (our *I*-operator). This is an *operative model* for the task solution. If we now count in the formula the number of schemes involved, we find at least six distinct symbolic/mental processes to be boosted with *M* and coordinated (we have underlined them in formula **f1**). This is the count for a complex Ravens item with four dimensions of variation. Items vary in the number of necessary dimensions. The item illustrated in figure 1.2, for example, could be solved by examining just two dimensions. However, all items require examination of rows and columns (and at times diagonals).

This is a sketch of operative task analysis conducted from within the participant's processing (metasubjective analysis). A complementary alternative way of doing analysis (meta-empiricist task analysis) would seek only to identify relational figurative features and their perceptual-rule patterns relevant in the test as a whole (and instantiated in one or another item). This would be an *objective* analysis focused on the relevant

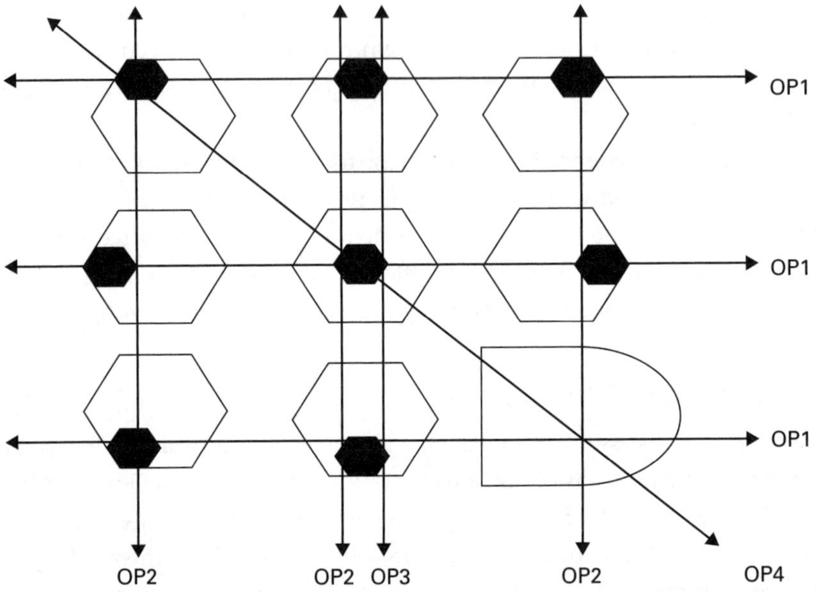

Figure 1.3
Dimensions of variation in solution of Raven's-like item. OP stands for an operative process to extract a dimension of variation across the figures. (From Pascual-Leone, J., & Johnson, J. [2017]. Organismic causal models "from within" clarify developmental change and stages. In N. Budwig, E. Turiel, & P. Zelazo [Eds.], *New perspectives on human development* [p. 82]. Cambridge University Press. Copyright 2017 by Cambridge University Press.)

figures of items, that is, a *figurative model*. Mindful use of these two complementary models helps participants to select the missing part of the matrix.

The TCO draws a sharp contrast between determinants of automatic/perceptual attention and those of mental attention—but this is a distinction among hidden operators, not between modules versus a central system. Piaget and other constructivist-learning researchers ignore mental attention as a distinct and independent cause of scheme activation. They attempt to explain mental (endogenous, focal, or executive) attention as being caused by the schemes' own assimilatory tendency (activation potency) and by cognitive learning. This formulation confounds processes of mental and perceptual or automatic attention. In contrast we see endogenous focal/mental attention as a main determinant in the working mind (working-with-memory) process—a determinant causally distinct from the repertoire of schemes.

Unlike Piaget, who only had one complex learning mechanism (i.e., structural learning), we have in our theory several categories of complex learning. We summarize these various learning categories now and provide more detail in chapter 5. The simple associative learning mechanism we call *C* learning (i.e., content-domain learning, playing the role of modules in Moscovitch's theory). We also have *LC* learning, which subsumes associative-structural learning—Piaget's logical-mathematical experiences when they are overlearned and occur in facilitating situations. *LC* learning takes place slowly, structuring together invariant associative-learning connections. The automatization (i.e., habits) that *LC* learning provides is useful in facilitating situations (when no misleading overriding schemes are found).

A selective and effortful dynamic synthesis of logical structures is necessary in conflict or misleading situations (when interfering schemes must be inhibited to solve the task); in this case, the power of mental attention (*M* and/or *I*) becomes necessary, as illustrated with the Raven task. This sort we call *LM* learning. There are multiple sorts of *LM* learning. They express different levels of *M*-capacity demanded by misleading tasks to be solved (see chapters 3 and 7). This requires progressively more *M*-capacity that grows maturationally with chronological age in normal children. Even when the needed executive know-how is available, a subject's level of mental (*M*) capacity must be commensurate to the task's *M*-demand, for the task to be solved easily. The *M-demand* of a task corresponds to the minimal amount of *M*-capacity needed to solve it. Because *M*-capacity has a maximum that increases with each developmental stage (the substages in Piaget's theory), each new stage enables children to cope with tasks of greater *M*-demand relative to previous stages. The maximum order/level of complexity-coping (or *M*-power level) that individuals' new *LM*-learning repertoire requires tends to increase with developmental *M-stage* levels (e.g., Arsalidou et al., 2010; Pascual-Leone & Baillargeon, 1994; Pascual-Leone & Johnson, 2005). These *M* levels determine effective complexity of truly novel tasks that an age group or person can solve—a major individual-difference variable.

Competition among schemes occurs in misleading or distracting situations. Piaget (1974, 1975/1985) talked about affirmations versus negations (i.e., opposite schemes) and about how these opposites must be numerous and in equilibrium to permit good adaptability (equilibration). This makes easier equilibration of dialectical opposites (such as subject's *M*-power versus task's *M*-demand, or schemes that contradict one another) within situations. As Heraclitus (2001 [46], p. 31), the old dialectical genius, said, "From the strain of binding opposites comes harmony."

2 Problems of Cognitive Developmental Theory

This is a discussion of problems and paradoxes that appear when constructivist organismic theories are not adopted (and schemes, as organismic process units, are not used). It refers to organismic-causal schemes and Kant's and Piaget's construal of them. We examine problems of constructivist theories that fail to recognize maturational processes in mental attention, raising the question of whether learning alone could explain development. Later we examine paradoxes and problems that traditional approaches may encounter: learning paradoxes; Plato's problem of low versus high cognition, still found in current psychological theories; Vygotsky's and Piaget's paradoxes; language and communication paradoxes; the meaning of *meaning*; and problems of emotions, cognition, and personal schemes.

> [W]e experience something that stands in opposition to us, that is different from us [Cassirer called it a "resistance"], and out of this opposition grows our consciousness of the object.
> —Cassirer, 1996, p. 140

> [O]verdetermined causality is multisided, reflecting and absorbing the mutual determinations of all the totality's [such as a developmental-stage] contradictions.
> —Cullenberg, 1996, pp. 137–138

> One tries to construct stages because this is an *indispensable instrument for the analysis of formative processes*. Developmental psychology seeks to examine mental functions in their construction, and stages are the preliminary method for analysis of these constructive processes.... But one must insist vigorously on the fact that stages are not by themselves the goal.
> —Piaget, 1956, p. 56, translation by JPL

Philosophers (e.g., Scheler, 1961) and psychologists (e.g., Hebb, 1949/1961; Luria, 1973; Piaget, 1948/1963) have emphasized that developmental growth in *mediational processes*[1] contributes much to a person's relative autonomy of performance vis-à-vis context/circumstances in which a performance occurs. Such processes of the brain's working-mind mediate between situations-as-experienced and the overt or covert responses of subjects. Mediational processes are inferred by the researcher (often via what Peirce called *abductions*) from objective or phenomenological/introspective data, task analysis, controlled observation of behavioral sequences, correlational methods, and so forth. In the best of cases they are tested by experiment. How these mediational processes (e.g., mental attention, functional structures, learning mechanisms, executive schemes) evolve with development, their characteristics, and causal modeling are issues in need of more detailed investigation.

Mediational processes are not just information-bearing processes of the working mind, like thinking/mentation, they are also hidden general-purpose and information-free resources founded on maturation. The need to assume such processes often is not recognized in psychology. When mediational processes are recognized (e.g., in the case of learning mechanisms of various sorts), they are not attributed enough diversity to provide a basis for causally explaining human performance and the information-carrying processes that we call schemes. This leads to conflation of performance-descriptive versus organismic-causal processes, causing epistemological/cognitive paradoxes, which cannot be explained without appealing to general-purpose processes in the brain's working mind. We discuss a number of these paradoxes below. However, first we comment on the main basic mediator of human processing: the scheme.

Kant's/Piaget's Schemas/Schemes versus Organismic Schemes as Mediators

Kant understood that complex, abstract categories or constructs cannot be directly related to more concrete experiential constructs they are meant to explicate, unless there is a suitable mediational process: that is, one or more mediators that can bridge over them. He called *schema* such a mediational process, which is abstract at one end but concrete and experiential at the other (Kant, 1929/1965; Pascual-Leone, 1998). Piaget called *scheme* a related sort of unit (e.g., Apostel, Mays, Morf, & Piaget, 1957; Piaget, 1959). In Piaget's definition, schemes are the learned coordinated relational-structures of all actions (or objects, presentations, mental expectancies, and representations) that a subject takes as relevant or interchangeable vis-à-vis a goal (i.e., in order to achieve some expected or intended results). Some simple examples of schemes are "sitting in a chair" or "seeing a given face" (irrespective of the concrete variations of the sitting act or facial expression). Kant and Piaget both were aware that schemes and schemas are different from the concrete instances (objects, activities) that they express. One thing is the concrete object, for example, a computer. Another is the more or less abstract set of schemes that correspond to the complex systems of coordinated functional interrelations that, in my mind, constitute this computational machine.

However, cognitive and development researchers often talk of functional-relational *stage* structures[2] as if they were unspecified concrete entities; that is, tangible substantive (not just relational) realities that either exist or do not. Consequently, performances that express these schemas or stages become hard to construe as they are: as products of a functional multiplicity—a mass of active diverse schemes that cause performance via overdetermination. Thus, the principle of Schemes' Overdetermination of Performance (*SOP*) becomes unintuitive and abstract. Abstraction is indeed involved, because each scheme retains or carries a few situated and repeatable (probabilistically "invariant") aspects of the experience in question. This constructivist abstraction is *epistemically reflective* (reflective in a sense that Peirce may have called iconic), because schemes capture recurrent aspects of the object's infrastructure[3] relative to the context of use (e.g., ripe McIntosh apples, but not just any fruit, are good for baking).

Because schemes are defined at multiple levels in various contexts (i.e., there are schemes of schemes of schemes), both performance and actual representations are overdetermined by nested schemes coming from different levels. As Valsiner (2006) has emphasized, the principle of theoretical parsimony in developmental science must be adjusted to this state of affairs: "in no case may we interpret an observable (i.e., emerged) outcome as being caused by a unitary lower level process (within the hierarchical network of processes), but always as a result of causal systemic processes that operate between levels" (p. 180).

Thus mediations/mediators are multiple and complex and must be analyzed with care. Perhaps for this reason, researchers often miss the fact (emphasized by neo-Piagetians, e.g., Morra et al., 2008) that human development is both dynamically overdetermined and uneven, as most dialectical processes are. It is uneven in the sense that different areas of development exhibit different degrees and density of schemes, because schemes are distinct functional processes and develop locally as a function of experience. Moreover, this is due not only to learning differentials but also to organismic holistic factors. These include Gestaltist principles (e.g., simplicity), effective-complexity limits, momentary organismic states, and other factors. These ideas are related to those put forward, to speak only of classics, by Goldstein, Koffka and other Gestaltists, Piaget, Wallon, Werner, Vygotsky and collaborators, and others. At the limit, this unevenness is such that individual development and individual developmental profiles become unique (e.g., Granott & Parziale, 2002).

Further, some areas of development are facilitating, in the sense that schemes in these areas are often compatible. In contrast, other areas are misleading, because individual schemes in them are in conflict, causing contradictions in the situation. Stable developmental stages are found in misleading situations, because mutually competitive processes exist in them, and mental attention is needed. In the common psychology literature, facilitating versus misleading situations often are confounded or misunderstood.

Problems of Constructivist-Learning Theories of Development

In chapter 1 we highlighted limitations of Piaget's theory and other current developmental theories. We now expand on these problems. We recognized two distinct groups of neo-Piagetian theories: the constructivist-learning theories group, which includes Piaget, and the group we called maturational attention theories. A further comparison between Piaget's constructivist-learning theory and our own maturational-attention theory (Theory of Constructive Operators, TCO) can illustrate problems and limitations of current developmental theories.

Recall that a theory is causal (not only descriptive) when its constructs both explain the theory's descriptive aspects and can be experimentally investigated independently from the descriptive constructs they intend to explain. Piaget's causal theory (**P**) is constituted by three key ideas, also found in different forms in more recent theories:

(**P1**) Knowledge is represented in the form of *schemes* or similar units (i.e., goal-driven blueprints of representations or operations). Schemes are self-propelling units of information or action; they are semantic-pragmatic functional systems that carry or mediate expectancies (e.g., after turning the ignition key, expect the car to start). Although he did not use the term expectancies, J. von Uexküll (1934/1957, 1982) introduced a functionalist/relational concept of *meaning* (possibilities of action or hindrances tacitly offered by objects and situations) to explain functional intertwining between animals and their own experienced milieu. von Uexküll's ideas are precursors of ethology and behavioral biology, and his meaning unit is very close to Kant's schema and to Piaget's concept of scheme.

In the context of starting a standard-transmission car, a Piagetian cognitive scheme may be expressed by *reaching-for-and-turning* the key in the ignition switch, *with the expectancy* that the engine will start. Another scheme could be expressed by pressing the gas pedal, when the person has an expectation of the motor revving or the car moving as a result; the latter is contingent on the current state of the clutch pedal (and this would be a condition in the scheme).

Schemes may incorporate various possibilities of actual praxis: that is, alternative courses of concrete (or mental) action planning for bringing about the expected result. Notice again that a scheme is not just an association between stimuli and responses: there is always (or nearly so) a conditional expectancy—anticipation of outcomes that follow the scheme's application. Furthermore, schemes are dynamic and self-propelling. They tend to apply and contribute to performance as soon as some releasing cues appear (this is Piaget's assimilation principle), unless other incompatible schemes interfere and prevent them from applying. Because schemes are self-propelling, someone who sees a key in the car ignition switch should get the meaning (if they have the scheme in their repertoire) and perhaps be tempted to turn the key. If people do not act in this manner, it is because they anticipate unwanted consequences and preempt or inhibit this tendency by using another scheme incompatible with turning the ignition key (e.g., doing nothing or removing the key). Schemes can produce psychological addictions (e.g., one may sit at the table and eat when not hungry, or check one's cell phone again and again, or rebel/submit to authority without really thinking or intending to), unless contradictory (i.e., opposing) higher-level schemes are used to prevent habitual schemes from applying when unwelcome. Such competition among schemes relates to Piaget's second causal idea, the equilibration principle.

(P2) The organism in its adaptation follows a *principle of equilibration.* Persons or animals adapt to a situation via the self-propelling application (Piaget's assimilation) of sets or clusters of relevant compatible schemes (cued externally or internally) that may be applicable. For instance, my reaction to what somebody says is influenced not only by the content of the utterance but also by my mood; his tone of voice, physical posture, and psychological attitude; the presence of listening others; and so on. The organism seeks satisfaction or internal balance/equilibration, and many schemes are directly or indirectly driven by affective needs, whether positive or aversive. Therefore, clusters of affectively and cognitively compatible schemes that successfully cope with the situation will tend to interconnect and are remembered in coordination, with connections (via associative and/or symbolic learning) that constitute *complex schemes.* We and others call the latter *schemas*, but Piaget called them psychological (functional) structures. Piaget's principle of equilibration assumes that the dynamic organism adapts and reorganizes its schemes. It forms flexible hierarchies (also called heterarchies) of schemes, to ensure that the three complementary aims of the equilibration function mentioned in chapter 1 are promoted by the internal change in the organism (Pascual-Leone & Goodman, 1979). Equilibration produces dynamic syntheses of here-and-now performances by adapting schemes to the situation's resistances to achieve intended performance goals. As mentioned, resistances (see the first epigraph) are aspects or constraints of the situation that help or interfere with agency or praxis (Gibson, 1979; Tolman, 1959, 1961).

(P3) In this manner, *flexible* nested functional hierarchies—often called *heterarchies*—of schemes and schemas get formed by constructivist-learning mechanisms Piaget called *reflective abstraction*. This is Piaget's third idea. The higher levels of these nested hierarchies stand for abstracted schemas of related schemes' coordination. Nested hierarchies tend to occur whenever schemes compatible with currently dominant flexible hierarchies apply together and codetermine useful performances. Thus, whenever schemes or schemas of any hierarchical complexity contribute together to successful performances (and do so maintaining certain repeatable mutual relations as they apply) the resulting functional order among applied schemes is internalized by the brain. This is a deeper meaning of reflective abstraction. This internalization of mutual relations, and their manifestation in tasks and tests, agrees with (and may serve to justify) current concepts of construct validity (Borsboom, Mellenbergh, & van Heerden, 2004; Krantz, Luce, Suppes, & Tversky, 1971; Ullmo, 1967).

In Piaget's sense, *reflective* has the double explicit meaning of a *mental* (conscious or preconscious) *reflection*/projection of meaning into a new higher-level scheme, which *epistemically reflects* (like a mirror's physical reflection) repeatable patterns of functional interrelation that schemes exhibit as they apply to produce performance. We call epistemic reflection this characteristic that Peirce may have called iconicity (Johansen, 1993). For instance, a photo portrait, or a suitable sunset, epistemically reflects (i.e., has iconicity with, is partial icon of) the person in the photo, or a very old person, respectively, because they share characteristics: respectively, some physical features of the body image or short time left before the light of life be extinguished.

In summary, reflective-abstraction processes construct or adapt new higher-level schemes and coordinate previously successful cofunctional and coactivated schemes, to epistemically reflect the mutual relations found among them, which were successful in agency or praxis. Schemes abstract and incorporate (schemes embody) encountered resistances of the task. For instance, a beginner driver of a standard shift car who takes a curve on an upward slope may eventually learn (by trial and error) to accelerate prior to the sloping curve and then change gears downward, to prevent the car from stalling on the slope. Learning often requires several trials; different schemes emerge in performances that preserve desirable functional relations with one another, in space-time, to ensure good results.

Can Learning Explain Development?

Piaget's theory and other cognitive causal theories attempt to explain development using only cognitive learning. However, Piaget's constructivist learning mechanisms (schemes, reflective abstraction, equilibration), even when explicated, may not suffice to account for development in high cognition (see chapter 1; Pascual-Leone, 1983,

1988; Pascual-Leone & Johnson, 2017). A pure learning account may suffice within facilitating situations, but not in misleading situations. (This is also a problem in many connectionist, information-processing, and memory theories.) Selective activation of task-relevant schemes is a function of the subject's past experience (their repertoire of schemes) and of the situation's cues. A *cue* is a match (an active functional correspondence) between a feature/resistance of a situation and a releasing condition (or cue-marker) in a subject's scheme. By definition, no scheme inconvenient for the task is "cued" within facilitating situations, where associative learning suffices and automatic (often perceptual) attention is enough to activate the needed schemes.

Misleading situations, in contrast, demand control of performance, which requires application of mental attention. Indeed, Piaget (1959, p. 47; JPL translation) wrote, "Certainly we could say that when two incompatible schemes tend to apply on an object, the one applying is that one compatible with a larger number of other schemes maximally activated at this moment (this is the formula that Apostel was proposing to us); but it remains to understand why these schemes reach precisely now a maximal degree of activation." Mental attention, executive schemes, and a few other organismic factors described in the TCO are a good answer to Piaget's perplexity.

Piaget's (and Other Researchers') Stage-Transition Problems

Baldwin (1894/1968), great pioneer of child psychology, thought of developmental growth as caused by some sort of cognitive learning. He contrasted *accommodation* (a problem-solving process reminiscent of both Piaget's accommodation and equilibration) with *habit* (driven by assimilation—a self-propelling disposition of schemes to apply, determining representations or performances). Similarly, James (1892/1961) contrasted *effortful* (voluntary-attention driven) *performance* with *habit*, as today we may contrast problem solving with learning. However, as far as we can verify, these researchers, including Piaget, interpreted (mental, effortful) attention and attentional growth as being a consequence, not a cause, of learning and habit.[4] They were, like many current researchers, learning theoreticians who see development as the long consequence of learning. This position was well expressed by Baldwin (1894/1968, p. 465):

> The organism thus acquires a *habit of accommodation* on a higher level. This is attention. … This attention-habit, this centralized function, is not all that attention is. The original excess function [Baldwin's name for an extra source of energy of operative/motor origin] must be kept in view. … Each new accommodation to idea carries a motor [i.e., operative] excess discharge of its own, and this also enters into the sense of attention, making each act of attention, and each sense-type of attention, different, as was said above.

Thus, for Baldwin or Piaget, attention is a conjoint product: an act of accommodation[5] that is followed by learning. Attentional acts for Baldwin express distinct sorts of automatic habits. The quote also shows that for Baldwin and Piaget attention can, in its accommodation moment, encompass a set of elements or schemes simultaneously. This is the *span of attention* or mental working space, which Piaget (1975/1985) called *centration* and expressed as a "field of equilibrium" (a functional semantic-pragmatic coordination) among the schemes involved. Nonetheless, neither Baldwin nor Piaget anywhere makes the claim that size of a mental centration set increases independently from learning as a function of maturation, nor do they think of mental-attentional growth as a major cause of cognitive developmental growth as we first proposed—the transition rule for developmental stages (Pascual-Leone, 1970).

In Piaget, stage transitions remained causally unexplicated, as he and his collaborators admitted (Garcia, 1980; Inhelder et al., 1977; Piaget, 1975/1985). This is a key problem of Piaget, originator of the current developmental-stage construct. For Baldwin, as for Piaget, Hebb, James, Wundt, and others, growth in size (centration set) of attention is just a consequence of learning and habit. They did not investigate other sources of attentional growth—like a mental-energy brain resource that increases maturationally with age in childhood. The construct of an independent mental-attentional capacity, possibly maturational and not reducible to learning or habit, may have first appeared around 1970 (e.g., Kahneman, 1973; Pascual-Leone, 1970; Pascual-Leone & Smith, 1969), unless G. Miller (1956) had considered maturational his magical mental-set number 7 (plus/minus 2), which arguably is not the case.[6]

Our attempt to incorporate mental attention into developmental theory and to interpret developmental stages as the progressive incrementation of complexity processing began the neo-Piagetian movement (Pascual-Leone, 1970; Pascual-Leone & Smith, 1969). Piaget's (1959, 1975/1985) interpretation of growth in his *field of centration/equilibrium* (maximum size of the person's mental set) as being caused by cognitive learning, leads to problems. A first step to correct these problems or paradoxes, some of which we discuss in this chapter, is to recognize dynamic adaptivity in the organism. This adaptivity seems to follow a principle of overdetermination—performances being determined by the dominant cluster of compatible determinants currently active in the organism, as we explained in chapter 1.

Such formulation of a multisided, organismic-overdetermination causality helps to understand the emergence of developmental (organismic, natural) *stages* of growth in humans, distinguishing them from states or steps of change. Piaget (1956) made a clear formulation of the stage construct: this is a model of developmental changes in the psychological organism, a probabilistic invariant of individual age groups (i.e., an expectable sequence of states/relative to situations) that is coupled with a causal model of these changes and their evolution. He indicated characteristics that set the stage construct apart from simpler descriptive constructs such as steps or states. He also emphasized (see the third epigraph) that stages are primordially descriptive theory, intended as instruments for sequential qualitative analysis, and necessary to clarify constructive processes behind developmental change.

According to Piaget (1956), proper stages (**stg**) must at least satisfy these characteristics:

(**stg1**) Stages must demarcate as descriptions a succession of states.

(**stg2**) This order of succession is probabilistically invariant in a context-sensitive way, or is constant.

(**stg3**) The cuts demarcated by stages are natural, in the sense that they epistemically reflect emergence of some new functional-structural landmark(s) in the organism.

(**stg4**) The structural characteristics of every stage are integrated into a structured dynamic whole (functional totality) that characterizes the stage-type individual—or can be so modeled.

(**stg5**) Each stage, by virtue of this integration, exhibits characteristics of a more-or-less stable functional totality (Piaget's "structure d'ensemble") that can be modeled with Piaget's logical models or by other means, to describe functional characteristics of the stage as a whole. Each stage is a new developmental moment of organismic equilibrium.

Steps are sequential descriptive-landmark moments in the changing performance, which may occur across ages and within or across individuals. As such, steps, although useful, do not imply distinctive organismic-process characteristics; they are purely descriptive and perhaps arbitrary. *States* stand for substantive or structural descriptors of process, at a given moment in time; they may not refer to proper developmental change but to processing.

Stages, we claim, are not reliably found in facilitating tasks. In these situations improvements are more or less continuous, and stable stage discontinuities fail to appear. Stages are found in misleading situations. Stage theoreticians, including Piaget, have not formulated with clarity the distinction between facilitating and misleading situations, nor can they explain how misleading situations eventually are overcome during development. Piaget's quote given at the end of the previous section demonstrates his perplexity on the matter.

Learning Paradox: Mental-Attention Growth Is Not Due to Habit

There are two forms of the learning paradox. The more famous (weak) *learning paradox* was formulated by Fodor (1981). A rationalist thinker, Fodor raised the issue that to learn a concept one must already have an idea of the concept to be learned. Thus,

Fodor concluded, to learn a truly new concept (i.e., one that is not already innately prewired) should be impossible. We know of course that this is not so: people do learn concepts truly new to them. This weak learning paradox is easily solved by recognizing that conceptual ("top-down") learning is not the only kind of learning: people and other animals often learn by way of trial-and-error associative ("bottom-up") learning, without prior conceptual ideas about what is to be learned (Pascual-Leone, 1987, 1991a).

There is, however, a *strong learning paradox* that we formulated in the 1970s, which is more difficult (Bereiter, 1985; Juckes, 1991; Pascual-Leone, 1976a, 1987, 1991a). This paradox aims to correct the mistaken belief that abstract and context-specific conceptual forms (e.g., symbolic schemes, complex schemes, mental structural frames), such as those underlying linguistic grammar, Piaget's conservations, or formal reasoning, can be acquired from everyday life by simple associative learning. This empiricist claim and expectation raises a strong paradox: to learn these abstract conceptual forms, a learner should produce in his or her working mind (spontaneously and without the help of prior learning) appropriate complex schemes stemming from diverse and often long praxis, or should be able to generate spontaneously (without enough schemes) the performances from which this direct learning of schemes could result. In either case, when a theory lacks suitable mechanisms for dynamically synthesizing performances in the absence of appropriate corresponding schemes, prior availability of schemes in the subject's repertoire of the schemes is necessary for conceptual learning to occur.

Thus, if the schemes in question are not innate (and they must not be if they involve specific abstract cognition that evolution could not explain), this is a real paradox. This is because abstract relational schemes cannot possibly be acquired by associative learning from the outcome of experiences, when the very emergence of such experiences requires application of the schemes in question. Current neonativists tend to assume (e.g., Renée Baillargeon, 2008; Spelke, 1994) that this sort of specific and abstract functional process (concepts or schemes) is innate. Such a neonativist view is epistemologically objectionable: evolution cannot offer to humans functional processes that could not evolve phylogenetically because they were not useful for activities and agency of other species, but there are more reasons (e.g., Kagan, 2008).

The situational specificity of these complex human schemes (often culture-based) leads to an *evolutionary paradox* when innateness is assumed, as neonativists do when interpreting their important infancy data (see chapter 3). It is incoherent to attribute innateness to such cognitive concepts, principles, regulatory dimensions, and skills that evolution could not have developed across species (Gould, 2002; Johnson & Pascual-Leone, 1989). Indeed, progressive phylogenetic emergence within evolution of such nonevolutionary schemes and processes may be impossible without the hand of God (Johnson & Pascual-Leone, 1989), an option outside scientific explanation. Such a strong learning paradox can be solved by positing innate but general-purpose mechanisms (e.g., brain general resources such as mental attention, neo-Gestaltist field factors, various sorts of learning, affects, motivation, Schemes' Overdetermination of Performance [*SOP*]) that can dynamically synthesize emergent experiences or performances; such general-purpose mechanisms can also exist in lesser animals (e.g., Pascual-Leone, 2006; Striedter, 2005).

Consider a well-known developmental example of strong learning paradox. Piaget and his team were the first to show the three distinct conservations of matter acquired by the child spontaneously at about 7 to 8 years (conservation of substance), 9 to 10 years (conservation of weight), and 11 to 12 years of age (conservation of volume). The order of acquisition and the intervals of about two years between these acquisitions are very well established, although usually not well explained. Piaget proposed an explicit psycho-logic model for substance and assumed that the same logical/structural model would apply to weight and volume (because content-domain material difficulty for substance, weight, and volume is progressively greater, although there is no logical difference among them). He called *horizontal decalage* the particular circumstance when two or more types of task exhibit the same logical structure in terms of psycho-*logical* modeling, but they are acquired with an invariant delay or "decalage" (Piaget, 1948/1963). He attributed this decalage in the three conservations to time taken by subjects to incorporate to their already acquired psycho-logical model the specific material or causal-situational factors of weight and volume.

For instance, an added constant weight must be intuitively attached to every one of the (intuitively imagined) "molecules"/particles of substance, to thus attain intuitive weight-conservation invariance. This "molecular" weight assignment results in an acquired psycho-logical grouping structure of weight (a mathematical group-like or "groupement" system-structure, Piaget's name for such model). This structure must be well coordinated with the psycho-logical grouping of substance. These developments take time, causing a horizontal decalage. Piaget gives a similar account for volume. The time taken by these logical-grouping coordinations explains horizontal decalages. He contrasted this sort of decalage with the *vertical decalages* caused (by definition) by difference in the complexity of tasks in their respective logical models. The models of Piaget, framed with his chosen (content-free and generic) psycho-logic, compelled him to believe that conservation decalages are horizontal. However, in our own theoretical modeling, using metasubjective task analyses (MTA) and including the tasks' nonlogical substantive "molecular" aspects, the decalages appear to be "vertical" (i.e., due to differences in computational complexity, see chapter 8).

Here is a more detailed account for those unfamiliar with the issue. When, in the conservation tasks, the child is shown two equal balls of Plasticine, presented as pretend candy, and has accepted that the two balls have the same in amount (i.e., eating the red ball would yield as much food as the blue ball), the tester then rolls one of the balls out to produce a long and thin "sausage." Nonconserver children (e.g., 5-year-olds) often believe that after being rolled into a sausage, the amount of substance decreases, because the ball seems perceptually to have more amount than the sausage. At about 6 years children often admit that if the sausage were rolled back into a ball it would be the same amount, but now there is less.[7] Two years later, at about age 7 or 8, children tend to answer that sausage and ball have the same amount, although the sausage looks thinner, because (children often argue) no matter what was taken away or added. These children distinguish well between an object's appearance (its perceptual/manifest facets that can change when we manipulate the Plasticine) and the matter that the real object contains.

Psychologists often call *proximal object* the functional totality of perceptual facets of an object. They call the object's actual functional (potentially displaceable or movable) totality of matter the *distal object* (a cognitively deeper spatial-temporal probabilistic invariant). The distal object is a functional invariant that always emerges during cognitive activities or transactions with the object—from one perceptual facet (or apparent presentation) to another, and back to the former facet or another, again and again (this is the enduring relational reality of the Real object). Reversible operations interconnect the many facets (perceptual presentations) into a coordinated system of variations, which is the distal object's functional totality. Thus, reversible operations transform one facet into another, preserving as invariant the distal object itself (Beth & Piaget, 1961).

Piaget modeled such a system of reversible sensorimotor transformations with one of his logical structures, the type of grouping he calls *grouping of vicariances* applied in the experiential/infralogical domain (Beth & Piaget, 1961; Piaget, 1975/1985). This vicariances grouping (which describes as a dynamic system the variation of the objects' perceptual facets—in Latin, *vicarius* means "a substitute") is a mental operational system that allows children to understand why any facet of an object is equivalent and can be substituted for another facet (within the distal object in question) and every facet can be reversibly transformed into another.[8] It should be clear that the job of this grouping of vicariances is just to coordinate facets (proximal objects) within their shared distal object (integrating all facets that have an invariant amount of matter, until some amount is added or taken away). This is a complex experiential symbolic structure, which is learned cognitively only when the subject has alternative ways to intuitively and perceptually coordinate facets that belong together, because they preserve an invariant amount.

This sort of perceptual relating and "hypothesizing" (Rock, 1983) makes the distal object emerge as a package of functional invariants that express the encountered "resistances," as the first epigraph suggests. According to Piaget's theory, such "hypothesizing" inference should happen when the child already has in his or her working mind an *infralogical* (purely experiential) grouping of vicariances. However, this reasoning is circular: *unless* the psycho-logical models are innate, they could be experientially abstracted only when the child intuitively extracts and coordinates the facets, guided by an empirically emerging functional invariant (as happens with the number 7 and its facets—its grouping of vicariances as discussed in footnote 8). But the functional invariants of Piaget's conservation, that is, the abstract concept of *conserved matter*, cannot be empirically appraised independently of its facets (the psycho-logical grouping). Thus, an empirical/experiential psycho-logic does not explain the new facet-coordination and the emergence of its invariant. Nonetheless, Piaget thought that conservations are the products of experience and cannot be innate. Notice that if conservations were posited to be innate (as neonativists may suggest), an evolutionary paradox would result because of the conceptual-relational abstractness of this specific logical structure. (Johnson & Pascual-Leone, 1989, argued this point with regard to language.) Piaget's account of conservations falls under a strong learning paradox: his grouping model could not be innate and can only be learned after conservations have been otherwise attained by the child (Pascual-Leone, Goodman, Ammon, & Subelman, 1978; see also chapter 8).

Solving the Learning Paradox in Conservation Tasks

For Piaget, children need a vicariance grouping to solve conservation tasks, but vicariance groupings cannot be learned without a child's ability to recognize the invariance of conservation (substance/weight/volume) in the distal objects' "molecules." A belief in conservation may emerge without using such logical grouping, however, if there were an endogenous booster of schemes' activation, a *mental attention* that grows with developmental maturation, and it (along with the *SOP* overdetermination resolution—see chapter 1) can coordinate the heuristically needed schemes/aspects of the task. According to our theory, this growth of mental attention induces a child of 7 to 8 years (who can coordinate a mental set of three task-relevant symbolic schemes) to infer substance invariance under reversible transformations, whenever the perceived object facets do not contradict the assumption of a single distal object. Such conservation (**Cons**) judgment (see figure 2.1) is contingent on three task-relevant schemes brought together in the child's working mind:

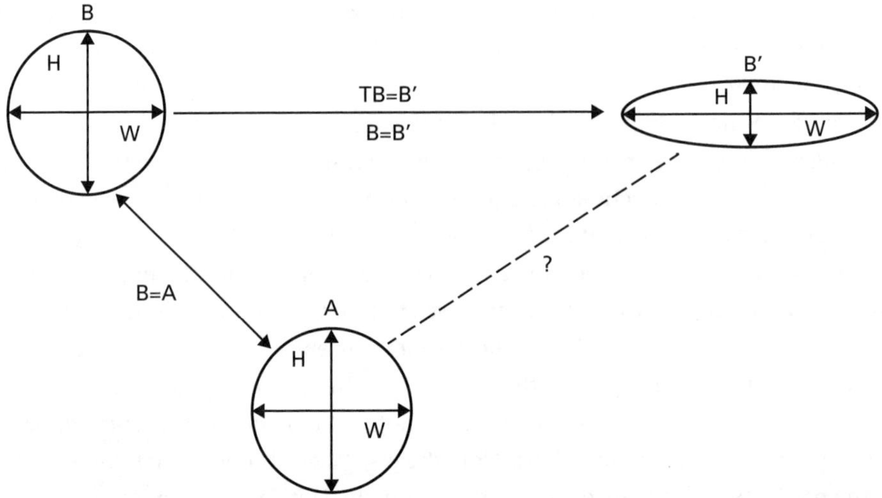

Figure 2.1
Neo-Piagetian depiction of substance conservation task. B and A represent balls of Plasticine.

(**Cons1**) The blue ball **B** was the same as red ball **A**.

(**Cons2**) **B** was transformed (T) into sausage **B′**, which is longer than blue/red ball, but also thinner (although a particular child may or not notice this contrast).

(**Cons3**) The operative transformation that changed blue ball **B** into sausage **B′** did not cause any loss or gain of matter.

With these three schemes in mind, a child may infer intuitively (nonverbally but symbolically), via overdetermination (*SOP*), that when red ball has equal amount as blue ball (because of **Cons1**), and blue ball has equal amount as blue sausage (because of **Cons2** and **Cons3**), then red ball must have equal amount as blue sausage (because of simplicity—the brain's internal field, or *F*, factor). This sort of inference (similar to "equal plus equal yields equal" or "the friends of my friends are my friends") is an instance of what Peirce called an abduction (Johansen, 1993).

For Peirce, abduction is a dynamic form of inference such that from a general proposition (e.g., major premise in a syllogism) and an intended conclusion, the thinker "infers the minor premise," that is, the particular assumption or hypothesis that makes the conclusion intuitively plausible (Apel, 1995, p. 39). For instance, having observed (in the conservation situation) that blue ball is equal to red ball (**Cons1: B = A**) and that blue ball is equal to blue sausage (**Cons2** and **Cons3: B = B′**), the child concludes by induction that in this situation all known relations among objects are equality relations;

this is a general proposition. Then by applying the general proposition onto the intended conclusion, which here is a question about whether **A** is equal to **B′** (**A ? B′**), the child reaches the intuitive causal belief of equality (*via implicit abduction*, i.e., a dynamic, not strictly logical, overdetermination, an *F-SOP process*). In chapter 8 we present this task analysis in more detail.

This is a heuristic, not a logical, deduction. The general proposition was inductively inferred, and it, with the intended conclusion, brings to mind the belief/hypothesis of equality. The child's experience can confirm (or falsify) later the so attained intuitive thought. Although Peirce does not always make a distinction, his abductions can be either explicit (the thinker fully conscious of the thought) or implicit (the thinker is unaware). We are assuming an implicit abduction. To avoid a learning paradox, this inference must occur to the child before she acquires the psycho-logical groupings for Conservation.

Notice that when conservation first appears the problem is truly novel: learning cannot be a cause. Rather, dynamic synthesis (sudden coordination) must produce conservation as an inference/hypothesis. Thus, learning paradoxes can be solved when the organism uses general-purpose resources or hidden operators that enable the inference. These organismic-causal abduction "packages" produced by general organismic models are context sensitive and individual-difference sensitive, like our TCO, and can make claims about Reality (i.e., ontological claims). These organismic causal claims are always conditional (i.e., *situated* or relative to situations) and intrinsically probabilistic. They exhibit a *multisided* organismic *causality* (many compatible overdetermining causes)—as, in the second epigraph, Cullenberg (1996) has asserted within sociology.

Low Cognition versus High Cognition in Plato's Problem: Paradoxes of Vygotsky and Piaget

A common paradox of cognitive-developmental theories is what we have called *Plato's problem* (Pascual-Leone, 1996a, 1996b). Plato recognized a radical difference (and functional complementarity) between two sorts of knowing. One is the world of things (or *world of becoming*), related to what Vygotsky has called *low cognitive functions,* largely made of automatic/unreflected sensorial and perceptual experience. The other is the *world of ideas or being,* related to Vygotsky's *high cognitive functions* and made up of conceptual-analytical (reflected and perhaps reflective) mentation. For Plato these two worlds had different origins. The former was a direct product of concrete experience, and the latter expressed gods' knowing, given to us by them. Today, we would say that the former entails bottom-up transmission and the latter top-down transmission. Plato did not see a problem in his formulation, but for scientific researchers gods cannot be a cause. High cognition, although radically distinct from low cognition, must also come from causal organismic interactions conditioned by experience and maturation.

How can low and high cognition emerge according to different formulas but be induced by same set of causal factors? This is Plato's problem. Like Plato, many current researchers in cognition and development fail to see the problem, and they may just replace gods by Innateness, Maturation, Evolution, Logic, Learning, Culture, Language, or Brain—one or more isolated abstract categories often written in capitals. Vygotsky and Piaget saw this problem. For Vygotsky, the synthesis of experience and conceptualization (the two worlds of Plato) results from adaptive/dialectical synthesis of *action* (goal-directed activity—agency/praxis) with *symbols and tools*. This synthesis gives meaning to the means-end relations between actions/procedures and reality. This is Vygotsky's *method of double stimulation*—stimulus-object (figurative schemes in our model) and stimulus-means (operative schemes in our model; R. Miller, 2011; Valsiner, 2006; Vygotsky & Luria, 1994). Vygotsky's formulation did grasp the importance of external intentional action to extract some causal models for the activity in question—internalizing them from experience. These are symbolic operative-and-figurative models: "what was an outward sign operation... is now transformed into a new intra-psychological layer and gives birth to a new psychological system" (Vygotsky & Luria, 1994, p. 155).

However, in Vygotsky's historico-cultural model of high cognition (in which high cognition emerges via internalization from external and intersubjective experiences), two unexplained problem-solving moments exist that paradoxically cannot be addressed within his theory. One is the act of dynamic synthesis (i.e., the invention in the idea-inventor's mind) of the to-be-transmitted novel ideas (e.g., a machine, symbol, method, procedure). Another is the act of dynamic synthesis within the recipient of the truly novel communicated idea, who must re-create the idea in his or her own working mind, to allow internalization and use. Some contemporary models of cognition and development, particularly social constructionism (Terwee, 1995), have the same problem. Even if historico-cultural theory could explain affective and social-cognitive choices that motivate the idea-inventor (and/or the recipient), a lack of explained mental mediation (of creative dynamic problem-solving processes) for these two moments makes historico-cultural transmission incomplete and unexplicated from an organismic-causal perspective.

Vygotsky's theory lacks explicit organismic-causal problem-solving mechanisms. It is a historico-cultural, cognitive-learning theory that hinders organismic explanation of these two moments. We call it *Vygotsky's paradox*, important because many current cognitive scientists, particularly those in social learning, computer-simulation/connectionism, or social constructionism, unwittingly suffer from this problem (R. Miller, 2011, gives a similar analysis). However, when a dialectical constructivist model of the organism is combined with historico-cultural theory, Vygotsky's and his followers' major insights about human development become free of paradox and appear extremely cogent.

Piaget investigated transition from low- to high-cognition using the method of developmental stages. This is a descriptive approach (see the third epigraph). He saw stages as marking discontinuities in developmental change. Yet, he knew that learning is continuous, as is often maturation. Multiple (dialectically interacting) deeper processes must exist in the organism to organismically cause developmental discontinuities in performance. Piaget called these deeper causal factors of transition *regulations*—the source of equilibration. Paradoxically, even though he understood that stages are only descriptive, he never investigated organismic factors ("regulations") responsible for them. Had he investigated organismic causes for stages, Piaget might have solved Vygotsky's paradox. Instead he followed Baldwin (1894/1968) and adopted a constructivist-learning theory to explain emergence of stages, leaving "equilibration" to be a name for the unexplicated problem (see chapter 1). Failure to clarify organismic processes makes both Piaget's and Vygotsky's theories (along with other current theories) fall under various instances of the learning paradox.

Language and Communication Paradoxes

There are two sorts of construction in psychology: *psychogenetic constructions*, which determine transitions of the human organism from one developmental *stage* to the next, versus *generative constructions*, which cause dynamic syntheses that produce new *states* of experiences or performances within particular situations. Language and sociocultural know-how are particularly important in generative constructions, because cognitive/affective situational interpretations are much influenced by them. Thought rooted in the present gets much power from language and this launches high cognition. In contrast, maturation[9] plays an important role in generative construction only when maturational (biogenetic) cognitive brain resources (such as mental attention) are insufficient to cope with the situation. In psychogenetic construction, the converse is the case: because general stages express brain-resource limitations, the role of maturational determinants is most important. To think otherwise, attributing to language a dominant role in psychogenetic constructions, as when language is seen as a cause and not a consequence of thinking, brings about problems and paradoxes.

The *language paradox* consists in the fact that semantic clarity and the very construction of language depend maximally on experiential nonlinguistic processes. Via linguistic communication uncommon nonlinguistic meaning (life-world or thought experiences) can be indicated symbolically only. Thus a nonlinguistic semantic process is needed to clarify the meaning. Such experiential knowledge (nonverbal schemes/schemas of all sorts, referring to present, past, and intentional future) serves as referential meaning domain. A simple illustration of these nonverbal roots is found in learning a second language. To enrich lexicon in the second language it is best to imagine the intended nonverbal, experiential meanings. Repeating the paired words is less useful. Rather, it is best to intuitively imagine the word-relevant objects, feelings, actions, emotions, and so forth that bring to mind experiences and life-world meanings.

Consider now the *communication paradox*. Education and public activity take place by way of communication. If communication is hindered, so also are public activity and education (including psychotherapy, in part an alternative form of radical education). To receive and understand any form of communication, *receivers* often must problem-solve and have in their repertoire suitable schemes congruent with the *producers'* intended meaning (expressing producers' own schemes). Thus, receivers can only truly understand what they, in some sense, already knew. How then is it possible to communicate anything truly new? This is the communication paradox; its solution may ease problems for the educational enterprise and public-exchange activities.

Karl Jaspers first formulated this communication paradox: "That which becomes clear to me seems then to be that which I have actually always known" (Jaspers, 1959, p. 65). Piaget, using more psychological language, expressed the same idea (tacitly alluding to reflective abstraction) by claiming, "But I would say that anything is only understood to the extent that it is reinvented" (Jennings, 1967, p. 830).

The solution to this paradox appears when we realize that there are two fundamentally distinct forms of communication, standard and radical. In *standard communication* the receiver already possesses schemes he or she needs in order to comprehend the true meaning of what is being said. In this case, communication is a quasi-mechanical transmission of information. This is the conception of communication that empiricists, and with them some journalists, economists, public speakers, fiction writers, and some educators seem to have. *Radical communication* is distinct, however, because here receivers do not have the schemes needed for deep understanding of what is said. Therefore, to succeed, the communicator and receiver must, consciously or not, spend time in preparatory communication (perhaps using Socratic or other form of dialogue/dialectics) that fosters in the receiver emergence of schemes/schemas needed to achieve deep understanding of the communicator's message. Concurrently, the receiver must spend time working through the message, reviewing the material to problem-solve in understanding. (For this reason, in this book we have presented themes repeatedly from various angles, to propitiate understanding and the emergence of these themes as functional invariants in the receiver's mind.) Much working-mind effort and communication time may be needed to achieve radical communication. This radical practice, frequent in good educational exchanges, solves the paradox.

Radical communication often is not recognized for what it is. A frequent presumption is that initial efforts to communicate are not so much directed to build within the receiver needed schemes as they are to produce clear-and-simple relevant messages that receivers can understand. Yet clarity and simplicity of messages are relative to the receivers' capacity to assimilate intended meanings, using their own schemes, and then internalize (learn) the obtained meaning in new schemes (reflective abstraction). Radical communication fails when simplification is done by distorting rather than *properly deconstructing* the intended message. The term *proper* here means that adopted and expressed simplifications can be later expanded (differentiated and explicated) to their full complexity, when needed, without causing contradictions or unwanted distortions. Failure to communicate (common in radical communication at first) can be due to prior-learning deficiency (lack of necessary schemes in the receiver) or to a mismatch between the mental-attentional demand of the task and the communication-receiver's mental-attentional capacity. A learning deficiency can be solved via radical communication, or with special content-based training, as schools usually do. A mismatch between child's *M*-capacity and the task's *M*-demand requires more complex interventions, to reduce the communicative task complexity to the point when task's difficulty (*M*-demand) may not exceed the receiver's comprehension mental-span (*M*-capacity; Arsalidou, Pascual-Leone, & Johnson, 2010; Pascual-Leone & Johnson, 2005, 2011; Pascual-Leone, Johnson, & Agostino, 2010).

Meaning of Meaning

Consider an actual life example of adult communication. It illustrates how meaning (in both language and communication) is embedded in personal history, contextualized, and largely nonlinguistic. Imagine that a man and a woman are walking on the street after leaving a subway station. It is softly raining. He carries an open umbrella; she does not. He asks, "Don't you have an umbrella?" "Yes," she replies, "but I don't wish to bring it out." So, the man asks, "Would you like to share my umbrella?" This exchange illustrates well how the meaning of language is intertwined with felt experience, prospective imagination, and thought, to synthesize a dynamic act of judgment—an instance of mental generative construction by the working mind. Indeed, it is raining, and he does not like to get wet, so he opens his umbrella to protect himself. She does not mind a little rain and decides that getting her umbrella out is not worth it. He notices they are getting wet and assumes that she, like him, dislikes that. So, he asks whether she has an umbrella, and from her reply he infers that she may not want to search for it in her bag. Thinking that she (like him) does not like getting wet, he offers to share his umbrella. The meaning of this exchange is clear, but not because of language.

Meaning is clear when language is inserted inside a nonverbal (subjective and objective) life-world communicative context. Notice that for an object, percept, action, or thought (as Piaget, von Uexküll, and Tolman appeared to believe) meaning is all the schemes that can bear/apply on the verbal-term/object/percept/action/thought in question, and all those schemes that cannot, including their interrelations. Language is necessary to make meaning explicitly clear to the Other, or to oneself, unless consequences of active praxis make it clear, but much (if not all) meaning is nonlinguistic. Its referential domain is the person's *life-world* (Husserl, 1970). Mentation, often nonverbal, interprets and justifies verbal exchanges, enabling meaningful communicative praxis with contextualized dialogue.

The need to situationally contextualize verbal exchanges to extract meaning occurs at all ages. Although babies may utter words at around 10 to 12 months and articulate some communicative language at about 18 to 20 months, their nonverbal communication with gestures and action is much more advanced than their language. As Ortega pointed out, babies learn to use language because they have meaning and have the urge to tell. In Ortega's (1935/1980) view humans have a "hyper-functional" communicative imagination when compared to other apes, and this richly flowing imaginative mentation presses to communicate, possibly causing the discovery, learning, and invention of spoken language (a view consistent with Tomasello 2008, 2018). *Intentionality* is the words' or images' or gestures' (any symbols) tacit semantic-pragmatic reference to an intended experiential referent of the speaker/thinker's nonverbal life-world. Brentano introduced the intentionality concept to express the internal object-directedness of thought (Audi, 1995). Current philosophers and linguists believe (e.g., Searle, 1983) that intentionality precedes emergence of language. These active cognitive processes, driven by affective goals (motives), may be causing performance via overdetermination (Piaget's equilibration, Vygotsky's dynamic synthesis), "which unites in complex cooperation and in complex combinations various separate elementary functions" (Vygotsky & Luria, 1994, p. 162). Internalization of language and logical forms is only possible because of maturational mental attention. It allows children to keep in mind and coordinate distinct relevant schemes and to inhibit those now irrelevant.

Early empirical evidence suggesting that psychogenetic constructions depend on maturational processes other than language is found in Piaget's qualitative research program with his three babies, some of which we discuss in chapters 3 and 9. This research called attention to the early emergence of expectancies and intentionality (as Tolman, 1959, 1961, had done earlier within behaviorist learning theory). Piaget investigated in detail the onset of intentionality prior to language and the problem-solving acts that this intentionality promotes. Piaget (1948/1963) Observation 133 describes his 9-month-old daughter, Jacqueline: "[S]he likes the grape fruit in a glass but not the soup in a bowl.... When the spoon comes out of the glass she opens her mouth wide, whereas when it comes from the bowl, her mouth remains closed" (p. 249). Her mother tries to trick her in various ways to take the soup, but without success. Two months later, Jacqueline does not even need to look at the spoon: she knows from the sound (of spoon hitting container) whether it comes from glass or bowl and closes her mouth with the bowl's sound. Jacqueline still refuses her soup at 10 months, but her mother tricks her. Without Jacqueline's noticing, her mother hits the glass and silently takes food from bowl, leading the baby to open her mouth and take the soup. (Piaget does not report Jacqueline's subsequent reaction!)

This example shows very well the 10-month-old's discriminating (nonverbal) intentionality as she is fed by her mother. Sound becomes a sign (a protosymbol) of soup or juice. Jacqueline's assertive operative choice reveals a complex intentional act—discriminative meaning that via willful praxis does not need language. We can symbolize the baby's "mental state" in this sensorimotor-inferential activity. Call TAKE and NOT-TAKE the two *operative* processes[10] that enable Jacqueline to take food into her mouth or to refuse and not to take it. Call *fruit, *soup, #glass-sound, and #bowl-sound, the four *figurative* processes (i.e., representational schemes) relating to the operative processes in question. The mental state is a competition between two alternative strategies (complex operative schemes) of the baby's working mind, each opposed to the other: the dominant one is that cued more strongly (sounds or visuals) by the situation:

TAKE (#glass-sound, *fruit) ➔ OPEN.mouth
Versus NOT-TAKE (#bowl-sound, *soup) ➔ CLOSE.mouth **(f1)**

The schemes #glass-sound and #bowl-sound are not actual sounds but the child's categorization of these sounds as predicative markers (code parameters or "discriminant stimuli" of the schemes); that is, respectively, the fruit and the soup schemes. Further, *fruit and *soup are not the actually present objects but imaginal anticipations of taste and so on, cued by the sound-marker (#) schemes. The *schemes are figurative expectancies of objects that prompt the actions TAKE or NOT-TAKE. This split of alternative formulas summarizes Jacqueline's decision moment as reported by Piaget in his Observation 133 [0:10(26)]. This is only the "decision-choice" step of Jacqueline's conduct in the task. In earlier sequences reported by Piaget [0:9(16)], when the child does not use sound but looks at the container, we can replace in the same formulas the sound schemes by the schemes #glass-container and #bowl-container, respectively. Notice that the prefix # in these schemes identifies *parameters* that function as "discriminant cues" for the action schemes that precede them in the formula. After the arrow (➔, which stands for overdetermination) we indicate the result of the action expected by Jacqueline, the mental state guiding the subsequent take or not-take act.

Note that these contradictory *strategies* have the same problem situation and complexity: in both cases three distinct schemes (one operative, two figuratives) must be coordinated to ensure a good (affectively motivated) outcome. If the three schemes are not jointly active and coordinated, the outcome (tasting the intended food) cannot be invariantly obtained. But because the schemes before the arrow are not cued by the external situation (perceptual attention does not evoke these schemes), they must be internally activated by the child's sensorimotor mental attention (which in chapter 3 we call Matt.sm). In chapter 3 we will show that during sensorimotor stage IV described by Piaget (i.e., from 8 to about 11 months) children can simultaneously coordinate three purely perceptual-motor (not fully symbolic but presymbolic) schemes. Thus, distal-object meaning and intentional acts of judgment need both schemes and mental attention to be dynamically synthesized.

Problems of Emotion, Cognition, and Personal Schemes

Emotions (affective *and* cognitive, hybrid schemes) are part of the schemes we call *personal* because they help to form the personality; they are often more visible in humans than "pure" cognition. One may argue that pure cognition does not exist, being always motivated and colored by affect/feelings/emotions. Some psychologists may treat affects/emotions as if they were not separable from cognition, although this is not the view of classic or information-processing theoreticians of emotion (Fraisse, 1963; Frijda, 1987). For instance, clinicians (e.g., emotionally oriented therapists, psycho-dynamic analysts, or cognitive behaviorists) as well as some neuroscientists (e.g., Pascual-Leone et al., 2015; Pessoa, 2013) do acknowledge within actual performance a close and frequent interaction between emotions and cognition, although perhaps questioning the complete distinctiveness between affects/emotions and cognition. Both aspects are often seen as conjoint and not separable (neurologically or psychologically) in organismic processing. Piaget in fact had a similar conception, because for him all scheme units are both cognitive and affective/emotive. For him cognition provides the descriptive (or causal) link of schemes with reality/Reality, and affect is the schemes' own self-propelling "energy"/motivation.

This is in partial accord with the frequent intertwining of four key sorts of entities (i.e, affect/emotions, cognitions, bodily reactions, and external reality/Reality or contextual circumstances). Such formulation leaves unanswered, however, why we distinguish among these different entities in our epistemic (intuitive psychology) theoretical analyses and why we assign to them distinct functional characteristics. If emotions/affects are functionally distinct from cognition they must also have, above and beyond their intertwining, differential neuroscience characteristics and dedicated neural sites. Essential differential characteristics, like those contrasting affects with emotions and affects/emotions with cognition, cannot just be epistemological (knowledge based) when they are *essential*. They must also be ontological (part of Reality and with distinct related networks in the brain). This ontological distinctiveness between cognition and affect is perfectly compatible with their epistemological intertwining in performance, both in psychology and brain processes.

To solve these problems, we call *affects* the innate-propensity causes (i.e., the vital-value category schemes—Edelman, 1989; or, in clinical, the organismic "forces" such as the "psychic energy" of Jung: Campbell, 1971). Affects are distinct schemes within the person's repertoire of schemes and have distinct tonalities (colors, modes) of feeling and emotion, which provide *vital* or life *value appraisals* and *propensities* ("psychic force/energy") to the persons' objective and subjective experience. We call *feelings* the pure, often subdued, expressions of affects in consciousness under certain circumstances, intertwined with essential bodily experiences and cognition (Pascual-Leone, 1990a, 1991b). We call *emotions* (e.g., Fraisse, 1963) the often strong feelings and expression (in body and mind) of negative or positive affects intertwined with corresponding cognitions.

Emotions bring unconscious affects into consciousness, often after a situational appraisal (Frijda, 1987) that evaluates the current situation as challenging or uncertain. Emotions are strong feelings (Damasio, 1999) acting like "forces" that can change the way intentionality, motives, cognition, and motor activity react and evolve in the situation. However, the force or "energy" they bring to consciousness is not their mark. Their mark is the emotional *appraisal* and subsequent affective tones and conscious or unconscious intentionalities they bring into the situation, as expectancies or evaluation of likely future consequences—a qualitative vital value attributed to the situation that may inform and codetermine action.

The problem is that neither the intensity ("force") nor the qualitative value (appraisal) can take place unless affects apply on cognition, which self-consciousness may then appraise. Cognition as such is a Real (truth function) mode of coping with life experience. It has no vital value or "force" of its own, other than the assimilatory tendency of cognitive schemes functioning in the brain and body activity. Cognitive processes tell only about contents and context of experience: their only evaluation is a probabilistic *truth value*, appraising, anticipating, or seeking a match between resistances encountered in Reality and the representation or planned action that cognitive processes can configure.

Thus, there is an *emotion formulation* problem. Experientially, emotion may appear as a feeling-process utterly distinct from cognition, but an emotion always unconsciously integrates three different but intertwined types of process. These are (1) some characteristic bodily experiences; (2) a personally challenging (more or less strong) negative or positive "psychic force" with its affective-and-cognitive appraisal that assigns vital values to the situation; and (3) a distinct cognitive description, interpretation/expectancy, or representation, assigning a truth value (true or false, if we simplify) to situations and contexts cognitively related or not to other (similar or not) experiences. Because neurological and psychological intertwining of the three components is strong, we can understand clinicians and neuroscientists like Pessoa (2013), who may be tempted to consider affect/emotion and cognition as a single category of processing. The problem is that this distinction is essential to analyze what is going on, and it should be maintained.

Critical constructivist process/task analysis also requires such distinction. The following example is an instance of emergent emotion generated from a real-life cognitive-affective symbolic experience that uses mental attention in a facilitating situation, an experience that could change the person's way of coping in life. A person suffers an accident as passenger (*A*) is in a car driven by *B*. In the rain, the car leaves the road, skipping out of control onto wet sloping grass, speeding as it moves and reaching the end of the slope at the riverbank. Then it flies 12 meters and falls into the river, where passenger *A* and driver *B* risked crashing into a huge rock. Although, surprisingly, they were unhurt, *A* kept for years a hard-to-control anxiety and a fear reaction whenever she was in a car driven by *B*, if driving circumstances seemed dangerous (e.g., passing or coming close to another car). This learned real-life emotional reaction could be dismissed as an instance of one-trial classical conditioning (an instance of associative learning). Doing so, however, obscures the fact that emotion-driven actual mentations (the mind's symbolic thinking processes) are involved and that this single experience was effortlessly and spontaneously (automatically) synthesized within *A*'s brain, possibly due to the strong negative feeling/affects. A result of this experience was a complex schema coordinating other simpler *shadow* (i.e., marginal but emotion-loaded and forceful) schemes into an overpowering anticipation of future danger. This soon automatized (signalic or symbolic) schema can be represented as follows:

> WHENEVER ([*A* is driven in a car] AND [the driver is *B*] AND [the present driving circumstances seem risky]), ANTICIPATE THAT [a life-threatening car accident is about to happen to *A* and *B*]. (f2)

Words in capitals indicate the overall semantic-pragmatic framework—an automatized *complex superordinate operative scheme* (at least initially unconscious) expressing a conditional affect-driven expectancy made possible by automatic-and-mental attention. The automatic schema coordinates four cognitive schemes demarcated with brackets [ ] and described in English, although they stand for purely experiential (nonlinguistic) knowledge. In the working mind of *A*, coactivation of this schema and its three subordinate cognitive schemes within the emotion-loaded facilitating situation automatically brings the strong expectation that an accident will occur. Because situations here (the original and subsequent) are facilitating, both automatic-perceptual and mental attention can conjointly boost this processing—making the signalic-and-symbolic emotional experience accessible to young individuals.

ANTICIPATE-THAT is the dynamic synthesis via overdetermination of the schemes preceding it (which in other formulas is represented by ➔), that is, a signalic/automatic and symbolic act facilitated by affect, which the person may eventually become aware of. These cognitive schemes must be coactivated, that is, coexist in the emotion-driven working mind as distinct and simultaneous, to evoke (via overdetermination) the actual fear of an impending car accident. The example illustrates how the working mind can, from affects blended within cognitive situations, create and maintain emotions. It also shows the epistemo-ontological problem of emotions. They are automatically constructed, albeit unconsciously, by blending affects with cognitions (i.e., vital-value schemes with truth-value schemes); they are products of, but not the same as, affects.

Perception and thinking usually are driven by both affect and cognition: an affective-and-cognitive criterion of cognitive relevance that may initiate perception or thinking (an implicit evaluation within the situation of the personal importance of actual here-and-now perception or expectable future). This is crucial in most sorts of cognition. The local coordination of schemes that can appraise cognitive relevance includes as dominant some affects. These tacit *affective goals* (or motives) convert into *cognitive goals* causing *motivation* and setting, via appraisal, a *criterion of relevance*. These processes may be related to activities in the brain's anterior cingulate, insula, and amygdala, sites of affect that we discuss in chapters 10 and 11.

Other Problematic Aspects of Cognition and Development

When one adopts a learning approach that does not recognize schemes and hidden brain resources/operators, many other (usually ignored) problems and paradoxes can be encountered. We mention very briefly three final paradoxes without giving concrete examples.

U-Shaped Curves of Developmental Growth

Strauss (1982) named this paradox, well known to neo-Piagetians who understand misleading situations. It consists in the finding that some tasks (from either low or high cognition) are solved easily early in development but become much harder for older children. This could happen when the task can be solved using different alternative strategies. In these sorts of tasks children could use a global (perceptual) strategy that is facilitating, although the same task becomes misleading when more analytical strategies are adopted. This sort of strategy-drifting tasks may exhibit a U-shaped curve in the developmental proportion-pass performance. Young children, who use low-cognition enabling facilitating global strategies, may obtain an early high performance. However, older children, using a high-cognition strategy and experiencing the task as misleading, would obtain worse performance, until finally they reach an age/stage level when performance is high again. This paradox is easily explained by our theory-guided finding (Pascual-Leone, 1969, 1989; Pascual-Leone, Johnson, Baskind, Dworsky, & Severtson, 2000; Pascual-Leone & Morra, 1991) that misleadingness can emerge at some chronological age (at developmental levels when misleading aspects/schemes are fully attended to, which influences performance), although it did not exist in the same task at an earlier age. Indeed, misleadingness often is caused by unconscious schemes, learned and automatized from activity/agency in other earlier situations. Then task difficulty will disappear later when the misleading strategy's *M*-demand (*Md*) can be controlled in older children with greater *M*-capacity (*Mp*). This is a case of *Mp/Md trade-off* in tasks (Pascual-Leone & Baillargeon, 1994), as discussed in chapters 1, 3, and 7. The existence of horizontal decalages, discussed above, can also be interpreted as an instance of this paradox.

Paradox of Truly Novel (Really Creative) Performances

These are never-learned performances that are not compounded from learned segments, nor can they be innate. Such performances (truly creative inventions are examples) appear as serendipitously synthesized. They require the use not only of schemes of all sorts but also of various general-purpose resource operators mentioned in chapters 1 and 7 and throughout the book. Truly novel performance results from implicit interactions among schemes in the person, aided by dynamic organismic processes of the working mind.

Paradox of Functional Totalities Without a Totalizer (Homunculus)

Complex creative people in the sciences, as well as in philosophy, human and social sciences, the arts, human performance, dance, advanced sport, and so on exhibit in their performances, each in his or her own way, a completeness and closure that gives the impression the work was produced by a functional totality: a flexibly organized and heterarchically unified organization. However, to avoid paradoxes of recurrent homunculi, this plausible functional totality cannot have a totalizer—a point raised by Sartre in his *Critique of Dialectical Reason* (Aronson, 1987). Many truly novel performances seem holistically organized. This is found in science (e.g., theoreticians) and is easier to see in the arts, where artists (e.g., writers, dancers, actors, musicians) can individually adapt doing/playing a given project or program, or improvise, and yet achieve together integrated performances that express meaning as totalities. Because such performances are often truly novel, they point to general-purpose regulations in the brain helping to produce these processes (i.e., resource operators and principles). Organismic general-purpose processes must creatively synthesize these performances as totalities without a totalizer.

Figure 3.1
Jasper feeding crackers to the screen image of his uncle Rafael. (Photograph by Antonio Pascual-Leone.)

3 Emergence of Mental Attention in Infancy: From Sensorimotor to Symbolic Processing

In memory of Dr. Nancy Benson-Hamstra

We examine cognitive development in infancy from a neo-Piagetian, dialectical-constructivist perspective. After defining developmental stages, we review mental-attentional (*M*-) capacity growth during infancy—a sensorimotor scale of *M* (*Me*). Various infancy tasks are examined via task analysis to illustrate sensorimotor mental processes. This *Me* growth enables emergence of the semiotic (signs) function, in particular symbols, and with it intersubjectivity ("Theory of Mind") and levels of consciousness. We then examine three experimental-developmental research programs that support our model of mental attention in infants: (1) EEG coherence data of Thatcher; (2) data of the Dimensional Inventory for Child Development Assessment by Silva, Mendonça Filho, and Bandeira; and (3) doctoral work on infants' mental attention, cognition, and language by Nancy Benson-Hamstra.

> Just look at any infant, any child, any young vertebrate: SEEKING [related to Piaget's active assimilation construct] lies at the foundation of all of their aspirations.
>
> —Panksepp & Biven, 2012, p. 496

> In the course of the first eighteen months ... there occurs a kind of Copernican revolution ... a general decentering process whereby the child eventually comes to regard himself as an object among others in a universe that is made up of permanent objects.
>
> —Piaget & Inhelder, 1969, p. 13

> We cannot explain the appearance of signs in children by postulating that they already exist; we must identify the cognitive and socio-cognitive mechanisms that generated them. Although reformulated, we still need Piaget's point of view.
>
> —Marti, 2012, p. 166

Piaget's constructivism and the developmental stages he described remain relevant in developmental theory (Carey, Zaitchik, & Bascandziev, 2015). In this chapter, we show

how Piaget's theory of infant development can be explicated (accepting his descriptive, but not his explanatory model) to account for transition between sensorimotor (sm) and symbolic (sym) processes. This transition underlies early emergence of relational abstraction and cognition. We discuss growth of *M*-capacity during infancy and the emergence in it of a limited symbolic function and consciousness; we conclude by discussing three sets of data—Thatcher's, Mendonça's, and Benson-Hamstra's—which support important aspects of the model.

It is not always appreciated that Piaget was a constructivist-learning, not an associative-learning, theoretician. Consider his concepts regarding reflective abstraction of experienced functional invariants and his self-propelling schemes, actively accommodating, when necessary, to bring about equilibration. Indeed, Piaget thought that functional information-bearing systems such as schemes and organismic regulations are process-structural categories that all humans share, as tools in knowing activities and action. Although Piaget never studied individual differences, he considered that they could be accounted for by variation in the strength of these process categories. He saw schemes as capable of generating expectancies, even in infants, more or less automatically. Piaget's intuition of scheme units resembles that of behaviorist E. C. Tolman (1959, 1961), whose favored psychological units (tested on rats!) were "means-end-readinesses," that is, inferential "expectancies" about what-leads-to-what. We explicate schemes in detail throughout this book, but particularly in chapters 1, 5, and 10.

According to Piaget, infants develop expectancy-bearing schemes very early, quite visibly after 2 months of age. However, he understated how much cognitive schemes are motivated and driven in development by affect and by infants' early emerging social-being processes. Infants' motivation (like that of older children and adults) consists in conversion of affective goals (often called dispositions, propensities, or motives) into cognitive goals (Pascual-Leone, Pascual-Leone, & Arsalidou, 2015). Positive or negative motivation, always present in healthy infants, drives development. A good example of these affectively driven cognitive goals (described in chapter 2) is Piaget's (1948/1963, p. 249) Observation 133 about his 9-month-old daughter, Jacqueline. It illustrates well the intertwining of affect and cognition (mental attention) in infant motivation. Talking about his daughter, he says, "She likes the grapefruit in a glass but not the soup in a bowl.... When the spoon comes out of the glass she opens her mouth wide, whereas when it comes from the bowl, her mouth remains closed."

Piaget treated schemes as essentially cognitive entities, even though schemes are products of a social being and often are affective-and-cognitive hybrids (i.e., *personal or psycho-social schemes*), which in our theory we call *B*-operator (social Being) schemes. This latter view is consistent with Vygotsky's school (R. Miller, 2011; Pascual-Leone,

1996b; Pascual-Leone & Johnson, 2004; Vygotsky, 1978). Schemes, cognitive and psychosocial, combine—generating *stages of development*. If a *step* is a state describing one stable moment in the developmental sequence of change, a *stage* is a sequence of descriptive states, each coupled with its corresponding causal process model. Note that proper stages are descriptive models coupled to organismic causal-process models.

The current school of neonativist researchers investigating infant concepts has made important contributions and is internationally admired (e.g., Renée Baillargeon, Scott, & Bian, 2016; de Hevia, Izard, Coubart, Spelke, & Streri, 2014; Spelke, 1994). These concept neonativists have enriched previous constructivist descriptions of infancy by demonstrating reasoning abilities (a working mind) in babies that apply to cognition and to psychosocial activities. These are abilities of a cognitively alert social being, endowed from the outset with a goal-seeking organization (a functional-totality capable of purposive behavior [Tolman, 1959]). Within facilitating situations the social infant easily develops refined expectancies about encountered encumbrances or affordances (in the sense of Gibson, 1979). Such functional ways are innately primed by pre-coordinations among sensorial/motor channels and emerging networks of the brain's functional totality. *Body schema* is the name often given to the more or less coordinated functional totality of figurative/representational and operative/transformational processes that, in the psychological organism, enables functional organization. This cross-modal functional-totality package or body schema is neurologically anchored and, in part, innate. This construct helps to explain how, in early weeks of life, a baby can imitate the facial or body movements of others, such as the protruding tongue of a friendly adult (e.g., Meltzoff & Moore, 1999), even when the model is no longer visible.

In a similar manner, functional-sensorimotor coordinations are found early in many domains. For example, at 2 to 3 months, and more so at 4 to 6, infants clearly demonstrate precursors of walking (Thelen & Smith, 1994; Thelen & Ulrich, 1991). An innate precursor of the body schema (global pre-coordinated innate schemes) must exist to explain the achievements of young infants within facilitating situations, including some achievements highlighted by concept neonativists. Many of these early (innately primed) precursor performances are less available or harder to elicit later on (often between 5 and 10 months, Bower, 1974), when emerging schemes/skills begin to compete with innate precursors. These precursor body-schema performances are too weak in competition with other schemes, until recreated and strengthened by the incipient willful mental attention.

Findings of experimental concept-neonativist researchers have encountered resistance among developmental constructivists like Piagetians or neo-Piagetians. These and other researchers question conceptual neonativism because the innateness of complex (conceptual, abstract, categorical) cognitive dispositions is often not a parsimonious causal interpretation for babies. One should not impute to evolution (innateness) concepts or dispositions that past animals in our evolutionary line cannot have acquired for their own survival (Johnson & Pascual-Leone, 1989). In this chapter we illustrate through examples how our theory can explicate these results in a way compatible with neuroscience and evolution. Our constructs are general-purpose (not specific), and mammals may have developed versions of them for survival. This chapter, like the book itself, is an empirically grounded theoretical essay, which offers an analytical general-organismic (from within) model. We analyze some neonativist tasks from our perspective as illustration but make no attempt to be comprehensive. In chapter 9 we analyze other infancy tasks.

We suggest that the brain's problem-solving mechanisms include (if formulated from within the subject's processes, Pascual-Leone & Johnson, 2017) three *categories of innate functional processes,* or resources, whose coordination with experience explains causally infants' cognitive/psychosocial evolving characteristics that neonativists emphasize. These categories are innately primed. They are (1) very active information-bearing innate or learned processes (i.e., *schemes*—simple or complex; see chapters 1, 4, and 5); (2) a number of specific regulation systems in the brain's functional organization (the *hidden operators*); and (3) a few general *organismic principles* that regulate general alignments (functional coordination) of hidden operators and schemes to produce activity or performances. These constructs can explain organismically infants' reasoning capabilities, without falling into an evolutionary paradox.

Maturation of *M*-Capacity in the Sensorimotor Period

The sensorimotor period spans from birth to about 35 months. During this time babies grow from purely sensorimotor, instinct-driven organisms, able to recognize and automatize content-based (nonsymbolic) states. These lead to expectancies and operative moves to serve affects and motives. During this time babies grow to become minimal symbolic processors; processors that have sensorimotor ("content") representations but also simple symbolic representations as well as simple mental acts. This onset of a *symbolic function* is enabled by emergence of mental/endogenous attention. There are many textbooks offering basic information regarding infancy, and we shall not present that here. Rather we present our model of mental attention growth in infancy. In table 3.1 we summarize Piaget's sensorimotor stages, to which we add the corresponding levels of endogenously growing mental attention that enables creative (i.e., nonautomatized) problem-solving acts and permits emergence of stages and the symbolic function. The first column of table 3.1 indicates the sensorimotor (sm) mental-attentional capacity. We call it *Me,* because this amount of sensorimotor capacity later is used to activate

Table 3.1
Me-capacity in the sensorimotor period

Me-capacity	Age (mos)	Piagetian substage
0	0–1	Use of some reflexes and/or innate schemes
1	1–3	Acquired adaptations and primary circular reactions; initial body schema
2	3–8	Beginning of secondary circular reactions (SCR) and procedures for making interesting sights last
3	8–12	Coordination of SCR and application of schemes to new situations; object permanence without visible displacement
4	12–18	Beginning of tertiary circular reactions (TCR) and discovery of new means by active experimentation; object permanence with visible displacement
5	18–26	Coordination of TCR; invention of new means by mental combinations; appearance of executive performance anticipations [Spontaneous construction of executive schemes demands four or five *Me*-units]
6	26–35	Transition to symbolic processes; initial working mind (mentation)
7	35–59	Early preoperational period; child mobilizes executive schemes and relates symbolic schemes; functional generic entities represented as generic objects (e.g., mother, father, brother, family, teacher, daycare); child has an *M*-capacity of $Me=7$, i.e., $Mk=e+1$

tasks' general *executives* (*e*) during symbolic (sym) processing after 12 months of age. The second column expresses the theory-predicted average age in months of children at each of these sm stages.

In the **zero *Me* stage** (0 to 1 month, the age estimates are approximate, but surprisingly stable), *Me*-capacity is not yet manifest: the baby's activity is controlled by innate reflexes and innate schemes. In the **first *Me* stage** (1 to 3 months) the baby exhibits *primary circular reactions* (PCR) and acquired adaptations in facilitating contexts. A situation or context is facilitating when it induces in the person only the desired behavior (i.e., only offers cues and elicits schemes facilitating it). For instance, when I touch the lip of a young infant with my finger, he or she may promptly grasp the fingertip with the lips, in a sucking action. This is a reflex facilitated or cued by the finger contact. When, at about 2 months, a mother sets her baby in her arms to start breastfeeding, the infant spontaneously turns his or her face, positioning it for the nipple that will soon arrive. The mother placing the infant and uncovering her breast constitute a facilitating situation, which cues (associative learning, *C* and *LC* learning) the baby's suitable head turning.

A PCR is a recurrent action or move of the infant in which he or she seems to make no distinction between the action itself and the object being engaged; both are conflated in the baby's "working mind." A common PCR is thumb-sucking. Perhaps accidentally, the infant may have brought the thumb to his or her mouth and begun to suck. Soon this becomes a PCR habit, because the end of a thumb-sucking action appears to prompt its reiteration; this is the functional characteristic of a circular reaction (a concept originally discovered in infants by Baldwin, 1894/1968). Another PCR is the often-interminable crying of young infants: the sound of their crying cues another act of crying, which makes crying contagious among babies. Act, action, and object are not differentiated during PCR.

In the **second *Me* stage** (3 to 8 months) *secondary circular reactions* (SCR) begin to appear. An SCR differs from a PCR in that infants now seem to differentiate functionally between their action/reaction and the object to which it is addressed, being able to act/adapt separately on each of them (reaction or object) to achieve the intended SCR goal. A clear example of this behavior is when we present infants with a desirable toy (e.g., small doll or rattle) that they want to grasp, and they reach their hand toward it. At this moment, the experimenter takes the desired object and displaces it to another site, reachable by the infant. More or less efficiently, the child may stop his or her hand movement toward the object's previous position and initiate another movement toward the object in its present position. This is easily visible at 5 months: "One observes retraction of the hand, often completely out of the visual field, followed by a second reach. In older infants, by contrast … the hand is not withdrawn to begin again but rather alters its trajectory in flight. There is correction on the act, rather than correction between acts" (Bower, 1974, p. 160).

This is a clear example of SCR: the goal-driven (intentional) reaction is to get the toy, but when this object is displaced, the infant in one or another way corrects the act. Object and reaction are separate in the infant's "working mind" and can be separately adapted. It is not uncommon, at this age, for parents to attach some device above the crib so that a string tied to the infant's foot can, when shaken, move the object and perhaps produce a sound. When infants are pleasantly surprised with the movement and sound of the rattle, they may try to repeat the effect and soon learn to shake the foot to cause the movement. This SCR is applied as procedure for making interesting effects last.

In the **third *Me stage*** (8 to 12 months), coordination of SCR appears. Several SCR schemes can be used together to attain a result. For instance, as Piaget first demonstrated, when an object that the infant wants to reach is unexpectedly blocked by the father's hand, or by an obstacle or cloth, the infant now is likely to apply another scheme in the repertoire (e.g., hitting or pushing away the obstacle) to get at the object.

Bower (see quote earlier in chapter) gave another example of SCR scheme coordination. An 8-month infant may, when the object he or she tries to reach has just been displaced, insert a different scheme within the same act of reaching, so as on the fly to displace his or her arm and align it with the new object location. Now there is correction on the act rather than between acts. The first level of Piaget's object permanence task, removing obstacle A to get at the object, is thus achieved. However, when, after some new screen B is added, and the object is visibly displaced behind A immediately followed by visible displacement behind B, children in this stage fail to retrieve the object; they instead persist in removing screen A, even though they saw the object displaced behind B. Piaget called this level of achievement *permanence of object without location* during visible object displacements; current researchers call it the A-not-B error. To avoid the A-not-B error, infants must overcome an added misleading factor, that is, the interfering habit created by the repeated prior act of removing A to obtain the object. A creative solution controlling this misleading habit would demand tertiary circular reactions (TCR).

In the **fourth *Me* stage** (12 to 18 months) TCR appear. Using Piaget's terminology, infants now can solve a *permanence-of-objects problem with location* (find the object placed behind successive screens, A, B, C…) provided that object displacements are visible to the child. The idea of searching behind multiple screens (thus overcoming the habit of going only to A) is a TCR creative idea. In our view, consistent with Piaget's analyses, TCR are tertiary because in them a different functional constituent appears to codetermine performances, one more term in addition to the action/reaction and the intended proximal object, the two constituents of SCR schemes. This new constituent is the distal object: a distinct (more or less complex) entity in the child's mind and the world, with its own characteristics and its functionality, which the child wants to know. The term was invented within psychology of perception: a distal object is contrasted with a proximal (sensorial-perceptual) object, as an object-model (i.e., coordinated package of interrelations among schemes) that emerges as a relational-invariant entity (i.e., as a hub for intercoordinated schemes) in the context of cognitive/personal activity. Now infants seem to have tacitly understood that objects are distal in the world and have their own multiple characteristics (are specific hubs of coordinated schemes) that can be known if one explores them attentively, thus exhibiting "l'experience pour voir" (i.e., experiments in order to see; Piaget, 1948/1963). Elsewhere, Piaget (1958) explicitly defined an object (a distal object) as being all the schemes that can apply on it, plus all the schemes that cannot.

Infants' discovery of objects as distal entities shows that accommodation is a key function enabling TCR in this sensorimotor stage. This includes all kinds of accommodation (Gibson's primary accommodation, see chapter 1, and the usual accommodation of

Piaget, which is secondary to assimilation). TCR directly motivate infants to discover and investigate distal objects in the world. Now infants apply schemes in problem-situations seeking to solve the problem and invent trial and error! In other cases, accidental happenings or spontaneous actions lead to a solution, enabling infants in this stage to solve the problem—and in the process learn the just-discovered new means to reach intended ends.

Infants in this stage synthesize a new operative scheme, a *means-end-readiness* as Tolman (1959) called it. Piaget highlighted examples of TCR leading to operative discoveries (new means discovered through active experimentation). The three key operative means that he discussed are (1) pulling a support (blanket, cushion) on which a desired object is found, to bring the object within reach; (2) pulling a string to which a faraway object is attached, to bring it closer; (3) using a stick to reach an object and bring it near. Notice that the infant's actions are often SCR schemes, but they become TCR schemes if and when the child uses the situation to discover/invent (and abstract) a new operative means that later will be used in other contexts.

In the **fifth *Me* stage** (18 to 26 months) infants clearly begin to have internal thinking (i.e., a working mind, mentation). They now begin to invent new operative means through *mental combinations*. Thus, in Piaget's object permanence problem, children now can solve a task with object location (behind screen A, B…), even when *displacement of the object was invisible* to the child. Another example of mentation leading to problem solution is shown by Piaget's daughter Lucienne (Piaget, 1948/1963, Observation 180). When Lucienne was 16 months old, after several days of exploring and playing with matchboxes (e.g., removing a chain her father had placed inside—with the matchbox-cover wide open), Piaget placed the chain inside a matchbox that he closed, leaving an opening of only 10 mm. After some trying, Lucienne succeeded in getting the chain out. Piaget placed the chain inside the box again and closed it, leaving an opening of only 3 mm. Lucienne tried to get the chain out but failed. At this moment, "[S]he looks at the slit with great attention; then, several times in succession, she opens and shuts her mouth, at first slightly, then wider and wider!…Lucienne by opening her mouth thus expresses, or even reflects, her desire to enlarge the opening of the box" (Piaget, 1948/1963, Ob. 180; translated by Flavell, 1963, p. 120). This is an embodied, motor mental image that represents what Lucienne is planning. Soon after, she "unhesitatingly puts her finger in the slit…and she pulls so as to enlarge the opening" (Piaget, 1948/1963, Ob. 180; translated by Flavell, 1963, p. 120); then she takes the chain. As Piaget emphasizes, this *representation* of a future plan embodied in a mouth-movement analogy tells Lucienne (in a symbolic mental synthesis of schemes) what she can do; leading her to invent a new means-end operative scheme, a mental tool for extracting enclosed objects from a box. Notice that this act of synthesis must control some misleading factors, such as a prior failed experience with closed boxes: thus, executive schemes must intervene. This is clearly a TCR but also a mental-attentional act of creative mentation. This problem solution illustrates the working mind and the power of a subject's internal symbols.

There are limits to what a child at this stage can comprehend, however. Jasper (our grandson) was eating crackers and interacting by video chat with his uncle Rafael. Rafael playfully invited Jasper to feed him, and Jasper, without hesitation, attempted to feed Rafael's screen image (see figure 3.1). Jasper was 17 months and 25 days old (see McClure, Chentsova-Dutton, Holochwost, Parrott, & Barr, 2018). An operative model of mental processing that likely led Jasper to attempt this feeding is symbolized in formula **f1**. Before looking at this formula, readers may wish to review our notation for task analysis. It is explained in the book's appendix and summarized in table 3.2. This notation is used throughout the book. Alternatively, readers could skip this and other task analyses on a first reading of this chapter. This, however, could reduce clarity, because task analyses add detail and precision to our infancy model.

FEED$_A$ (HAND(#what food: #to.whom Rafael[*com.image*]:
#where mouth[Rafael])) ➔ RESPONSE [Rafael is fed] **(f1)**

In this metasubjective formula we use a logical *observer's language* to represent from within a semantic-pragmatic action (nonverbal processes). We write operative schemes in capitals (FEED, HAND) to indicate that these schemes generate action. We represent figurative (i.e., representational-object) schemes in lowercase. Operative schemes apply to their right on figurative schemes to bring about semantically expected action. HAND stands here for an operative process: the hand maneuver or action that brings the food to Rafael's image. This HAND operative has three ordered schemes on which it applies (notice that schemes/entities are separated by [:] to signify that each of them corresponds to distinct interrelated, semantic but not-yet-linguistic, categories [#what, #to-whom, #where]). They correspond to three functional aspects (*observer's parameters*), which here are just category names for the schemes. Parameters (indicated by #) are pragmatic conditions, attributes, or dimensions of variation, by which we specify conditions or functional-structural aspects that the child must consider in his operation or action: in this case, #what, #to whom, and #where. Notice that in formula **f1** the computer image (*com.image*) may be construed by the child as a perceptual part of Rafael, perhaps not having task-significant characteristics different from Rafael's. Notice that the sequence of schemes in this formula begins on the left side with FEED and ends on the right with mouth[Rafael]. As this book's appendix suggests, rightmost schemes in the formulas are more concrete, so schemes to their left are progressively

Table 3.2
Summary table of codes used in metasubjective task analysis (MTA)

Equal by definition	:=
Operative scheme (OP)	uppercase NOTATION
Figurative scheme (fig)	lowercase notation or *
Parameter	# appended to the scheme
Scope of scheme	(…) where "…" stands for a scheme
Semantic aspect of scheme X	X[…]
Knowledge scheme	[K:…]
Scheme for an actually present, true entity	$…dollar symbol precedes scheme
Language scheme	λ…lambda precedes scheme
X is semantically conjoined with Y	X.Y X and Y are connected schemes
X leads to Y	X::Y
X,Y,Z follow each other	X:Y:Z
X controls Y	X:.Y
Y leads to local emergence of X	X←Y
Result of preceding operation	X …→Y
Y caused by schemes' overdetermination	X … ➔ Y
Separation of schemes	(,)
L-structured dominant scheme	X^{L1} post-superscripted L1, L2, etc.
L-structured subordinate scheme	$\{\ldots\}_{L1}$
Boosted by hidden operators	$\{\ldots\}_{LC, F}$
Boosted by current situation	$\{\ldots\}_{sit, F}$
Violation of expectancy	… !
Boosted by Mental Attention	Underline
Not task relevant/misleading	^…
Dialectical relationship	@…
Dialectical/contradictory relationship	X…<@> Y…
Uncertainty	… ?

at a more (>>) abstract level: FEED >> HAND >> food >> Rafael-image >> mouth. In contrast, the whole process is controlled (:>) in its unfolding in the formula from left to right: FEED:>HAND:>food:>Rafael-image:>mouth. However, control is primordially exerted by operative schemes (FEED, HAND) over figurative schemes.

Underlining denotes distinct schemes/schemas that the child must boost with sensorimotor mental attention (Matt.sm), to synthesize a RESPONSE. Notice that there are five underlined schemes, and consequently the *Me*-demand of this task should be, congruently with Jasper's age, *Me*=5, (i.e., the mental capacity first available at about 18 to 26 months, *Me*-stage 5; see table 3.1). How long should it take for Jasper, or another child, to correct this error of performance? To do so, the child must recognize that *com.image* is an object distinct from Rafael, a functional totality with different characteristics, which cannot be fed even though it shows Rafael's image. Thus, *com.image* should be boosted with *Me*-capacity separately from Rafael. This brings the *M*-demand to *Me*=6 (past 26 months).

In the **sixth *Me* stage** (26 to 35 months), true symbolic processing of the working mind fully appears (see chapter 4). This is mentation: the mind's hidden processing of (perhaps conscious) ideas, symbols of symbols. The **seventh *Me* stage** (35 to 59 months) is not just a sensorimotor stage. The child has now a mental capacity of *Me*=7, which corresponds to the first level of the symbolic processing scale Matt .sym (which we abbreviate as *Mk* scale: $Mk=e+1$). Thus, this seventh *Me*-stage marks the transition from the sensorimotor (Matt.sm) scale of mental-attention measurement (*M*-measurement) to the symbolic (Matt.sym) scale usable at and after 3 years of age. The Matt.sym scale keeps growing in capacity till 15 to 16 years. Its *Mk* units are not sensorimotor. However, both sorts of processing, sensorimotor and symbolic, are often used together by the child, particularly within facilitating situations that lack misleading or distracting aspects. Within the seventh *Me*-stage children appear to have attained (via congruent schemes' coordination) conscious representations of familiar objects as functional totalities relevant to their life, such as "mother," "father," "family," "person," "television," and so on. This abstraction of distinct cognitive and psychosocial entities is perhaps expressed by the child's new creative capacity to draw images of familiar persons that use ovoids standing for human bodies, and dot eyes, stick arms, and legs. Notice that behaviorally *M-capacity energy* is measured in terms of the number of task-relevant schemes that can be internally activated by *M*, to enable task performance (*M-measurement*, see chapters 7 to 9).

The essential schemes are demarcated using Metasubjective Task Analysis (MTA), called metasubjective because analysis is done from within the subject's task-process perspective. The method does not make any assumption about whether all essential schemes require the same amount of *M*-energy; the organism will automatically allocate *M*-energy as needed. The cognitive-developmental stages described by the growth of *M*-capacity are essentially Piaget's. We call them *Me-stages* to emphasize that mental attentional growth, not just learning, produces them (Arsalidou & Pascual-Leone, 2016; Pascual-Leone, 2012a; Pascual-Leone & Johnson, 2005). Our (TCO) stages of development are probabilistically tied to chronological age, because growth in *M*-capacity (the "transition rule" for stages—Pascual-Leone, 1970) is maturational. This is in contrast to Piaget's theory, where stages are tied to cognitive learning, not to age. In this book (and numerous published papers) we present task analyses to show concretely how growth of mental attention could produce stage transitions (Pascual-Leone, 1970).

Symbolic Function and Consciousness during Sensorimotor Period

Although maturational *M*-growth produces stages, its most obvious empirical consequence is the transitioning of children's thinking from concrete/particular and local to abstract/generic and general (or semantically universal). This transitioning causes mentation, the complex and internalized working mind. The cognitive-developmental *Me*-stages lead to the emergence of two major, cognitive and personal (i.e., cognitive, affective, psychosocial, and sociocultural), modes of processing—two landmark achievements: the *symbolic function* and *consciousness*. We describe symbolic function in more detail in chapter 4; and discuss consciousness, from a broader perspective, in chapter 12. In this section, we will briefly explicate these qualitative functions and show how they can be derived from combined working of mental attention, executive processing, and constructivist learning.

Prior to the emergence of a symbolic function, mammals and babies only have signals to inform them about the environment. A *signal* is an automatic associative link connecting a present indicator, or *token sign*, with a perhaps distal entity (e.g., object, action, relation, process) to which signals refer (more or less automatically). This entity is the *referent*, and a *sign* (for instance, a signal) can bring its meaning to mind. All associative connections among schemes can constitute signals, even when they include symbols (see below). Signals and their referents are *functionally attached* to one another: the signal indicates the actual, here and now, presence and meaning-relevance of the referent. For instance, in adults or older children, smoke is a signal of fire in this very sense: the present actuality of fire is announced by smoke. Similarly, in adults, the loud scream of "*FIRE!!*" in a movie theater may induce people (a signal-driven reaction) to think that fire is a here-and-now relevant reality. Signals, unless innate, are conditional to prior experience and learning, but their functional characteristics are the same in mammals, babies, and adults.

A much more flexible connection of meaning emerges in human development with the symbolic function at about 12 months of age, and it keeps developing into adolescence and adulthood. A key for the symbolic function is the *symbol* (Nöth, 1990). A symbol is *functionally detached* from its referent and its context (i.e., its object and actual situation). For instance, a car was a "gaga" for our just-symbolic grandchild, but so was a bicycle, a truck, or any big moving object in the streets. A symbol is an abstract invariant, a concept motivationally driven by affects (*A*) or by tacit consensus among social beings (*B*). It emerges from the systemic coordination of four distinct and complementary category-schemes: <**SEM(cxt, sgn, ref)**>. These aspects differentiate symbols from signals (signs that are not symbols).

Our model of a symbol is a *semiotic* (i.e., referring to signs) *system* characterized by the formula <**SEM(cxt, sgn, ref)**>. This is a pragmatically evolved coordination of generic schemes, explained as follows. **SEM** stands for a semantic operative scheme that carries meaning and coordinates three category parameters within the semiotic system: the token **sign**, its **context** or sense (one or several sense values), and its **referent** (one or many values); each parameter is distinctly detached and carried by its own scheme. Such distinctiveness allows some concrete specific contexts to evolve their own appropriate sense (activity-related specific meaning) relative to a given referent. Together the four semantic components (functionally integrated by the superordinate **SEM**) produce the situated (i.e., context sensitive), flexible semantic functioning typical of symbols. For instance, when an 11- or 12-month-old infant sees her mother putting on her hat and coat, she may begin to cry. Mother's actions mean to the infant that her mother is going to leave, that is, will disappear and no longer will answer the baby's call ("mother leaves!" reaction of the baby). Piaget (1948/1963, p. 250) described this observation about his daughter Jacqueline (11.5 months); we task-analyze it in chapter 9. Clearly, such reaction is caused by a symbol, because with a different sign token (the mother putting on a different hat or coat), or a different context (the mother in the bedroom trying her hat on in front of the mirror), the baby's alarm reaction may still occur; when the housekeeper puts on her hat in the family room as mother does, the baby, if attached to her, may infer (an operative expectancy) that she will leave too and perhaps show some reaction.

Because symbols are functionally detached and the symbolic function is recursive (there can be symbols of symbols of symbols), symbolic-processing growth eventually generates consciousness: distinct symbolic (detached) representations of others, objects, or situations; and symbolic self-representations. It is thus useful to examine consciousness evolving during infancy. Damasio (1999, 2012) and Dehaene (2014) have interesting neuroscientific views about consciousness. Our model independently converges with Damasio's (2012), adding a developmental perspective that complements his work. Damasio (2012, p. 167) defines consciousness as *"a state of mind in which there is knowledge of one's own existence and the existence of surroundings."*

Early Development of Consciousness

In our view (Pascual-Leone, 2000a), the mind is the encompassing functional totality of schemes and operators (expressed by the brain's neuronal networks, their regulations and connectivity) that produce via overdetermination actions, feelings, representations, learning, problem solving, memories, and so forth. Consciousness is a particular state of the mind, in which the embodied subject/Self and its external context can be felt as distinct and functionally detached (Damasio's *primordial feelings*) and also experienced as situated, adapted/adaptable to a here-and-now situation. This experience of consciousness is done from within the subject's own processing, adopting a first-person perspective (Pascual-Leone, 2013; Pascual-Leone & Johnson, 2017). "Ordinary consciousness is the aptitude of an organism to have a *distinct* representation of some constituents of its own thinking (cognition), feeling (affects and emotions) and/or willing (deliberate intentions)" (Pascual-Leone, 2000a, p. 242). We say *distinct* because to be conscious a representation must be different (in its processes) from the processes/schemes that it represents (i.e., its object or referent) and be distinct also from the experiencing self-subject. Another constituent of true self-experience, the reflective agent/Self who does the representation (the knower or knowing self) appears developmentally later, at the end of infancy; this is the onset of reflective/conceptual self-consciousness (Legerstee, 1998). This reflective self-consciousness leads to what Damasio (2012) calls autobiographical self.

From a constructivist neuroscience viewpoint, Damasio (2012, Fig. 8.1, p. 192) distinguished three levels (stages) of processing: protoself, core self, and autobiographical self. Respecting these important distinctions, we expand and explicate them to give a clear account of their developmental emergence, which Damasio does not provide. A key developmental factor is, in our view, the growth of mental-attention capacity, which we have outlined in table 3.1 with distinct *Me*-stages. Our developmentally expanded, explicated version of Damasio's consciousness (**CONSC**) model follows.

(**CONSC0**) There is, already before birth, a *sentience*, some sort of global sensorial perception and organismic reactivity vis-à-vis the environment (Pascual-Leone, 2000a), that is, primordial feelings that Damasio (2012, figure 8.1, p. 192, figure 8.2, p. 202) subsumes under the *protoself* and attributes to brainstem-level processes.

(**CONSC1**) After 3 to 4 months and up to 8 months of age (*Me*-stage 2), an *incomplete subject's awareness* appears (some representation of the object of experience as distinct, without representation of the self-subject's own experience). Damasio (2012) has called *protoself* this incomplete level of consciousness, in which the self-subject is not cognitively distinct and explicitly identified but is instead conflated with the object of experience. Developmental research shows that protoself awareness emerges and matures earlier within affective, psychosocial, and interpersonal relations than in purely cognitive processes (Legerstee, 1998), possibly because primary affects, some of them psychosocial, are innate (e.g., a mother's general schema, as Lorenz demonstrated famously with his ducks). Babies at about 5 or 6 months can recognize themselves in the mirror: "It was when they observed their own face that they began to coo and smile" (Legerstee, 1998, p. 637). This observation shows that a distinct object of experience (here their own face) is recognized and reacted to appropriately (Pascual-Leone, 2000a).

In mirror self-recognition, the *object* is both the mirror and the face—face that mediates (or signals) the experienced *self-subject*. However, the important point is that object and subject cannot be simultaneously distinguished at this stage. To see the mental demand of this task, consider how babies could recognize their own face in the mirror (as a familiar "object"). This demands the coordination of two (without practice, three) distinct schemes, highly activated together in the child's working mind. These schemes are as follows. (1) The baby's scheme of the mirror and its physical characteristics is denoted as mirror* (postscript * indicates a figurative scheme). (2) The figurative scheme of the baby's face and its changing appearance under movements is connected to a purely experiential (i.e., perceptual and unreflective) baby's self-representation conflated/confused with the object, which we call *self0*. We denote this complex scheme as baby-self0-face*. (3) The operative/procedural scheme that serves to recognize an object is here symbolized as RECOGNIZE (an action blueprint developed in contact with people and mirrors). To recognize his- or herself, the baby must allocate mental attention that forces RECOGNIZE to apply to both (interconnecting/interdifferentiating) baby-self0-face* and mirror* to bring about the actual recognition. This is symbolized in the following logical formula (Pascual-Leone, 2000a):

RECOGNIZE(baby-self0-face*, {mirror*}$_{sit?}$) ➔ baby-self0-face*-in-mirror **(f2)**

In this formula, mirror* is placed inside braces to indicate that it does not need to be boosted with mental attention if (but only if) situational factors (*sit*) such as content and associative learning (our *C*- and *LC*-operators, as well as perceptual salience, our *F*-operator) are likely (*sit?*) to be boosting it; this depends on prior practice. Thus, 3- and 4-month-olds with enough experience can do this recognition task. Task solution could be delayed to 5- or 6-months, however, if additional experience with mirrors must accrue. If mirror* has to be boosted with mental attention (because sufficient experience with mirrors is lacking) three schemes would be involved, and then 8 months of age would be needed to succeed (*Me*-stage 3).

A demonstration that babies at level **CONSC1** do not have distinct concepts of their self-subject as different from the object (possibly they cannot focus simultaneously and distinctly on self-subject and object) is their failure to pass a misleading mirror test designed by Gallup (1982), who showed that adult chimpanzees pass the test. In this task, after the baby is familiarized with the mirror and his or her reflection in it (a purely perceptual self, self0 or protoself), the experimenter surreptitiously marks the child's face with rouge (forehead, ear, or nose). Babies pass this test when, upon looking at the mirror, they bring a hand to the rouge mark, tacitly demonstrating they know that their own face is physically distinct from its image in mirror*.

At level **CONSC2**, at or after 18 months of age (late *Me*-stage 4, or *Me*-stage 5), a *proper consciousness* emerges in the baby, when the child has a simultaneous distinct awareness of both the object and the self-subject of experience. In fact, the earliest time when such fuller consciousness has been found is 14 months (see Meltzoff, 1990, p. 42ff). This self-subject, distinct and physically separate from objects, was demonstrated using the just-mentioned classic mirror task of Gallup (Legerstee, 1998; Meltzoff, 1990). When children pass the rouge-mark mirror task, one concludes that they are fully conscious of the distinctness and then interrelations between the self-subject (face) and object (mirror) of consciousness. This achievement we call *self1*, and Damasio calls it *core self*.

In the two steps of formula **f3**, we present a process analysis of this achievement. In this formula, we call mirror* babies' scheme for the mirror reflection and call, respectively, baby-face* and face-self1* their schemes for any baby's face and for his or her own face. Symbols written in capitals represent operative schemes: NOTICE or RECOGNIZE symbolize perceptual procedures for, respectively, noticing a feature in a recognized pattern and recognizing one pattern as being an instance of a known conceptual category (e.g., a face). The solid arrow (➔) symbolizes the dynamic process of Schemes' Overdetermination of Performance (*SOP* principle); this process causes the infant to synthesize the representation that there is red in the mirror's face. In these formulas, underlined schemes are boosted in their activation by mental-attention activation (using the sensorimotor *Me*-operator). The following formula (Pascual-Leone, 2000a) models in two steps the emerging baby's *consciousness* that the red marks are on her own face, which she sees in the mirror image:

NOTICE(baby-face*, mirror*) ➔ red-in-face/mirror* **(f3.1)**

NOTICE(RECOGNIZE(face-*self1**, mirror*), red-in-face/mirror*) ➔
red-in-face-*self1** **(f3.2)**

In step **f3.1** infants notice a red mark on the face of the mirror image, without recognizing themselves. Then, in step **f3.2**, they recognize the face as their own, noticing the red mark on their face reflected by the mirror. In this more complex step, five schemes must be activated with mental attention: mirror*, face-*self1**, RECOGNIZE, red-in-face/mirror*, and NOTICE. Because growth of *Me*-capacity in infancy does not generally permit simultaneous hyperactivation of five distinct schemes before 18 months (table 3.1;

Pascual-Leone & Johnson, 1991, 1999), the resolution of this task at 18 months confirms the prediction that *Me*-capacity is needed to enable proper consciousness (a *self1* that can distinctly and concurrently represent the object and self-subject of knowing, which is not yet proper self-consciousness).

The next level of consciousness, **CONSC3**, usually occurs at or after 35 months of age (late *Me*-stage 6, or *Me*-stage 7, which is the same as *Mk* stage $M=e+1$). Only at 3 years of age, when a child can simultaneously boost (strongly activate) seven distinct sensorimotor schemes (Pascual-Leone & Johnson, 1991, 1999), does reflective *self-consciousness* appear. We call this stage *self2.1*, and Damasio might call it *autobiographical self* (he also calls by that name later stages of self-consciousness). However, for us the autobiographical self probably begins at about 5 years of age with what we call *self2.2* (**CONSC4**, see later), when mental capacity is at least $M=e+2$. This is usually the time when first memories from infancy remain in people (memory of experiences before 5 years of age seem not to survive). With the *self2.1* there is a true *self-concept*, because the child is simultaneously aware of three "entities": *the object* of experience, *the self-subject* who knows, and the *reflective Self* (i.e., *Self-concept*) that contemplates object and self-subject as distinct. Only with this achievement is there a full (simple but adult-like) mental self-representation. Children often exhibit this new level of consciousness by drawing a picture of a person, perhaps symbolizing their dad or mom. Formula **f4**, from Pascual-Leone (2000a), models this achievement in the context of drawing daddy's picture (first accessible to *self2.1*, 3-year-olds, albeit with a rough representation). This process-landmark is the true onset of reflective self-consciousness.

SYMBOLIZE(DRAW(pencil*, paper*, IMAGINE(daddy*, body*))) ➔
drawing-image-of-daddy **(f4)**

To symbolize him, the child imagines daddy's body and draws it with pencil on paper. Imperfections of this drawing (e.g., a head with arms and legs) exhibit mental-attentional limits in a 3-year-old's mind (e.g., Morra & Panesi, 2017). Also at this time children are able to pass DeLoache's (1989) dual representation task. Here, children are shown a toy-object and witness it being hidden somewhere in a small model room, and then they are taken to a life-size version of the model room and asked to find the hidden object there. At 3 to 4 years (*Me* stage 7) children can find the object, which demonstrates that they now are fully symbolic. They differentiate between their self-subject and their operative self-concept (relational, abstract, invariant across all their concrete manifestations); similarly, they differentiate between the model room, the real-life-room, and their shared room-concept, which enables the child to mentally transfer location of the hidden object from the model room to the real room.

This mental decentration also appears in the psychosocial domain: At this stage (3 or 4 years of age) children pass, for example, the Sally–Anne false belief theory-of-mind (ToM) test. Here, the child (the *Sub*ject) sees a child or doll, Sally, hide a desired object (e.g., chocolate) in a location (a basket) and then leave the room. When Sally is gone, another child or doll, Anne, moves the object to a different location (e.g., a box). Then Sally returns and wants to get the chocolate: where will Sally look? Children younger than 3 to 4 years consistently assume that Sally will look where the chocolate is located now, tacitly attributing to Sally knowledge that only the child and Anne have. This is a failed ToM response (Wimmer & Perner, 1983). When children pass this test, usually soon after 3 or 4 years, their expectation will instead be that Sally goes where she originally had placed the chocolate. These children distinctly differentiate between their own mind's expectation (or "theory") and Sally's own. When we task-analyze from within the subject's processing perspective, we find (see below) that its mental-attentional demand is seven sensorimotor schemes (equal to $M=e+1$ in the symbolic scale). Our theory (see chapter 7) predicts that this capacity is not accessible to children until about 35 months of age. As Gordon and Olson (1998) claimed, mental attention of the symbolic variety is a necessary organismic factor in ToM performance.

During the next level of consciousness, CONSC4, usually at and after 5 years of age (*Mk*-stage 2, i.e., $M=e+2$), reflective consciousness (*self2.2*) has reached what Damasio (2012) calls an *autobiographical self*. From this point on, according to our theory, consciousness keeps growing to later levels of (autobiographical) self-consciousness, thanks to growth of mental attention and accrual of experience throughout adolescence (when mental-attention maturational growth stops at $M=e+7$). Later stages of self-consciousness in adult life are due solely to the growth of experience (constructivist learning) and wisdom. We call the developmental series of autobiographical-self stages up to adolescence by the names of *self2.2* (at about 5 to 6 years, $M=e+2$), *self2.3* (at about 7 to 8 years, $M=e+3$), *self2.4* (at about 9 to 10 years, $M=e+4$), *self2.5* (at about 11 to 12 years, $M=e+5$), *self2.6* (at about 13 to 14 years, $M=e+6$), *self2.7* (at about 15 to 16 years, $M=e+7$). For all these *M*-stages or levels, please refer to chapter 7.

Development of Symbolic Function

Table 3.3 shows in a synthetic manner emergence of the symbolic function during the sensorimotor period. We elaborate on these *Me*-stages of symbolic development below.

(*Me*=0; 0–1 month): There is no *semiotic function* (i.e., no learned and produced signs).

(*Me*=1; 1–3 months): The first nonsymbolic signs (sgn) with meaning (SEM—semantic) appear, because of innate schemes (e.g., a mother/human schema). They

Table 3.3
Symbolic function during the sensorimotor period

$M=Me+Mk$	Age (mos)	Symbolic-Function Formula	Piagetian Substage Description
$Me=0$	0–1	No S-F Formula	Development and coordination of innate schemes, **CONSC0** (*sentience*)
$Me=1$	1–3	$\underline{\text{SEM}}\{\text{sgn}\}_{sit/inn} \rightarrow \text{ob}[\text{SEM}(\text{sgn})]$	PCR, adaptation of initial body schema and other innate schemas, *sentience*
$Me=2$	3–8	$\underline{\text{SEM}}(\underline{\text{sgn}}) \rightarrow \text{ob}[\text{SEM}(\text{sgn})]$	Weak SCR, semi-intentionality, object conscious,**CONSC1** (*protoself*)
$Me=3$	8–12	$\underline{\text{SEM}}(\underline{\text{sgn}}, \underline{\text{ref}}) \rightarrow \mu_0\text{: ob}[\text{SEM}(\text{sgn, ref})]$	Strong SCR, intentionality, proto-symbols ("mother-leaves" task), signal-words, *self0* or *protoself*
$Me=4$	12–18	$\underline{\text{SEM}_1}(\underline{\text{ctx}}, \underline{\text{sgn}}, \underline{\text{ref}}) \rightarrow \mu_1\text{: ob}[\text{SEM}(\text{ctx, sgn, ref})]$	Weak TCR, symbols, strong intentionality, one-word holo-phrase (symbol words), **CONSC2** (*self1* or *core self*)
$Me=5$	18–26	$\underline{\text{SEM}_2}(\underline{\text{ctx}}, \underline{\text{sgn}}, \underline{\text{ref}} : \underline{\mu_1}) \rightarrow \mu_{21}$	Strong TCR, symbolic executives, mirror self-recognition, initial two-word sentences (symbolic), quasi-verbal communication, proper consciousness **CONSC2** (*self1* or *core self*)
$Me=6$	26–35	$\underline{\text{SEM}_3}(\underline{\text{ctx}}, \underline{\text{sgn}}, \underline{\text{ref}} : \underline{\mu_{21}} : \underline{\mu_3}) \rightarrow \mu_{321}$	Mentation (weak symbolic processes). At about 35 months weak self-consciousness: **CONSC3** (*self2.1*)
$Me=7$	35–59	$\underline{\text{SEM}_5}(\underline{\text{SEM}_4}(\underline{\text{ctx}}, \underline{\text{sgn}}, \underline{\text{ref}} : \underline{\mu_{321}}) \underline{\mu_4}) \rightarrow \mu_{54.321}$ $\underline{\text{SEM}_7}(\underline{\text{SEM}_6}(\underline{\text{ctx}}, \underline{\text{sgn}}, \underline{\text{ref}} : \underline{\mu_{54.321}}) \underline{\mu_6}) \rightarrow \mu_{6.54.321}$	Passes "draw-a-person" task and DeLoache model-room task. At 60 months reaches strong self-consciousness: **CONSC4** (*self2.2, autobiographical self*)

are prompted by the situation (*sit*) and by inner states (*inn*), which here we represent as $\underline{\text{SEM}}\{\text{sgn}\}_{sit/inn}$. In this formula, sgn is being boosted in its activation by *sit* (i.e., by situational activation factors *C, LC, F, A,* etc.) and/or by *inn* (i.e., inner, perhaps innate, activation factors). Hence, sgn does not need boosting from *Me*-capacity. This SEM-activation of sgn causes emergence, in the babies' internal field of activity, of an internal object (ob) with the meaning and characteristics (ob[…]) carried by $\underline{\text{SEM}}\{\text{sgn}\}$. An example of this sort of sign, perhaps innate, is the mother/human schema that newborn babies have, which is expressed by babies' preference for human faces, newborns' imitation of tongue protrusion, and their imitation of facial gestures such as mouth or head movements—only imitated from humans (Legerstee, 1991; Meltzoff & Moore, 1999).

(*Me* =2; 3–8 months): Now the baby has two sensorimotor *Me*-capacity units, used to separately boost activation of SEM and of sgn (underlines in the symbolic-function formulas of table 3.3 indicate boosting by *Me*). This *Me*-boosting gives them functional distinctiveness and allows emergence of the secondary circular reactions (SCR) explained earlier. This *Me*-stage corresponds to **CONSC1** discussed earlier. Active imitation begins, as does self-recognition in the mirror (see formula **f2**). At 8 months or later, the baby can adopt procedures to make interesting sights last, for instance, shake the leg to get the rattle sound, and so forth.

(*Me* =3; 8–12 months): Full intentionality (e.g., indicative pointing—pointing to things of interest) and strong SCR appear (e.g., pulling a tablecloth to bring near the toy that is on it, so as to reach and get the toy). There are also coordinations of secondary schemes in cognition and in psychosocial activities (e.g., the "mother leaves!" reaction). Now the baby can focus on three schemes simultaneously: $\underline{\text{SEM}}(\underline{\text{sgn}}, \underline{\text{ref}})$. This semantic processing creates, via overdetermination (➔), a meaning that here we call μ_0 (μ_0: ...). This meaning is constituted by sensorimotor (quasi-) representations, that is, mental/internal objects that we call *obs*, which function as *protosymbols*. They are protosymbols because they have semantic function, sign, and referent but are missing a context parameter (cxt) that would add location relativity (μ_0: ob[SEM(sgn, ref)]). An example of this semantic-processing level is when children recognize themselves in the mirror (see formula **f2**). In this case, SEM is the recognition operative, sgn is the mirror figurative scheme (including baby's face reflection), and ref is the face-*self0* figurative scheme (which is part of the baby's evolving body schema). Notice, as we mentioned before with formula **f2**, that this protosymbol can be achieved with an *Me*-power of 2, when scheme mirror* is made salient by the situation (associative learning).

A spontaneous gesture (semiotic act) that for Tomasello (2008) marks the beginning of intersubjective communication (a precursor of language) is *indicative*/communicative *pointing*. When the baby notices an exciting object and wants to share this interest (declarative pointing), or wants the Other (mother, father, or other adult) to give the desired object (imperative request), and so forth, the baby points with the finger to the target object. Tomasello (2008, p. 112), in an excellent book, claims that, "Surprisingly, no one knows where pointing comes from ontogenetically." We believe we know. For us, the origin of indicative pointing is clearly threefold: (1) It comes from the excitement and affective motivation of an interesting object, which the child points to, driven only by its charm. We call this initial phase *expressive pointing*. Notice, however, that Tomasello's question referred only to indicative pointing, not expressive pointing. (2) The pointing action soon acquires the functions of an operant: it begins to be used to call the attention of Others, to share a target object with (or request it from) intimate Others. Indicative pointing occurs only when all requisite schemes are available in the baby's repertoire. (3) All these schemes can together be boosted in activation by baby's sensorimotor mental attention (Matt.sm), because this is a facilitating situation, and so they overdetermine production of a pointing RESponse to the actual ($) object ob (i.e., RES(sub.*Points*: $ob)). Our metasubjective task analysis of indicative pointing is:

$$\underline{\text{POINTING.sub}}^{L1?}(\#\{\underline{\text{NOTICE![Other: ob]}}\}_{L1?,A.B?},\ \$\underline{\text{ob.ref}}\text{: }\$\underline{\text{Other}}) \rightarrow \text{RES(sub.}\textit{Points}\text{: }\$\text{ob)} \qquad \textbf{(f5)}$$

In other words, the subject (sub) enacts the indicative finger-pointing action, which may or may not (?) already be chunked (superscript *L1?*) with his or her attempt to call the Other's notice, leading them to look at the object (ob). This operative POINTING intends (#) to call Other's attention (NOTICE!) to the actual ($, i.e., truly there) object-referent ($ob.ref), which causes (➔) the pointing act. The mental demand of the task is the number of underlined schemes in its formula: four when NOTICE is not chunked with POINTING, or three when it is already chunked (perhaps because the baby's parents playfully trained the child at this task). Thus, indicative pointing will be accessible to children with *Me*=3 (8 to 12 months) or *Me*=4 (12 to 18 months). This model of mental-attention demand in indicative pointing answers Tomasello's (2008) perplexed question "Why don't three-months-olds point?" (p. 136 ff), but 9- or 12-month-olds do?

(*Me*=4; 12–18 months): Proper symbols appear in the baby's conduct. The meaning of the signs now is context sensitive; that is, it varies in detail with the context, being nonetheless semantically invariant across contents. The formula for a proper symbol is SEM_1(ctx, sgn, ref), a process that produces a meaning (➔ μ_1) that leads to symbolic mental objects or representations (:ob[SEM(ctx, sgn, ref)]). A good example of this sort of symbolic processing is the rouge-mark mirror task, analyzed in formulas **f2** and **f3.1**. In the key formula **f3.2**, we find all the elements of a symbol in the superordinate inference: SEM is the operative scheme corresponding to NOTICE the red-in-face; the subordinate act (RECOGNIZE face-*self1* in the mirror) is the sign of the main inference; red-in-face/mirror (i.e., the red seen in the mirror) is the context; and red-in-face-*self1* is the referent. This example shows that referents in complex processing can be symbolic mental objects or representations. As a consequence, when TCR appear, cooperative actions become possible, and language begins to develop.

(*Me*=5; 18–26 months): Now, with an additional unit of *Me*-capacity, the infant can easily coordinate previously synthesized schemes (prior meanings) using a scheme that is semantically synthesized. This is expressed by the formula SEM_2(ctx, sgn, ref : μ_1) ➔ μ_{21}. Here, the previously acquired meaning μ_1 is being coordinated within the synthesis/formation of a new symbolic scheme SEM_2. This symbolic coordination produces the more complex meaning μ_{21}. An example of this sort of symbolic coordination of schemes appears in the formula sequence **f3.1** and **f3.2**. Another example is found in formula **f4**, in which the product of imagining the father's body (IMAGINE [daddy*, body*]) corresponds to μ_1, which serves as referent for the act of drawing daddy with pencil and paper, generating the new symbolic scheme μ_{21}, daddy's drawn picture. With this new ability of mentally synthesizing coordinated symbols (symbols of symbols of symbols), a true symbolic *executive function* appears, and with it, soon after, mentation (the symbolic working mind), during the next two stages. Notice that executive schemes (or *executives*) are generic operative schemes and scheme systems that describe possible situated procedures and plans of action. Executive function is the collection of different processes that can generate executives. We discuss executive schemes and processes in chapter 5.

Stages **(*Me*=6; 26–35 months)** and **(*Me*=7; 35–59 months)** are functional expansions of *Me* stage 5; in them the symbolic function and the executive function come to their first full expression. This is shown in tasks such as draw-a-person, DeLoache model-room task, or false belief theory-of-mind tasks. It is also shown in the rapid growth of language and in the appearance of Damasio's autobiographical self. To illustrate in some detail how these, perhaps complex, sequential mentations take place (at least in our theory), we summarize a task-analytical process-model of the ToM task previously described, that is, the Sally–Anne false belief task.

ToM results from the totality of a subject's schemes for mental intersubjective decentration or mind reading; this is a key paradigm illustrating the emergence of a working mind doing intersubjective inferences (e.g., Gordon & Olson, 1998; Tomasello, 2008; Wimmer & Perner, 1983). Table 3.4 presents this process/task analysis in four coordinated steps. The task analysis models the cognitive/affective skills of ToM that may control this task (a misleading task, because of the child's tendency—*F*-factor—to attribute his or her own mental representations to Sally). We paraphrase in English the schemes represented in table 3.4. In **step 1**, the child-subject witnesses that the Sally doll places (PLACE) a bar of chocolate in a certain location (Loc1). Thus, Sally knows (K: ...) that the chocolate is in Loc1; this is her representation resulting from Sally's act of placement. Then Sally leaves the room, and Anne, in **step 2**, moves the chocolate to location Loc2. As a consequence, the child-subject (Sub) knows (K: ...) that the chocolate is now in Loc2. **Step 3** expresses the subject's spontaneous observation that Sally,

Table 3.4
Metasubjective task analysis of false belief theory-of-mind task

Step 1	PLACE(Sally : chocolate : Loc1) → Sally[K: chocolate : Loc1]
Step 2	PLACE(Anne : chocolate : Loc2) → Sub[K: chocolate : Loc2]
Step 3	NOT.SEE(Sally : PLACE(Anne : chocolate : Loc2) → Sally[NOT.K: chocolate : Loc2]
Step 4	WHEREGO{TOGETchoco}$_{sit}$ (^{Sub[K: choco : Loc2]$_I$}$_{sit}$, #Sally[NOT.K: choco : Loc2] #Sally[K: choco : Loc1], Sally) → RES[GO.GET(Sally : choco : Loc1)]

Note: Step 4 coordinates information/schemes attained in steps 1, 2, and 3. Step 4 constitutes a misleading situation, because some schemes are contradictory. This step, therefore, is an act of conflicted mentation: a judgment task in which the correct response (RES[…]) can be reached only by boosting relevant schemes with *M*-capacity, and concurrently inhibiting the misleading schemes with *I*. Whereas steps 1 and 2 are purely sensorimotor, steps 3 and 4 are clearly symbolic. This task illustrates well the transition from the *Me*-scale (sensorimotor) to the *Mk*-scale (symbolic). Indeed, $Me = 7$ corresponds strictly to $Mk = e + 1$.

being absent, could NOT SEE Anne moving the chocolate. Thus, Sally does not know (NOT.K…) that the chocolate is now in Loc2. This is clearly not a sensorimotor step because the observation and its conclusion (→) are purely symbolic-mental. The same can be said of **step 4**, which synthesizes the conclusion (→) that Sally will go to Loc1 to get the chocolate.

This is a valid inference reached in a misleading situation. The situation is misleading because the subject *Knows* that the chocolate is now in Loc2. Such belief may induce the child to assume that Sally will go to Loc2. This response is produced by a simplistic semantic idea: <since the chocolate is in Loc2 and she wants it, she will go to Loc2>. This erroneous inference is caused by the brain's internal-simplicity field factor or *F*-operator, caused by cortical lateral inhibition. The child-subject's knowledge scheme (Sub[K: choco: Loc2]$_I$) induces the expectation that Sally will go to Loc2 (because *F* leads the child to conflate Sally's knowledge with his or her own). Thus, this scheme is negative (unsuitable for the task), which we represent in table 3.4 with the prefix ^. This scheme is not boosted by *M*-capacity; it is boosted (which we symbolize with {…}$_{sit}$) by *situational* factors (*sit*) such as content (*C*) and structural-relational learning (*LC*), affective empathy (*A*), or sociocultural habits (e.g., an expectancy that she will do what is "logical" in the situation). This negative scheme must be actively inhibited (*I*-interrupted). Active interruption requires bringing the schemes into mental attention; for this reason the scheme is underlined in step 4, with a subscript *I* indicating inhibition. Such misleading scheme is contradictory with the scheme (Sally

[NOT.K: choco: Loc2]), but because the latter is consistent with other activated schemes and can cluster with them, the overdetermination principle (➔) reaches the correct response, provided that the child applies attentional inhibition (*I*) to the misleading scheme (^{$\underline{\text{sub[K: choco : Loc2]}}_{I}$ }$_{sit}$).

The formulas have underlined schemes that require attentional boosting from *M*-capacity. As table 3.4 shows, step 4 demands *M*-boosting of seven schemes (*Me*-scale). Thus, the minimal age for solving this task would be, in our theory, 3 or 4 years of age. This is the chronological age that marks a developmental transition from purely sensorimotor mental attention to a symbolic mental attention. In the current literature these two forms of mental attention are often conflated and confused with executive function/controls. However, the literature recognizes a clear important transition in executive controls between before 4 years of age and after 4.5 years of age, which tacitly agrees with our concept of two distinct levels of mental attention: sensorimotor and symbolic (Espy, 2016).

Mental Process in Facilitating Tasks versus Misleading Tasks: Sensorimotor versus Symbolic Mental Attention

As discussed before, there are two scales for measuring mental attention: the sensorimotor scale *Me* or sensorimotor mental-attention (Matt.sm), which is most useful in facilitating situations, versus the symbolic scale *Mk* or symbolic mental-attention (Matt .sym), useful in misleading and complex situations. Total mental-attentional power (*Mp*) is a combination of both scales; that is to say, quantitatively, $Mp = Me + Mk = e + k$ (where *e* and *k* are the *M*-capacity estimates corresponding, respectively, to *Me* and *Mk*). Two related but distinct methods of task analysis (MTA_{sm} versus MTA_{sym}) are used to estimate task demand in terms of these scales of *M-measurement*. It is important to recognize when to apply the one or the other, because the scales cannot be combined; their estimates of *M*-capacity have different units (*sm*-units versus *sym*-units). Indeed, schemes in the sensorimotor period are simple, and the amount of *M*-capacity likely needed to boost each scheme is small. Using the metaphor of measuring water, in the *Me* scale the sm-unit measure is like a cup of water (*M*-energy); in the *Mk* scale, in contrast, a sym-unit is like a pail of water. This difference is shown in the rate of development found for sm-units versus sym-units acquisition. They cannot be summated. However, we may assume that at least three sm-units from the sensorimotor period (those at 12 months, 18 months, and 26 months), because they occur with or after the symbolic function, could be used in symbolic tasks to boost schemes that are facilitating (not misleading), thus aiding in the number of total *M*-units available (this is only a theoretical inference or hypothesis).

As a first essential step, any task analyst has to decide what mental-attentional scale (*Me* or *Mk*) should be used; whether the task in question needs only *automatic*, driven by affectively immediate, *attention* (Matt.sm, i.e., *Me* scale) or needs *effortful and executive-driven symbolic* mental attention (Matt.sym, i.e., *Mk* scale). A decision can be reached by using two complementary task-analytical checks. **Check 1** uses the **facilitating/misleading task criterion** to determine whether task performance is *affectively immediate* (i.e., driven only by affect or habit in a facilitating situation) and can thus be readily solved if enough *M*-capacity is available, or it is misleading. In the latter case, a subject-controlled problem-solving solution must be *executive mediated* (i.e., executive driven because it requires a prospective plan, or the situation is complex, or strongly misleading). Notice that executive mediation can occur optionally in facilitating situations, when automatic and executive processes are mutually compatible. However, executive mediation is mandatory to solve the task in misleading situations due to the competition and interference among schemes. **Check 2** involves evaluating whether the required task performance can be carried out without executive schemes, boosting with sensorimotor *Me*-units. Here we examine the **number of essential schemes** that would have to be boosted. Given the child's age, does he or she have enough *Me*-capacity to match the minimum task-required number of sensorimotor or automatic schemes? Otherwise *Mk*-units of older children may be necessary. If he or she does not have enough *Mk*-units, training (*LC* learning, chunking) may be needed to solve the task, segmenting it into few scheme components. Misleading/symbolic schemes can be boosted only with symbolic (Matt.sym, *Mk*-scale) *M*-units, unless they are already automatized via associative (*LC*) learning.

Should we assume that if a task's performance is executively mediated it needs the symbolic, Matt.sym, scale? Not necessarily, because when executives are automatized, (overlearned) schemes could be activated and apply without help from mental attention. To clarify the two important checks, we examine four examples.

Example 1

The baby follows objects that move from one side to the other in front of his or her eyes. To apply criterion **Check 1** (i.e., affectively immediate or automatic *vs.* executive-mediated tasks), one should ask whether automatic (innate or acquired) reactions, or affects, exist that propitiate the behavior and whether there are task-misleading schemes elicited by the situation. In this example the answer is yes to the first issue and no to the second. There is a reflex or innate reaction that follows with the gaze interesting moving objects, when they can be discriminated well. This behavior would be affectively immediate, because the child experiences direct pleasure in keeping the eyes on interesting objects, and this positive affect itself propitiates and boosts intended performance. Consider now criterion **Check 2** (estimate number of distinct sensorimotor *Me*-units needed to produce performance, when executive schemes are not used). To perceive a moving object with clarity infants must be able to distinctly separate in their mind the eye motor action from the representation of the moving object, that is, at least two sensorimotor (*Me*) units. This *Me*-capacity is developmentally available first at about 3 or 4 months. After this age normal children can easily notice and track a moving object, so their innate positive-affect factors (curiosity, agency, desire to maintain an interesting situation) as well as Gestaltist perceptual-field mechanisms (e.g., contiguity, good continuation—this is our *F*-factor) should help the child achieve this task.

Example 2

The child counts objects properly from 1 to 10. Let us adopt first the perspective of **Check 1**. This task clearly is not affectively immediate; instead, it is executive mediated. The very young child does not initially care about counting objects, even when these objects have been nicely arranged in a row very close to each other. But the child does care about pleasing her mother, or the cordial teacher or tester, who motivates the child to do or imitate what the adult has done and is telling the child to do. Because the task itself does not immediately elicit the positive desire, it is not affectively immediate but mediated by executive schemes. With the perspective of **Check 2**, it seems that without symbolic executive schemes, and using only very concrete sensorimotor/automatic schemes, this task cannot be learned. This is because there are three jointly needed aspects in the task of counting: the row of objects (X1, X2, X3 ...), the verbal sequence of number names (N1, N2, N3 ...), and the complex COUNTING process, a process that young children mark by bringing the finger to the object (X1, represented by scheme x_1, saying "one!"; N1, represented by verbal numbering scheme n_1), and then repeating the act of pointing for each of the other objects in a perceptually facilitated order, voicing the corresponding next number, until the row of objects has been numbered by pointing and voicing (we symbolize this pointing and voicing as $x_1.n_1$). This task may be attempted by using only sensorimotor means, but in this case at most three or four objects could be counted, never ten objects. To see this limit, consider that without using symbolic executives the subject must carry out a learned (associative learning) sequence such as:

$$\underline{\mathrm{COUNTING}^{L1}\ (\mathrm{POINTING}_{L1}}(\underline{x_1 \cdot n_1}:\underline{\mathrm{GO\ NEXT}}[\leftarrow \underline{x_2 \cdot n_2} \leftarrow \underline{x_3 \cdot n_3}])) \quad \text{(f6)}$$

Formula **f6** shows how by functionally coordinating pointing and counting with the two series X and N young children may count three objects. In our notation words in capitals stand for operative schemes, and lowercase words or letters for figurative/object-representational schemes. Sensorimotor counting should require to have in mind the requisite schemes (if there is no symbolic executive processing and no executive-driven recursive procedures). GO NEXT is an operative telling the subject to repeat the operation with the X-N pairs inside the square bracket [...]; the inverse arrow (←) says that the child is mindful of this repetition—replacing the pair $x_1.n_1$ with $x_2.n_2$, and so on, provided that he or she can keep the additional pairs in mind. If we count the schemes involved (underlined in the formula), there are five (i.e., *Me*=5) with just three pairs, and six (i.e., *Me*=6) with four pairs. We can assume that schemes COUNTING and POINTING are chunked (*L*-structured—indicated by superscripted/subscripted *L1*) because they were taught before and used together, possibly learned by practicing counting with an adult. The operations and object schemes have to be learned, and this takes time, but they are facilitated by a situation that cues this process if suitable prior training was given. Thus, sensorimotor *Me*-units suffice to handle the tasks. A capacity of *Me*=6 exists at about 26 months. If the chunking of COUNTING and POINTING had not taken place, the total demand of four pairs would be seven, only available at 35 months. No more than three or four pairs could thus be repeated, because the lack of abstract/symbolic recursion compels children to keep in mind all units in order to repeat the operation.

For this reason, counting to ten is a different story. The child gets tired of counting (a misleading affective factor), and the two long X and N series are very hard to memorize in order and require recursion. Counting to ten asks for motivation and symbolic executive schemes, to maintain repeated counting (via recursion) across all available similar objects. A task analysis of this symbolic-executive COUNT procedure is summarized in **f7**.

$\underline{\text{symCOUNT}^{L1}\ (\#\{\textit{MATCH all}\ \text{X: N}\}}_{L1,}\ \{\text{ob[X-series]}\}_{sit,F,LC,LM}\ \{\text{number[N-series]}\}_{LC,LM}$ **(f7)**

The procedure is not a simple sensorimotor COUNTING effort, as it was before, but a symbolic recursive operative COUNT process (notice the rather abstract parameter #). This symbolic process anticipates the sequence of finger pointings to the X-objects, and voicing of their matching numbers (this could only happen when parents and tutors have worked enough with the child so that she has learned the X- and N-series and the procedure). The symbol # indicates that this scheme is a parameter of (i.e., constraining condition to) COUNT, stipulating how it is to be done. By repeated practice of the procedure indicated by **f6**, tutoring of parents or mentors may have led the child to learn the recursive COUNT procedure, and possibly may have chunked together (*L*-structured) the motor COUNT procedure with instructions about how to do it (parameter #). The object series (ob[X-series]) that is in front of the child and the number series (number[N-series]) that is in her mind need not be boosted with mental attention because of prior practice/training with the procedure **f6**. Both series are well learned via content learning (*LC,* automatic associative structuring), as well as symbolic mental-attention learning (*LM,* effortful, more or less conscious explicit learning). In addition, the object series X has its activation boosted by perceptual salience (our neo-Gestaltist field factor *F,* see chapters 1 and 4), which the row of objects has when placed in front of the child. Under these conditions *F* and *L* are facilitating factors in this task, monitored by a symbolic/recursive executive.

Thus, the task *M*-demand in symbolic mental-attention (Matt.sym) will be the *Me*-power at the end of sensorimotor period, which we denote *e,* plus one symbolic *Mk*-unit (i.e., $Md = e + 1$; see chapter 7). This amount of *M* is the measured *M*-capacity of 3- or 4-year-olds. In this manner the counting to ten can be completed, because the child uses a recursive executive that, together with symbolic processing, minimizes *M*-demand. This is a symbolic mental-attention process and its *M*-demand is represented in formula **f7**. The schemes that need *M*-boosting are underlined; all other schemes are being boosted by alternative hidden operators elicited by the situation (*sit*) such as *F, LC, LM,* and so forth. If the child had not received required training, mental demand of this task would be much higher, perhaps $Mp = e + 2$ (the *M*-power/capacity of 5- or 6-year-olds) or $Mp = e + 3$.

Example 3

Indicative pointing means the child points to some object with the finger to call attention to it. We task-analyzed indicative pointing in **f4**, but we return to this behavior to highlight the complexity difference between indicative pointing and purely expressive pointing within sensorimotor facilitating situations. Let us adopt the perspective of **Check 1**. This sort of task is affectively immediate because the object itself calls positive affect-driven attention from the infant-subject (sub), and he or she points to express this positive affect (e.g., curiosity, novelty). Pointing to an exciting object is innately facilitated, through positive affects (tendency to approach what one likes) coupled with the internal-field factor, which promotes this sort of S-R (stimulus-response) compatibility action (Proctor & Reeve, 1990), approaching with the finger the exciting object. Unavoidably, this pointing becomes, later on, a way of calling others' attention to the object, as formula **f4** has shown. Formula **f8** is a model of how the sensorimotor infant invents expressive (but not yet indicative) pointing behavior within an affectively immediate situation.

$$\{\text{NOTICE![ob]}\}_{sit,A,F,LC}\ (\{\underline{\text{POINTINGsub}}\}_{A,F}\ (\$\underline{\text{ob}})) \rightarrow \text{RESP(sub.}\textit{Points}\text{: \$ob)} \quad \textbf{(f8)}$$

Initially the child notices the object and finds it interesting or charming (*A*, affective factor). The reaction is also potentiated by perceptual salience (*F*, field factor), and some family-mediated practice in noticing objects (*LC* – Logical-structural Content-learning). These three activation-boosting factors promote noticing of the object (NOTICE). Such act of noticing induces activation (because of positive affect and S-R compatibility) of the POINTING scheme as expressive action. It expresses (in presymbolic ways) the child's charmed interest. Mental demand of this expressive pointing is no more than *Me* = 2, as indicated by number of schemes underlined in **f8**.

Note that POINTING requires *M*-boosting because it is a nonautomatized motor action at first. Soon the child discovers that pointing induces other persons to look at the object, as modeled in formula **f4**. The child may then produce POINTING with the explicit intention to share this excitement with the Other. Then NOTICE is directed to the scheme of the Other, whose scheme becomes a parameter of POINTING. This new state of affairs was represented in formula **f4**, with a mental demand of *Me* = 4 (accessible for the first time to 12-month-olds). However, if in a child (because of prior familial learning) the scheme NOTICE is already chunked with the operative scheme POINTING, mental demand will become *Me* = 3 (accessible to 8- or 9-month-olds). We have thus fully addressed the developmental perplexities of Tomasello mentioned in connection with formula **f4**.

Example 4

Consider now the sort of tasks so cleverly used by neonativists. Renée Baillargeon and colleagues (2016) reviewed this sort of data, examining with care psychological reasoning in infancy. From this and other reviews it can be inferred, as our theory has long proposed, that for as long as the situation is basically facilitating and there are opportunities for learning, the baby's learning will not be too constrained by developmental stages. Stages express mostly mental-attention capacity (mental attention of the sensorimotor variety in the baby, i.e., *Me*). However, there are no strong stages in mostly facilitating situations, in the sense that learning could make the child's performance move up in such situations. This is much harder to do in misleading symbolic situations. The review of Renée Baillargeon and colleagues (2016) tacitly supports this point, although the authors do not draw our conclusion. The new findings provided by neonativists are not inconsistent with a constructivist neo-Piagetian approach like ours. What has to be investigated in their tasks is whether sensorimotor (Matt.sm) mental-attention demand (*Me* demand) is always congruent with our predictions of an infancy *Me*-scale for mental attention. We think so. To back up our claim we now analyze two well-known tasks: the drawbridge screen task of Renée Baillargeon, Spelke, and Wasserman (1985) and the small-number estimation task of Wynn (1992).

Drawbridge Screen Task In the drawbridge screen task (Renée Baillargeon et al., 1985), 4- and 5-month-olds ($Me=2$) were trained (habituation) to see a wooden screen, movable like a drawbridge, move back and forth (the baby's direction) from vertical to flat (180 degrees). Babies then were presented with a salient yellow box of about half the height of the vertical screen, placed behind the floor-flat screen (see figure 3.2). The experiment consisted, for all children, in two task versions presented in alternation: the possible event and the impossible event. The *possible event* showed the flat screen with the box close behind it so that when the screen was moved back toward the vertical, the box became invisible; and when the screen was moved further (toward 180-degree flat) it would encounter the box as an obstacle at 120 degrees. Then the screen moved again forward to its flat position near the baby showing again the box. The *impossible event* showed a similar sequence including the close-behind-the-screen box. The important change was that now, when the screen moved back beyond the vertical, it did not stop at 120 degrees but went on to a flat position. And yet, when it returned to the forward flat position, the box could still be seen behind the screen base. Findings show, and they have been replicated, that babies looked, as if surprised, to the impossible event more than to the possible one. Thus, they had some notion of the impenetrable solidity of the wooden screen and the box and had some notion of object permanence (perhaps without location in the sense of Piaget).

Our task analysis, which we explain in English below, appears in formulas **f9**. Our text in the formulas is only an observer's language to describe with clarity the babies' nonverbal knowledge state.

$$\textit{(Possible Event)}\text{: }\underline{\text{FALL}}^{L1}(\#\underline{\text{obstaclebox}}^{L1}, \#\{\text{EXPECT[screen.fall.120]}\}_{L1},\{\text{screen}\}_{sit} \rightarrow \text{RES[\$screenFall 120]:>}\{\text{MATCH [expected Fall=found Fall]}\}_{L1} \quad \mathbf{(f9.1)}$$

$$\textit{(Impossible Event)}\text{: }\underline{\text{FALL}}^{L1}(\#\underline{\text{obstaclebox}}^{L1}, \#\{\text{EXPECT[screen.fall.120]}\}_{L1}\ \{\text{screen}\}_{sit} \rightarrow \text{RES[\$screenFall 180]:>}\{\text{MISMATCH [expected120=found180]}\}_{L1}\text{:>Surprise!}_{A} \quad \mathbf{(f9.2)}$$

The notation used is explained in table 3.2. We modeled the possible event in **f9.1** as the baby may have experienced it: the baby observes the screen FALL (sensorimotor operative scheme attributed to the external screen movement). FALL has a condition (parameter #) saying that the screen has an obstacle (i.e., a visible box). Because these are facilitating situations not bound to stages, prior learning with solid objects and obstacles will have led the baby to learn and automatize (*L1*) in everyday situations that the obstacle should impede the screen's full fall; such knowledge

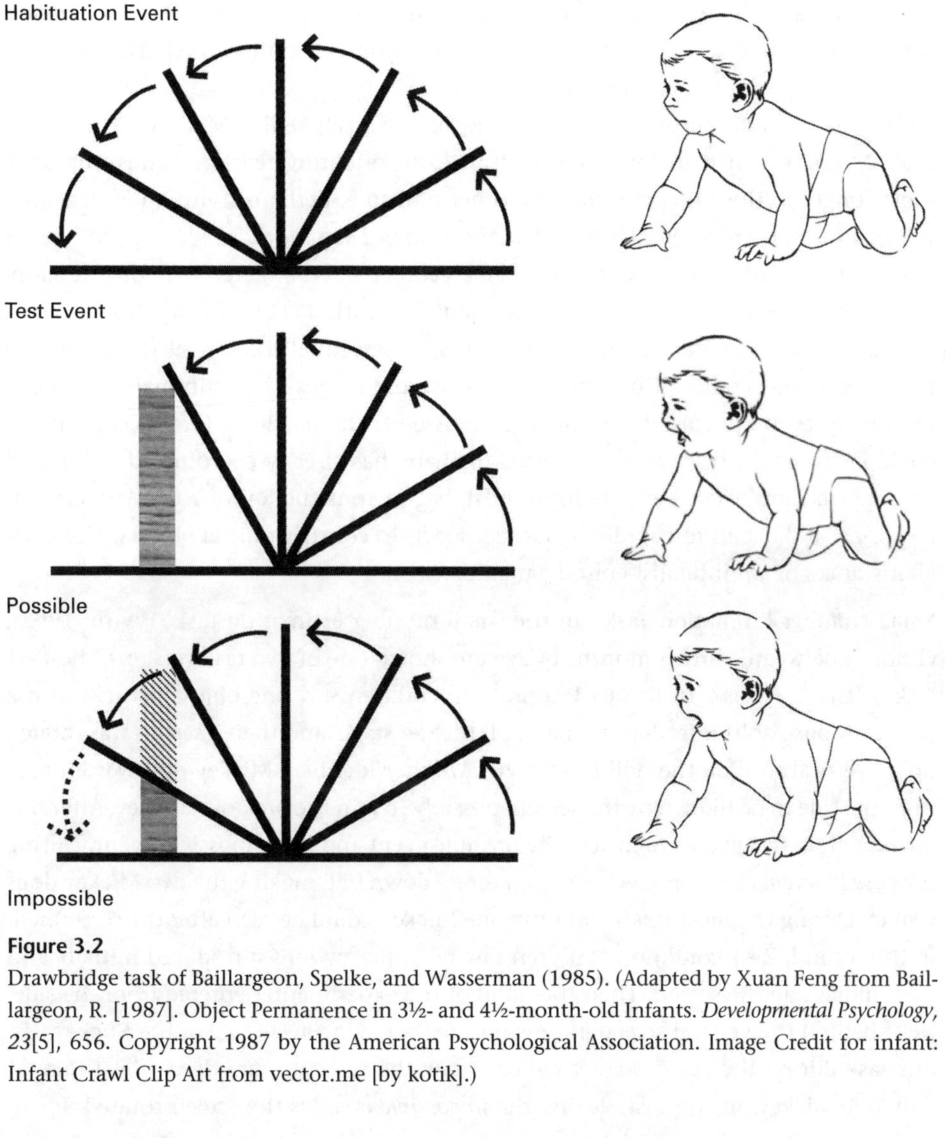

Figure 3.2
Drawbridge task of Baillargeon, Spelke, and Wasserman (1985). (Adapted by Xuan Feng from Baillargeon, R. [1987]. Object Permanence in 3½- and 4½-month-old Infants. *Developmental Psychology, 23*[5], 656. Copyright 1987 by the American Psychological Association. Image Credit for infant: Infant Crawl Clip Art from vector.me [by kotik].)

is overlearned (*L*-structured or chunked). The infant EXPECTs the screen to fall only to about 120 degrees (qualitatively appraised!). Note that screen does not have to be *Me*-boosted here, because it is activated by situational (*sit*) facilitating factors (i.e., *C, LC, F*). What happens next, causally overdetermined (➔) by the preceding events, is the result (RES[$screenFall 120]). This result (screen falling 120 degrees) is congruent with the baby's qualitative expectation (our symbol $ indicates that Fall 120 degrees is empirically/actually true). This dynamic result forces his or her mind (:>) to a comparison between the expected and found result. The comparison is a MATCH, and so no surprise or orienting reaction is elicited.

However, in **f9.2**, the obtained results (RES[$screenFall 180] :>{MISMATCH …) force a MISMATCH, which in turn causes surprise, an orienting reaction. Thus, the baby looks longer to this outcome than the other outcome, perhaps trying to understand. In both formulas mental-attention demand is *Me*=2 (i.e., Matt.sm capacity of babies after 3 or 4 months). Results of Renée Baillargeon et al. (1985) are, therefore, developmentally explained by our constructivist mental-attention (TCO) theory. Notice, let us insist, that expectation of an obstacle (#{EXPECT[screen.fall.120] in **f9.1**), as assumed in this example, could be available to 3-month-old babies as a "minimalist" sensorimotor expression of episodic memory (see Russell & Hanna, 2012, for this concept). It could be available, because at 3 months the baby has already coordinated vision and hearing (turning his or her gaze to sounds), begun construction of secondary circular reactions, and begun to coordinate facets/aspects to construct distal objects. These are all instances of a minimalist episodic memory.

Small-Number Estimation Task In the small-number estimation task (Wynn, 1992), children between 4 and 5 months of age are shown one of two task versions: the **1+1** task or the **2−1** task. In the **1+1** condition, children saw one object (let us call it a Mickey Mouse doll) introduced on a small theater stage, and then a screen was rotated up to vertical to hide the doll from view. Another identical Mickey doll was entered from the side to be hidden by the screen, possibly to join the previous Mickey. After this introduction, two events followed: the possible event and the impossible event. During the *possible event* children saw the screen rotate down flat, making the two Mickey dolls visible. During the *impossible event* only one Mickey could be seen after the screen fell. In the second, **2−1** condition, children saw two Mickey dolls introduced initially and then hidden by the screen. Then the hand of the experimenter entered from the side, went behind the vertical screen and reappeared, carrying away one of the Mickeys. In this task during the *possible event* children saw the screen rotate down flat to reveal only one Mickey. In contrast, during the *impossible event* as the screen rotated down, children saw two Mickeys.

Thus, implicitly, the impossible events exhibit the calculations **1+1=1** or **2−1=2**. Results showed that children looked significantly longer at the impossible events, as if they knew the calculation was wrong. Can babies do small-number arithmetic? Let us look at our task analysis, in which we model only the **1+1=1** impossible event. It is easy to see that similar analyses yielding the same *Me*-demand could be done for the possible event and also for both events in the **2 – 1** task. We begin when the first Mickey has been introduced to the stage and the screen is being rotated up.

Small-number estimation task (1 + 1 = 1):

{SCREEN-UP}$_{sit}$ ($Mickey1) → EXPECT$_{A,LC}$ [SCREEN[$Mickey1]] **(f10.1)**

{ADD}$_{sit}$ (#K:SCREEN[$Mickey1], Mickey2) → EXPECT$_{A,LC,F,T}$ [SCREEN[$Mickey1, $Mickey2]] **(f10.2)**

{SCREEN-DOWN}$_{sit}$ (EXPECT[SCREEN[$Mickey1, $Mickey2]]) → RES[$Mickey1] **(f10.3)**

COMP$_{A,LC,LM,F,T}$ (EXPECT[SCREEN[$Mickey1, $Mickey2]]: RES[$Mickey1]) → RES[*MISMATCH* :>surprise!]:> OR **(f10.4)**

If we describe in English these four steps, the process is as follows. **(f10.1)** Mickey1 is on the stage when the screen goes up and hides him, creating the knowledge-based (K:) expectancy (SCREEN[$Mickey1]) that he is behind the screen. **(f10.2)** Mickey2 is added and goes behind the screen, inducing the expectancy that both Mickeys are now behind it. **(f10.3)** The screen is now brought down, and the baby expects to see both Mickeys, but the result (RES) is that only one Mickey appears. **(f10.4)** Babies automatically (*A, LC, LM, F, T*) compare (COMP) what they expect (EXPECT …) with the obtained result (RES). A mismatch is found, which causes surprise, releasing an orienting reaction (OR) with longer inspection time in the impossible event.

Notice that our task analysis essentially corresponds to the explanation offered by Simon (1997), emphasizing not an arithmetic calculation but a visuospatial qualitative match between expectations and obtained results. What we add to Simon's account is a finer formulation of causal factors (such as the *T*-operator, automatic-perceptual attention, or sensorimotor-*Me* mental attention), promoting this matching strategy and the orienting reaction that results. Consistent with our theory, when the tasks are presented in a symbolic manner, with verbal instructions and responses, passing age is 2 to 3 or 3 to 4 years (Houdé, 1995), in contrast with Wynn's looking-time results with 4- to 5-month-olds. This difference occurs because in Houdé's case the task is no longer sensorimotor and facilitating, but it is symbolic and somewhat misleading.

Some Research Data with Relational-Developmental Patterns That Support the TCO Model

We now summarize the work of three researchers who discovered surprising relational-developmental patterns that support our theory of infancy, and within this theory, the

Me (sensorimotor mental-attentional capacity) model. On first reading, the reader may choose to skip this section and move directly to Conclusions. The patterns bear on three distinct but related predictions **(Pr)**.

(Pr1) Thatcher's research: Our theory claims two different scales of *M*-measurement, one for sensorimotor mental attention (Matt.sm, i.e., the *Me*-scale) and the other for symbolic mental attention (Matt.sym, i.e., the *Mk*-scale). The theory also claims that *M*-capacity is a maturational construct. Thus, the growing brain should exhibit two distinct periods of growth, a fast one for sensorimotor (Matt.sm) processes and a slow one for symbolic (Matt.sym) processes.

(Pr2) Mendonça's research: If these fast-growth (Matt.sm) and slow-growth (Matt .sym) scales are maturational, they should be manifested behaviorally across most important and distinct content domains. As we showed above, and in chapter 7, time intervals of stage transitions in the two scales are very different. Transitions occur about every 4 or 5 months during the first year, versus 6 to 9 months in the second year (these are Matt.sm transitions), but they occur about every two years after the third or later years of age (these are now Matt.sym transitions). Progressive lengthening of these transition intervals suggests that more mental-attentional energy is required for *M*-units emerging in later stages, on the assumption that rate of growth in *M*-energy (whichever its brain causal substratum) is probabilistically constant throughout.

(Pr3) Benson-Hamstra's research: If **Pr1** and **Pr2** were correct, then it should be the case that, across distinct content domains, tasks developed (with *M*-theory) to measure Matt.sm all exhibit the age-predicted values of *M*-capacity.

(Pr1) Thatcher's EEG Coherence Data

Thatcher (1997) used the theta frequency band from left lateral frontal-parietal regions to examine mean percentage EEG (electroencephalography) coherence (i.e., correlations of neuronal firing) in children from 6 months to 16 years. The results yield clear support to our model. Pascual-Leone et al. (2010) discussed Thatcher's data. We also discuss it in chapter 10, reproducing one of his figures. A different set of data reported by Thatcher (2010) is also interpretable as supporting our maturational *M*-capacity model, with its two distinct scales of growth (*Me* and *Mk*). In this study of beta waves, with 458 participants ranging from 2 months to 16.67 years, Thatcher and collaborators examined developmental change of EEG coherence and EEG phase differences for cortical pathways of various lengths (6 cm, 12 cm, 18 cm, and 24 cm) in left (LH) versus right hemispheres (RH) and for different directions (anterior-posterior versus posterior-anterior). We focus on Thatcher's data for short-distance connections (6 cm) and for the anterior-posterior direction. This is the most relevant direction for us, because short-distance connections are simple and likely to increase with mental attention without requiring many learning opportunities. Thus, the two main sources of mental-attentional activation (scales *Me* and *Mk*) may be expressed by these data.

In Thatcher's (2010) data, coherence of short-distance interelectrode combinations (local-connections of 6 cm length from anterior-posterior direction in both hemispheres) increased rapidly during the first year of life, reaching a measure of about 35 (at about 2 years) and 30 (at about 3 years). There was a sudden reduction of coherence between ages 3.5 and 4.5 years, when the coherence measure was below 25 (Thatcher, 2010, figure 2, p. 88). Then again, after the age of 4.5 years, there was a continuous yearly increase in coherence until age 16, which was considerable in the anterior-posterior direction (from a coherence measure of about 20 at ages 3.5 to 4.5 to about 55 at 16 years). A linear fit of the mean coherence as a function of age for all electrode pairings showed (Thatcher, 2010, table 1) that these data exhibited a positive slope of the linear fit to age (slope 1.81 for LH and 1.75 for RH, with correlations of .88 and .90, $p < .0001$ in both cases).

This is our interpretation. Semantically, EEG coherence expresses coordination of neuronal processes, coordination of schemes. Coherence within short-distance interelectrode combinations expresses local and simple schemes' connection as a result of everyday experience, possibly potentiated by the activated *M*-capacity, both sensorimotor (*Me*) and symbolic (*Mk*). Thus, high coherence prior to 3 to 4 years of age should express predominantly use of the *Me*-scale of mental attention. The sudden coherence reduction between 3 and 4 years may express termination of *Me*-scale growth. Finally, the progressive new growth of coherence until 16 years (oldest age in the sample) should be expressing use of the *Mk*-scale. Thus, reduced coherence between 3.5 and 4.5 years of age strongly suggests that these scales of mental attention (sensorimotor and symbolic) are distinct, with partly different causal processes.

Such interpretation is also supported by Thatcher's analyses of EEG phase differences in the same data and by developmental ultraslow oscillations in phase differences over the age span. Semantically, EEG phase differences express change toward differentiation (i.e., progressive distinctiveness) of schemes being formed—the second key aspect of human development (coordination versus differentiation). Differentiation creates more complex schemes, likely to be involved in distant connections that integrate separate sets of local schemes. Thatcher's data (2010, figure 3 and table 2) show that long-distance phase differentiation (of 24 cm interelectrode span) grows developmentally much more than short-distance (6 cm) phase differentiation. In fact, short-distance interelectrode pairings in these data exhibit no developmental change in phase differences and tend to have negative slope and correlation. In these curves (Thatcher's figure 4) one can see a transition break in phase-difference measure between data prior to 3 years and after 4 years. Developmental oscillations over the age span show (his figure 4) a phase shift between 3 years and 4 years of age: Prior to 3 years short-distance phase differentiations are somewhat dominant, whereas after 4 years the long-distance differentiations greatly dominate—the cross-over being at about age 3.5 years.

Thatcher commented on this issue: "There appears to be a competitive type of dynamic between the short and distant connections with an approximate 1 year shift between the short- and long-term systems between approximately 3 and 4 years of age" (2010, p. 91). In our terms, long-distance phase differentiations are indexing differentiation of complex schemes that express developmental growth in *Mk* mental-attention symbolic units, which have the high boosting capacity required for complex (long-distance) differentiation of schemes. Thatcher (2010) concluded, "The most likely explanation is that a common genetic factor is responsible for the regular rhythms and slopes of change in mean coherence and mean phase over a 16-year period" (p. 100). We propose that this genetic factor may be developmental growth of mental attention.

(Pr2) Mendonça's Research on Maternal Report of Developmental Landmarks

Another source of data supporting our model was obtained in Brazil by Euclides Mendonça Filho (2017; Silva, Mendonça Filho, & Bandeira, 2019) for his master's thesis. We learned of it when he did a research stay in our lab. It was part of a larger study, Construction and Validation of the Dimensional Inventory for Childhood Development Assessment (IDADI) (translated from the Portuguese). Silva et al. (2019) analyzed psychometrically a refined scale of items adapted from six well-known infancy instruments (such as *Battelle Developmental Inventory*, *Bayley Scales of Infant and Toddler Development*, *Vineland Adaptive Behavior Scale*, and *Griffiths Mental Developmental Scales*). These items were transformed into specific questions, asked to mothers of children from birth to 6 years of age. For each item, mothers responded whether their child did the described behavior, did not do it yet, or no longer did it. The scale was named *Dimensional Inventory for Child Development Assessment* (IDADI; *Inventário Dimensional de Avaliação do Desenvolvimento Infantil*; Silva et al., 2019).

Items were grouped into seven age bands (all expressed in terms of months): 0 to 6, 7 to 12, 13 to 24, 25 to 36, 37 to 48, 49 to 60, and 61 to 72. Finally, these items were purified psychometrically using Rasch analysis and other methods. The complete measure covered seven domains: cognitive, socioemotional, fine motor, gross motor, receptive language, expressive language, and adaptive behavior. Brief descriptions of these domains appear in table 3.5. A sample of 1,336 mothers in Brazil completed the IDADI.

Items were administered in their presumed developmental order, with each mother completing the band suitable for her child's age, plus the bands one level below and above. Here we focus mainly on the one hundred six–item cognitive development subscale.

Within each domain, data were submitted to Rasch analysis. The Rasch model accounted for 72% of variance in the cognitive subscale responses. Figure 3.3 shows a scatterplot of the transformed cognitive-development Rasch score (i.e., [Rasch-score $\times$ 15] + 100) against the child's chronological age. The overall correlation between age and cognitive-development Rasch score was high, $r = .88$, $p < .001$. Inspection of the scatterplot shows a nonlinear relation between age and the cognitive score, which resembles a logarithmic function. Within-age-group regression lines (shown in figure 3.3) are much steeper for the first two years of age. They are flatter after the second year, looking like a linear function from 3 years onward. This is clearly what our model of two *M*-scales (*Me* vs *Mk*) predicts. Essentially the same logarithmic-like pattern was obtained in all seven domains where alternative scales were created (see table 3.5). This is entirely consistent with our model of mental attention.

(Pr3) Benson-Hamstra Doctoral Dissertation

Nancy Benson (1989), later Dr. Benson-Hamstra, completed a very creative doctoral thesis under Pascual-Leone's direction. Her work used our task-analytical methods to construct various *Me*-tasks and obtained results in infancy that our theory predicted. The methods and results, however, were unexpected in the research field at the time. Perhaps for this reason, initial attempts to publish the results in major journals were unsuccessful, and her premature death prevented further attempts. Here we summarize this doctoral work, focusing on two tasks: a linguistic-complexity task created by her (semantic complexity task—SCT) and a motor imitation task (imitation sorting task—IST). Both tasks were developed in Pascual-Leone's lab, the second by Pascual-Leone and Alp (Alp, 1988, 1994). Both tasks yield very close mental-attention (*Me*-) capacity mean estimates per age. These, and other tasks used in her dissertation, confirm experimentally the *M*-capacity of infants as predicted by our theory, contrasting children's *M*-power against the tasks' *M*-demand across tasks and age (*M*-) levels. This age-expected quantitative relation between infants' mental capacity (*Me*-power, *Mp*) and tasks' mental demand (*Me*-demand—*Md*) predicts children's mean performance and their differences in various *M*-measurement tasks, across domains.[1] Importantly, this *Mp/Md trade-off* yields the same results for visuospatial, motor-imitation, and language tasks.

Other tasks used in Benson's doctoral work were her versions of Piaget's object permanence task (Uzgiris & Hunt, 1975) and of Robbie Case's balance beam task for young children (Case, 1985, 1992); plus a semistructured and spontaneously elicited

Table 3.5
Description of domains and subdomains of Dimensional Inventory for Child Development Assessment

Domain (Subdomain)	Description
Cognitive	Conceptual and perceptual development, i.e., attention, symbolization, abstraction, processing speed, visuospatial processing, problem solving, memory.
Socioemotional	Understand one's own and others' feelings and emotions. Behavioral and emotional regulation, empathy, attachment. Ability to establish and maintain social relationships.
Motor	
(Fine)	Ability to reach, grasp, and manipulate objects. Precise coordination of small muscles of the body, i.e., feet, hands, fingers, wrists, lips, eyes, and tongue.
(Gross)	Ability to walk, sit, run, stand, balance, and coordinate physical activities involving major muscles of the body.
Language and Communication	
(Receptive)	Comprehension of meaning in spoken language including words, phrases, verbal expression. Understanding different tones of voice, gesture recognition, and forms of nonverbal communication.
(Expressive)	Ability to use spoken language for verbal expression, conveying information and instructions for social interaction. Gestures and nonverbal expressions for communicative purposes.
Adaptive Behavior	Ability to perform daily tasks for personal and social autonomy. Personal care, perception, establishment and maintenance of relationships. Communication of needs and feelings.

From de Mendonça Filho, E. J. (2017). "Evidências de Validade Relacionadas à Estrutura Interna da Escala Cognitiva do Inventário Dimensional de Avaliação do Desenvolvimento Infantil" [Validity Evidence Related to the Internal Structure of the Cognitive Scale for Dimensional Inventory of Infant Development] [Unpublished master's thesis]. Universidade Federal do Rio Grande do Sul, Porto Alegre, Brazil. Reprinted with permission.

request-to-mother task (i.e., request elicitation task) that she created. We shall not present these last three tasks in detail but provide empirical evidence for the close quantitative interrelations of all these *M*-measures, in particular semantic complexity (SCT) and imitation sorting (IST). To repeat, the significance of Benson's results is the predicted *Mp*/*Md* trade-off effects (see chapter 7 and Pascual-Leone & Baillargeon, 1994; Pascual-Leone & Johnson, 2005). Readers can consult Benson's dissertation through

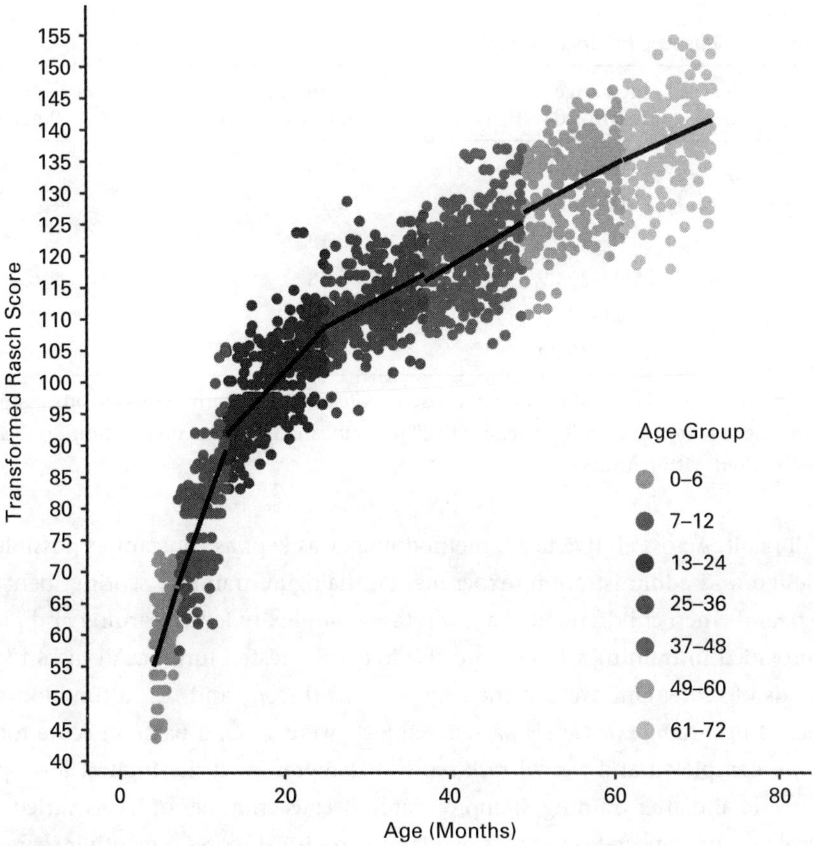

Figure 3.3
Association between child age and transformed cognitive-development Rasch score on Dimensional Inventory for Child Development Assessment. (From de Mendonça Filho, E. J. [2017]. "Evidências de Validade Relacionadas à Estrutura Interna da Escala Cognitiva do Inventário Dimensional de Avaliação do Desenvolvimento Infantil" [Unpublished master's thesis]. Universidade Federal do Rio Grande do Sul, Porto Alegre, Brazil. Reprinted with permission. See color plate 1.)

York University Library (library.Yorku.ca) to study her full results. Much of what follows is adapted from Benson and Pascual-Leone (1997).

Children from 10 to 36 months of age were assigned to six age-level groups, each 5 months apart (table 3.6). Each group was comprised of twelve carefully selected normal children, about half of them girls. Participants had no developmental delay or disorder, had English as their first language, and came from a middle socioeconomic background. Within each task, items were arranged in progressive order from simple to

Table 3.6
Age groups in Benson participant sample

Group	Age Range (months;days)	Mean Age (months;days)	Predicted *Me*
1	10;00–10;29	10;13	3
2	15;00–15;28	15;17	4
3	20;01–20;29	20;19	5
4	25;02–25;29	25;18	5, 6
5	30;01–30;29	30;18	6
6	35;01–35;29	35;16	7

Adapted from Benson, N. (1989). "Mental capacity constraints on early symbolic processing: The origin of language from a cognitive perspective" [Unpublished doctoral dissertation]. p. 112. York University. Used with permission.

more difficult. Across all five tasks, methodology was kept as constant as possible (i.e., task design and administration, experimental dialogue, training, scoring, behavioral coding), and when children failed a given task-complexity level, learning and practice were provided in training trials before children were tested further. All tasks received two kinds of score. One we call the *empirical (item) score,* and it is a function of the number of *item-complexity levels* passed. Subjects were given a pass/fail score for each task level completed and a final task score that represented the highest level passed (both before and after training, if appropriate). Because number of levels varied across tasks, there is no expectation of task overlap in empirical scores. The other score is the *M-complexity score,* which corresponds to the *M*-demand of the highest level of performance the child exhibited on each task. *M*-demands were predicted a priori on the basis of task analysis. In the *M*-complexity score, performance on each and every task is measured on the same *Me*-scale, and consequently the various tasks should yield similar *M*-complexity curves across age. Notice that empirical (item-level) scores express empirical performance of subjects across items, the highest level of items that subjects can pass. In contrast *M*-complexity score expresses highest *M*-demand of items that subjects can pass, after items have been task-analyzed for their *M*-complexity; this is an empirically grounded *theoretical* score.

Semantic Complexity Task (SCT) This task uses five key semantic constituents or categories of action semantics in language (Bloom, 1973; J. Miller, Chapman, Branston, & Reichle, 1980): verb (**V**), agentive (**A**—or agent), genitive (**G**—or possessive), objective (**O**—or object), and locative (**L**—or location), which gives the most complex sentences the semantic structure **VAGOL**. The category titles come from Fillmore's (1968) case grammar. The verbal comprehension task required children to enact with suitable toys sentences that differed based on such semantic-complexity categories. The complete set of sentences appears in table 3.7. The lexical items chosen were selected from high-frequency words reported in children's early vocabularies (e.g., Anglin, 1977; Nelson, 1974). Suitable toys and playthings were provided for acting out the sentences. Task procedure presented items to children verbally and in random order within levels of complexity. The tester said, "Would you like to play a game? Look some toys! Show me this [the tester voices one item of SCT, e.g., "*The baby dances on the book*"—a three-term (VAL) item]." After the child's response (i.e., enactment of the sentence), the tester asked, "What was that?" We present data for the children's enactment (i.e., comprehension) of the items. Children's understanding of lexical items (single words) was tested in advance of items with higher-order semantic relations. All children received levels 1 through 5. Beyond level 5, testing ended when the child failed two consecutive levels of complexity (see table 3.7). Total testing time lasted from five to fifteen minutes.[2]

Appendix 3.A.1 presents task analyses of SCT items from complexity levels 1 to 5. Here we simply summarize task-analytical conclusions without presenting the analyses. Figure 3.4 shows the number of children per age group passing each of the first five SCT levels. No children passed items beyond level 5.

Level 1, identify a single object such as the "dog" (O): The task analysis shows mental (*M*-) demand of this item to be $Me=4$. Such mental-attention capacity is theoretically attained at or after 12 months of age (table 3.1), so this level should be passed by the 15-month age group.

Level 2, show (e.g.) "The ball on the book" (OL): Here the child is asked to act out the phrase. The mental-attention complexity of this item is $Me=5$ (theoretically, at or after 18 months) and should be passed by the 20-month age group.

Level 3, show (e.g.) "The dog eats the cookie" (VAO): The child reaches and grasps the cookie (1), then locates the dog (2), and (3) mimics the action of the dog eating the cookie. Here $Me=6$, because this is the number of schemes to be held in mind in step (3), the most complex step of task analysis (see appendix 3.A.1). This level should be passed by the 30-month group and possibly at 25 months.

Level 4, show "The baby eats the cookie on the book" (VAOL): After grasping the cookie (1) and locating the baby (2) and the book (3), the child (step 4) enacts the baby doll eating the cookie on the book. *M*-demand is $Me=7$, that is, the *M*-capacity of 3-year-olds ($Mk=e+1$), so should be passed only by the 35-month group. However, in

Table 3.7
Comprehension items for Benson's semantic complexity task (SCT)

Level	Relation	Test Items
1		baby dog ball book blanket cookie banana hat shoes eat sleep kiss
2	AL	The baby on the blanket.
	VA	The dog dances.
	VO	Kiss the baby.
	OL	The ball on the book.
	GO	The baby's shoes.
	VL	Sleep on the blanket.
3	VAL	The baby dances on the book.
	VAO	The dog eats the cookies.
	VOL	Eat the banana on the blanket.
	VGO	Kiss the baby's hat.
	GOL	The dog's hat on the book.
4	VAGO	The dog kisses the baby's shoes.
	VAOL	The baby eats the cookies on the book.
	VGOL	Kiss the dog's hat on the blanket.
5	VAGOL	The dog eats the baby's shoes on the blanket.
		The baby kisses the dog's hat on the blanket.
		The dog kisses the baby's hat on the book.
6	V-VAGOL	The baby dances and kisses the dog's hat on the book.
		The dog sleeps and eats the baby's hat on the book.
		The dog dances and kisses the baby's shoes on the blanket.
7	VA-VAGOL	The dog dances and the baby eats the dog's hat on the blanket.
		The baby sleeps and the dog kisses the baby's shoes on the book.
		The baby dances and the dog eats the baby's hat on the blanket.
8	VAO-VAGOL	The dog eats the cookies and the baby kisses the dog's hat on the book.
		The baby kisses the ball and the dog eats the baby's shoes on the blanket.
		The baby eats the banana and the dog kisses the baby's hat on the book.
9	VAOL-VAGOL	The baby eats the cookie on the book and the dog kisses the baby's shoes on the blanket.
		The dog eats the banana on the blanket and the baby kisses the dog's hat on the book.
		The dog kisses the ball on the blanket and the baby eats the dog's shoes on the book.

Adapted from Benson, N. (1989). "Mental capacity constraints on early symbolic processing: The origin of language from a cognitive perspective" [Unpublished doctoral dissertation]. pp. 62–63. York University. Used with permission.

Benson's study some children (25%) pass at 25 months and 66% pass at 30 months (see figure 3.4). This earlier passing can be explained by intratask learning (*LCLM* learning, see below and chapter 5), because these are facilitating situations.

Level 5, show "The dog eats the baby's shoes on the blanket" (VAGOL): In step 4, which carries the highest *M*-demand, the child simulates the described action. Its mental-attention complexity is $Me=8$, corresponding to $Mk=e+2$, that is, only accessible theoretically to 5- or 6-year-olds. In Benson's study, however, 33% (four children) of her 35-month-olds pass this level, possibly because of intratask learning caused by the items' presentation in progressive order of difficulty (see figure 3.4).

Rate of passing showed a marked linear increase with age (see figure 3.4). On average, consistent with the predictions (see appendix 3.A.1), 15-month-olds (but not 10-month-olds) identified a single object (level 1 "dog"); 20-month-olds were able to enact a two-term relation (e.g., "the ball on the book" [OL]); 25-month-olds enacted a three-term relation (e.g., "The dog eats the cookie" [VAO]); at 30 and 35 months children tended to be able to enact a relation with four constituents (level 4, "the baby eats the cookie on the book" [VAOL]). Only four children from the oldest age group passed level 5 (VAGOL).

Panel A of figure 3.5 shows mean empirical scores for SCT. An analysis of variance yielded an effect for age group ($F[5,60]=48.27$, $p<.001$); and a nonsignificant effect ($p<.05$) for sex. Post-hoc comparisons revealed that all adjacent pairs of age groups differed significantly from one another, except the 30- versus 35-month-old comparison (Benson, 1989; Benson & Pascual-Leone, 1997). For all groups performance was very close to theoretically predicted (see figure 3.5).

Imitation Sorting Task (IST) This task was designed as an *M*-measure by Pascual-Leone and Alp in our lab. It was first implemented in Alp's doctoral thesis (1988, 1994) to study mental-attention development from 1- to 3-years of age. In this task, infants must imitate tester's placement of a set of objects into two canisters and repeat this performance across a number of trials with progressively more objects (from one to eight). The child's assigned goal is to sort objects (by dropping them into the canisters) in the same two clusters as the tester just did, and not to group or classify objects according to a criterion of form or function (e.g., Bruner, 1964; Gopnik & Meltzoff, 1992). The apparatus consisted of a yellow wooden platform with one or two canisters (24 cm high and 15 cm diameter), inserted at each end of the platform. Seven different toy sets, comprising small (e.g., 7 cm long) colorful plastic toys, were prepared for each of the nine difficulty levels. Levels were indexed by the number of toys to be placed in either of two canisters: one toy in level 1, two in level 2, and so on. A level 0 with only one canister was included to determine whether the infant could drop one toy into a single container.

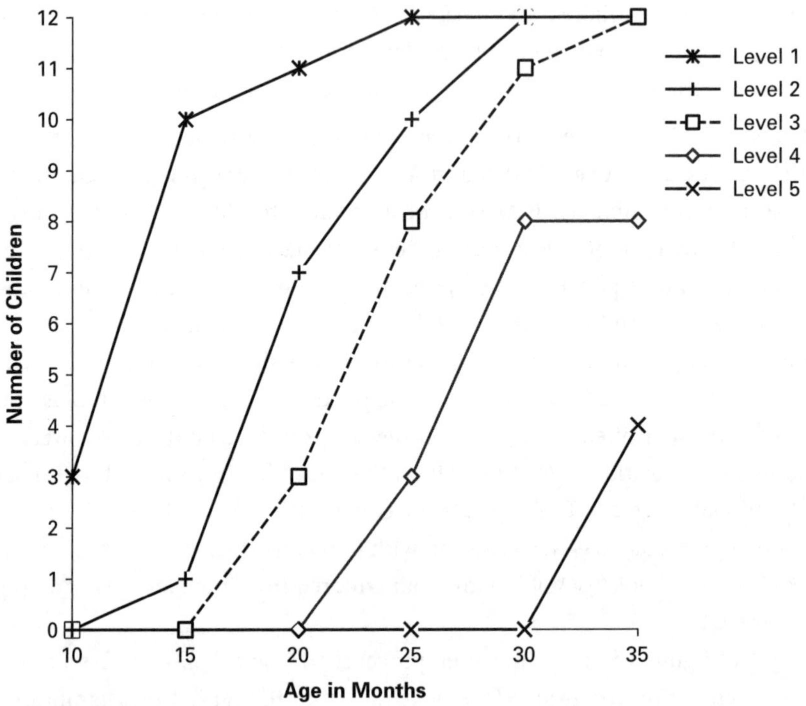

Figure 3.4
Frequency of children passing semantic complexity task (SCT) levels per age group.
Note: Points at 12 on the y-axis indicate the highest level that all children at a given age pass. Points at 0 on the y-axis indicate the lowest level that no child at a given age passes. (Adapted from Benson, N. [1989]. "Mental capacity constraints on early symbolic processing: The origin of language from a cognitive perspective" [Unpublished doctoral dissertation]. p. 172. York University. Used with permission.)

Alp's (1988, 1994) procedure was modified to include not only testing but also training and retesting items. (a) During *testing* there were two warm-up trials with the platform and only one canister. The tester held up a toy in front of the child (who sat in front of the platform) and said, "Look a [here the toy name was given—let us call it *T*]! Watch the *T*. *T* goes here!" (Tester drops the toy in the canister.) The tester made the toy hop three times on the way to the canister (to make this course more perceptually salient) and dropped it in, where it could not be seen. The tester then retrieved the toy and placed it on the platform in front of the child saying, "Now [child's name, e.g., *C*] you do it." The child was expected to respond. The procedure was repeated with another toy (and if needed, depending on the outcome, with a third toy); a passing criterion of at least two out of three correct trials was used.

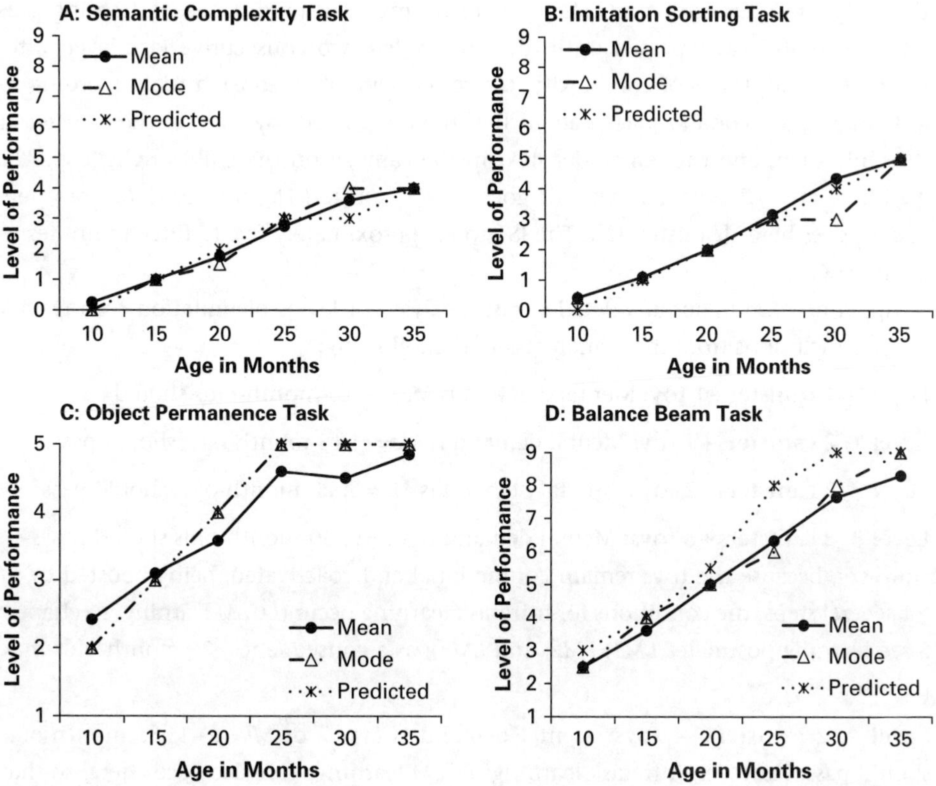

Figure 3.5

Empirical scores and model predictions across ages, for four sensorimotor *M*-measure tasks.
Note: For each task, the vertical axis reflects the possible range of level scores. (Adapted from Benson, N. [1989]. "Mental capacity constraints on early symbolic processing: The origin of language from a cognitive perspective" [Unpublished doctoral dissertation]. p. 278. York University. Used with permission.)

If the child passed level 0, then level 1 (one toy, two canisters) was administered. Otherwise, two training trials were conducted. (b) During *training,* the tester held up a small hand puppet and said, "Hi! My name is Friendly, and I'm here to help you play this game. We'll play with the toys together." The tester then hopped the toy to the canister ("*T* goes here"), returned the toy to the center of the apparatus, saying, "Now *C* you do it," and then used the puppet to help guide the infant's hands to enact the correct response. (c) Two or three *re-testing* trials (pass = 2/3) were administered next, and the exact procedure used in testing trials was followed with novel toy sets. (d) The above procedure was repeated for the next and subsequent levels of complexity (e.g., sorting one toy into one of two canisters, sorting two toys into two canisters, sorting three toys into two canisters). The cycle of testing trials (and if needed training plus retesting trials) was repeated until the child failed two consecutive task levels after training. For all levels of complexity, the tester demonstrated each trial according to a standard protocol and placed about half the designated toys first in the canister at the child's left, and the remainder next in the canister on the child's right (e.g., "*T1* goes here and *T2* goes here and *T3* goes here" [canister#1]. "But look! *T4* goes here and *T5* goes here" [canister #2]). The IST took approximately five to fifteen minutes to administer.

Appendix 3.A.2 contains detailed task analyses of levels of imitation sorting task (IST). Conclusions from these analyses are the following:

Level 0, 1 canister + 1 toy: Mental demand is $Me=3$; 10-month-olds should pass.

Level 1, 2 canisters + 1 toy: Mental demand is $Me=4$; 15-month-olds should pass.

Level 2, 2 canisters + 2 toys: Mental demand is $Me=5$; 20-month-olds should pass.

Level 3, 2 canisters + 3 toys: Mental demand is $Me=6$; 30-month-olds should pass. However, because the toys remain cofunctional and coactivated, being boosted by *M* for several steps, the conditions for intratask learning occur (*LCLM* learning, in chapter 5 see learning postulates *LM1, LM2,* and *LM3*); as a consequence, 25-month-olds may pass this level.

Level 4, 2 canisters + 4 toys: Mental demand is $Me=7$ or $Mk=e+1$; 35-month-olds should pass. However, intratask learning (*LCLM* learning) can also occur here, so that at 30 months children may pass this level.

Level 5, 2 canisters + 5 toys: Mental demand: $Me=8$ or $Mk=e+2$; 60-month-olds should pass, although because of intra-task learning 35-month-old children may also pass.

Thus, when intratask learning (*LCLM*) is not considered, the basic *Me*-demand of all items is equal to $\mathbf{3+n}$, where n is the number of toys in levels with two canisters. The detailed analyses in appendix 3.A.2 show that the number 3 of *Me*-demand is made up by the following processing schemes: a move-and-drop operative, the scheme of the object to be dropped (ob-grasped), and the scheme for the canister. Figure 3.6 shows the number of children per age group passing each IST level. Performance is consistent with predictions, particularly when intratask learning is assumed at the higher levels.

Responses on the IST were scored 0 when the child failed to place a toy in a single canister (i.e., failed level 0) or 0.5 if level 0 was passed. Children received scores of 1 to 8, respectively, when they passed levels 1 through 8 (sorted one to eight toys into two locations). As panel B of figure 3.5 shows, improvements in this empirical score with

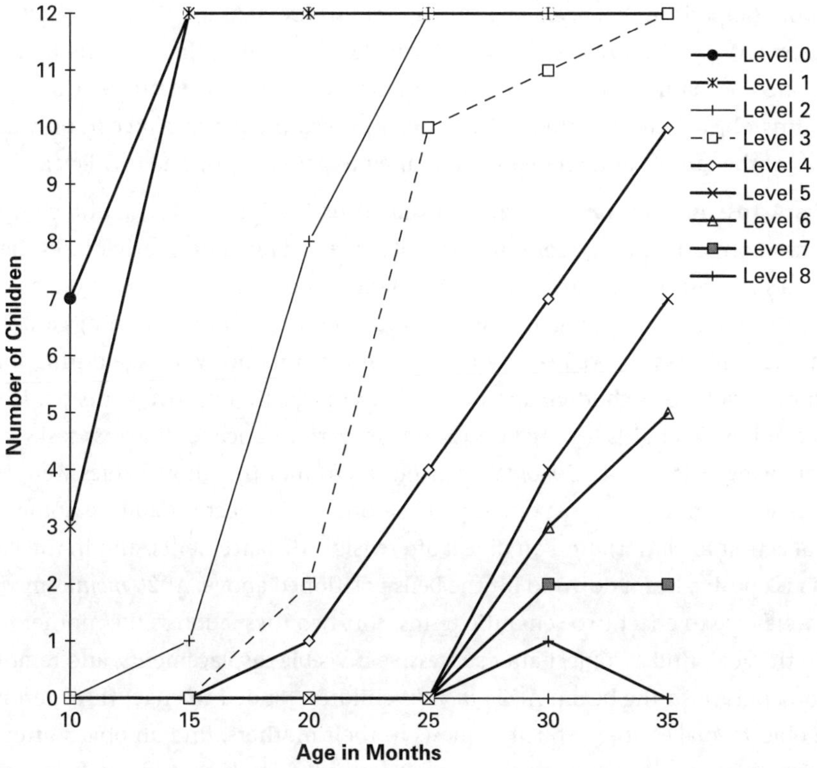

Figure 3.6
Frequency of children passing imitation sorting task (IST) levels per age group.
Note: Points at 12 on the y-axis indicate the highest level that all children at a given age pass. Points at 0 on the y-axis indicate the lowest level that no child at a given age passes. (Adapted from Benson, N. [1989]. "Mental capacity constraints on early symbolic processing: The origin of language from a cognitive perspective" [Unpublished doctoral dissertation]. p. 190. York University. Used with permission.)

age were strong, and very close to theoretical predictions (figures 3.5 through 3.7 show IST performance following training, when needed). The average number of toys sorted increased by approximately one toy with each successive age group, as predicted.

A Group (6) × Sex (2) × Time of Testing (2; testing versus retesting trials) analysis of variance performed on the empirical scores yielded effects for group ($F[5, 60] = 31.29$, $p < .001$) and time of testing ($F[1,60] = 20.30$, $p < .001$), but not sex and no interactions ($p > .05$). The level of children's performance tended to improve following training, and scores obtained during testing and retesting trials both increased as a function of age.

Post-hoc comparisons between adjacent age groups revealed significance for all but the youngest (10 vs. 15 months) and oldest pairs (30 vs. 35 months). In general, training functioned primarily to help children maintain necessary task strategies (for instance, separating objects into locations). Training improved the performance in only about a quarter of the sample and tended to be limited to gains of just one task level.

Results Across All Five Tasks Although our focus has been on semantic complexity and imitation sorting tasks, congruent results were obtained from other tasks (Benson-Hamstra's request elicitation, object permanence, and Case's balance beam; see figure 3.5 for the latter two). Taking all the tasks together, key summary findings are the following: (1) All tasks showed steady increase in performance with age during infancy, less dramatically so as children approach 3 years of age. (2) Improvements tended to be similar in boys and girls. (3) Age transitions in performance level across tasks yielded the following pattern: *at 15 months* (but not at 10 months) most babies were able to enact one semantic term, sort one toy, name one goal object (or address some instrumental action toward it), find an object after visible displacements, and in the balance beam task push a beam down to hit the bells positioned above. At *20 months* most children were able to enact two semantic terms, sort two toys, address the mother to help obtain the goal, find an object after successive invisible displacements, and remove the supports plus push the beam. At *25 months* children tended to enact three terms, sort three objects, make more explicit requests to their mothers, find an object after invisible displacements, and remove the supports plus lift the beams in the balance beam task. At *30 or 35 months* the majority of children could enact four terms, sort four toys, find an object after invisible displacement plus rotation of screens, and place a weight plus remove the supports to shift the position of the beam.

These results suggest a consistent/invariant pattern of complexity improvement in performance across domains, featuring increase of roughly one task component (e.g., semantic relation, toy, action) with each successive age group during infancy, in agreement with our task analyses (done prior to obtained results). After the results were known, the TCO model was enriched by adding the prior theoretical assumptions about intratask *LCLM* learning (see Pascual-Leone & Goodman, 1979, and chapter 5). These two prediction models are empirically contrasted below.

To compare across tasks, we must put them on the same scale, and this is what the *M*-complexity (*Me*) score does. Recall that the *M*-complexity score for a task is the *Me*-demand of the highest level passed by the child. Figure 3.7 shows in the left panel the obtained mean *Me*-score for the two main tasks (semantic complexity and imitation sorting), as well as the other three tasks used by Benson (request elicitation, object permanence, balance beam). These *M*-complexity scores were assigned using

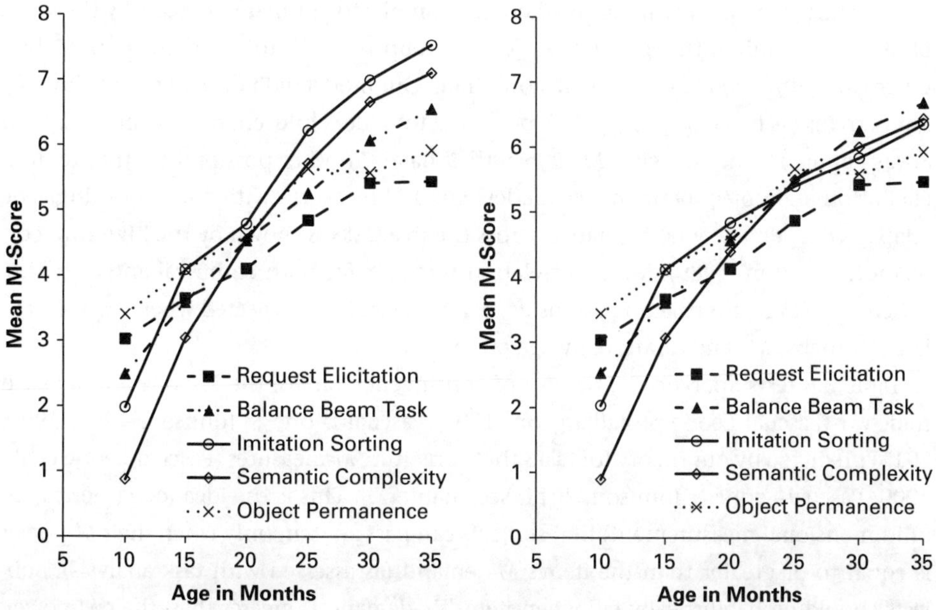

Figure 3.7
Mean *Me*-complexity scores as a function of task and age group. *Left panel* based on original *Me*-demand estimates; *right panel* based on intratask (*LCLM*) learning adjusted estimates. (Adapted from Benson, N. [1989]. "Mental capacity constraints on early symbolic processing: The origin of language from a cognitive perspective" [Unpublished doctoral dissertation]. pp. 276 and 282. York University. Used with permission.)

M-demand estimates from the a priori task analyses, and without considering the likely contribution of intratask learning (*L*-structuring, *LCLM* learning, see chapter 5). All scores increase with age yielding maximal homogeneity at about 20 months (when the middle-range item complexity is accessible). Intratask learning could reduce the *M*-demand of task items. In revised (a posteriori) task analyses, Benson (1989) applied the theoretical *L*-structuring principle and found that it lowered *M*-demand estimates in higher-level items.

The right panel of figure 3.7 shows the resulting *L*-learning (*LCLM*) adjusted mean *Me*-scores plotted as a function of task and age group. These adjusted *Me*-scores yield group means that are more consistent across tasks. The four structured measures (semantic complexity, imitation sorting, object permanence, and balance beam), after adjustment, yield the following across-task average *Me*-scores for age groups 10 through 35 months, respectively: 2.20, 3.70, 4.54, 5.44, 5.83, and 6.25. The corresponding purely theoretical expected *M*-values from ages 10 months to 35 months are 3, 4, 5, 5/6, 6, and 7. Thus (consistent with our predictions) complexity estimates increase by the value of about one unit with each successive age group from 10 to 25 months and subsequently continue to increase, but at slower rate. Our model did not predict this slowing, but perhaps (as Pascual-Leone, 1970, predicted for older children) older babies may not always apply all the capacity (*M-reserve*) they have. Another possibility is that request elicitation and object permanence are less good *M*-measures with a lower ceiling. The relative equivalence of performance across the five tasks is shown by the high intercorrelations (Benson, 1989). Mean correlation was 0.78 for both empirical and *M*-scores. When age was removed, correlations decreased severely, as expected when their shared latent-variable *M* is maturationally age bound.

Table 3.8 tests analytically, by way of contingency tables, the *M*-power/*M*-demand trade-off (Pascual-Leone & Baillargeon, 1994; Pascual-Leone & Johnson, 1991, 1999, 2011) predicted by our theory for tasks that constitute *M*-measures (Pascual-Leone et al., 2000; Pascual-Leone & Johnson, 2011; see chapter 7). This is the idea (congruent with unidimensional measurement) that subjects can pass an item only when their *M*-power is equal to or greater than the item's *M*-demand (as assessed with task analysis). Subjects should probabilistically fail when items' *M*-demand is greater than their *M*-power. The five contingency tables in table 3.8 show in columns the subjects' theoretical *Me*-capacity. The rows exhibit the items' estimated *Me*-demand, based on the (*L*-learning) revised task analyses. The diagonal boundary lines mark the partition between data in which *M*-power (columns) is equal or greater than *M*-demand (rows), which is above the diagonal; versus data below the diagonal, where *M*-power < *M*-demand. Our *Mp/Md* trade-off hypothesis predicts that all passing scores must be above the diagonal, and cells below the diagonal should be empty, which is probabilistically the case in these data.

To statistically test such filled/empty-cells contrast, Benson used prediction logic procedures of Hildebrand, Laing, and Rosenthal (1977). A prediction logic model assigns weights of one to the critical cells that theoretically should be empty, and zeros to other cells in the matrix; this is the most conservative weight assignment with respect to our theory. In addition, whenever the theoretical prediction from task analysis could accept disjunctively two contiguous values of *M*-demand (e.g., 3 or 4, 3/4 in table 3.8) the most conservative test of predicted values (e.g., 3 for prediction 3/4) was adopted. Cases falling in these disjunctive predictions are indicated by boxes in that appear below the diagonal. Prediction-logic statistics evaluate the probability of finding by chance the obtained pattern match of filled/empty cells. A significant Del statistics indicates that a match between predicted and obtained results cannot be due to chance. Benson found a significant Del statistic for all five tasks' contingency tables, indicating support for the predicted *Mp* vs. *Md* trade-off (semantic complexity: Del = 0.48, SE = 0.10, Z(Del) = 4.93,

Table 3.8

Relation between age and complexity of response as a function of sensorimotor *M*-task

Complexity Or Mental Demand of Item	Age in Months: 10	15	20	25	30	35	
	Theoretical Mental Capacity of Participants: 3	4	5	5/6	6	7	}*Mp*
Semantic complexity task							
Md							
—							
Fail	9	2	1				
3/4	3	9	4	2			
5		1	4	2	1		
6			3	8	11	8	
7						4	
Imitation sorting task							
Fail	5						
3	4						
4	3	11	4				
5		1	7	8	5	2	
6			1	4	5	5	
7					2	5	
Object permanence task							
3	7						
4	5	11	5				
5		1	7	5	6	2	
6				7	6	10	
Balance beam task							
2	6	2					
3	6	1	1				
4		9	4	1			
5			7	5	2		
6				6	6	5	
7					4	7	
Request elicitation paradigm							
2/3	30	16	5		1	3	
3	20	4				1	
3/4	13	27	15	6	2	5	
4	10	31	28	19	13	16	
4/5			23	14	8	3	
5			4	17	17	12	
5/6				5	1	2	
6				7	17	23	
7				1	13	10	

Note: Within task, the line marks the border of the trade-off between task *Me*-demand and child theoretical *Me*-capacity. For the first four tasks, numbers above the line indicate children performing as theoretically predicted (i.e., $Md \leq Mp$); numbers below the line indicate those performing above the predicted level (i.e., $Md > Mp$). For Request Elicitation, numbers correspond to responses, not individuals. (Adapted from Benson, N. [1989]. "Mental capacity constraints on early symbolic processing: The origin of language from a cognitive perspective" [Unpublished doctoral dissertation]. p. 284. York University. Used with permission.)

$p < .001$; imitation sorting task: Del = 0.58, SE = 0.10, Z(Del) = 5.64, $p < .001$; object permanence: Del = 0.50, SE = 0.11, Z(Del) = 4.65, $p < .001$; balance beam task: Del = 0.64, SE = 0.10, Z(Del) = 6.62, $p < .001$; request elicitation paradigm: Del = 0.59, SE = 0.05, Z(Del) = 12.57, $p < .001$).

In summary, Benson's doctoral thesis shows probabilistically that across very different tasks (different in content and context) the maximum *M*-complexity scores achieved by representative children of 10, 15, 20, 25, 30, and 35 months of age are most likely to be, respectively, 3/4, 4, 5, 5/6, 6, and 7. Very rarely did children succeed beyond their theory-predicted *M*-capacity, even though when needed they received tester-mediated training. This is strong support to our TCO because the retraining *learning loops* (Pascual-Leone, 1976b; Pascual-Leone et al., 1978) built into the testing procedures by Dr. Benson-Hamstra (for imitation sorting, object permanence, and balance beam), go counter the most frequent alternative explanation to a model of *M*-capacity developmental growth, that is, a constructivist-learning interpretation of development.

Conclusions

The analyses and data on infancy reviewed in this chapter suggest that infants, after 2 or 3 months of age, have a sort of working mind that develops using constructivist and mental-attention-driven (not just associative) learning. This learning is mediated by self-propelling dynamic functional units, the schemes that embody functional invariances abstracted from experience. Schemes produce inferential expectancies about future or conditional outcomes, and infants' problem solving is often directly driven by affective and psycho-social motives. These inferential expectancies are made possible by interactions among schemes and content-free, general-purpose organismic-causal dynamic factors (hidden operators such as maturational *Me*-capacity and other operators, see chapters 1, 5, 6, and 7). Our theory of sensorimotor mental processing models the infant's emerging mind as deeper and different, but still consistent, with descriptive theories of Piaget, Case, and other neo-Piagetians such as Demetriou (Demetriou & Spanoudis, 2018). Infants exhibit creative cognitive activity, because of four intertwined organismic factors: (1) They possess a basic repertoire of innate schemes geared to agency and even praxis, causally oriented toward implicit biological or psychosocial goals. (2) The learning situations of infants are basically facilitating. (3) Infants have enough (sensorimotor) *Me*-power, and other hidden operators, to easily synthesize new concrete-scheme invariants by coordinating cofunctional and coactivated schemes, with the help of organismic operators and the principle of overdetermination (*SOP*).

(4) They also have enough opportunities for protracted constructivist learning because of their facilitating mutuality with parents and caretakers. Theories of self-organizing dynamic systems in infancy are compatible with these organismic factors, although self-organizing dynamic approaches do not entertain the growth of a maturational mental-attention resource (e.g., Kelso, 2016).

Piaget did not fully anticipate in infants the sort of constructivist mentation they exhibit, which research by neonativists has established. Our task analyses and theory can explicate these new findings without assuming proper concepts within the infant's mind.

Appendix 3.A.1

Task-analysis notation is explained in more detail in chapters 8 and 9, together with the book's appendix. Table 3.2 provides a summary of notation conventions. As appendixes to this chapter we "translate" into English our *metasubjective task analyses* of the semantic complexity task (appendix 3.A.1) and the imitation sorting task (appendix 3.A.2).

Task Analysis of Semantic Complexity Items

1. **Identify single object "dog" (O).** *Me*-demand = 4.
   (0) LAN(λdog) ➔ dog
   (1) SHOW(#dog, POINT($dog)) ➔ $dog.pointed
2. **Show: "The ball on the book" (OL).** *Me*-demand = 5.
   (0) LAN(λball: λON: λbook) ➔ ON[ball : book]
   (1) SHOW(#ON[ball : book], REACH+GRASP($ball)) ➔ $ball.grasped
   (2) SHOW(#ON[ball : book], LOCATE($book)) ➔ $book.located
   (3) SHOW(#ON[ball : book], MOVE.ON($ball.grasped: $book.located)) ➔
   ➔ $ON[ball : book]
3. **Show: "The dog eats the cookie" (VAO).** *Me*-demand = 6.
   (0) LAN(λdog: λEATS: λcookie) ➔ EAT[dog : cookie]
   (1) SIMULATE(#EAT[dog : cookie], REACH+GRASP($cookie)) ➔
   ➔ $cookie.grasped
   (2) SIMULATE(#EAT[dog : cookie], $cookie.grasped, LOCATE($dog)) ➔
   ➔ $dog.located
   (3) SIMULATE(#EAT[dog : cookie], EAT(#dog.mouth,$dog.located: $cookie.grasped)) ➔ $EAT[dog : cookie]

4. Show: "The baby eats the cookie on the book" (VAOL). *Me*-demand = 7 ⇔ *Mk* = *e* + 1

....

(4) SIMULATE(#EAT[baby : ON(cookie : book)], EAT(#baby's.mouth, $baby.located: ON[$cookie.grasped : {$book.located}$_{sit}$])) ➔
➔ $EAT[baby : ON[$cookie.grasped : book.located]]

5. Show: "The dog eats the baby's shoes on the blanket" (VAGOL). *Me*-demand = 8 ⇔ *Mk* = *e* + 2; or *Me* = 7 ⇔ *Mk* = *e* + 1

(0) LAN(λdog: λEATS: λbaby's: λshoe: λONtheblanket) ➔
➔ $EATS[dog : shoe[baby's . ONblanket]]

....

(4) SIMULATE(#EATS[dog: shoe[baby's . ONblanket]]: EATS(#dog.mouth, $dog: $shoe[baby's. ONblanket])) ➔
➔ $EATS[dog : shoe[baby's . ONblanket]]

Level 1, identify a single object "dog" (O): LAN stands for linguistic task-specific operative schemes, that is, procedures for language processing, decoding, encoding, and communication. We use λ as an affix referring to a suitable language scheme of the subject. In this case, LAN refers to the identification of the word λdog (which the tester has voiced) as being a dog-toy ($dog) out there in reality. It is encoded as a figurative scheme (using lowercase without an affixed star *, for simplicity). In step (1) the child wants to SHOW (behaviorally demonstrate) the object-referent of the word λdog. This intended object is #dog, written with # because it functions as a parameter/condition of SHOW. The showing action is realized by pointing (POINT) with the finger to the actual toy dog ($dog). This is the entity the child perceptually sees (we signify with the affix $ that here "dog" stands for the actual thing here-and-now perceivable, true). The result of this pointing is $dog.pointed. The underlined schemes are those that effortful attention, *Me*-power, must boost. This gives a count of 4 as the mental (*Me*-) demand of this item.

Level 2, show "The ball on the book" (OL): Here the child is asked to demonstrate the phrase. In step (0) this phrase is linguistically decoded to the experiential knowledge item ON[ball : book]. The square braces [...] indicate some characteristic of this ON scheme: the operative ON refers to the relational coordination of ball (object) and book (location). In step 1 the child reaches for and grasps the ball. In step 2, he or she locates the book; then, step 3, moves the ball onto the book. In all steps the underlined schemes are those being boosted by *Me*-capacity. The step with a highest number of underlined schemes is the one that specifies the *Me*-demand of the item. In this case, it is step 3, which has five schemes in need of *Me*-boosting. Thus mental-attention complexity of this item is *Me*=5.

Level 3, show "The dog eats the cookie" (VAO): The child reaches for and grasps the cookie (1); then locates the dog (2) and simulates the action of the dog eating the cookie (3). Here *Me*=6; that is, the number of schemes underlined in step 3, the most complex step.

Level 4, show "The baby eats the cookie on the book" (VAOL): After steps 0 to 3 here omitted, the baby (step 4) simulates by enactment that the baby doll eats the cookie on the book. *M*-demand is *Me*=7, which corresponds to (*Mk*=*e*+1).

Level 5, show "The dog eats the baby's shoes on the blanket" (VAGOL): In step 4, the one with highest *M*-demand, the child simulates this action. The mental-attention complexity is *Me*=8, corresponding to *Mk*=*e*+2 (i.e., only accessible theoretically to 5- or 6-year-olds). However, if the schemes $shoe[baby's. ONblanket] are fully chunked or *LC*-learned (i.e., $shoe[baby's . ONblanket]) then the *M*-demand would be instead *Me*=7 ⇔ *Mk*=*e*+1.

Appendix 3.A.2

Task Analysis of Imitation Sorting Items

Level 0: 1 canister (can)+**1 toy** (ob). *Me*-demand=3.

(0) EX.PLACE($ob1: $\{\$can\}_{sit,LC,B}$) ➔ $ob1[IN.can]

(1) $\{IMITATION.PLACE\}_{sit,A,B}$ (#ob1[IN.can], $ob1: $can) ➔ REACH+GRASPob

(2) $\{IMITATION.PLACE\}_{sit,A,B}$ (#ob1[IN.can], REACH+GRASP($ob1)) ➔ $ob₁.grasped

(3) $\{IMITATION.PLACE\}_{sit,A,B}$ (#ob1[IN.can]^{L1}, **MOVE+DROP** (**$ob1.grasped)** : $\{\$can\}_{sit,B,L1}$) ➔ $ob1[IN.can]

Level 1: 2 canisters (can)+**1 toy** (ob). *Me*-demand=4.

(0) EX.PLACE($ob1: $\{\$canA\}_{sit,LC,B}$) ➔ $ob1[IN.canA]

(1) $\{IMITATION.PLACE\}_{sit,A,B}$ (#ob1[IN.canA], $ob1: $canA) ➔ REACH+GRASPob

(2) $\{IMITATION.PLACE\}_{sit,A,B}$ (#ob1[IN.canA], REACH+GRASP($ob1)) ➔ $ob1.grasped

(3) $\{IMITATION.PLACE\}_{sit,A,B}$(#ob1[IN.canA],**MOVE+DROP($ob1.grasped):$canA**) ➔ $ob1[IN.canA]

Level 2: 2 canisters (can)+**2 toys** (ob). *Me*-demand=5.

(0) EX.PLACE($ob1: $\{\$canA\}_{sit,LC,B}$) ➔ $ob1[IN.canA]

(0) EX.PLACE(# $ob1[IN.canA], ($ob2: $\{\$canB\}_{sit,LC,B}$)) ➔ $ob2[IN.canB]

(1) {IMITATION.PLACE}$_{sit,A,B}$ (#ob1[IN.canA], #ob2[IN.canB]) ➔ REACH + GRASPob

(2) {IMITATION.PLACE}$_{sit,A,B}$(#ob1[IN.canA],#ob2[IN.canB],REACH + GRASP($ob1)) ➔ $ob1.grasped

(3) {IMITATION.PLACE}$_{sit,A,B}$ (#ob1[IN.canA], #ob2[IN.canB], MOVE + DROP ($ob1.grasped) : $canA) ➔ $ob1[IN.canA]

(4) {IMITATION.PLACE}$_{sit,A,B}$(#ob2[IN.canB],$ob1[IN.canA],REACH + GRASP($ob2)) ➔ $ob2.grasped

(5) {IMITATION.PLACE}$_{sit,A,B}$ (#ob2[IN.canB], $ob1[IN.canA], **MOVE + DROP ($ob2.grasped) :$canB**) ➔ $ob2[IN.canB]

(6) IMITATION.EVAL ($ob1[IN.canA], $ob2[IN.canB]) ➔ RES[*STOP*]

Level 3: 2 canisters (can) + **3 toys** (ob). *Me*-demand = 6.

....

(1) {IMITATION.PLACE}$_{sit,A,B}$ (#ob1[IN.canA], #ob2[IN.canB], #ob3[IN.canA]) ➔ ➔ REACH + GRASPob

....

(5) {IMITATION.PLACE}$_{sit,A,B}$ (#ob2[IN.canB], #ob3[IN.canA], $ob1[IN.canA], **MOVE + DROP ($ob2.grasped) :$canB**)) ➔ $ob2[IN.canB]

....

(7) {IMITATION.PLACE}$_{sit,A,B}$ ($ob2[IN.canB], #ob3[IN.canA], $ob1[IN.canA], **MOVE + DROP ($ob3.grasped) :$canA**) ➔ $ob3[IN.canA]

(8) IMITATION.EVAL ($ob1[IN.canA], $ob2[IN.canB], $ob3[IN.canA]) ➔ RES[*STOP*]

Level 4: 2 canisters (can) + **4 toys** (ob). *Me*-demand = 7 ⇔ $Mk = e + 1$.

(1) {IMITATION.PLACE}$_{sit,A,B}$ (#ob1[IN.canA], #ob2[IN.canB], #ob3[IN.canA], #ob4[IN.canB]) ➔ REACH + GRASPob

....

(9) {IMITATION.PLACE}$_{sit,A,B}$ ($ob2[IN.canB], $ob3[IN.canA], $ob1[IN.canA], #ob4[IN.canB], **MOVE + DROP ($ob4.grasped) :$canB**) ➔ $ob4[IN.canB]

....

Level 5: 2 canisters (can) + **5 toys** (ob). *Me*-demand = 8 ⇔ $Mk = e + 2$.

....

(1) {IMITATION.PLACE}$_{sit,A,B}$ (#ob1[IN.canA], #ob2[IN.canB], #ob3[IN.canA], #ob4[IN.canB], #ob5[IN.canB]) ➔ REACH + GRASPob

....

(11) {IMITATION.PLACE}$_{sit,A,B}$ ($ob2[IN.canB], $ob3[IN.canA], $ob1[IN.canA], $ob4[IN.canB], #ob5[IN.canB], **MOVE+DROP** (**$ob5.grasped**) :**$canB**) → $ob5[IN.canB]

....

Note: Prior to implementing *LCLM* learning, the *Me*-demand = 3 + *n*. That is, the three schemes bolded in the formulas plus number *n* of toys.

The notation is as before, with a few added features. In **level 0 step 0**, the experimenter places (EX.PLACE) toy ob$_1$ inside (IN) the canister (can). Notice that ob$_1$ and "can" carry affixed to them the symbol $. This affix marks the scheme/object in question as *true, in a logical sense* (i.e., as existing in the present actual reality); it is not just known but is physically present and acted upon. In step 0, the experimenter's (EX) placement of the object inside the canister produces the true ($) result object-inside-canister (ob_1$[IN.can]). Seeing this, the child proceeds in **step 1** to imitate and place (IMITATE.PLACE) ob_1$ inside the canister, while keeping in mind the imitation parameter or condition (#) to place the object inside the canister (#ob$_1$[IN.can]). To do so, he or she must first (→) reach for and grasp the object (REACH+GRASPob). Next, in **step 2** the grasping occurs, changing the object to a grasped one (ob_1$.grasped). Finally, in **step 3** the object is moved to the canister ($can) and dropped into it (ob_1$[IN.can]). Schemes that need boosting by mental attention (*Me*-activation) are underlined in the formulas. Some schemes do not need *M*-activation, because they are being boosted by other organismic causal factors: *situational* factors (*sit*, e.g., prior Content *C*—or relational/structural *LC* learning, or perceptual field *F* salience). Other sources of schemes' activation are affects (*A*) or *B*-schemes derived from interactions with other persons; that is, schemes relating to psychosocial Beings, and so on. These other causal factors are postsubscripted to the braces {…} to signify that schemes inside these braces are not *M*-boosted but rather activated by these factors. We also assume in **step 3** that schemes #ob$_1$[IN.can] and $can are chunked; that is, are connected by Logical structural relations (*L*) so that the former scheme in the relation can boost activation of the latter, which is signified by superscript/subscript *L1*.

The same process repeats in **level 1**, but now with two canisters, A and B, and one toy. Because now there are two canisters and a choice must be made, in **step 3** $scanA (or $scanB, we use A in our example) cannot be boosted by #ob$_1$[IN.canA], and must be boosted by *Me*. **Level 2** exhibits the same procedure, now applied on two objects/toys. Here, we assume there is an imitation-evaluation **step 6** that confirms completion and stops the response sequence (RES(*STOP*)).

As the number of toys to be placed increases with the following levels (there are eight toys), the procedure remains the same, repeating this process with each and every toy, as summarized in the task-analysis formulas. The mental-attentional demand (*Me*-demand) of each level is given by the formula $Me = 3 + n$, where n is the number of toys at the corresponding level, and 3 represents the recurrent (process invariant) schemes that we have marked in bold at the step that is the most *Me*-demanding in the level. These three schemes are the Move-and-Drop operative, the scheme of the object to be dropped (ob-grasped), and the scheme for the canister. However, as mention above, intratask (*LCLM*) learning can be applied to this task to cause (see *LCLM* learning in chapter 5) a higher performance per each *Me*-level, because this intratask learning lowers the item's *Me*-demand. Comparison of the two panels of figure 3.7 shows the difference that these *LCLM* learning assumptions make in the modeling.

4 Meaning, Mental Attention, and the Symbolic Function

This chapter introduces signalic and symbolic meaning (semiotics) and analyzes from a dialectical constructivist perspective the functional characteristics of meaning and signs. We examine three dimensions of variation (i.e., signalic vs. symbolic, arbitrary or indexical vs. iconic, figurative or representational vs. operative or procedural) and investigate emergence of these dimensions within organismic schemes, which first occurs at about 12 months of age. Discussions of meaning refer to Peirce, Cassirer, Saussure, Piaget, and von Uexküll. Mental-process task analyses of signs (both signals and symbols) give metatheoretical (epistemological) and evolutionary reasons for these dimensions of variation. We explain ontogenetic development of the semiotic functions (signalic, symbolic) in humans and discuss their presence in problem solving of chimpanzees.

> Meaning is a decisive factor in nature; it appears always, often in novel and surprising ways.
> —J. von Uexküll, 1982, p. 77

> Meaning is *invisible*, but the invisible is not the contradictory of the visible: the visible itself has an invisible inner framework, and the in-visible is the secret counterpart of the visible, it appears only within it.
> —Merleau-Ponty, 1968, p. 215

> [T]he course of human knowledge leads from [intuitive] "representation" to "signification," from the schematism of perception to the symbolic grasp of pure relationships and orders of meaning.
> —Cassirer, 1940/1996, p. 111

In chapter 3 we suggested that infants' transition from perception to higher cognition is made possible by growth of mental attention, of more complex schemes, and emergence of the symbolic function. Because this function is often unexplicated but is essential for high cognition, we will now explain semiotic processes—the organismic generation of signs (i.e., signals and symbols).

Meaning is a decisive experiential factor in animals, whose organisms carry meaning via information-bearing brain processes that are largely learned, context-sensitive, and organized in flexible hierarchies (heterarchies). To help clarify meaning in the organism one should adopt *metasubjective*[1] units applicable at different levels of meaning-defining hierarchies. Organismic schemes/schemas are an example of such units. Schemes satisfy the basic reflective-abstraction properties of G. Miller, Galanter, and Pribram's (1960) TOTE (Test-Operate-Test-Exit) units and can explicate developmental growth in complexity processing with age. Scheme units can dynamically coordinate into more complex systems, which could be seen as organismic analogs of probabilistic computational production systems and related methods.

Classic learning theorists have adopted a meta-empiricist approach (a sophisticated outsider's perspective), failing to propose a workable psychological unit. The exception was Tolman (1959), a great precursor of cognitive learning. Tolman thought of these units as means-end-readinesses ($s_1r_js_2$, i.e., stimulus$_1$ & response$_j$ $\rightarrow$ stimulus$_2$, which can be assimilated to our operative schemes). He contrasted these units to representational units of the form $s_1 - s_2$ (stimulus$_1$ <*connects to*> stimulus$_2$), which, in our terms, are empiricist precursors of figurative schemes. In the TCO we adopt schemes, or schemas, as basic information-carrying psychological system-units (which logically correspond to self-propelling *semantic-pragmatic conditionals*—conditional to situational antecedents).

Piaget (Apostel, Mays, Morf, & Piaget, 1957) defined the *meaning of an object* as constituted by all schemes that can apply to it, plus all schemes that cannot. Related definitions were offered independently by Peirce (founder of modern semiotics—the science of signs) and von Uexküll (precursor of modern ethology). The constructs of sign and symbol often are seen as externally identifiable or interpreted and as incarnate or embodied within the interpreter's organism. A *sign* is something whose presence, in the mind of an interpreter, stands for (intends or announces) something else. A *symbol* is a conceptually explicit or implicit functionally detached sign. Cognition should be grounded on organismic/brain processes intertwining cognition with affect and motivation. However, the "cognitive revolution" of the mid-twentieth century, with its pure (decontextualized and organismically disembodied) cognition, failed to do so. It produced signs/symbols and structures that are purely abstract, decontextualized, and *amodal* (Barsalou, 2008)—amodal in the sense of excluding all modalities and modes of processing that intervene in an actual contextualized communication among subjects/interpreters, or *within* subjects.

Using Morris' (1946) distinction between *semantics* (relations between signs and their referents/objects), *syntactics or syntax* (relations among signs), and *pragmatics* (relation between signs and their use by interpreters), we can clarify a key difference between amodal and modal signs/symbols: whereas amodal ones are focused on syntactics and some semantics, *modal signs/symbols* emphasize pragmatics and semantics (Apostel, 1967) and can be interpreted as schemes. Within a constructivist developmental framework, pragmatics functionally contributes/causes semantics and syntax, because it expresses activity of interpreters. To explain pragmatics, however, one needs a general theory of action/praxis (Apostel, 1967). Our theory of schemes and cognitive operators is one such theory of semantic-pragmatics.

We proceed to review some epistemological (meta-theory of knowledge) reasons for making signs and their science necessary for a deeper understanding of cognitive processes. Readers interested only in signs as products of psychological organizations may wish, on first reading, to skip this section and go to the next section.

Meaning and the Ways of Signs

As discussed in chapter 2, Plato's problem (Pascual-Leone, 1996b) consists of his having recognized two distinct forms of knowing: the world of sensorial/experiential becoming (world of things) versus the world of beings (world of forms or conceptual ideas), although he was unable to explain how the latter emerges from the former. The former refers to "bottom up" (i.e., purely experiential) knowing and the latter to "top down" (i.e., conceptual) knowing. The modern problem is to explain how conceptual knowing emerges from experiential learning via constructively abstracting across situations. This issue remains a problem for cognitive psychology and neuroscience as much as for philosophy. Kant (1929/1965; Jaspers, 1962) gave us a more sophisticated version of the same basic distinction when he talked of three forms of knowing: intuitive sensibility, understanding, and reason. *Sensibility* refers to contact with the concreteness of experience, which is organized by the "pure forms" of space and time (in current neuroscience, the dorsal versus ventral prewired brain networks). Kant's sensibility corresponds to Plato's world of becoming. Kant's second and third ways of knowing, understanding and reason, jointly refer to Plato's world of being or ideas. *Understanding* is Kant's name for intelligent perception and local intellection. *Reason* is his name for truly high cognitive processes, which attempt to fully integrate sensibility and understanding with memory of past experiences and creative imagination (fully rational or intellectual processes). Schelling and Hegel (Beiser, 1993; Lukacs, 1978) criticized Kant for having a problem. They thought that sensibility, understanding, and reason, being radically different modes, could not be integrated unless they shared a common functional substratum (i.e., in the current view, shared organismic/neuronal characteristics). This common substratum would underlie an equivalence or "identity" between the subject (interpreter or agent of understanding/reason) and the object (source of sensibility—the experiential-input processes of the subject). This sort of "identity" (which for Hegel made possible the viable or true knowing of objects) can be explained in a current idiom by referring to Peirce's iconic and indexical functions (Apel, 1995; Buchler, 1955; Johansen, 1993), as discussed below. Advanced semiotics, the science of symbolic signs (pioneered by Peirce, Saussure, Cassirer, Vygotsky, Piaget, and others) is very important to comprehend communication in high cognition and help to clarify Plato's problem.

Let us examine what Hegel meant by "identity" of subject and object. Subject and object of knowledge maintain a dialectical "identity" relation with each other, because they maintain an informational correspondence. In our terms, a subject's initial experience of an actual object is expressed by specific encountered resistances (affordances or encumbrances/obstacles) manifest when the subject interacts with the object. *Informational correspondence* is expressed by either of two functional characteristics that may underlie Hegel's obscure concept of subject-object "identity." (1) The *iconic function* (Peirce's *icon*) is obtained when a subject's sign (a representation token, perhaps an object, sound, or figural unit) has some important aspect like, or resembling[2] in its configuration, the intended object's appearance (the referent). For instance, a photograph or a portrait is an icon sign (iconic representation) whose configuration reflects important aspects of the object as experienced by the subject. This iconic function explains why a referent often can be spontaneously recognized in, for example, a photo or painting. As Peirce claimed, images, diagrams, and metaphors (among others) are icons by the same relational definition. (2) Peirce called *index, indices,* or indicators another sort of informational correspondence. According to Peirce,

> indices may be distinguished from other signs, or representations, by three characteristic marks: first, that they have no significant resemblance to their objects; second, that they refer to individuals, single units, single collections of units, or single continua; third, that they direct the attention to their objects by blind compulsion. … Psychologically, the action of indices depends upon association by contiguity and not upon association by resemblance or upon intellectual operations (Buchler, 1955, p. 108).

Two Main Semiotic Functions

Thus, for Peirce, there are two organismic functions: the iconic (association by resemblance or similarity, due to stimulus-response compatibility or *F*-factor pattern matching) and the indexical, which produces indices (due to association by contiguity, associative learning). The two together explicate Hegel's dialectical concept of subject/object "identity." Both iconic (similarity) and indexical (associative) functions stem from a nonlogical but experiential active bias in the subject. Subjects tend to notice (be biased toward) what is relevant for their agency or praxis, and what is pragmatically relevant is highlighted and made salient. The subject's current purposive activity designates what is relevant in "informational correspondence" alignments. Driven by agency/praxis, a common dynamic substratum (shared brain processes) interconnects the cognitive representations (and the indexical function) to the subject's sensorial/perceptual experience of objects (and the iconic function). Hence, emergence of high cognition from low cognition becomes possible (e.g., objects, procedures—progressively more complex at various levels), which eliminates Plato's problem. The brain's organismic-functional characteristics and connectivity that a subject's top-down cognition and bottom-up sensorial perception share account for Hegel's "identity" of subject and object. Subjects automatically construe objects and procedures appropriately for their agency and survive. A signaling system alone, functionally independent from agency/praxis (goal-directed activity addressed to the environment) would not work. As Jorna (1995, p. 236) said, representation not only involves "standing for" but also "interpreting as" or "interpreting within" a given sort of agency or praxis.

Heinz Werner formulated in psychology his *orthogenetic principle* with the same idea. Werner and Kaplan (1984, p. 7) described this principle: "[D]evelopment is a constitutive moment of organismic functioning. We assume that organisms are naturally directed toward a series of transformations—reflecting a tendency to move from a state of relative globality and undifferentiatedness toward states of increasing differentiation and hierarchical integration" (geared toward agency/praxis). This orthogenetic principle serves to "characterize development as distinct from other types of change over time" (Werner & Kaplan, 1984, p. 7). The orthogenetic principle assumes with neuroscience that our organism is very active—endogenously driven to adaptation, development, and growth, via flexible hierarchies (heterarchies) that a symbolic function makes possible. Such adaptive tendency converges with Piaget's optimizing equilibration and his reflective abstraction, concepts discussed in chapter 1.

Major qualitative-functional differences exist between initial levels of processing (perceptual sensibility, low cognition) and more advanced levels of processing (understanding, high cognition, reason). Such differences (in the view of Cassirer, 1940/1996, and others) are largely due to the functional difference between signals (meaning-bearing processes functionally attached to their intended referent) versus symbols (meaning-bearing processes functionally detached from their referent). We discuss these two sorts of signs below.

Why is the symbolic function important? In addition to play, imitation, and dreams there are four major psychological processes that the symbolic function enables: (1) emergence of language; (2) reflective abstraction (i.e., progressive structural integrations that begin from concrete/intuitive cognitions to create symbols progressively referring to more complex and abstract knowledge, with wider scope of applicability); (3) executive processes, which anticipate the future and prospectively organize behavior to serve plans and expected future outcomes; and (4) effective communication, objective, subjective, and intersubjective, even when language is not used (e.g., gestures, drawings).

Signs as Products of the Working Mind

Peirce (Buchler, 1955; Johansen, 1993) thought that objects of knowledge are cognitively grasped and re-presented to the person's own mind by way of signs.[3] For Peirce, a sign (which we designate as **SGN**) is the product of a habit, that is, a durable mental coordination (often a complex scheme or schema—a functional structure) made of four meaning-generating constituents. (**SGN 1**) First is the *token-sign* itself or *sign vehicle*, which Peirce called "representamen" (Saussure called it "signifier"; Nöth, 1990), often manifested in acts of communication as when the French word "homme" is uttered or written as signifying man. (**SGN 2**) Second is the *context* or *ground of meaning* for which the token sign stands; Frege called it "sense" and Saussure called it "signified." An example is the awareness that "homme" is a given word with meaning in the French language. Frege's example is the morning star and the evening star, which express the same referent (the planet Venus) using two different contextual senses or grounds of meaning. (**SGN 3**) Third is the *referent* or actual *object* for which the sign stands; in the case of "homme" a two-legged rational animal, and so forth. (**SGN 4**) Finally, the *semantic operative* or *functor* (the operative process generating the semantic function) coordinates the meaning of the three previous constituents to combine them into a single semantic-pragmatic entity, that is, the meaning of man in the example given.

Peirce's concept of a semantic "ground" seems to refer to the operative process of the semantic/semiotic function or to the sense (or context) of meaning. Indeed, Johansen (1993) pointed out that Peirce often collapsed our constituents **SGN 2** and **SGN 4** into a single term called *interpretant*. This triadic relationship among sign (or symbol), its referent/object, and their interpretant (or interpreted meaning, i.e., semantic operative plus context) often is emphasized in semantic logic. According to Peirce, what enables an *internal-meaning sign* to function as interpretant of another *external sign* is the equivalence of representative functions that relate each of the signs ("homme" and "man" in our example) to the referent/object (Johansen, 1993); this is the subject-object "identity" in Hegel's sense that we have discussed. Thus, *semiotic function* can be seen as triadic, although, its formulation in terms of the four SGN-terms just given is much more revealing in cognitive development.

Cassirer (1944/1966) emphasized that symbol is a sign without unique (fixed or rigid) meaning, the sign remaining semantically detached from both its referent and its context, so that meaning can change with a change in context, and so on. For instance, when a baby begins to use language, one of the first words he or she learns is "mom," which is used in relation to the mother (Pascual-Leone, 1995, 1996a; Pascual-Leone, Escobar, & Johnson, 2012; see figure 4.1). At first, the sign in question is not a symbol but rather a signal, that is, an intuitive sign that is attached to its referent or context. Thus, the uttered "mom" sign has a fixed meaning irrespective of context. For instance, if the baby learned to say "mom" as the mother tried to feed baby, the sign in question may have the fixed meaning of "feed me"; if learned in the context of playing, this sign may have acquired the meaning of "raise me in the air," and so on. At this point the infant may be 8 or 9 months. When the child grows older, after 12 months of age,

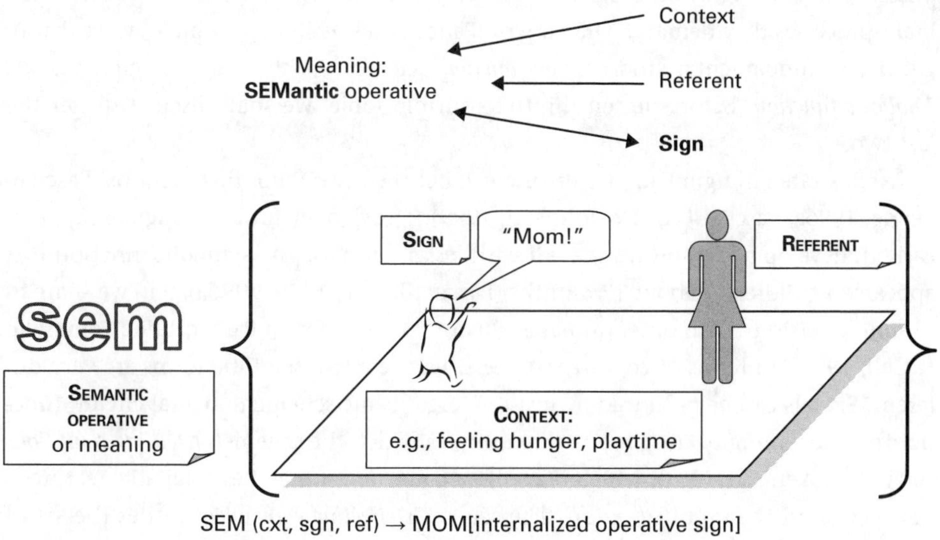

Figure 4.1
Schematic representation of constituents in a sign with illustrative example of baby calling his or her mother. (From Pascual-Leone, J., Escobar, E. M. R., & Johnson, J. [2012]. Logic: Development of logical operations. In W. Hirstein [Ed.], *Encyclopedia of human behavior* [2nd ed., p. 539]. Elsevier. Copyright 2012 by Elsevier.)

the sign "mom" will have changed into a symbol, carrying the meaning of "feed me" in the context of feeding, but the meaning of "raise me in the air" in the context of playing, and so on.

Notice that symbols and signals, if compatible, can replace each other as functional processes. However, as Cassirer (1940/1996) pointed out, signals and symbols often may contradict each other, creating misleading situations. For instance, under emotional stress or with habitual practice, symbols can functionally turn into signals. When an anxious, panic-prone person hears the shout, "Fire!" inside a movie theater, he or she may not take it as a symbol suggesting the possibility of actual fire but rather may take it as a signal and run in panic to the door. Whereas a symbol conveys a possibility (for instance, "perhaps there is a fire in the theater"), a signal conveys the actuality here and now of its referent.

There is a vast literature on signs/semiotics that we need not review, since Nöth (1990) and other excellent books have done so. Instead we attempt to generalize some basic concepts from the literature with a fourfold classification table for the main sorts of signs. Our classification has emerged by taking seriously the idea (in various ways assumed by Peirce, von Uexküll, Cassirer, Saussure, Piaget, and others) that there are different sorts of signs, ontologically and developmentally, that differ in terms of the brain processes they demand. Drawing on Peirce and Cassirer's terminology, we distinguish four fundamental kinds: *iconic function, indexical function, signalic function, and symbolic function*. Before presenting this fourfold table, we shall discuss further the last two.

As illustrated in figure 4.2, we distinguish between two semiotic functions (Pascual-Leone, 1996a; Pascual-Leone & Johnson, 1999). The signalic function (signals) appears early in development, and we share it with lesser animals. The symbolic function first appears in children at about 12 months (Pascual-Leone, 1995, 1996a), and we share it, in simple versions, with other primates (Russon, Bard, & Taylor Parker, 1996; Tomasello & Call, 1997). In terms of cognitive processing, these two semiotic forms are very different. Signals can be produced by just one organismic scheme (a neural circuit, functioning as meaning-bearing processing unit), in which the scheme's *releasing conditions* carry the token-sign (**sgn**), which becomes a cue when it matches a suitable feature(s) or aspect(s) in the input situation. The scheme's *effecting component* carries the set of effects resulting from the scheme's application to or assimilation of the situation. Finally, the scheme's *functional component* embodies the gist of the scheme's functioning, and it often carries a semantic intention (intentionality) toward the signal's referent (**ref**), anticipating the scheme's application outcome. Signs result from application of individual schemes, and signalic meaning is the consequence of initially applying

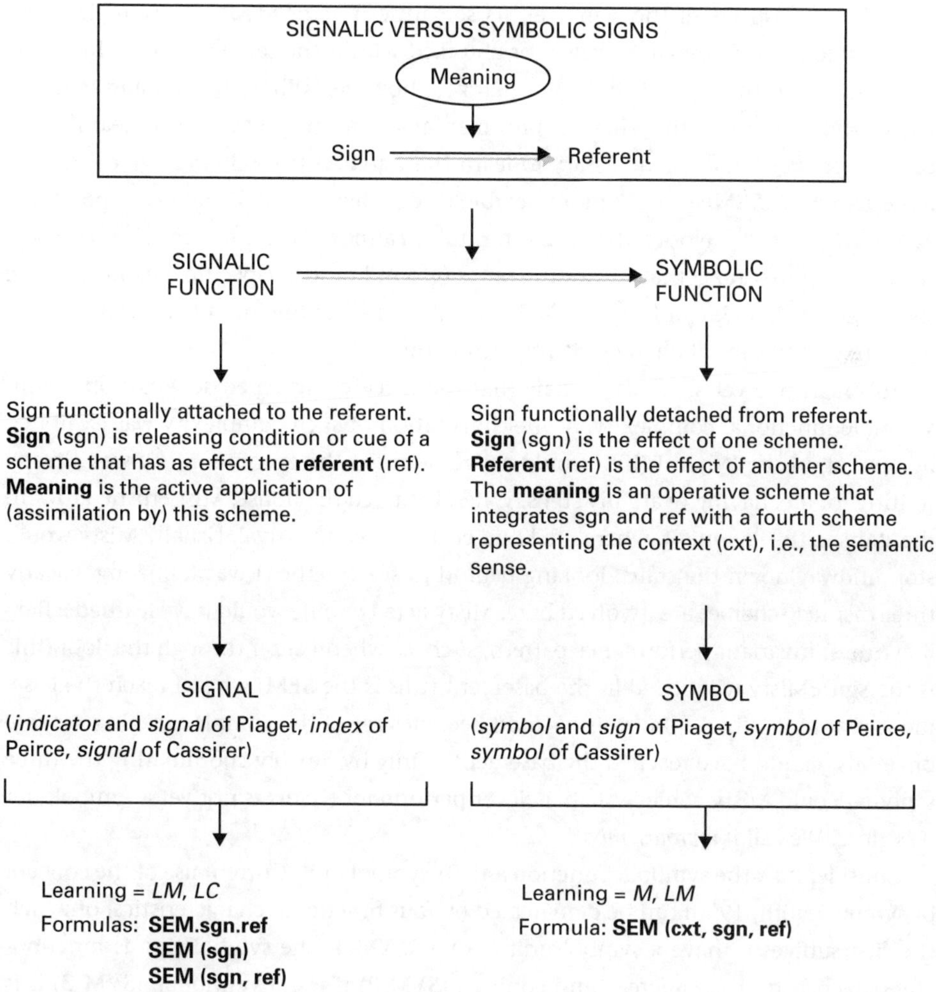

Figure 4.2
Signalic versus symbolic signs.

just one scheme. The learning of signals is largely the product of associative learning, which we call *content* (*C*) *learning* when it is simple and not relational. We call it *LC* learning (*Logical*-schemas resulting from automatizing *C* learning) when it includes more or less complex schemes or schemas (Pascual-Leone & Goodman, 1979; Pascual-Leone & Johnson, 2005, 2011).

When fully automatized, such a signalic scheme is a coordinated logical system or "package." We could represent it as **<SEM.sgn.ref>**, to indicate that the three essential sensorimotor aspects of the sign (i.e., its semantic functor **SEM**, its token sign **sgn**, and its referent **ref**) are all packaged or chunked within the same scheme. This is the case only when the signal is fully automatized, however. When it is not automatized, two or three sensorimotor schemes (plus mental attention and associative learning, to coordinate them) may be necessary to learn the signal. A two-scheme signal could be represented as <**SEM(sgn)**>. A more flexible three-scheme signal could be represented as <**SEM(sgn, ref)**>. Notice that these formulas cannot stand for symbols, because a semantic context (**cxt**) that carries the sense (ground or local/specific meaning) of the sign is not included. Unlike symbols, in signalic schemes the intended specific meaning is fixed and cannot change with the situation.

Although not yet symbols, signals that use three coordinated schemes offer considerable functional complexity. A brief illustration of such complexity was exhibited by our cat, Misty. When I (JPL) would arrive home in the evening, Misty was already waiting for me, having heard my steps. When I entered the house, Misty rushed toward the stairs to the basement, where I fed him customarily on arrival. Usually Misty would stop midway down the stairs, looking behind to see whether I was following. Clearly three distinct schemes are involved here. Misty acted *as if* he would have learned a flexible rule or invariant performance pattern, such as, when I enter through the door (this is the **sgn**), Misty will be fed in the basement (this is the **SEM**) when I reach the basement (I am initially **ref**, but Misty's food becomes **ref** of the complex feeding scheme once Misty and I have reached the basement). Only by flexibly coordinating the three schemes could Misty achieve such a clever performance. This is not yet a symbol, but it is close. We call it a *protosymbol.*

Consider now the symbolic function and its symbol unit. Current use of the concept of symbol (Nöth, 1990) can be demarcated by four functional characteristics, of which the first suffices to have a symbol unit (**SYM**): (**SYM 1**) the symbol/sign is functionally detached from its referent and context; (**SYM 2**) it is conventional; (**SYM 3**) it is generically abstract; and (**SYM 4**) it is motivationally driven by affects. The four together clearly differentiate symbols from signals. For instance, when a symbol-processing 18-month-old baby calls "Mom," the sign's meaning can change with the context (**SYM 1**). A symbol is conventional (**SYM 2**) in that it often is related to its referent by way of more or less arbitrary associative links with contextual schemes (e.g., "connection rules" shared and adopted, tacitly or explicitly by all interpreters, e.g., the flag as a symbol of a nation, the math parameter Pi as symbol of the ratio between circumference and its diameter).

This is not always the case, however: iconic symbols relate first via similarity. Peirce, who emphasized only the cognitive-associative aspect of symbols, insisted on their conventional and rule-governed aspect (e.g., "the signification I attach to it, that of a conventional sign, or one depending upon habit"; Buchler, 1955, p. 113). A symbol is generically abstract (or general, **SYM 3**) in that it usually corresponds to a class (or type) of "objects" and has many distinct particular referents that are equivalent in terms of the symbol's meaning (or semantic dimension—highlighted by the purposive activity that the symbol supports or relates to, e.g., "food," "writing").

Consider again the characteristic **SYM 1** of functional detachment (which Cassirer, 1944/1966, emphasized). In principle, every symbol (symbol etymologically means "thrown together with") is semantically pointing to a set of possible individual referents and not just one of them. Thus, we should define symbol as a semiotic sign in which the token sign (or sign vehicle) is functionally detached from its current referent. We know that a sign is functionally detached from its referent when the meaning attributed to the symbol changes with the context (as already illustrated with the baby-Mom example). Such functional detachment of the symbol's meaning is often called "situated" or with "context sensitive" meaning (e.g., Pribram, 1971). It generates true conceptual thinking (at times called mediation or reflective abstraction), which explains why these symbolic representations give mobility and cross-situational generality to thought (often constituting mental operations).

Simple (not complex—not compound, nor hierarchically organized) schemes are functionally indivisible processing units (embodied in basic meaning-bearing neural circuits—Pascual-Leone & Johnson, 1991, 2005, 2011). For this reason, when a single simple scheme (perhaps a single neural circuit) carries simultaneously the semiotic meaning, the referent, and the context of a token sign, these parameters must remain functionally fixed as constituents; consequently these semiotic units function as signals. It follows that to have symbols (whose meaning or sense changes with the given context-situation) organisms must carry the various semantic constituents in distinct coordinated schemes. Symbols emerge when four distinct schemes can function in coordination, each carrying either a semantic operative (SEM) or one of its semantic parameters (token-sign, context, or referent).

We can represent in a semantic formula the symbol's semantic scheme as follows: **<SEM(cxt: sgn: ref)>**. **SEM**, a semantic operative scheme, coordinates three dimensional parameters: context/sense (one or several sense values), token-sign, and referent (one or many values). Each parameter is detached in a different categorial scheme. This separation ensures that each specific (actual or real) context can elicit the appropriate sense (the context-relative specific meaning) corresponding to the given referent. Together these four distinct semantic components (functionally integrated into a complex—superordinate—semantic scheme) produce the flexible semantic functioning typical of symbols.

To illustrate how this formula for symbols can be used in a concrete task analysis, consider the pretend-play symbolic imitation of "talking by phone," which children may practice after 12 months, and more often between 18 and 24 months. Reinecke and Fogel (1994) describe an example. To paraphrase their account: Hannah's mother had tried for some time, without success, to get her daughter to pretend-play talking by phone. But one day, when Hannah was beyond 12 months of age, she spontaneously picked up the phone and offered it to her mother. Her mother took the phone, put it to her ear, and said, "Hi, Grandma!" Hannah smiled and reached for the phone; she put it to her ear (first time she did that) and said, "Ha-o!" Hannah looked and again offered the phone to her mother for some more pretend talk. This observation illustrates well a symbolic act of deferred imitation. Notice that imitation is largely driven by innate affects—a disposition to imitate well-liked persons in the surround, parents in particular (this may be an innate root of intersubjective communication—see Nadel, Guerini, Peze, & Rivet, 1999; Trevarthen, Kokkinaki, & Flamenghi, 1999).

When we apply to this case our formula for symbols (see figure 4.3), the semantic operative (SEM) appears as an operative scheme induced by positive affect: a TALK-to-people-BY-PHONE scheme. Context or sense (cxt) of this operative is the representational scheme *play-with-mom. As usual, we symbolize operative schemes by writing them in capitals, and figurative/representational schemes by writing them in lower case preceded by *. The token-sign (sgn) is the figurative scheme *visual.object/seen.

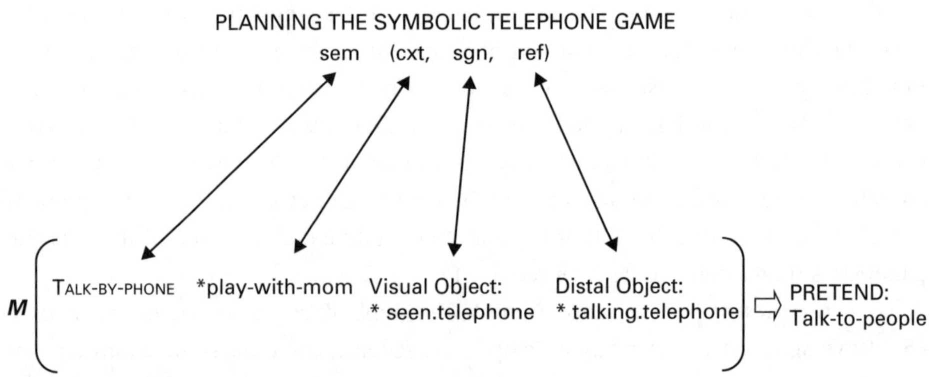

Figure 4.3
The telephone game. Paraphrase of Reinecke and Fogel's (1994) observation:
At about 12 months (*Me*=4) Hannah picks up the phone and offers it to her mother. Her mother takes the phone, puts it to her ear, and says, "Hi, Grandma!" Hannah smiles and reaches for the phone; she puts it to her ear (first time she does it) and says, "Ha-o!" Hannah looks and again offers the phone to her mother for more pretend talk.

telephone. The referent (ref) is the representational scheme *distal.object/talking.telephone. Notice that one must distinguish between proximal object (here visual) and distal object. An *object* is an entity or "package" of Reality resistances (represented by a complex representational/figurative scheme), which appears in cognition in a given context and is a probabilistic invariant. This object-invariant is a coordination of two sorts of schemes (e.g., actions, perceptions): (1) *positive schemes* that could assimilate or apply to the "package" of possible resistances elicited by the given real object and (2) *negative schemes* that cannot assimilate these resistances. The latter schemes demarcate the distal object in question and make it more clearly distinct. Such definition is a paraphrase of the meaning of an object as formulated by Piaget (Apostel et al., 1957). It is similar to Peirce's (Apel, 1995; Buchler, 1955) and to J. von Uexküll's (1982) object meaning. As discussed in chapters 2 and 3, a proximal object is the object-invariant that emerges in the context of purely perceptual processes (here and now). A distal object is the object-invariant that emerges in the context of intentional activity across situations, where one's own actions encounter invariant resistances, often to be controlled or eliminated while pursuing intended goals.

There is another distinct source of often unconscious meaning that a distal object (or distal act) can elicit: the affective or affectively loaded personal/emotion schemes elicited by (or seeking expression in) the object. Following Jung's terminology in a very liberal manner, we call these strong unconscious schemes *shadow schemes* to highlight their frequent unconscious presence, and perhaps haunting expression, particularly within psychosocial or emotional situations. A shadow scheme relevant to the present talk-by-phone example is the one presented below in formula **f2**. This is a baby's compelling scheme of exploring a small object by taking it to his or her mouth. Such shadow scheme makes it hard for the baby to synthesize the pretend play "talking by phone," unless she or he has enough mental attention (*Me*-capacity). Shadow schemes contribute much to the individual's personal style and the form or content of art and literature. Notice that these shadow schemes often are signals under our definition, although as the subject becomes aware of them, they could evolve into symbols.

Using this terminology, the symbolic formula for the pretend play of "talking by phone" can be stated thus:

<u>TALK-BY-PHONE</u> (*cxt*: <u>*play-with-mom</u>, *sgn*: <u>*seen.phone</u>, *ref*: <u>*talking.phone</u>) **(f1)**

According to this model, the child must be able to coordinate simultaneously four distinct sensorimotor schemes to understand the pretend game and thus have the idea of playing. Mental attention grows with chronological age in children, as described in chapter 3. Simultaneous activation with mental attention of four distinct sensorimotor schemes cannot occur before 12 months of age; three such schemes can be activated at about 8 months, but not before (Alp, 1988, 1994; Benson, 1989; Pascual-Leone & Johnson, 1991, 2005). In this way we explain why flexible signals (<**SEM (sgn, ref)**>) are not often found before 8 months; and the symbolic function begins in most children at 12 months. This developmental timing is well recognized in the literature, but it remains without explanation, which we have now provided.

Having dynamically synthesized in her mind (via some concrete instances) the mentally real equivalent of this symbolic formula, the child can have the idea to pretend-play talking with her grandma. But actual implementation-in-action of this idea (or executive plan) requires other schemes as well. Notice that such implementation is hindered (interfered with, i.e., dialectically contradicted) by a misleading factor: the affect-driven (*A*) habit of exploring physical objects by taking them to the mouth—an overlearned shadow-scheme automatism of children at this age, which is strong at 8 months and earlier. The corresponding formula for this *misleading factor/strategy* is:

EXPLORE.PHYSICALLY^{A}(*sgn*: seen.phone, *ref*: solid.phone) ➔ phone-in-mouth **(f2)**

To shorten the formulas, henceforth we eliminate the * in front of figurative schemes. This formula corresponds to a flexible signal as described earlier, which induces the child to explore the phone physically and take it to her mouth. This signal, which initially demands simultaneous activation of three schemes, becomes automatized and turns into a habit. Notice that in this formula, after the arrow, we represent the performance that results, via overdetermination, from the schemes' application. When the resulting scheme is automatized, it does not use mental capacity. A habit such as mouthing the phone creates a misleading situation for babies who would try to produce the symbolic performance: this habit interferes with implementation of the plan "talking to grandma." Symbolic processing, which ensures that all necessary distinct schemes are separately boosted with mental attention, is then needed to overcome the interference. When this is done, a symbolic performance emerges, which we model as follows (Pascual-Leone, 1996a):

TALK-BY-PHONE^{L,A} ({#GRASP.TAKE-TO}$_{L,A}$ (*where*: ear, *subject*: baby/mother, *distal.object*: talking.phone)) ➔ phone-at-ear-talking **(f3)**

In this formula the baby, following an operative TALK-BY-PHONE partly boosted by affect (superscripted *A*) and already overlearned as a logical-structural procedure (superscripted *L*), moves (GRASP & TAKE TO) the telephone to the ear for talking. This operative scheme (i.e., GRASP.TAKE-TO) functions here as an operative parameter (#) of TALK-BY-PHONE and is already overlearned and chunked with the main operative, indicated by the subscript *L, A* matching the superscript of the logical-structural procedure. Activation of the main operative scheme transfers automatically to its overlearned subordinate operative parameter. Thus, the total number of distinct sensorimotor schemes to be boosted with mental attention (to ensure production of the overt response "place **phone-at-ear** and begin **talking**") is four at most, which makes mental demand of implementing this task the same as the demand (shown before) of planning the gist of the task and its executive. Although there has been much research conducted on play and imitation, there does not seem to be a developmental organismic theory that explains, as our model does, growth of imitation and play induced by maturational growth of the child's mental attention. In this sense the model we have just outlined fills a gap in the literature.

The misleading-factor (shadow scheme) strategy created by the three schemes for taking the phone to the mouth (see formula **f2**) is an interference factor that can be overcome by boosting with mental attention the task-relevant strategy, made up by four other schemes. Because the four *M*-boosted schemes in this symbolic strategy are stronger as a functional totality than the three misleading automatized schemes (which concurrently are being interrupted by attentional inhibition), the relevant symbolic strategy dominates and applies resolving the conflict situation. This conflict resolution illustrates how all organismic schemes apply locally and compete for application when a situation offers suitable releasing features, which is why effortful symbolic processing must be used. Indeed, signalic/intuitive experiential processing tends to bear on salient elements of a situation, whereas symbolic processing tends to deal with more complex aspects that involve coordination of scheme units from both the situation and long-term memory.

Semiotic Relations and Cognitive Processing

We have discussed four semiotic functions: iconic, indexical, signalic, and symbolic that can be crossed to generate types of signs. Table 4.1 presents the resulting fourfold semiotic table. We use *sign* as the general name or generic designation for all types of semiotic relation. This is the generic extension of a term found in Peirce, von Uexküll, and Cassirer, although this denomination is not universal (for instance, Piaget used the term *sign* to denote what our table calls *conventional* or *symbolic indexical* sign). The columns of table 4.1 present two semiotic modes (signalic vs. symbolic) bearing on the degree of a sign's functional detachment from its referent. The rows present two modes (iconic vs. indexical) bearing on degree of configural relation (iconicity versus lexicality) between sign and referent.

Table 4.1
Signs: Fourfold table of semiotic relations

	SIGNALIC FUNCTION: Functionally attached; Cassirer's signal sem (sgn) Mechanism: *C, LC, S, T, F, E*	SYMBOLIC FUNCTION: Functionally detached; Cassirer's symbol sem (cxt, sgn, ref) Mechanism: *M, LM, I, F, E*
ICONIC FUNCTION: configural relation; Peirce's icon Mechanism: *SOP, F, E*	**configural signal** (perception) Peirce's icon Saussure and Piaget's motivated sign	**universal symbol** poetic symbol Freudian/Jungian symbol Piaget's "symbol" Saussure's motivated symbol Peirce's iconic symbol
INDEXICAL FUNCTION: arbitrary/non-configural relation Mechanism: *C, LC, LM, I, E*	**arbitrary signal** (Pavlov) Piaget's "indicator" Peirce's "index"	**conventional symbol/sign** much of language, logic, math Piaget's "sign" Peirce's "symbol"

To each of these four semiotic functions corresponds a distinct way of processing information. In the case of signalic processing the main mechanism is associative learning: both content-based learning (*C* learning) and logical-structural (relational) learning caused by automatization (Hebb, 1949/1961; Pascual-Leone & Goodman, 1979; Tolman, 1959), which we call *LC* learning. For instance, consider our cat, Misty, discussed before: he had learned with practice to interpret as cue my entering the house through the main door. This cue (largely via *C* learning) served to anticipate that Misty was about to be fed in the basement, an expectancy acquired via *LC* learning. Other mechanisms may intervene in the emergence of signals, such as executive processes (the *E* organismic factor), the spontaneous spatial and temporal structuring of information (what we call *S*- and *T*-factors—occipito-parietal and occipito-temporal cortical processing), or Gestalt-field saliency (*F*-factor). Iconic examples of signalic processing are, for example, the marks of steps in the snow, a photo, or the standard sign for a ladies' washroom. Examples of arbitrary signals are the sound of rain at the window's glass, or smoke as sign of fire, or the bell of mother calling for dinner (when it evokes thoughtless hunger and desire to go to eat).

In the case of symbols (symbolic function) organismic mechanisms are mostly problem-solving acts via mental-attentional activation (working memory processing), which we call *M*-processing, accompanied by a logical-structural learning that is enabled by this *M*-processing, which we call *LM* learning. However, other mechanisms such as executive processing (*E*-operator), attentional inhibition or interruption (*I*-operator), or Gestalt-field saliency (*F*-operator) may also intervene.

The iconic function of Peirce (which generates icons—i.e., configural or perceptually "motivated" signals or symbols) occurs because of the brain's disposition to express/adjust itself, in cortical processes of representation or action, to the configural/relational characteristics of sensorial/perceptual input processes. By *configural signals or symbols* we mean signs that in their own structural constitution express configurally (i.e., are *in-formed* by, in the sense of "injecting form into") some salient or overriding structural aspects of the referent or input-situation itself (as perceptually represented). For instance, when we refer to a configuration of points in the data graph as a "cloud of points," the term *cloud* is used as an icon configurally expressing actual arrangement of points in the page (as synthesized in perception).

We think that iconic processing, whether in language (e.g., metaphors) or cognitive performance (e.g., mimetic representation, imitation, dancing to music), is made possible by what Gestalt psychologists and others have called internal field effects, which we call *F*-operator/factor (also known as simplicity or "minimum" principle of representation and as "stimulus-response compatibility"; Hochberg, 1964; Pascual-Leone, 1989; Proctor & Reeve, 1990; see chapters 1 and 3). If compatible, schemes may apply together to codetermine the ongoing process or performance. Other organismic factors such as mental attention (*M*-operator) or executives (*E*-operator) possibly intervene in this semiotic function. Iconic symbols (at times called universal symbols, because they can be understood without prior learning) are illustrated by experiences captured in linguistic descriptions such as "the sunset of life," "dawn of civilization," or "cloud of points" mentioned. Every metaphor, verbal or not, has as its kernel an iconic symbol.

Finally, the indexical function, formulated by Peirce in contrast to the iconic, is the product of associative learning (*C* and *LC* learning), although factors already mentioned in chapter 1 (e.g., *M, LM, I, E, SOP*) may also be relevant. There are two distinct products of the indexical function: *indexical signals* (often called arbitrary or conventional signals) and *indexical symbols* (often called conventional signs or arbitrary symbols). Examples of arbitrary indexical signals include, as mentioned, the sound of rain on the window or smoke as sign of fire. Examples of arbitrary indexical symbols are a country's national flag, the number of stars in a hotel's rank, the key to the city that exceptional visitors might receive from the mayor, or the bell of mother calling for dinner (only when it evokes mother's love, the dinner table, or mother's demand not to be late for dinner, and so on).

Because the two semantic dimensions of variation in table 4.1 (degree of functional detachment vs. degree of iconicity between sign and referent) can be crossed, we obtain the four main ways of semiotic processing: *configural* motivated *signals; arbitrary* or indexical *signals; universal*, configural, motivated *symbols*; and *conventional symbol* or *sign*. The four types result by combining compatible processing mechanisms for each semiotic mode. For instance, universal configural symbols may result from the combined use of *M*, *LM*, *SOP*, and *F*. Signalic processing often is referred to as mental *habits*, useful in facilitating situations, whereas symbolic processing may be described as *mentations* or metacognitions, thinking, mental mediations, or problem solving (particularly in misleading situations).

Complex systems of signs may be combined into *mental models*. These models may stand for complex entities of reality (e.g., universities, airports, cities, transportation organizations—internalized *figurative* or *object models*) or for complex blueprints of action (internalized *operative* or *executive models*). They are also found as mixed models, combining these two aspects (e.g., *causal models*). Symbolic models often are organized in levels of relational representation (levels of reflective abstraction) that become progressively more abstract.

A different important sort of symbol that our fourfold table does not include is the *personal* (i.e., cognitive *and* affective) symbol. When personal symbols are complex and strongly boosted by affects, they may generate spiritual, *numinous symbols* that clinical depth analysts, and in particular Jung (Campbell, 1971), have emphasized. Such numinous symbols could be explained by powerful unconscious "packages" of compatible and complex personal (affect-driven) schemes. They may be universal expressive symbols produced unconsciously by our brain's innate organization, which all humans share, as the brain encounters and reacts to many existential categories of situations that we all encounter. Because the symbolic function is universal in humans, the resulting expressive (perhaps numinous) universal symbols should be similar across humans for different cultures.

Notice, however, that the essential aspect of symbols is that they have their sign-token, referent, and meaning functionally detached and contextualized. It does not matter much for cognition whether the symbols are expressive, universal, or numinous. And because this detachment and contextualization is brought about by mental attentional syntheses, any animal whose brain processes resemble ours, and has enough mental attention, could exhibit symbolic processing. This is the case of the other high primates.

Symbolic Function in Chimpanzees

Preschoolers and older children produce anticipatory symbolic processing, and so do our biological cousins, adult high primates. However, mental processes of nonhuman primates lack mobility, reversibility, and combinability, relative to preschoolers, perhaps because the mental-attentional capacity of apes only compares to that of

26-to-35-month-old humans. The following discussion is adapted from Pascual-Leone (2006).

Consider the chimpanzees' use of multiple tools (hammer-stone, anvil-stone, wedge-stone to stabilize the anvil) for cracking nuts, discovered by Matsuzawa and Yamakoshi (1996; Pascual-Leone, 2006; Yamakoshi, 2004). The logic of action in this behavior includes two nested functional components. First is an operative scheme (PLACE) for positioning the nut (*nut-food) on the anvil stone (*anvil) during nut cracking. Second is another operative (CRACK) that uses the hammer-stone (*hammer) as tool and the nut as target (*nut-target). Although the schemes *nut-food and *nut-target refer to the same actual object (same semiotic *referent*), in our method of task analysis they are distinct schemes. This is because they belong to different *contexts* of use and thus have a different sense—distinct functional activities or practical skills (this rule is an application of postulate of schemes **Sch4** given in chapter 5). To crack a nut, chimpanzees must coordinate simultaneously six sensorimotor schemes:

CRACK (*hammer, *PLACE* (*anvil, *nut-food), *nut-target) **(f4)**

First the chimpanzee, with the intent (*operative expectancy*) to crack the nut, places the nut-food on top of, and supported by, the anvil. Then, always maintaining the nut in place, he cracks the nut by using the hammer-stone. This is symbolic processing. The situation is misleading (e.g., prior habit-schemes should lead chimpanzees to take the nut to their mouth and ignore the stones), and so performance requires chimpanzees to boost all relevant schemes with endogenous mental attention (*M*), and inhibit or interrupt (*I*-operator) interfering scheme habits. Specifically, six distinct schemes must be simultaneously boosted in activation using mental attention. This is the most complex symbolic coordination of schemes that great apes seem capable of (six or seven perceptual-motor schemes), which places them on par with the mental attentional capacity of 26-month-old or 3-year-old children (Pascual-Leone & Johnson, 2005; see chapters 3 and 7).

This example of problem-solving performance shows chimpanzees' symbolic anticipatory processing, as well as their limitation. Processing is essentially perceptual-motor, cued by facilitating situations, and somewhat lacks mobility. Once habits are formed, great apes have a hard time in overcoming them within misleading situations. Although it is a sensorimotor symbolic operative coordination, this performance marks the peak ceiling of their symbolic processing. Lack of more mental-attention capacity (which includes both *M* and *I*) prevents them to produce more complex and hierarchically deeper mental symbolic processing. In contrast 2- or 3-year-old children, who achieve equally complex symbolic processing, are just at the start of their symbolic-complexity growth ladder, which in due course brings them to the symbolic-processing ability of human adults.

Conclusions

We have shown how signs can be signalic or symbolic. High primates, and particularly humans, are symbolic because they have enough mental attentional capacity to synthesize symbolic schemes. This symbolic function opens cognition to *intentional mentation*, the capacity of the working mind for analysis and agency in future and possible problems. This working mind creates the psychosocial and sociocultural domains, as well as domains of arts and science. But this is beyond the scope of this book.

Because meaning and signs are unavoidably constituted by schemes, the next chapter examines in more detail organismic schemes, their development via dialectical-constructivist learning, and their expression in performance.

II Theory of Constructive Operators (TCO)

5 Schemes/Schemas and Their Causal Constructivist Learning

The constructs of organismic schemes and schemas are examined in detail as suitable psychological processing units for the neuropsychological organization. Expanding on Kant's original insights, we show how schemes mediate between hard Reality (the encountered resistances) and the organismic information processing that decodes experience and enables thinking. We describe the principles of schemes, their kinds, and their various modes of learning mechanism.

> Thus an intelligent functioning (that is, a successful functioning in the reaching of a goal) is, as I see it, an expectancy on the part of the organism, aroused by that part of the stimulus layout which is immediately presented, to the effect that such and such performances or behaviors (if carried out) would be successful in reaching such and such goal.
>
> —Tolman, 1945, p. 165

Jakob von Uexküll (1982; T. von Uexküll, 1982) considered signs and meaning to be part of animals' dynamics. They were essential for the constitution of a *subjective life-world*, that is, an interpreted immediate-environment or context/situation that he called *Umwelt*. He attributed a distinct Umwelt to every species and individual, with similarities expressing their functional biological proximity. He emphasized that "a sign never exists in its own right; it is always part of a circular process [i.e., a functional system or totality] in which a receiver (receptor) receives stimuli, codes them into signs, and responds to them as such" (T. von Uexküll, 1982, p. 9). We call *organismic schemes* the functional systems expressed by signs.

Schemes are elementary functional-structure units of meaning that Kant, Piaget, Tolman, and von Uexküll pioneered using different terms. Our organismic schemes, congruently with Tolman's "expectancies" (Tolman, 1959), von Uexküll's "functional/circular systems," and Piaget's "schemes of assimilation," can be seen as meaning-bearing dynamic functional units with which to interpret the life-world. Life-world is an animal's construal of its surrounding context, other animals, and itself; the immediate external reality experienced by the subject from within (Husserl, 1970). In this surrounding context animals encounter resistances to their own agency. This sense of resistance (with lowercase r) is psychological, because it expresses constraints (affordances or obstacles/encumbrances) that Reality imposes on the animals' senses or acts. We use the term Reality (with capital R) to mean the immediate environment, prior to being represented or perceived.

Call Resistances (with capital R) the potential constraints for a species, within any given unrepresented or unknown environment (this is Reality as experienced). Resistances affect animals as they sense or act, which induces them to adapt or accommodate. When experienced by the animals "from within," we call them resistances (with lowercase r; this is now an epistemo-psychological perspective) to indicate that the Resistances are now semantically interpreted. In figure 5.1 we illustrate this

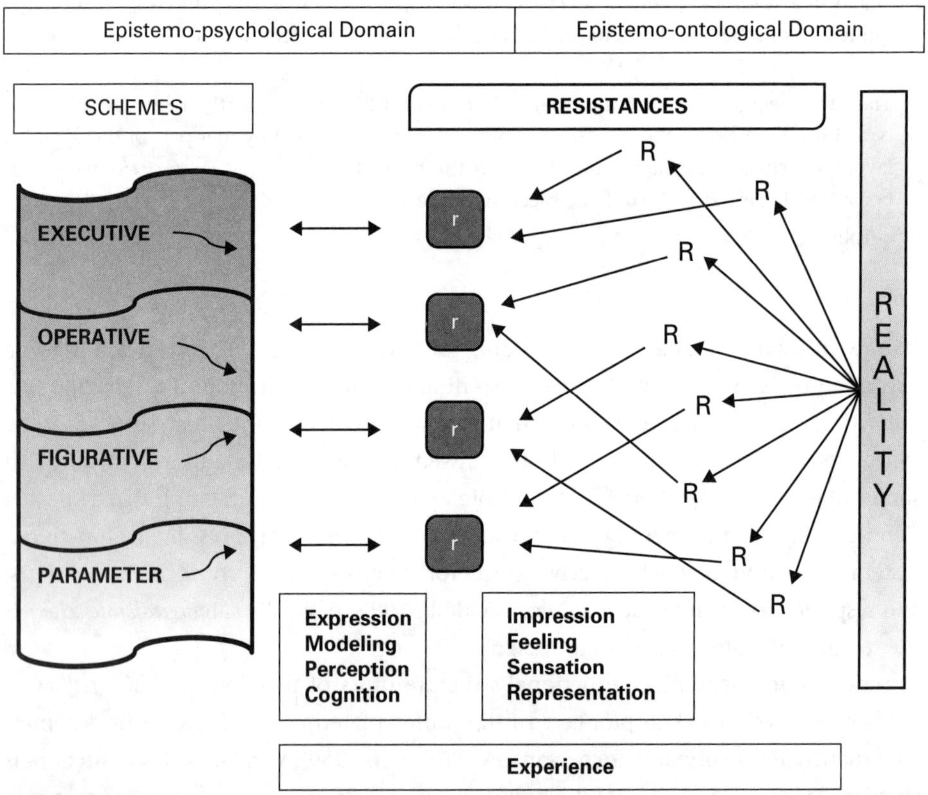

Figure 5.1
Model of organismic perception of reality/Reality.

Resistance → *resistance* transition. The two sides of figure 5.1 symbolize the two transitions that in classic psychology differentiate *sensations* (i.e., Reality stimulating the receptor organs; the right side of the figure) versus *perceptions* (sensations coded or modeled into percepts or cognition, the left-hand side). The center of the figure is where *resistances* emerge, incorporated to perceptual/cognitive processes in the making. This transition is at first (near the right side) a very *weak interpreted presentation* (only *C*-coding, impressions, feelings, sensations) of the input. Then, on the left side, they are transformed and incorporated into real *representations* (expressions, coding or modeling, perception, cognition). This second transition (strong representation) of the interpreted input is done by suitable schemes.

The schemes come in four kinds: executives, operatives, figuratives, and parameters. These four kinds resolve the dynamic interaction between the subjects' interpretive activity and Reality Resistances. This dynamism differentiates and incorporates into schemes four epistemo-ontological types of functionality: abstract and generic *pros* (procedures), action *pros*, action *obs* (internalized object representations), and relational and generic *ads* (adjunct information). There exist in the knowing subject many different concrete instances of these four basic functionalities. The various types are discussed in detail below.

Examples of resistances commonly occur in perception. When I look at the computer screen on my desk, raise it, explore it with my hands, and so forth, I unavoidably experience resistances that inform the schemes I construct. I attribute these resistances to Reality, that is, the encounters between my schemes (mobilized and differentiated when I act on the Real screen) and the screen's Resistances. These resistances (coded by active schemes) provide *cues* (matches between scheme conditions and Reality features). Reality constraints are resistances from an epistemo-psychological perspective and Resistances from an epistemo-ontological perspective. Perceptual illusions illustrate well the dynamic interactions between reality resistances and cognitive processes (see our discussion of the Kanizsa illusions in chapter 6).

Any resistance can be positively helpful (affordances) or negative/unhelpful (encumbrances or obstacles). Notice that the sign of resistances, positive or negative, appears relative to the goals or agency/praxis that the person pursues. Indeed, as such, Resistances of Reality do not have a sign; the sign appears in the resistances, that is, at the psychological not the epistemo-ontological moment/level. For instance, as Tolman (1961) emphasized, a major property or feature of chairs is their sit-on-ness (the aptitude of allowing us to sit on them easily). Solidity, shape, and size are major physical properties constituting the sit-on-ness of chairs, and when one is tired, the sit-on-ness with its three physical properties becomes a positive package of resistances. However, when the agency/praxis changes, for instance, when we wish to pass a big chair through a door, its solidity, shape, and size can become negative resistances—an obstacle or encumbrance. The sign of resistances is only within and relative to agency/praxis being pursued. These resistances eventually are internalized within schemes, which are our meaning-bearing units. Such "transmission of information" from Resistances to resistances to schemes is illustrated in figure 5.1.

A scheme is "an organized set of reactions that can be transferred from one situation to another by the assimilation of the second to the first" (Piaget & Morf, 1958, p. 86). Piaget also defined schemes as the common functional structure of all reactions (including signalic, symbolic, and agency/praxis functions) that a subject takes as interchangeable in pursuing a given affective or cognitive goal.[1] The concept of schemes or schemas, as internal mediators of knowledge processing, goes back to Kant. Interpreted in the language of current psychology, Kant (1929/1965; Pascual-Leone, 1998) saw the scheme/schema as a knowledge-processing bridge between our organism and its situational context (the Resistances of Reality), expressing actual constraints or resistances (both encumbrances/obstacles and affordances) of the current situation, as figure 5.1 symbolizes. However, schemas/schemes of Kant or Piaget are neither organismic (i.e., embodied in the brain and behavioral processes) nor always situated (i.e., contextualized to a situation). These authors do not show a relation of schemes/schemas with organismic constructivist processes in evolutionary epistemology (Pascual-Leone, 2006).

Human activity can be understood as a goal-directed interaction (intentional or not) with situations, with the key aim to control and understand objects, persons, or situations (Dormashev, 2010; R. Miller, 2011). It ultimately can be expressed as conscious or unconscious agency, whether praxis or not.[2] Goals always involve an affective component (appetitive/positive or aversive/negative), expressing the organism's history or innate internal constraints (Pascual-Leone & Johnson, 2004). We think of reality-out-there (Reality) as a universe of species-specific Resistances potentially affecting our goals: for instance, externally determined cognitive constraints or (often relational) sensorial-perceptual features of the experienced situations and outcomes. These resistances emerge during individual activity within a given context or situation. For instance, when I hit the table with my knuckle, I hear the sound (resistance1) and simultaneously feel the blow (resistance2) on my finger. These are resistances causally expressing two different "objects" that clash. As in this example, resistances often are found to have dependency relations with one another. Reality is populated with packages of interdependent potential Resistances that are relative to each species but also (when converted into resistances) related to the personal history of each individual.

Any physical object is (epistemo-ontologically) one such sort of package, which functionally may change with the subject's species and history. An example from Kagan (2002, p. 72) can serve as illustration: "a person who has just cut his finger on a knife and watches the blood ooze over his palm has no uncertainty about the existence of objects that can cause blood to flow, and is certain that he feels different than he did moments earlier." The finger and the knife are Real packages of resistances. They are distal objects—objects out there—construed by us to represent, and coordinate vis-à-vis life contingencies and action, the interdependent external Resistances (shape, size, color, utility, sharpness, and other characteristics). The cutting action (a key resistance of the knife) here serves to expose a conditional characteristic or resistance of the finger (i.e., when cut it will bleed). Although this construct of *Resistances/resistances* is our solution to the key problem of how thought can be effective in Reality (and how Reality can affect thought), some idea of Reality resistances opposing a person's agency is found (with different terminologies) in thinkers throughout history—albeit not stated as an epistemo-ontological explication of Reality.

Resistance packages can be interpreted, without falling into empiricist excess, as indexing Real (i.e., actually external) functional and probabilistic invariants (cf. Gibson, 1979; Nozick, 2001; Ullmo, 1967). They are recurrent relational aspects of Reality that animals can and must learn and cognize, to re-present these Resistances to themselves, as resistances by way of schemes. Motivation (which functionally intertwines affect/emotion, cognition, and Reality) leads higher animals to internalize these packages and their interdependencies, thus learning the *causal texture of the environment*[3] (Tolman & Brunswik, 1935). Notice that *internalize* means to learn by constructing schemes that epistemologically reflect (*epireflect*[4]), that is, express or attempt to control, important resistances and their affective value for us. They often do so iconically (see chapter 4) because the schemes epistemically reflect important aspects of the Reality at hand. From this perspective, it is appropriate to recognize that internalization (a conscious or unconscious epireflective abstraction) of reality's functional packages results from our own activity and occurs in three distinct but interrelated modes of invariance, three different sorts of packaged Resistances/resistances:

(1) First are packaged resistances of Reality that can serve as *targets* of the person's agency, which underpin the "*ob*jects" of experience, whether external or internal/mental experience. When these packages of resistances are represented psychologically, we call them figurative schemes (or internalized *model obs*). From a psychological semiotic perspective, these *model obs* correspond to simple or complex figurative (including parameter) schemes that have the function of signs (relational signal or symbols), together with their expectable meaning, their context or sense, and their referent.

(2) Second are packages that stand for *patterns of action* or *mental operations*, which can change *obs* or relations among *obs* in expectable ways. Epistemologically we call these packages *pros* (from procedures). When internalized, we call them *operative schemes* (or *model pros*), which are substrata for constructing procedures and operative processes. From this perspective, *pros* correspond to the actual operative or transformational possibilities of action offered by *model obs* (figurative schemes) as they become activated. Or vice versa: figurative schemes or objects are definable, as Piaget emphasized, by the set of operatives that can apply on them and those that cannot.

(3) Finally, there are packages of Resistances, simple or complex, that give *ad*junct information about *obs*, *pros*, and situations in which they usefully can be applied. We call these packages *ads*. When internalized, we call them *model ads* or *parameters* (a sort of relational figurative scheme). They describe properties or relations pertaining to *obs*, or conditions (i.e., contextual parameters) that *pros* need to satisfy to be applicable to particular *obs*. From this perspective, *model ads* (parameters) often correspond to signalic or symbolic (iconic or indexical) cues. *Ad*jectives, *ad*verbs, the meaning of relative clauses, and *ad*vertisements all have this category of *adjunct-information* as their reality foundation.

In the quote of Kagan above, the finger, the knife, the blood, and the palm are each represented in the person's brain as *ob*-indexing processes or models (figurative processes). The category description "objects that can cause blood to flow" is an *ad*-indexing process that causally relates *obs* such as knives to parts of the body (e.g., fingers) and to blood. The brain representation of the knife's action, which caused the blood to flow, stands for a *model pro* (and operative). By reflectively abstracting these *obs*, *pros*, and *ads* from experience, intelligent animals can internalize descriptive dynamic representations, and causal relations, relevant for life within their environment. These three sorts of resistances are psycho-logical processes, in Hegel's or Piaget's sense of (dialectical) logic, that is, schematized effective processes that mentally mediate amongst Reality, life experience, and action.

Pros, when internalized as action moves or procedures, serve as blueprints for people's actions or transformations, expressed as constituents of *operative schemes* or *operatives* (generators of scheme *effects*, see below). In contrast, *obs* and *ads* are internalized as descriptions of states and relations; combined, they make up units that we and other developmental constructivists call *figurative schemes* or *figuratives* (i.e., declarative or representational knowledge, implicit or explicit). Psychologically (i.e., when represented) a main difference between *obs* and *ads* seems to be motivational: *obs*, but not *ads*, can usually serve as targets for the person's praxis. Notice further that reflective abstraction and internalization of *obs*, *ads*, and *pros* cannot be made in a piecemeal manner. They emerge all together, coordinated in the context of activities within situations, as distinct functional aspects of the internalized total activity. They are *dialectical trios* abstracted together in coordinated packages, which produce organismic schemes. They are carried in the brain by distinct unitized neuronal circuits distributed over the brain. These circuits are cofunctional (vis-à-vis certain activities) and often coactivated in some tasks. Functionally, the unitized circuits are organismic schemes, definable within functional totalities of psychological brain processes.

Organismic Schemes

Schemes are situated (i.e., situation specific) and self-propelling. They are causal factors that together overdetermine manifest performance when they are cofunctional and often coactivated (Piaget's assimilation principle tacitly embodies this assumption). Every scheme **(Sch)** has a dual reality. **(Sch1)** It appears as a *package of qualitative functional characteristics* that informs (i.e., injects form into) experience and action when it applies. **(Sch2)** It has a variable *quantitative parameter* (i.e., *activation weight* for its neural network area), which conditions probability that the scheme can become highly activated and apply in a given context (perhaps as a synchronized neural circuit) to cause performance. Schemes are like recursive procedures (can produce schemes of schemes of schemes) and can be learned. It is common to have coordinated flexible hierarchies of schemes (i.e., heterarchies) with a context-sensitive ranking of dominance for the schemes. Complex schemes are often called functional *structures* or *schemas* (such as the coordinated "scripts" that govern our behavior in restaurants or within already-familiar complex problem-solving situations). Complex schemes express coordinated neuronal circuits or networks (Arbib, Érdi, & Szentágothai, 1998; Fuster, 1995; Kagan, 2002; Mars, Sallet, Rushworth, & Yeung, 2011).

Long-term memory results from the repertoire of schemes/schemas sorted in kinds: *executive* schemes (prefrontal), *operative* schemes (frontal), *figurative* schemes (i.e., perceptual or representational or linguistic; occipital, parietal, or temporal), and coordinated packages of more or less automatized *operative/motor* schemes (frontal hemispheres, basal ganglia, and cerebellum). There are also *affective or emotion* schemes (connected with the limbic system and the medial cortex), *analytical* problem-solving schemes produced with mental effort (initially in the left hemisphere, inferior and dorsolateral prefrontal), *holistic or automatized* schemes (right hemisphere), *complex analytical thinking (mentation)* schemes (frontopolar networks), and so forth.

From a functional-structure perspective, a scheme can be understood as expressing a unitized, well-learned coordination of three subsystems or components: *functional, releasing, and effecting* components. The *functional component* (*fc*) embodies the gist of the scheme's functional organization (scheme goals and task-relevant model configuration of its *pros* or *obs*). It embodies implicit or explicit expectancies about future outcomes of the scheme's application(s), as well as the gist of its whole functional activity. Further, the *fc* possibly has within it a *terminal state* or *stop rule* (Pascual-Leone, 1970) that tells the scheme in question when it has achieved its expected results. Notice that expectancies carried by the *fc* of schemes are essential to enable the inner appraisal (top-down and perhaps bottom-up) of a scheme's relevance for a given task or situation. The executive processes can use this information. When an affective need or goal induces executive schemes to bring about this affective-goal satisfaction, the fc of schemes, with their expectancies, can provide inner cues semiotically, telling the currently dominant executive that the scheme in question is or is not task-relevant here and now. This is why when one is thirsty, any potential source of water becomes more salient; and mailboxes become more noticeable when one needs to mail a letter. If they are congruent with intended cognitive or executive goals, *fc*'s of schemes are boosted with mental attention by executive schemes, increasing the perceptual sensitivity for task-relevant schemes cued by the situation. Thus, effectiveness of situational cues is potentiated by the figurative and operative expectancies in a subject's schemes.

Second is the *releasing component* (*rc*), that is, a set of conditions (perhaps indexed by schemes—models of *ads*, *obs*, or *pros*) that activate or release the scheme. Scheme conditions that satisfy current features or resistances within the immediate (external or mental) situation become cues. Thus, cues are never only external: they express matches between releasing components of schemes and the external resistances/Resistances or features of the actual reality/Reality.

Finally the *effecting component* (*ec*) is a set of effects (often expressed by models of *pros*, *obs*, or *ads*) that follow from the scheme's application to reality experience. For instance, I encounter my spouse in the kitchen holding a knife. In this context, the *rc* of the knife scheme is its visual characteristics and the way it is held; the *fc* is my anticipation in this context that she may be preparing dinner; the *ec* is my operative knowledge of how to use a knife to chop or slice food.

A scheme can be conceived as a dialectical dynamic system <: *fc* : *rc* : *ec* :> constituted by the three components, each dialectically coordinated (:) with the other two, and all constituting an integrated dynamic triplet that cannot be divided without breaking the scheme as a unit of processing. It cannot, because schemes emerge developmentally as functional and probabilistic (dialectical/dynamic) trios—the viable functional invariants from interactions among these three components extracted from regularities of actual experience or thought.

Although most schemes express *pros*, *obs*, and *ads* as models, the actual relative importance of these distinct semantic-pragmatic constituents changes with the sort of scheme. Some schemes can be said to function mostly as models for *pros* or *obs*, others mostly as models of *ads*. We cannot show it here, but our formulation of a scheme's functional structure is congruent with the processual units that Vygotsky pioneered with his organismic conception of semiotics (Dormashev, 2010; R. Miller, 2011; Pascual-Leone, 1996a, 1996b; Van der Veer & Valsiner, 1994; Zinchenko, 1985). Our scheme construct also is congruent with the implicit units (often called, e.g., chunks, structures, moves, responses, options, schemas) formulated within current reinforcement-learning theories of neuroscience (Mars et al., 2011).

To be well defined, these organismic carriers of signs and operative activity must, in our view, satisfy some other basic postulates of schemes **(Sch)**:

(Sch3) Schemes are recurrent (probabilistically invariant) functional constituents of *agency or praxis*. Therefore, they must be defined relative to the totality of agency (activity, actions, or mental performance) that prompted their emergence. For instance, relative to the task of shopping for food, necessary complex scheme constituents are <getting dressed to go out>, <walking to the store>, and so on. From a neurological perspective schemes are functional-activity component systems of brain processing that carry active information (are subjective operators of the brain "software") relevant for some sort of agency/praxis, which can organize functionally various content domains (e.g., visual, auditory, motor, affective, psychosocial, cognitive-relational). As information carriers, schemes are qualitative, functional-relational (*intentional* or *intensive*) entities with specific activity dispositions or propensities. Schemes have no here-and-now *extension* until they are applied (i.e., no set of concrete instances, objects, or operations exist related to them prior to the schemes' activation and application). Every scheme is self-propelling (Piaget's assimilation) under releasing conditions (cues), and an applied scheme imposes attributes, restrictions, or constraints to the external or internal performance, unless other contradictory schemes prevent its application. Competition among schemes often precedes performance.

(Sch4) Functionally, schemes are *indivisible units*. Distinct aspects of experience that are not cofunctional must be kept within separate coordinated schemes if they are to be used separately. Thus, schemes must be internally consistent to be formed and congruent to be coordinated. This constraint implies that whenever two objects, or two aspects of the environment, or two action moves, are functionally different in terms of agency/praxis, they have to be expressed organismically as two distinct schemes—even when they are similar in other respects. Otherwise, the organism cannot perceive or treat them as distinct and cannot use them differently (this is postulate SC1 of Pascual-Leone, Goodman, Ammon, & Subelman, 1978).

(Sch5) A scheme is a sort of *recursive function*: "the conditions and/or effects of any scheme could in turn be constituted by schemes" (Pascual-Leone et al., 1978, p. 254). Schemes can be defined at different epistemic or knowledge levels within functional hierarchies, with schemes functionally nested (usually into heterarchies). The conditions or effects of higher-level schemes are constituted by partial "copies" of other lower-level schemes that remain functionally autonomous.

(Sch6) Organismic schemes substantively correspond to cofunctional and often coactivated neuronal circuits that carry shared semantic-pragmatic meaning. As with the "symbolic forms" of Cassirer (1923/1957) or the schemes of Piaget, *"the form of a scheme's structure is the structure of its functioning"* (Pascual-Leone, 1984, p. 183). From this perspective, schemes or schemas may be seen as semantic-pragmatic systems of processing for the performances they help to produce; performances being overdetermined by sets of compatible schemes that can apply together and achieve intended results. Notice that each scheme embodies a distinct, necessary, and invariant functional constituent of the strategy of action (agency, praxis, or practice) from which it is abstracted. However, the functional structure of any scheme tends to be minimalist, in the sense that, in the process of a scheme's psycho-logical formulation (i.e., reflective abstraction by the organism), this functional structure remains as general or generic as possible. In constructing a scheme, and vis-à-vis the intended performance, the organism always takes as much error (variability, features, or aspects omitted from the prescription and left undefined) as it can afford, relative to the intended ideal agency or praxis. In other words, for as long as the abstracted scheme is viable, no further differentiation or refinement of the scheme may take place. This progressive degree of abstraction facilitates generalization of schemes to similar or related situations (i.e., transfer of learning).

(Sch7) To emerge or be formed, schemes must be abstracted from performance/experience or constructively learned by the subject. A high enough *ecological frequency* of situations (and affective goals) to promote formation/learning of these schemes must therefore exist in the life experience.

(Sch8) Schemes are key organismic-causal determinants of agency/praxis and experience, but they do not suffice to explain truly novel or creative performance. Agency, praxis, and causal abstraction use processes of the brain "hardware" together with the schemes' "software." These "hardware" brain resources (e.g., hidden operators like mental attention, space or time factors, internal Gestalt-field factor), along with organismic principles such as Schemes' Overdetermination of Performance (*SOP*, see chapters 1 and 2), Piaget's equilibration, and so forth, apply on schemes to generate truly novel *dynamic syntheses* that create probabilistically transitory organismic states that can induce and orient a subject's performance and may lead to emergence of new schemes. These transitory states can synthesize performance (via "creative" coordinations) in ways not previously expressed in already learned schemes and can do so without assuming a *homunculus* (i.e., a "little man" inside the organism that produces its performance). These novel and ephemeral dynamic functional structures could, with repetition, be learned, generating new schemes or schemas.

There are interesting epistemological convergences of our process-qualitative, developmental approach with current dynamic field/system theories and with computational-rationality probabilistic-process models (Friston, 2010; Gershman, Horvitz, & Tenenbaum, 2015; Schöner, 2014; Spencer, Perone, & Buss, 2011). They are all dynamic-process theories of various sorts that produce performances by activating locally ("in situ") multiple processes; in our case, schemes/schemas and operators, which with their interactions make affectively motivated (internal or external) performances emerge. Schemes/schemas (together with TCO's organismic hidden operators and principles) could perhaps serve as a qualitative-process, general causal account that may add an intuitive macro perspective to dynamic or quantitative probabilistic-process models of human performance.

(Sch9) The *assimilatory strength*, or total weight, W_T, of a scheme (Pascual-Leone & Goodman, 1979, p. 321) "is a combination of all the weights received by it from the various organismic boosters." The boosters usually are cue salience, hidden operators, or innate propensity to apply. A tentative mathematical sketch for this sort of weight combination was offered by Pascual-Leone and Goodman (1979), which we present (modified) to add epistemological clarity to the issue:

$$W_{\mathrm{T.i}} = (W_{E.i} + W_{A.i} + W_{C.i} + W_{L.i} + W_{F.i} + \ldots - W_{I.i})\, W_{cues.i}$$

In other words, the total activity weight of a scheme h_i is a function of the sum of all the factors (e.g., hidden operators) that boost this scheme, minus the weight of deactivation factors such as attentional inhibition (I_i). The result of this algebraic combination applies on ("multiplies") the weight of the cues that activate this scheme ($W_{cues.i}$). Notice that *cues* in the TCO are matches between releasing conditions in the *rc* of the scheme and features, characteristics, Resistances of the input (immediate Reality).

(Sch10) The main mechanism for causal abstraction is the formation and application of schemes, which are the most important dynamic causal constructs of the organism.

Action Schemes

In TCO's terminology, *action schemes* can directly produce performance. In contrast, executive schemes need action schemes to do so. Action scheme is a category that includes cognitive (operative, figurative, and parameter) as well as affective and personal schemes. Application of action schemes directly causes actual experience or performance (activity in the sense of Vygotskian psychologists).

The two main superordinate kinds of action scheme are affective and cognitive. *Affective schemes* (Greenberg & Pascual-Leone, 1995, 2001; Antonio Pascual-Leone, Greenberg, & Pascual-Leone, 2009; Pascual-Leone, 1990b, 1991b; Pascual-Leone & Johnson, 2004; see also chapter 6) add biological and life evaluation feelings, *vital values,* to the experience or situation at hand (feelings such as good/bad, like/dislike, and seek/avoid). They set up the life or vital-value appraisals (*values* for short) for experiences and evaluate situations (external cognitive states) and inner states, in terms of vital feelings they elicit.

Affective schemes introduce tendencies, which we call *affective expectancies* and *affective* (abbreviated as *A*) *goals,* which boost (*A-boosting*) activation of cognitive schemes congruent with the dominant affective tendency in question (e.g., if in fear, escape; if in love, approach). The *fc* of a purely affective (*A*) scheme stipulates the qualities or propensities of a given affective tendency. Its *rc* contains the cognitive states (signals or symbols; models of *obs*, *pros*, or *ads*) and the intuitive mental/experiential configurations that release the affect(s) in question. Its *ec* contains the effects occurring when the affective scheme applies: affect-related physiological manifestations (e.g., paling with fear, warming with love), related cognitive memories, and an affective activation weight (*A*-boosting—sometimes called *conation* or *psychic energy*) that applies on cognitive schemes suitable for the affect's affective goal—to bring about affect-valued outcomes or experiences. Purely affective schemes easily combine with cognitive schemes to constitute affective-cognitive hybrids, often called *personal* or *emotion schemes* (including emotions, character/moral schemas, psychosocial and sociocultural schemes, and others).

Unless they are personal (i.e., hybrid) schemes, *cognitive schemes* are distinct from affective schemes, because they do not contribute feeling-appraisals of vital value to the organism. Instead, figurative cognitive schemes provide epistemic *truth appraisals*: correct/incorrect, right/wrong, agreement/disagreement between perception or thought and actual experience. Truth can be appraised in agreement or disagreement with reality/Reality in four different ways: direct appraisals of correspondence with reality/Reality, appraisal of the schemes'/schemas' internal consistency, appraisal of pragmatic utility, or via consensual validation. These are tests of the schemes as signs, testing their relation to their referents and to one another.

To recapitulate this important point: within the initial biological complex system that von Uexküll first outlined (the system formed by an organism vis-à-vis its own interpreted life-world or immediate environment), affective/personal schemes yield feelings or vital-value appraisals, whereas predominantly cognitive schemes that are figurative produce truth-value appraisals. The latter evaluate whether the interpretation/representation embodied by a scheme corresponds to resistances offered by this aspect of Reality that for the scheme is referent (referential *intentionality*).

As mentioned, cognitive schemes split in two complementary sorts: figurative and operative. *Figurative schemes* are processes that assign predicates of all sorts, simple or complex (including relations, here construed as multiterm predicates), and more complex compound predicates (at times temporally structured in sequences and epistemic levels), which may generate models in the brain. Such models are distal objects (*obs*) or complex procedural descriptions (description of *pros:* strategies for doing "jobs" of one or another sort) or else complex adjunct information (*ads*). When figurative schemes apply, they together generate a mental state (phenomenal representation), which is an intuitive synthesis, or configural experience, bearing meaning and perhaps accompanied by figurative or operative expectancies. *Figurative expectancies* are associative anticipations of "what" is related to something else, contingent on some action. For instance, given an object examined from one perspective (my home, my pet, my street), I can anticipate what I can see or experience when I act and take other viewpoints: for example, when I go and look at the back of such object or mentally seek parts of the object that are hidden but accessible after using appropriate operative-scheme steps.

Operative schemes (*operatives* for short) are mental or external transformations or actions. Their application induces emergence of *operative expectancies* (Tolman, 1959, aptly called them "means-end-readinesses"), that is, anticipations of results when applying one or another operative scheme. A transformation changes a figurative state into another, distinct and possibly different, figurative state. For instance, I hit a glass, and it will fall from the table, breaking into pieces. This anticipation is an operative expectancy; the action itself is a transformation. Encompassing across situations, and more abstract, is the special sort of operative stipulating plans for action that are called executive schemes. This term, now very popular, was first introduced in psychology by Koffka (1963/1935) before its importance could be understood.

Executive Schemes

Much current literature has an overly broad definition of executives and executive functions. Executive functions/controls often are seen to subsume all planning and control of cognitive/personal processes, together with the brain resources (e.g., mental-attention activation or inhibition) needed for functioning. Generally, the literature recognizes two kinds of executive control: before 4 years of age and after 4.5 years. This is congruent with our sensorimotor mental attention versus symbolic mental attention (Espy, 2016; see chapter 3). In our organismic formulation we restrict the term *executive* to refer to a person's currently dominant set of compatible executive schemes. We restrict *executive functions* (often illustrated by the functions of shifting or updating[5]) to refer to certain executive-driven schemas caused by learned complex packages of executive schemes (*LC*- or *LM*-structures, see below). Our theory recognizes other organismic factors in executive-driven performances, such as mental attention (activation and inhibition or attentional interruption). These are distinct brain resources that executives use and control (Im-Bolter, Johnson, Ling, & Pascual-Leone, 2015), thus acquiring power to activate or inhibit action schemes of any sort. Other organismic factors, like affective/personal schemes, are the initial mobilizers of executives, sometimes controlling them (Pascual-Leone, Pascual-Leone, & Arsalidou, 2015).

Functional heterarchies and the nesting of schemes are illustrated in the dynamic interactions among executive schemes and ordinary action schemes. *Executive schemes* are higher-level operative units that embody plans or blueprints for action and strategy. Within situations, *executives* (i.e., currently activated, compatible, and dominant executive schemes—often known as *the executive*) define tasks and can regulate application of general-purpose brain resources. These resources are hidden organismic capacities—complex and innate subcortical and cortical controls (e.g., activation and inhibition functions of mental attention, various sorts of learning, field factor). They are applicable to action schemes to change their functioning, in a context-sensitive manner. Executives are induced or mobilized by (often implicit) affective goals or motives. Affective goals are brought up by currently dominant affective and personal schemes, and their conversion into cognitive goals, often via executives, is what we call *motivation*. This sense of motivation (i.e., conversion of affective goals into cognitive ones) has specific processing sites in the cortex, such as the insula and anterior cingulate cortex; at least in sufficiently complex cognition (Arsalidou, Pawliw-Levac, Sadeghi, & Pascual-Leone, 2018; Pascual-Leone et al., 2015).

By controlling organismic resources, executives can monitor application of suitable action schemes to the task at hand. To have the relevant action schemes strongly activated when needed, executives can mobilize mental-attentional (*M*) capacity and/or attentional inhibition (*I*). With practice, however, a chunking or associative-learning process may take place, which we call *logical*-structural *content* (or *LC*) *learning*, which transforms the flexible hierarchical package of schemes into a single, more or less flexible, complex schema or chunk (which, in our theory, we call *LC-structure*, an automatized and coordinated complex processing package). Once formed, executives and mental energy (*M*-capacity) can treat and boost the *LC-schema* as if it were a single unit. The semantic ground (or meaning) of well-established signs often corresponds in the organism to *LC*- or *LCLM*-schemas/structures (see below).

There are two sorts of executive schemes, which often are confounded: *plans* and *controls* (Pascual-Leone et al., 1978). As plans, executive schemes are in charge of selecting a task strategy and organizing processing for and within the task (i.e., schemes that must be simultaneously activated at each task moment, and the sequencing of task steps). As controls, executive schemes (in the service of current affective/cognitive goals) can mobilize and control application of brain resources such as mental attention, to ensure that essential task-relevant schemes become not just activated but dominant, to carry out the intended task performance. This would be effortful executive-driven (or executive-mediated) processing.

Executive-driven processing requires executives to "recognize" schemes relevant for the situation at hand and notice schemes that are misleading and should be interrupted (attentionally inhibited). How do executive schemes know which action schemes to recruit or inhibit? This important question should be answered at two distinct levels: At the level of action schemes (stipulating what they can do for the task) and at the level of the type of task (estimating degree of relevance/irrelevance of one or another scheme, in terms of its task effects). At the level of action schemes, the functional component (*fc*) of schemes is extremely important, because it carries expectancies about what a scheme can do or bring about, and the gist of its action. The functional components of schemes can provide information to, and are activated by, executives. This enables the current task executive to qualitatively appraise potential relevance of action schemes.

A quantitative (probabilistic) appraisal of degree of relevance can also be provided at the level of type of task. Such probabilistic appraisal would constitute a *criterion of relevance* (Pascual-Leone, 1984). This criterion may emerge via some sort of heuristic rule like the following (this is a hypothesis): schemes that in the past were frequently highly activated (or boosted by M; i.e., inside the field of mental centration or working memory) at moments when a suitable performance in this type of task was produced are regarded by the executive as relevant for producing this performance (proportionally to the frequency of occurrence).[6] Other schemes that had low activation (or were absent from mental centration) during past performances in this type of task are taken to be irrelevant by this criterion. Such criterion of relevance is similar to those adopted in neuroscientific computational models of reinforcement learning. This similarity is illustrated by the scientific model of Cockburn and Frank (2011, p. 319): "the pre-SMA [the pre-Supplementary Motor Area of the medial cortex] eventually learns to refine its selection of candidate actions based on the probability of having selected these actions in the past for a given context."

In misleading situations, executive-driven processing is more likely to succeed when executives are mental and symbolic, that is, functionally detached from action schemes applicable in the task (see chapter 4). However, when executive schemes are only sensorimotor and signalic (usually automatic), ability to cope with misleading situations depends on associative connections to task-relevant (local) operative schemes, which is only possible when the task is not novel, and the goal-relevant associative learning already has taken place. Furthermore, signalic executives may not be able to control ongoing task processing due to functional attachments (associative chunking) to other irrelevant action schemes—a common source of errors and performance inadequacies.

A clear distinction and diagnosis should be made between these two sorts of executives (i.e., mostly sensorimotor/signalic vs. fully mental/symbolic executives), because their respective strategies and demands on brain resources are very different. One difference tends to be in *complexity*: a greater number of distinct (executive and action) schemes is involved in mental-symbolic executives, which demands mental effort, and where relevant, schemes must be activated jointly or in the prescribed sequences. However, an important, more subtle difference is found in *their relation to affective goals* that motivate or induce the task. As discussed in chapter 3, sensorimotor/signalic executive tasks are always motivationally immediate (usually affectively or cue driven), because task activity (agency, praxis) is automatically cued by the actual immediate situation, which activates most or all needed task schemes (via *C, A, B,* and *LC* boosting). In contrast, symbolic-processing (high mental) tasks involve affective goals that must use mediation by symbolic executives.

In these executive-mediated tasks, the necessary operative processes are not directly related to current affective goals (e.g., working on a math assignment because my mother wants me to is executive mediated, if I do not much like working on math). Instead, the needed operative processes are activated and kept alive in a mediated manner by mental attention and other brain resources. For instance, when I take my car to visit my friend to get a book from him, the current goals/schemes activated for driving, getting to his place, parking, and so on, are unrelated to the symbolic executive (i.e., the main operative process of getting the book). These goal-schemes are kept alive in my thinking by both automatized/signalic executives (by prior LCLM learning) and mental attention (i.e., effortful activation or inhibition). Another example more relevant for development and education is when a primary-school teacher designs a task and asks children to do it. A child's dominant affective goals may not be doing the task but pleasing the cordial, friendly teacher and the child's parents, who encouraged his or her attendance and value his or her work. In this case, the tasks themselves are not affectively immediate but mediated by executive schemes boosted by these personal affective goals (e.g., pleasing teacher and parents, doing what the child feels he or she must do—an "ethical/moral" injunction). Even when the child finds the task tiring or boring, executive mediation may motivate the child to do the task.

Because distinction between signalic versus symbolic executive processes is so important, we illustrate signalic executive processes again using a sensorimotor (affectively immediate) task, which is complex and motivating enough to induce in babies an executive-mediated activity. Take the case of a 9-month-old who spontaneously reaches for a rattle to shake it because she likes its sound. Her sensorimotor behavior emerges as an overdetermined sequence of dialectical/dynamic syntheses in performance: coordination and application of several figurative schemes, operative expectancies, and transformation operatives. Table 5.1 symbolizes the stream of processing elicited by this task when it is solved via signalic/sensorimotor performance. There are two aspects to the description in table 5.1. The first is the likely activated repertoire of schemes that a child may need to generate a stream of processing that can solve this task. The second aspect is our descriptive model for this stream of processing.

The first three schemes listed in the activated repertoire are figuratives (i.e., predicates) for three aspects of the same distal-object (*ob*) rattle: the visual rattle, the touched-and-handheld rattle, and the sound-producing rattle. When, with child development, the figurative schemes of three "perceptual rattles" are coordinated, these figuratives contribute to constitute the symbolic distal-object "rattle." We name these schemes with meaningful abbreviations and, after the equal sign (=), we give more detailed description of each. We have numbered the schemes (i.e., … :1; … :3; … :5; … :2) to simplify description of the complex schemes they constitute, representing these constituents by their number (e.g., *GRASPING*:2.1.3). In a purely signalic processing, as we are now modeling, the three perceptual aspects of distal rattle may not be integrated, and each may represent a distinct perceptual object. Notice further that, as usual, we represent figurative schemes by writing their names in lowercase (we omit the prefix * used elsewhere to symbolize figurative schemes).

Recall that any scheme has three dynamically coordinated components that define it. Let us illustrate these three components in the simple schemes just mentioned. In the scheme vis.rattle:1, the functional component may correspond to the description [*fc*:= visual pattern signaling the distal rattle that can produce rattle sound when shaken, and can be touched]. This stands for a purely experiential, nonverbal associative description in the child. The releasing component corresponds to [*rc*:= sensorial-perceptual features that match conditions characteristic for a distal rattle]. Its effecting

Table 5.1
Secondary circular reaction: reaching for and shaking a rattle

Activated Repertoire of Schemes
vis.rattle:1 = "visual rattle:1"
touch.rattle:3 = "touched-and-handheld rattle:3"
sound.rattle:5 = "sound-producing rattle:5"
GRASPING:2 = "reaching and grasping:2"
SHAKING:4 = "shaking:4"
***GRASPING*:2.1.3** = "GRASPING:2; visual rattle:1; touched-and-handheld rattle:3"
***SHAKING*:4.3.5** = "SHAKING:4; touched-and-handheld rattle:3; sound-producing rattle:5"
APPLY.THEN: 2.1.3-4.3.5 = "APPLY *GRASPING*:2.1.3 **AND THEN** *SHAKING*:4.3.5"
Stream of Processing (Model of Processing Steps in Performance)
{vis.rattle:1} → {*SHAKING*:4.3.5, vis.rattle:1} →
{APPLY.THEN: 2.1.3-4.3.5, *SHAKING*:4.3.5, *GRASPING*:2.1.3, GRASPING:2, vis.rattle:1} →
{*SHAKING*:4.3.5, SHAKING:4, touch.rattle:3} → {sound.rattle:5, ...}

component could be [*ec*:= the actually experienced pattern of a visual rattle, which, when shaken, yields its specific sound, and so on]. The overall distal rattle is constituted as a probabilistic invariant by experiential coordination of the three distinct schemes: vis-rattle, touch.rattle, and sound.rattle.

The next schemes in the list of the repertoire (table 5.1) are two operatives: GRASPING:2 and SHAKING:4. We indicate operative schemes with CAPITALS. Next, we have written two operative expectancies, identified by writing their names in italics: *GRASPING*:2.1.3 and *SHAKING*:4.3.5. Again, digits identify heterarchical constituents or subordinate schemes. For instance, in the first case the three subordinates are the operative transformation GRASPING:2 that applies on vis.rattle:1 to bring about touch.rattle:3. The last scheme in this repertoire is APPLY.THEN:2.1.3-4.3.5. This complex operative is temporally structured (i.e., 4.3.5 applies after application of 2.1.3) and embodies an operative expectancy (i.e., *APPLY.THEN*), which is not explicitly represented because it is part of the same scheme. As an example, consider briefly how the full scheme <:*fc*: *rc*: *ec*:> of the operative expectancy *GRASPING*:2.1.3 could be represented: <:[*fc*:= WHEN GRASPING is applied, THEN vis.rattle turns into touch.rattle] : [*rc*:= vis.rattle] : [*ec*:= touch.rattle]:>.

Let us now examine the second section of table 5.1, the stream of processing model. In this representation, schemes included within the same braces {...} are activated within the same attentional centration (focus of attention or attentional act). Centrations separated by arrows occur sequentially in the given order. Within each centration, the schemes are arranged in a functional hierarchy: operative schemes at the left apply on and control schemes on their right. An English paraphrase of the flowing sequence of actions described in this stream of processing model follows: the child sees the visual rattle [{first centration}]. Then [→] seeing-this-rattle activates the operative expectancy of shaking the handheld rattle, to produce the rattle's sound [{second centration}]. Next [➔] to implement this expectancy, a grasping-the-rattle expectancy is activated as an intermediate step to the actual grasping act, and grasping of the rattle occurs [{third centration}]. To add clarity, we describe again, very concretely, this third centration of the sequence: the operative APPLY.THEN controls and mobilizes expectancies *GRASPING* and *SHAKING*, which in turn mobilize the operative GRASPING:2, which applies on vis. rattle:1 to produce an expected touch.rattle:3 feeling (not represented but appearing in the next centration). The mentioned sequence occurs within a single centration, within a short time or simultaneously. Finally, the shaking expectancy can be applied, and this application mobilizes the operative SHAKING:4 upon the rattle being touched [{fourth centration}]; which causes [➔] the rattle's sound [{fifth centration}].

Such sequence may result from signalic processing. Each scheme to the left in the sequence should induce activation of schemes on its right because of associative (*C*, *LC*—see below) learning; the process begins with the cues from vis.rattle:1. However, this or similar sequences may also occur (without much associative learning) under symbolic processing with the help of mental attention (*M*-activation, attentional inhibition). In this latter case, a symbolic executive scheme (not represented in table 5.1) would exist, activated to anticipate and "prescribe" the set of schemes, which mobilizing mental attention, could produce dynamically the sequence of the third centration in the "stream of processing" model of table 5.1. Use of mental attention, and presence of control executives, explain the functional difference between symbolic versus signalic processing, as Cassirer (1929/1957, 1944/1966) may have recognized. In signalic processing the sequence of applied schemes is known only a posteriori, being induced by here-and-now affective goals and by circumstantial local cueing of one scheme or another. In contrast, sequences produced by pre-existent, now activated, symbolic executives are controlled or prescribed. This makes such sequences more robust to local circumstantial variations, enabling individuals to anticipate sequences more easily.

Diagnosis of Schemes in Concrete Situations

For many researchers, an objection to the scheme-unit construct is difficulty in finding criteria to identify schemes unambiguously. Schemes are constructivist concepts, and so they often appear in nested flexible hierarchies, because many abstracted cognitive processes are complex. We now sketch a useful approach for the diagnosis of schemes that may underlie a performance. The method uses constructs and definitions already given. We investigate each term in the scheme sequence, using five analytical-test steps.

Diagnostic step 1: Relative to the characteristics of the explicitly formulated target task and chosen strategy we recognize and itemize the *totality* of schemes that may cause the *task's final performance*, given a chosen strategy. This deconstruction of the performance into schemes (plans, controls, and courses of action) is a preparatory step—an inference of plausible causal processes that may determine (overdetermine) the intended performance. We often begin by demarcating the intended final task performance (and final cognitive goal or goals), analyzing backward from this expected performance to infer distinct "jobs" that, combined together, can cause it. Finding these component "jobs" requires that we enter a problem-solving process prior to task analysis.

We can illustrate this prior problem solving with a simple example: when a tourist in a new city wants to attend a certain concert in the city's concert hall, she has to configure the final task performance of arriving to the hall to purchase a ticket and enter. One job needed for getting there is to find the location on a map. A second job is to decide the means of transportation. If it is near a metro/subway station, this way will be cheaper; if not, a taxi may be the choice. Assume that the choice is the metro. Where is the nearest station? One must find out and get there, and so on. Notice that this and many final task performances become progressively configured and clarified as we encounter possible encumbrances or obstacles (Reality resistances) and find suitable ways of resolving them while obtaining means-end affordances needed. Schemes are concise representations of these psychologically internalized obstacle solutions and obtained affordances, as we explain below. This sort of final-task-performance problem-solving helps to clarify what sort of schemes would be needed for a given task prior to a detailed task analysis.

In the task modeled in table 5.1 the intended final cognitive performance is "getting the rattle sound," cued initially by the sight of the rattle and prior experience with rattles (*LC* learning). To reach the intended performance, a sequential operative-expectancy scheme available in the subject's repertoire (i.e., "the rattle can be grasped and then shaken") must be present to induce an affective goal that activates, in our example, the scheme APPLY.THEN associated with this motive (*LC* learning). The grasping of the rattle causes new stimulation, configured by application of the figurative scheme touch.rattle. It is upon this particular touching-the-rattle facet, a partial *ob* of the distal object (i.e., the total *ob* rattle), that the operative SHAKING is cued to apply. The "stream of processing" model in table 5.1 shows the result of this backward analysis—a sort of analysis that Peirce would have called a chain of *hypotheses* or *abduction.*[7]

In this table the operative schemes within each attentional centration (i.e., demarcated by braces) tend to be ordered: operatives on the left side applying to schemes on their right. This is an iconic way of representing pragmatic relations holding among schemes in an attentional act. Notice further that different distinct processes that influence the task dynamics must be represented as separate schemes, cued by the situation or retrieved from memory. Without separating distinct processes within distinct schemes their distinctiveness would be lost.

Diagnostic step 2: When the totality analysis and potentially relevant schemes have been demarcated, we must address the issue of whether affective goals can, by activating this functional set of schemes, directly cause the overt or covert intended performance. Affective goals (*A*- or *B*-schemes applying their pragmatic tendencies) work either directly in an immediate manner (as could be the case in affective, motivationally immediate tasks) or via mediation by executives, in executive-driven or mediated tasks. Diagnostic step 2 must determine which of these two sorts of motivation applies to the task being analyzed. Motivationally immediate tasks typically are facilitating, and here signalic processing may suffice. Then slow associative learning may lead to automatization with practice, even without mental attention being mobilized. In executive-mediated tasks, however, symbolic processing is necessary, because such tasks often are misleading. They demand use of mental attention (attentional inhibition and *M*-activation), which brings about much faster learning. (As we discuss later in this chapter, this is *LM* learning and its automatization via *LCLM* learning.)

The use of this effortful (presymbolic or symbolic) processing is constrained by the individual's mental-attentional (*M*-) capacity limit, a capacity that develops from infancy to adolescence in a graded manner (see chapters 3 and 7). For instance, if the process symbolized in table 5.1 were to be done by a child using mental attention, complex schemes not easily acquired via associative learning should have to be synthesized and boosted with *M*-capacity. These potentially M-boosted schemes have been underlined in the stream-of-processing section of table 5.1. If we take the third attentional centration {...} of this stream, the schemes underlined are APPLY.THEN and the two operative expectancies *GRASPING* and *SHAKING*. A 9-month-old infant should have already automatized the operative GRASPING and the object-facet vis.rattle, which therefore should not need boosting from mental attention. There are only three schemes to be boosted jointly with mental attention (here symbolized by underlining the schemes). This amount of *M*-capacity is generally available to 8-month-olds (Pascual-Leone & Johnson, 2005; see chapter 3). Thus, 9-month-olds could solve the task presymbolically (creatively inventing the task solution without much prior practice).

Diagnostic step 3: To refine and clarify the schemes entertained in steps 1 and 2 a task analyst should examine again, more abstractly, the schemes found in step 1. For each of these schemes the analyst should identify them as carriers of the pragmatic functions (*obs, pros,* and *ads*) demanded by the task (i.e., the target cognitive goals, the strategy). This is done to see whether these functions are expressed in the schemes chosen in step 1: figuratives, figurative expectancies, operative transformations, operative expectancies, or executive schemes (such analysis may validate results of step 1, clarify the need for other schemes, or show redundancy in some chosen schemes).

Diagnostic step 4: This step attempts, like the previous step, to investigate further; and when necessary revise, schemes chosen in step 1. To do so one can re-examine qualitative information-processing in the intended performance. This can be done from two complementary perspectives. (1) Investigate the component-structure of the analyzed task schemes, by concretely formulating the schemes' functional component *fc* (gist and outcome expectancy of the scheme), concretely formulating their releasing component *rc* (concrete conditions that activate the scheme), and formulating concretely their effecting component *ec* (effects resulting from the scheme application). (2) If there is a doubt about the function of chosen task schemes, investigate how these schemes (particularly figurative schemes, and figurative or operative expectancies) should be classified *semiotically*. This involves analyzing schemes in terms of four semiotic functions (see chapter 4): symbolic (with detached sign and referent), signalic (with attached sign and referent), iconic (with configural similarity between sign and referent), or indexical (with no similarity but an arbitrary relation between sign and referent).

Diagnostic step 5: Complex schemes are often semantic-pragmatically nested, organized in flexible hierarchies that may change adaptively with the situation. We can investigate the chosen schemes further by determining the different *levels* of processing (different epistemic "grains" of analysis) in the concrete task where they appear as functional invariants. For instance, in table 5.1, within attentional centration three, we find three distinct necessary levels. At the most abstract we find APPLY.THEN, which during signalic processing functions as a sensorimotor executive (or complex signal). At the middle level we find the two operative expectancies, *SHAKING* and *GRASPING*, which stipulate how to implement the most abstract operative. At the lowest concrete level we have GRASPING and vis.rattle, the two concrete operative and figurative schemes that actually implement the task.

Theory of Schemes Applied: Organismic Explanation of Hierarchical Reinforcement Learning

There are two ways in psychology to formulate change: *development* proper and *learning*. Development proper brings change within the organism; for instance, change of schemes and of hidden operators or brain resources. It results from maturation patterns carried by genes (including, for mental-attention resources, its growth in infancy up to adolescence and then decay during old age). Learning (neuroplasticity) is a change within the organism resulting from life experience and training.

A dominant view in North American psychology and education emphasizes learning and often conflates development and learning to reduce the former to the latter. However, these two processes are distinct and complementary, interacting to effect growth in performance. As De Houwer, Barnes-Holmes, and Moors (2013) have emphasized, different formulations exist about learning that should be explicated (note that Tolman raised a similar point eighty years ago). We retain here the important distinction between learning as a *functionalist descriptive construct* that formulates *from an observer's perspective* "regularities" of behavioral and neurological change (i.e., what we have called functional invariants, recurrent invariances); versus the *organismic-learning* construct, aiming to explain changes in the psychological organism that *cause emergence* of these regularities. This distinction is useful, as shown by two established classifications of learning, namely: (1) *associative learning versus cognitive learning* (Holyoak & Cheng, 2011; Shanks, 2010) and (2) within associative learning, the distinction between *Pavlovian (i.e., classical) conditioning versus instrumental (i.e., operant) conditioning.*

Except Skinnerians, most researchers today see the associative-learning concept as expressing the two distinct (often coordinated) aspects we just mentioned: descriptive-functional changes in performance (i.e., behavioral/neural probabilistic regularities) versus change in the underlying organismic-causal processes, which along with external factors cause visible performance and brain changes. De Houwer et al. (2013) emphasized that the two aspects are needed and must be distinguished: regular (probabilistically invariant) changes in performance can index learning, irrespective of whether organismic-causal theories of associative or cognitive learning are correct. Currently the consensus (Gallistel & Matzel, 2013; Holyoak & Cheng, 2011; Mars et al., 2011; Shanks, 2010) is that associative and cognitive learning are complementary brain processes, jointly required for a complete account.

The second necessary classification of learning constructs is between Pavlovian (i.e., classical, respondent) conditioning versus instrumental (i.e., operant) conditioning. From the perspective of performance and adopting a functionalist descriptive viewpoint, these two sorts of learning are fully distinct. *Pavlovian conditioning* involves contingency relations between two figurative schemes (first and second stimuli), S1 and S2.*O* (e.g., the sound of the bell and the sight and taste of food—the latter experienced as reward because it carries a desired outcome *O*, i.e., feeding, in Pavlov's dogs). *Instrumental conditioning,* in contrast, involves contingency relations between an operative scheme R and a figurative scheme S1.*O* (for instance, the dog's motor act of standing on its rear legs *and* the food it gets as a reward). There is also in instrumental conditioning a later-acquired "discriminant stimulus" (to use Skinner's term): the cue S2, whose meaning is to signal that in this situation the association R-(S1.*O*) is likely to succeed. Notice that the example just given of appetitive (positive-outcome) learning is just one case. Aversive (negative-outcome) learning is another alternative case, not mentioned here for the sake of simplicity.

Note that, as Tolman (1959) emphasized, this kind of outcome learning tends to be *discriminative* in the sense that as the learner learns positive outcomes (with their contingency conditions and context), he or she also acquires, in a probabilistically contrasted fashion, learning of negative outcomes (and vice-versa) when contingency conditions fail to materialize and there is no expected outcome. Tolman called these (positive- or negative-outcome) learned connections S2-R-(S1.*O*) *expectancies,* or learned "means-end-relations" ("sign-Gestalts" or "means-end-readinesses"; Tolman, 1959).[8] Although rarely mentioned, Tolman's purposive/cognitive learning theory pioneered current reinforcement-learning theories (Klein, 1987). From this perspective, the two main descriptive forms of associative learning can be seen as two distinct forms of expectancy: the Pavlovian *figurative expectancy* [S1-(S2.*O*)] and the instrumental *operative expectancy* [S2-R-(S1.*O*)]. To enable effective performance, associative learning requires that figurative and operative schemes, together with suitable motivation (affective goals, affective schemes), be available to the learner.

As De Houwer et al. (2013) suggested, recognition of this "two-process" functional-descriptive theory of reinforcement learning (i.e., Pavlovian vs. instrumental conditioning and its corresponding organismic-causal theory) must be maintained. Indeed, such "two-process" formulation is universally admitted from a functional-descriptive perspective (Mars et al., 2011; Rescorla & Solomon, 1967; Talmi, Seymour, Dayan, & Dolan, 2008). However, from an organismic-causal perspective, the "two-process" distinction could be reduced to an organismic multiplicity of distinct causal processes common to both types of descriptive learning. Thus, Pavlovian and instrumental learning are different situation-paradigms, which have in part common organismic-causal determinants. This is supported by the finding that Pavlovian processes and instrumental processes affect one another in their manifestation (Liljeholm & O'Doherty, 2011; Rescorla & Solomon, 1967; Talmi et al., 2008).

The best-known illustration of this interaction occurs in the research paradigm called *Pavlovian-instrumental transfer* (PIT). This paradigm is achieved by training participants (animals or humans) on two distinct schemes or relationships: (1) a perceptual stimulus S1 associated with a rewarding outcome O_1; and (2) an instrumental action

R associated with a rewarding outcome O_2. These two rewarding outcomes can be distinct and have different affective goals, ag_1 versus ag_2 (symbolized in the $_1$ and $_2$ of O_1 versus O_2). After training the two schemes, participants exhibit increased instrumental responding during presentations of S1, relative to the responding found when S1 is not present and a neutral alternative stimulus S′ is present. These increases occur even when O_1 has a different affective goal than O_2. The PIT effect disappears when animals are satiated, which shows basic motivational determinants in PIT—an effect that hierarchical reinforcement theory cannot explain (Liljeholm & O'Doherty, 2011). Notice the importance of the PIT phenomenon. It shows empirically that instrumental and Pavlovian conditioning share the same organismic processes. TCO can explain why.

Metasubjective (TCO) Model for the PIT Effect

Unlike the S-R or S-S associations of reinforcement learning, constructivist schemes are self-propelling units applying to codetermine performance under minimal conditions of activation: A assimilates B, when B is construed as part of or related to A. This is an asymmetric "psycho-logical" relation between A and B, if they do not assimilate one another (Piaget's *reciprocal assimilation* or coordination), producing a schema or *L-structure* (a chunk). This mutual assimilation principle has a consequence: when a situation activates multiple schemes (as often happens), all activated compatible schemes tend to apply together to overdetermine performance (*SOP*). If they are not compatible, there is a competitive conflict and the stronger (most activated) *cluster of compatible schemes* will apply, suppressing noncompatible clusters. Affective schemes (either appetitive—affectively positive—or aversive—affectively negative) are also part of the SOP competition. This TCO principle explains the PIT (Pavlovian instrumental transfer) and other reinforcement-learning phenomena (e.g., latent learning, outcome selective transfer, conditioned inhibition).

To explain PIT effects, the needed schemes come from prior training: **(tr1)** The first training with S1 (scheme *S1) produces the scheme *$[S1 \rightarrow O_1.\ ag_1]$. In our notation star * indicates that this is a figurative scheme, and italics indicate an expectancy. Finally, square brackets with a sequence-forcing marker [→] indicate the expected consequence brought by scheme *S1; that is, outcome O_1 (serving affective goal ag_1) follows S1. In our formulas the symbol "→" indicates the act of *forcing* into actuality or bringing about O_1. **(tr2)** The second training produces R_1 with its operative (*OP*) expectancy $OP[R_1 \rightarrow O_2.\ ag_2]$. This second training also produces a discriminant stimulus for this *OP* expectancy, that is, the stimulus S2 (scheme *S2) that incorporates aspects of the context where the scheme in question can work (*S2cxt). The performance PIT results from the self-propelling (*SOP*) concurrent application of all these schemes. This is symbolized in formula **f1**:

R_2 [#*OP*[*S2cxt: $R_1 \to O_2$. ag_2*], #*[*S1→O_1. ag_1*], *S2cxt, *S1] ➔ PIT **(f1)**

That is, when *S1 and perhaps *S2cxt are present in the situation (and the animal is suitably hungry), scheme Response R_2 applies, because it has been mobilized by the two parameters (or necessary conditions #), which express the two learned schemes: relevant figurative expectancy (*[*S1*→ ...) and relevant operative expectancy (*OP*[*S2cxt: R_1→1* ...). The response application R_2 produces (➔) the performance PIT. Notice that ➔ stands for the *SOP principle* (which causes R_2). Thus, all available compatible schemes (in this example, *S2cxt, *S1, ag_2, ag_1, and the two expectancies) will summate (in a dynamic synthesis) their strength, combining compatible characteristics to cause the result of R_2. There is no misleading scheme in this situation, and training is provided until participants have learned all schemes. Thus, this is simple learning based on signalic (i.e., sensorimotor) processing, which after slow habit learning yields fast and effortless performance. No symbolic mental attention is needed; rats or babies of about 8 months with a mental attention of *Me*=3 (see chapter 3) could learn and perform this paradigm. Content-structural learning, however, may occur at very different levels of processing complexity. Consider our theory's classification of this complexity.

Developmental Levels of Affective-and-Cognitive Processing Complexity: *A, C, LC, LM, LCLM*

Affective (*A*) Schemes

We distinguish the *A*-operator from *A*-schemes as follows: *A*-operator is a general affective booster, a quantitative potency or power. *A*-schemes, in contrast, are specific sorts of affects, qualitatively distinct, which *A*-operator can boost as they are activated by the external or internal context. *A*-schemes implicitly or explicitly produce affective goals (*ag*) in tasks, which influence choice of strategies, planting seeds for *motives*. They also actively boost (*A*-boosting) relevant cognitive schemes active (or inactive) in the situation. *A*-boosting of cognitive schemes comes earlier and is possibly weaker than the *M*-boosting (mental attention) that cognitive schemes may receive under the direction of executive schemes. Cognitive schemes mediate information about world, experience, self, own or Other's capabilities, and so forth. They emerge in interaction with *A*-schemes whose affective goals boost or interfere with these cognitive schemes (including executive schemes). Thus *A*-processing (affective goals and their contextual cues) is a determinant of the *cue-function* (salience, induced activation) of experiential-content schemes (*C*-processing) and schemes at other higher levels (executives and knowledge/parameter schemas). Above the primal *A*-level are cognitive schemes, often organized in flexible levels of control (heterarchies) that work together by overdetermination. Cognitive schemes emerge developmentally as probabilistic functional regularities of situated experience or behavior.

Constructivist Learning and Schemes

We now discuss key levels of complexity from a causal-organismic perspective, classified by determinants that may cause their emergence (starting from local and concrete schemes). This presentation elaborates on Pascual-Leone and Goodman (1979), who attempted to formulate principles of associative-and-cognitive learning that result from dynamic interactions among features, habit learning, and inferential processes. These levels of complexity are causal-organismic but congruent with explanations found in current learning theories (Mars et al., 2011; Shanks, 2010). The learning principles are empirically grounded functional distinctions within our theory, which could help to explicate findings and heterarchies found in current reinforcement-learning theory, as illustrated above.

However, before discussing different basic forms of learning, we should clarify how we understand the very complex notion of *content*, which to us appears as an encoded initial expression (interpretation) of what has been called matter. Matter is Reality prior to any interpretation. Our sense of content corresponds to what neuroscience may call the input and its initial processing, first occurring in primary brain areas. It also relates, we think, to the most concrete substratum (the relational resistances) of diverse content domains in psychology and in our culture; the basic differential concrete substratum for (qualitative or quantitative) sensorial-perceptual and action processing of any sort. These are the initial differential constraints (resistances of all kinds) imposed by Reality to reflective abstraction and elaboration in any content domain (visual, auditory, motor, spatial, verbal/linguistic, social, or causal—dependency relations that predict change). These domains (undetermined in number) emerge when our reflective abstraction (thinking and learning) applies to various sorts of content for various kinds of praxis. For instance, Demetriou and Spanoudis (2018) demarcated some six to eight distinct domains (called *specialized capacity systems*) especially important for the study of intelligence, although they are not the only ones. We agree with the importance of these differentiated capacity domains but omit their discussion in this book. For us, they result from application of hidden operators and principles, reflective abstraction, and so forth on schemes generated by different sorts of agency/praxis. Our book is exclusively concerned with causal (and neuropsychological) processes of the working mind's context-sensitive strategies for agency/praxis.

Content Learning: C-Schemes As suggested, content (*C*-) for us is what neuroscience may call the received input and its immediate processing in the primary brain areas and limbic system. *C*-schemes are initial mediators of the content; they are the basic concrete substratum of experience and processing. *C* learning is the first cognitive learning level. It corresponds to what Piaget, Wallon, Werner, or J. J. Gibson may have called *local* (bottom-up) scheme *differentiation*. It also relates to what Piaget called accommodation of schemes to "physical experience," which causes *empirical abstraction*. These schemes represent or embody local functional invariants, that is, probabilistically recurrent and situated context-specific features or moves, concrete figurative or operative patterns. *C*-schemes, the products of *C* learning, express innate organismic (neurological) determinations for sensorial, simple perceptual, and simple motor processes and express local experiential resistances, driven by sensorimotor neurological readiness. According to Piaget and others, when experience violates expectations built into a scheme's prescribed invariant (i.e., violates regularities or lawfulness that this scheme represents) the scheme adapts or accommodates. This is what in chapters 1 and 3 we have called secondary or Piaget's accommodation; in it, schemes modify themselves when failing to assimilate experience, to bring the represented invariants in closer agreement with Reality resistances.

There is, however, another complementary sort of accommodation: primary or Gibson's accommodation (which he called *differentiation learning*). This is the neurological internalization of invariants via bottom-up neural adaptation or coordination (to encountered resistances) of relevant circuits. This primary accommodation may become incorporated into more complex circuits/networks and may enter the psychological organism (appearing as *C*- or *LC*-schemes that eventually could reach consciousness). During this accommodation process, the *fc, rc,* and/or *ec* of schemes change their characteristics to increase adequacy and viability of the schemes vis-à-vis local resistances of the attended-to reality/Reality. Functional aspects, conditions, or effects changed during *C* learning may themselves be psychological schemes, but they often constitute physiological (sensorial, motor, or associative) central processes never represented in the psychological organism, as explained in chapter 1 (this would be a primary accommodation).

For us, any psychological organism or metasubject includes regulations (hidden operators and principles) plus the repertoire of cognitive, affective, personal, or action schemes. Whenever accommodation change consists in coordinating two or more preexistent schemes (an instance of Piaget's secondary accommodation), *C* learning appears as an *asymmetric relation*. A main scheme h_i comes to *assimilate* (incorporate, making associative bonds with) another scheme h_j, which thus becomes subordinate to

h_i (h_i changing into $h_{i[j]}$). This change is likely to occur when this contingency repeats, and the change renders more efficient the application of h_i. This case corresponds, using a different notation, to both S1-S2.O_i learning (Pavlovian conditioning) and S1-R-S2.O_i learning (instrumental conditioning). In the second case, when *C* learning is not yet psychological, just a purely physiological process, that is, not yet present in the brain's psychological organization, the modified scheme h_i (i.e., $h_{i[j]}$) may come to embody conditions or effects that are *truly novel* for the metasubject. This is the intuition behind Gibson's concept of *differentiation* (practice without feedback, like in repeated wine tasting, can create possible discriminations and judgements that were previously inaccessible). It also helps to understand the centrality of trial-and-error learning, as well as practical learning (learning with practice and often without feedback), products of a purely experiential (not mentally reflective) know-how.

Our purpose in this chapter is not to review current work on learning or have proven rules for each level of learning. We wish to differentiate clearly *neuroscientific* (epistemo-ontological) *levels of learning*, formulating plausible differential rules (hypotheses), and show how change from associative learning to constructivist (cognitive-inferential) learning is a graded transition. The graded transition is caused by changes in the differential balance of organismic determinants (e.g., hidden mental-attentional operators), which all sorts/levels of learning share (Pascual-Leone, 2012a; see chapters 1 to 3). Learning is a constructivist *internalization*—a durable acquisition by the brain of repetitive, replicable patterns (external or mental) of performance constituents, whether qualitative or quantitative. These repeatable patterns constitute probabilistic functional invariants that can facilitate learners' "purposeful" activities. The various compatible levels of learning that we now discuss are complementary ways of pursuing internalization.

There are three conditions jointly needed for *C* learning to take place. **(C1)** First is a *high assimilatory strength* (i.e., a high degree of activation and propensity to apply; Piaget's concept of assimilation) of the main scheme h_i. This high activation and propensity to apply may be due to activated affective goals (*A*-schemes) or to sensorial salience, motor readiness, and so on. **(C2)** Second is a *lawful relation of coactivation* between the main scheme h_i and the subordinate scheme h_j (or, if h_j is not yet formed, its purely physiological processes that h_i may incorporate). Coactivation occurs between two or more schemes when their activation is physiologically interconnected, tending to occur simultaneously or in immediate succession, as a probabilistic invariant. **(C3)** Third is a *lawful relation of functional service that relates h_i with h_j* (or with purely physiological processes), meaning that within a given context the same affective/cognitive goals are served by both schemes. When application of h_j invariantly facilitates (or at least does not hinder) application of h_i, the first has a relation of functional service with the second. When such relation is mutual, we talk of a *cofunctional relation*.

Logical (*L*) Learning The *L* stands for *Logical* in the psycho-*logical* sense of embodying functional, relational structures (i.e., schemes, schemas) that emerge as generic, probabilistic invariants from one or other kind of external or mental experience. These emerging *L*-structures express some experiential (external or mental) functional characteristics abstracted as thinking proceeds. Notice that this concept of functional structure (i.e., scheme, schema) is very different from the quasi-architectural, passive concept of structure often found in cognitive psychology. This neo-Piagetian concept of functional structure, our *L*-structure, is expressing or abstracting an active way of functioning. As R. Miller (2011, p. 410) put it: "the term 'functional-structure' could be replaced by 'structure/way/form of functioning.'" It embodies coordinations or integrations of *C*-schemes or *L*-structures into more encompassing structures, often seen as more-or-less-complex packages of interrelated associative-learning schemas. Notice that *L learning* is like a recursive function: it can apply to coordinate its own products. The resulting *L*-structures/schemas are complex, perhaps superordinate, schemes that epistemically reflect integrated patterns of coordination found in agency/praxis across compatible schemes. Recurrent simultaneous or successive patterns of coactivation existing among schemes in the repertoire (i.e., long-term memory) are abstracted into *L*-structures when they recur sufficiently to generate probabilistic invariants in performance. Piaget explained this case of learning as *reciprocal assimilation* or *coordination* among schemes involved.

The *rc* of an *L*-structure has as conditions "copies" of releasing components in its different subordinate scheme constituents, and perhaps also patterns of coactivation among them. The *ec* of an *L*-structure is the various compatible and coordinated effects found among its scheme constituents when they apply. The *fc* should have as a functional totality its overall script (e.g., the summarized gist of a musical piece or a play, as intuitively felt by a working musician or writer). With the help of *L*-structures, the metasubject may achieve an internal *functional model* of schemes applicable to performance in frequent types of contexts and situations, progressively reaching a situated functional totality. *L*-structuring processes explain developmental emergence of cognitive-psychology constructs such as semantic networks, representations of distal objects (our *ob*-structures), procedural systems or subroutines (our *pro* complex schemes or structures), knowledge structures (our *ad*- and *ob*-structures of all levels of reflective abstraction), control processes (task and organismic-resource executives), complex scripts or operative/procedural models of performance, and so forth.

There are four necessary conditions for *L* learning. **(L1)** First is a *high assimilatory strength* of every one of the constituent schemes involved in the structure to be learned. **(L2)** Second is a *nearly equal assimilatory strength* among the constituent schemes involved. **(L3)** There must be a *lawful relation of coactivation* among the constituent schemes. **(L4)** Finally, an optional relation that seems to facilitate *L* learning (and may be necessary for *LM* learning, see below) is *a lawful mutual relation of functional service* among or between schemes involved. Thus, *L* learning is facilitated when schemes involved are cofunctional, and this relation is mutual among the schemes involved, that is, it is functionally *symmetric* (in contrast with *C* learning). Consider, as example, health food, such as ripe avocados, to replace butter on your bread. The beginner is shown how to open the avocado and how to use a knife to get some ripe flesh to spread on bread. If the person likes it, after few trials his or her feeding schemes will get changed: the scheme "eating bread with butter" is replaced by "eating bread with avocado flesh," and the expectancy of taste is replaced by the creamy taste and flavor of avocado. Soon this scheme h_i "eating bread with avocado flesh" strongly coordinates with the sensorimotor scheme h_j "sight of avocado and its flesh." Coordination of these two schemes is an instance of *L* learning, "reciprocal/mutual assimilation." Thereafter, when the person feels like having a snack, and the operative h_i (eat bread with avocado) becomes activated, he or she experientially anticipates taste and flavor of avocado flesh; conversely, when the person sees avocado or its flesh there is a temptation to eat it, which people with addiction may find difficult to avoid.

There are at least three compatible ways of generating *L* learning. **(L-learn1)** Overlearning/automatization of *C*-schemes leads to habituation and automatized patterning, cause of *L*-structures, as Tolman (1959) and other classic learning theoreticians (Harlow, 1959; Hebb, 1949/1961) pointed out, using different terminologies. We call this form of learning *LC learning,* because in it an *L*-operator (or logical/structural-learning process) applies on *C*-schemes. **(L-learn2)** When a set of cofunctional and often coactivated schemes is repeatedly boosted in activation with mental attention (i.e., our *M*-capacity resource or *M*-operator) the postulates **L1** to **L3** or **L4** are fulfilled, and they generate *L*-structures much faster. We call this form of learning *LM learning,* because it results from *L* applying onto *M*-processes. **(L-learn3)** Finally, a set of cofunctional and coactivated schemes may be strongly boosted in activation by affective (*A*-) schemes, that is, affective goals/motives. Such *A*-boosting could lead to overlearning, giving origin to an *L*-structure via *LA learning*. This case is more common in emotionally exceptional circumstances (e.g., personal crises, car accidents) or in people with difficult lives, great affective sensitivity, or emotional problems. We shall not discuss *LA* learning further but elaborate on the other forms of *L* learning.

Logical content (*LC*) learning (L-learn 1): This is *LC* learning or *L*-structuring via overlearning. If someone encounters many times the situations that coactivate a cluster of cofunctional schemes, which bear on affective goals, conditions will be set for *C* learning (as prescribed by **C1**, **C2**, and **C3**). Such learning occurs much without conscious control, particularly when this contextually grounded invariant is too complex to be consciously analyzed (e.g., Bach's musical style for a music lover; or the many cues helping a wood cutter to find his way through a forest, without trails, on the way to a certain cottage; or the know-how coming from much book writing or business practice). Repeated exposure to the same practice will tend to make coordination of *C*- and *LC*-schemes progressively stronger, causing emergence of probabilistically invariant patterns, which leads to *LC* learning.[9]

There are four important characteristics of *LC* learning. **(LC1)** Learning is quite slow and incremental. **(LC2)** The learning process is often implicit or tacit, that is, may take place latently and without mental effort. The resulting *LC*-structures correspond to what is often called *chunks,* but these are *effortless chunks*. **(LC3)** Because of the packaged (or associative-compound) nature of the *LC*-structure, when a package as a whole is relevant for a task, its activation will tend to become fast and effortless (because habit-driven), and its subjective (phenomenal) experience will be holistic, relational, and Gestalt-like. **(LC4)** Interference occurs whenever constituent substructures of an *LC*-package must be separately activated, because other schemes from this package are inconvenient for the task at hand. Indeed, constituent schemes/schemas of one *LC*-package may be functionally interlocked with unwanted contextual or context-driven schemes (cognitive or affective), connected to this *LC*-package during some learning process. As a result, *LC*-structures cannot be transferred easily across contexts or situations, because interference occurs easily. These structures make up much of the experiential knowledge or empirical experience of many individuals.

Tolman (1959) was first to point out that *LC* structural learning (called associative learning) occurs more easily under conditions of free-time repetition and low-to-average affective arousal. That is: to maximize *LC* learning it is better to satisfy in advance the subject's needs. Tolman worked with rats, and cognitive maps of rats are more likely to be developed when they are placed in a maze when well fed and anxiety free (analogically this is also true in humans). This makes sense, because under emotional disarray or stress not all schemes involved may be equally activated (violating postulates **L1** and **L2** given above), making *LC* learning more difficult.

***L*-structuring via *M*-boosting, or *LM* learning (L-learn 2):** Mental-attentional energy or effortful activation (i.e., *M*-capacity allocation, or working-memory's attentional centration) can produce rapid *L* learning when applied onto a set of cofunctional and coactivated schemes. This is because *M*-energy can boost the level of activation of schemes to reach a maximum level, which immediately brings all activated cofunctional schemes to a high and equal level of coactivation.[10] Such state of affairs fulfills postulates **L1**, **L2**, **L3**, and **L4** given above, producing rapid *L*-structuring. The time duration of *M-allocation* (or attentional *centrations*) required to cause such effect is a parameter in need of experimental investigation using *LM* learning paradigms. In our postulates of *LM* learning, we hypothetically estimate such time duration "qualitatively" in terms of the number of protracted *M*-centrations required to cause *L*-structuring of two schemes (more of them may take longer). We define a *protracted M-centration* as being a self-paced effortful mental-processing step.

We describe ten processing rules for *LM* learning. **(LM1)** If two cofunctional schemes are boosted by *M* during two (or three) successive and protracted *M*-centrations, they will form a superordinate *L*-structure that embodies or reflects them in their functional coordination. The learning concept of a very effective rehearsal (when rehearsal is not construed necessarily as verbal) is congruent with this principle. This *LM*-structure can then functionally replace the two original schemes in mental processing within the required *M*-centration(s), lowering the *M*-demand. That is, the needed expenditure of effort or *M*-energy will be reduced from two units to one. This postulate expresses the principle of *effortful chunking*. **(LM2)** The **LM1** process cannot be applied at once to more than two schemes, but it can be reapplied recursively until all schemes involved have been incorporated into the *LM*-structure, provided that at each two-schemes step there is a consolidation-process time that is cumulative. **(LM3)** *LM*-structures formed by the process **LM1** could be ephemeral unless they enter a consolidation process. Consolidation requires some amount of effortful re-attention to the formed *LM*-structure. In the context of process-task analysis, we have found that this consolidation process must be equal to one or more effortful and protracted *M*-centrations to have stable *LM*-structures. **(LM4)** The *LM* process may also occur with a single scheme or structure. In that case, the *LM* process may strengthen the structure formed (or create "replicas" of the scheme/structure in question, which increases its assimilatory strength and prevents forgetting).

(LM5) The ease of *LM* learning, or even its possibility, depends on the *M*-capacity available to the subject, relative to *M*-demand of the *LM*-structure in question (indexed by number of distinct separate schemes to be effortfully coordinated). When *M*-demand is small, relative to the subject's *M*-capacity, *LM* learning may be all-or-none or very fast. Thus, vis-à-vis *LM* learning there is a *trade-off* between the subject's *M*-capacity and the task or item's *M*-demand, which determines meta-subjective item/task difficulty. Subjects will tend to pass items when *M*-capacity is equal to or larger than *M*-demand, and tend to fail otherwise (Pascual-Leone & Baillargeon, 1994).

(LM6) This form of learning brings awareness of mental effort (James, 1892/1961; Kahneman, 1973) and tends to use conscious learning strategies with executive schemes monitoring it. Once learned, the resulting *LM*-schema will still need mental effort (i.e., *M*-boosting and/or *I*-interruption) to be fully activated and apply. However, because the *LM*-schema already has in it the appropriate interrelated constituent schemes, strategic application of this *LM*-schema will tend to be errorless and relatively fast, because selection by executives of the task-relevant scheme constituents is no longer needed.

(LM7) Figurative *LM*-structures (and their operative counterpart, complex operatives) are functionally autonomous, that is, discrete and usable across contexts with little interference from other schemes. Sequences of *M*-operations monitored by executive schemes are necessary to constitute and to apply flexible hierarchies of *LM*-structures and complex operatives. For instance, symbolic/mental executives require *LM* learning to emerge. This contrasts with signalic/sensorimotor executives that can be produced by *LC* learning, with the help of human mediation. **(LM8)** Interference with contextual schemes is minimized within *LM*-structures (and within complex *LM*-operatives). Indeed, because only *M*-boosted schemes are coordinated within *LM*-schemes/structures, and other nonboosted contextual schemes are automatically interrupted or inhibited, the latter are excluded from these *LM*-schemes. Thus, this sort of learning creates flexible and mobile schemes relatively robust to interference. High-cognition rational processes, responsible for the good problem solving and judgment in adults (what Piaget called "formal" mental operational structures) emerge in this manner, via *LM* learning. Piaget called *logico-mathematical experience* this sort of learning but failed to formulate the causal processes. With practice *LM* processes become overlearned and in part automatized.

Automatization of *LM*-structures or *LCLM* learning: Because they are automatized (overlearned), these *LM*-structures no longer need mental attention to be activated and applied. They do not involve mental effort (*M, I*). This level of learning allows emergence of mental operative models and refined cognitive maps, within high cognition. It requires two further processing rules. **(LM9)** Our three main types of learning (i.e., *C, LC,* and *LM* learning) are not only mutually compatible (each can apply on the products of the others), but they are often developmentally intertwined. They function as constructive operators of diverse types, which apply on each other recursively to generate hybrids such as *LC, LA, LCA, LMCA, LMLC, LCLM,* and so on. (Each resource-operator here applies to those on its right.) *LCLM* marks a fourth level of learning. As mentioned, it consists in automatization (*LC*) of *LM*-schemas/structures, supported by diverse scheme-learning types (*C, LC, LCA, A,* etc.). Because they are automatized,

LCLM-schemas no longer have to be boosted by *M*-power to apply; they are now effortless. This automatization may include the executive processes (*E*-schemes) that planned and regulated the *M*-operations now turned into *LM*-structures. Such process can form complex automatized high-cognitive *LCLM* functional totalities. This is a fourth level of processing because it enables peak performance of the working mind. This produces an *encompassing reason* that can take a cognitive "distancing" (Sigel, 1993; Werner & Kaplan, 1984) or *cognitive decoupling* (Evans & Stanovich, 2013) of high *LM*- or *LCLM*-schemes from more concrete, lower-level schemes that semiotically/symbolically relate (but are not essential) to the *LCLM* functional totality in question. This is needed to enable a distinct re-representation of equivalent meanings at different orders of relation (different levels of processing).

The constructive levels of processing expressed by our heterarchy of learning and scheme types (*C, LC, LM, LCLM,* etc., with their dynamic combinations) explain with organismic causal mechanisms currently recognized, important modes or ways of thinking (Evans & Stanovich, 2013; Kahneman, 2011). The types (or "systems") of processing, Type 1 and Type 2, are often characterized as *automatic* (*LC* or *LCLM* learned) versus *controlled* (i.e., *attentionally effortful* and executive driven). Evans and Stanovich (2013) do not formulate these two types as polar dichotomies, but rather as a sort of functional-structural continuum—two modes of processing that can be combined gradually as demanded by current cognitive goals. This useful cognitive characterization is a descriptive model, in need of a matching organismic-causal model to explain its causal determinations. The constructivist-learning theory just summarized, and explicated further in the next two chapters, generates organismic causal models of task processing consistent with the descriptions of Evans and Stanovich and of Kahneman.

(LM10) Because *LM* learning requires use of a mental attention, which increases with chronological age in children (Pascual-Leone, 1970; see chapters 1, 3, and 7), we should find as many distinct sorts of *LM* learning as there are levels of *M*-capacity. Notice further that the individual's obtained *M*-level is a combination of his or her age group *M*-capacity limit (i.e., the age-group's *reserve or structural M*—Pascual-Leone, 1970) and the individual's own ordinary or common *M*-capacity level of functioning (the individual's *functional M*). The various sorts of *LM* learning (and their resulting schemes/schemas) differ in the *M*-demand of tasks that can be solved with them (this *M*-demand is estimated via metasubjective task analysis). The developmental *M*-capacity/*M*-demand trade-off principle (mentioned in **LM5** and **LM10**) suggests that mental complexity (*M*-level) of the *LM* learning accessible to a subject changes with his or her developmental stage because of developmental growth in the *M*-reserve (e.g., Arsalidou et al., 2010; Pascual-Leone & Baillargeon, 1994; Pascual-Leone & Johnson, 2005). These developmental levels of *LM* learning correspond to substages of cognitive development, consistent with those of Piaget or Case, among others (see chapters 3 and 7). Such modeling of *M*-levels offers a quantification method for other current cognitive-complexity theories (e.g., the Type 1 vs. Type 2 modes of thinking), which lack developmental quantification and methods of process/task analysis.

LM-structures cannot emerge in an individual unless he or she has, in addition to the requisite *M*-capacity level, executive schemes needed to organize the complex processing required. Suitable task-relevant executive schemes are necessary to monitor sequential *LM* learning. These sequential executives are particularly needed because of scarcity: the limited power of subjects' *M*-capacity, relative to the complexity of high cognition. When *M*-capacity and good-enough executives are available, sequential *LM*-processes can be formed rapidly (which tend to incorporate few associative bonds with unwanted schemes). The result is relatively independent abstract structures (*LM-*, *LCLM-*) that remain detached from context and unwanted details; thus keeping a decoupling or "distancing" from other related schemes. This option, made possible by executive control, is very important in misleading situations (see chapters 6 to 8). Finally, because *LM* learning is recursive and can be applied repeatedly to create schemas of schemas of schemes, hierarchical reinforcement learning (Mars et al., 2011) can be reinterpreted as a by-product of schemes' constructivist learning in the organism.

Types of Reflective Abstraction That Schemes and *C*, *LC*, *LM* Learning Generate

Abstraction is the process of extracting from experience generic aspects (functional invariants) possibly useful for future acts of agency and praxis. So defined, abstraction is at the heart of learning. Abstractions that are not arbitrary are informed by, or express (epistemically reflect—epireflect), contents and relational characteristics of actual experience, indexing resistances encountered by the knower. Piaget called this epireflective abstraction *reflective abstraction* (see chapters 1 and 2).

There are three complementary senses or modes of reflective abstraction, which correspond to three dimensions of variation (and dialectical moments). The three modes are mutually compatible, complementary, and often interacting. Together, they characterize the working mind's processing complexity. These three modes of reflective abstraction are listed in table 5.2. This table also gives hypotheses about key regions in the cortex that, along with their subcortical interconnections, are important for the mode of abstraction in question. We discuss them successively.

Table 5.2
Modes of reflective abstraction and hypothesized key cortical regions

Modes of Reflective Abstraction			Brain Projection
Reductive (or Symbolic) Abstraction	Conceptual: Logological (or *LM*-structures)	⟷	Left Hemisphere
	Mereological	⟷	Left and Right Hemisphere
	Experiential: Infralogical (or *LC*-structures)	⟷	Right Hemisphere
Constructive Abstraction	Intellectual	⟷	Quaternary Area
	Intellective	⟷	Quaternary and Tertiary Areas
	Intelligent Perceptual	⟷	Tertiary and Secondary Areas
	Sensory Perceptual	⟷	Secondary and Primary Areas
Causal Abstraction	Executive	⟷	Prefrontal Lobes
	Operative	⟷	Frontal Lobes
	Figurative	⟷	Parietal, Temporal, and Occipital Lobes

Reductive Abstraction: High/Symbolic versus Low/Signalic

Empiricist thinkers often see abstraction as a construction of informational units (i.e., schemes and *L*-structures) that ignore or reduce task-irrelevant aspects of the experience and retain only relevant aspects. The more aspects of experience excluded from abstraction, the higher the level of reductive abstraction. Piaget called *logical abstraction* the high-reduction levels of abstraction, in which a selected-few relational invariants extracted from experience are retained (and many are omitted). This is the mode of high abstraction often identified with high conceptual processing. This *high-reductive abstraction* could also be called symbolic, because the symbolic function (not just signals) is needed to accomplish such abstraction form (which demands *M*-processing and *LM* learning). This sort of high-reductive abstraction comes in three distinct forms. One is pure *logological* abstraction, named to highlight that language emerges as a symbolic vehicle for communication by using this sort of high-reduction abstraction. Another form of high-abstraction is not directed to language but to explicate functionally invariant, generic aspects of objects, procedures, and adjunct information abstracted from experience; via visuospatial, temporal, causal, and more or less generic representations (experiential cognitive maps and intuitive models). This is what has been called *mereological processes* (from the Greek *meros* "part"—part-whole relations within objects and physical entities; Mieville & Vernant, 1996). By using logological structures adapted to stand for and represent experiential objects, procedures, and so forth, the mereological processes internalize (in more or less generic ways—e.g., object-centered models, cognitive maps) useful functional experiential invariants of objects and situations.

Finally, Piaget called *infralogical abstraction* (or empirical experience) the low-level, *low-reductive abstraction* that may be called *experiential processing* (if we exclude high-abstraction mereological schemas, which are also experiential but complex). Here, many particulars from concrete experience are internalized into scheme(s) or *L*-structure(s). This sort of experiential abstraction could be done by *LC* learning. Principles of learning discussed above make clear that reductive abstraction varies in a graded manner, with its logological or mereological poles demanding both *M*-processing and *LM* learning, and its infralogical pole demanding less of both *M*-processing and *LM* learning.

Table 5.2 suggests that deliberate, perhaps symbolic, conceptual/logological or complex mereological processes (involving mental attention, *LM*-structures, and complex schemes needed in misleading situations) tend to be stored preferentially in the brain's left hemisphere, in right-handed people. In contrast, automatic and often signalic, experiential/infralogical/mereological processes (involving automatic-perceptual attention, *LC*-structures, and usually automatized operative schemes—efficient in facilitating situations) tend to be located in the right hemisphere of right-handed people (Arsalidou et al., 2018; Pascual-Leone, 1989).

Constructive Abstraction

Constructive abstraction internalizes functional regularities that may have a wide scope of relevant experiences: complex probabilistic functional "invariants" (various levels of resistances, or ways for coping with them) abstracted across types of situations (across space and time, perhaps across conceptualizations). Whereas degree of reduction is a key mark of reductive abstraction, scope of referential domain is the keynote of constructive abstraction. The *scope of the referential domain* of a scheme is the set of different sorts or types of referent from which the scheme/structure in question has been derived as a functional invariant.

Congruently with distinctions in the literature under different names, we distinguish four modes or levels of constructive abstraction **(CA)**: **(CA1)** *sensorial-perceptual processes* deal with easy information, schemes activated in the here and now (i.e., the *immediate present*) or the very-close present-related events (practically, the "saliently given" present). This is basically the processing level that Piaget called *sensorimotor stage*, found in children and also in adults. **(CA2)** *Intelligent perception* deals with information closely related or linked to present events but stored in long-term memory (the *inferred present*).

This is basically the level of Piaget's *pre-operational stage.* **(CA3)** *Intellection* (intellective processes) deals with information bearing on both the present and the future. These are schemes or structures activated either by the present situation or by mediational processes (the *empirically possible,* remembered or expected present of executive and action schemes; Edelman, 1989) stored in the repertoire. Piaget's *concrete-operational stage* expresses this level of processing. **(CA4)** Finally, *reason or intellectual processes* deal with all sorts of events (present, future, and *formally possible*—noncontradictory). These are schemes of high-cognitive levels that may encompass or synthesize experiential-and-conceptual results from schemes with diverse modes and modalities of processing.

Table 5.3 summarizes these levels of constructive abstraction and shows which their key cortical regions may be, relative to the classic category modes or levels of processing: *primary* (initial areas of cortical reception/afference and of efference), *secondary* (locally coordinated, domain-specific, receptor or effector areas), *tertiary* (regional multimodal areas of cortical processing synthesizing information across domains, integrating some of their specific sorts of processing), or *quaternary* (also called *high tertiary,* the highest and most comprehensive multimodal areas, no longer specific).[11] It should be clear that, broadly speaking, as brain areas progress from primary to quaternary, the *M*-demand (*M*-levels) of the corresponding *LM*- and *LCLM*-schemes and structures

Table 5.3
Levels of constructive abstraction

→ **Conceptual** — Concrete Decentration

Luria: Brain Areas	**Neo-Piagetian Levels (TCO)**	**Phenomenological Categories**	**Piaget: Stages**
Quaternary	Intellectual →	Possible future →	Formal operational
Quaternary / Tertiary	Intellective →	Empirical future →	Concrete operational
Tertiary / Secondary	Intelligent perception →	Inferred present →	Preoperational
Secondary	Perceptual →	Given present	Sensorimotor
Primary ←	Sensorial →	Immediate present	Sensorimotor

Abstract Decentration — **Experiential** ←

tends to increase, up to the maximum *M*-capacity that the age group in question can afford. Thus, initiation of Piagetian stages may be related (in its stage-characteristic processes) to these brain areas, without of course excluding lower areas. Note that in table 5.3 levels of processing are assumed to be nested from low to high, in the sense that higher (more conceptual) processes are controlling, but largely founded on, the lower (more experiential) levels. Higher levels are functionally built and dependent on lower ones. Another implication of this model is that, relative to the age group of normal children, the higher a brain processing level (from primary to quaternary in table 5.3), the more errors in thinking/cognition, within misleading situations, could be related to the *M*-capacity limitation and executive deficiencies. When processing levels increase, task demands with regard to *M*-level, *LM* learning, and executive know-how should also increase, unless processes have already been automatized via *LCLM* learning.

Notice further that levels of processing suggested in table 5.3 result from dynamic/dialectical interactions among the three forms of abstraction indicated in table 5.2: reductive abstraction, causal abstraction, and constructive abstraction. These three forms of abstraction, in their intertwining, co-construct the working mind (Vygotsky's gradient from low cognition to high cognition).

Causal Abstraction

This is abstraction of functional invariants related to *change* (i.e., what leads to what; figurative or operative expectancies), as well as prediction of consequences when the needed antecedents are given. If reduction and scope are, respectively, the characteristic markers of reductive and constructive abstraction, the marker of causal abstraction is *control and prediction of change*. This modal dimension can be grasped (at least in complex situations) only when temporal sequences of action and change are coordinated with the situation and its changing spatial arrangements. Three sorts of schemes play basic, distinct functions in this mode of processing: figuratives (configural representations, state and process descriptors like our parameters), operatives (blueprints for and dynamic representations of action), and executive processes (encompassing processes of coordination). Causal learning can recursively apply at any level of constructed complexity as it dialectically coordinates with reductive- and constructive-abstraction processes to synthesize functional invariants that can control, predict, and/or postdict the change. Cognitive psychologists have recognized existence of causal abstractions but have usually failed to see this mode of processing as epistemically distinct from the other two modes, even though reduction and scope have very different processual characteristics than control of change. Indeed, control of change involves critically constructive/reductive abstraction of temporal sequencing—related to what we describe as the *T* (time) hidden operator in the next chapter.

6 Automatic Attention: Effortless, Perceptual, and Personal

Automatic attention processes and automatisms are examined, together with organismic processes that in normal individuals codetermine them. Such processes differ in facilitating versus misleading situations, as shown by brain dynamics. We look at organismic factors affecting automatic attention and automatism, such as the Schemes' Overdetermination of Performance (*SOP*—neuronal spreading of activation and convergence), the brain's "internal field" factor (*F*-operator), learning factors (*C, LC, LCLM, LA*), attentional inhibition, affective and personal factors, and automatic spatial and temporal resource-factors. When automatic and mental attentional processes combine harmoniously, a mental flow and efficiency appear in performance. We discuss contemplation and meditation as ways to optimize this flow in mental processes and performance.

> Attention as an activity implements a relation of a human to oneself, rather than to one's environment.
> —Dormashev, 2010, p. 309

> [B]y impression, and not by comprehension, is man primarily open to himself and other things.
> —Zubiri, 1966/2001, p. 195, translated by JPL

> [I]n the state of flow the goal of each action is the following action. *Action and awareness merge.*
> —Dormashev, 2010, p. 313

Attention is defined in the *New Oxford Dictionary of English* as "the mental faculty of considering or taking notice of someone or something: [e.g.,] *he turned his attention to the educational system*" (p. 108). This definition and example suggest that within common sense two assumptions are made: (1) the object-content of attention is always outside the person and (2) attention itself is always willful and mentally conscious. We disagree with these two assumptions, as do many psychologists and philosophers

(Bruya, 2010). As Dormashev's first epigraph states, and contrary to the first assumption, attention is best understood as applying activation to internal information-bearing processes, which we call schemes. Attention, as a source of added activation (we call it a "booster" function), should be studied "from within" the person. This is the constructivist perspective we have called metasubjective (Pascual-Leone, 2013; Pascual-Leone & Johnson, 2017). Further, "considering or taking notice" is an organismic function that precedes mentation, because it can be found in lesser mammals and birds, including animals that seem to lack a preconscious or conscious "functional organ"—a mind for mediated thinking.

This evolutionary perspective and the realization that attention must exist prior to consciousness (making possible, eventually, the emergence of both consciousness and self-consciousness) indicate that there must be two sorts of attention. As the reader knows, we distinguish *automatic/effortless attention* versus *mental/effortful attention*. The term "effort," introduced by William James (1892/1961), refers to the subjective experience of requiring careful focusing, together with the notion that there is some source of brain energy that functions as a scheme's activation booster. Both James and Pierre Janet (1889) identified this function with the Will, which heightens activation of whatever "idea" the person is pursuing (induced by affects or executive schemes). *Volitional attention* is James' name for the intentional act of applying this effort. "This strain of the attention is the fundamental act of will. And the will's work is in most cases practically ended when the bare presence to our thought of the naturally unwelcome object has been secured" (James, 1892/1961, p. 319). James later adds, "It is in one word an *idea* to which our will applies itself.... *Consent to the idea's undivided presence, this is effort's sole achievement*" (James, 1892/1961, p. 319).

For James, as for many classic philosophers beginning with Plato (e.g., Leibniz, 1966), "idea" refers to a unit of meaning (in our terms, a scheme or compatible cluster of schemes) stored in the person's knowledge repertoire. Thus, James' voluntary attention comes very close to the construct of mental attention (many would wrongly call it "working memory") effortfully mobilized by the executive (i.e., the currently dominant set of executive schemes) and allocated to suitable action schemes. James distinguished explicitly between primary or perceptual attention, which is effortless and more or less automatic, versus volitional attention. This distinction is the key early precursor of our own distinction between *automatic attention* versus *mental attention*. Another precursor was Pierre Janet, who adopted a dynamic developmental (psycho-genetic) perspective and influenced Piaget's theorizing. In his classic doctoral thesis book, Janet (1889) described dynamic variations of a person's functioning, as it is affected by two mutually complementary but contradictory organismic resources: *automatisms* versus *mental effort* (and mental energy). According to Janet, mental effort can expand the person's *field of consciousness* (Piaget's "mental centration," related to current notions of working memory) and so boost various distinct situation-relevant schemes (Janet's schemas) that together can bring about novel or "creative" dynamic syntheses, generating good performances and, in normal people, regulating unwelcome automatisms.

Automatisms (James' habits, good or bad relative to the person's life-coping or conscious intentions) are instincts, strong habits, affects, or subconscious sources of activity and agency—all usually self-propelling and difficult to regulate by the person's own conscious control. When persons are immersed in their automatisms, levels of active mental energy and effort tend to be low. However, performance, good or bad for the person, may still occur as a result of instincts, habits, or behavioral states occurring under low vigilance (sleep, somnambulism, catalepsy, narrowing of field of consciousness—e.g., amnesia, distraction). Claparede (1915/1946), the medical doctor and educational psychologist mentor of Piaget, explained the relationship between automatisms and consciousness as follows: consciousness does not usually occur until adaptation of the individual can no longer be achieved automatically, when a difficulty arrives that requires effortful attention. He then formulated a law of "taking into consciousness" ("lois de prise de conscience," 1915/1946, p. 206), saying that the more a person has automatized and used automatically (unconsciously) some knowledge relations, the longer it will take for the person to take it into consciousness.

It is important to emphasize that automatized processes, habits in particular, which may cause automatic attention, can be either good or bad for the person or animal's life and living. An example of good habits appears in what Claparede somewhere called the "paradox of the Will": when one needs the Will in praxis, one does not have it; and when one already has the will, one does not need it in praxis. One does not because, with practice, good executive and operative habits will have been developed that facilitate this particular activity, which may even cause a "flow" of performance, making exertion of Will unnecessary.

Bad habits are illustrated by psychological addictions. A nonuseful habit is illustrated by our cat, Luna. Luna previously was mistreated by a male owner. When we received Luna into our home, she was fearful, in particular of men, and she took more than a month to feel at ease and close to her owner, Janice. Soon after her arrival to our home, she developed the habit of jumping through the stair railings to escape (initially in panic) every time Juan climbed to the second floor and she was sitting at the top of the stairs. As Juan came close, she ducked behind him and jumped through the rails to run down the stairs. Now, after eight years, she is well adjusted and no longer afraid of Juan. However, she still feels compelled, when he climbs the stairs, to calmly stroll through the rails and walk down the stairs. Luna cannot correct this old bad habit.

Generalizing and simplifying a concept of Jung, we call unconscious and compelling habits such as this one *shadow schemes*. Shadow schemes have three characteristics: they are initially unconscious, they are compelling (often haunting us), and they often are distracting or misleading (although they can be facilitating in certain tasks). Shadow schemes may interfere with, bias, or prevent, executive-driven processes. Notice that conscious misleading schemes will remain shadow schemes if they are not fully mastered. Perhaps executive-driven processes of mental attention appeared in evolution to control or correct performance errors, biases, and difficulties induced by automatized shadow schemes. Already Janet and James seemed to contrast habits and automatisms to volitional attention and intentional processes, assuming the latter may control the former.

Ideas explicating aspects of James' and Janet's notions of attention have been articulated in neuroscience by Posner and his collaborators (e.g., Petersen & Posner, 2012). Posner distinguishes among three sorts of attentional control (or attention): alerting, orienting, and executive. His alerting and orienting attention are, together, instances of what we call *automatic* (perceptual, personal, or cognitive) *attention*, which often is effortless. His executive attention is what we call *mental or focal attention*, which is effortful. We prefer to call it mental and not executive attention (although executives do control this sort of attention as soon as they appear in ontogenetic human development), because mental attention seems to exist in babies after 2 or 3 months of age, whereas executives do not emerge until about 12 months (see chapter 3). Thus, mental attention already exists before becoming executive attention.

An example of an alerting task is cue detection used with humans, monkeys, and rats (Petersen & Posner, 2012). Such a task can separate effects of alerting/warning signals (telling *when* something is going to happen) from orienting signals (telling *where* it will happen). As described by Petersen and Posner (2012), to illustrate this difference one uses three different cue conditions presented prior to a target when a rapid response is required. These are an alerting cue (inducing general attention arousal), a non-cue, or an orienting cue condition (inducing modality- or location-specific arousal), prior to the response-releasing target. Brain activity corresponding to alertness (attention vigilance) and to orienting attention involves much participation of right hemisphere (e.g., Petersen & Posner, 2012), in particular in tonic (background, repetitive, persistent) processing that becomes overlearned. This is important, because it is likely that right hemisphere is where automatic processes tend to be stored and mobilized (see chapters 10 and 11; Arsalidou, Pawliw-Levac, Sadeghi, & Pascual-Leone, 2018). This finding agrees with our claim that Posner's two forms of nonexecutive attention are instances of automatic personal/perceptual attention.

In contrast, mental-attentional (or executive attention) processes may take place predominantly in the left hemisphere, where truly novel performances, particularly in problem solving, are synthesized (Pascual-Leone, 1989). Schemes difficult enough to be synthesized via problem solving may emerge as *ephemeral schemes*, that is, not-yet-schematized but momentarily synthesized or coordinated in neuronal activation patterns, in the left hemisphere. However, if the situation repeats and the ephemeral scheme becomes learned, it would make an *LM*-scheme in the left hemisphere; and then, with more repetition, when overlearned/automatized, it would appear in the right hemisphere, because processes of automatization occur more easily there (Arsalidou et al., 2018; Pascual-Leone, 1989; see chapters 10 and 11). The biological distinctiveness of automatic personal/perceptual attention versus mental attention is shown by another difference: whereas automatic attention processes use the neurotransmitters norepinephrine for alerting and acetylcholine for orienting (two neurotransmitters involved in processing external inputs), mental attention clearly uses dopamine as the dominant neurotransmitter (Petersen & Posner, 2012), which is more involved in feelings and mentation. We present a constructivist semantic model of these neurotransmitters in chapter 11.

Notice that the key function of automatic (cognitive, perceptual, personal) attention is to promote arousal in contexts that signal affective or perceptual-cognitive situations[1] (as when an event releases an orienting reflex; Howard, Johnson, & Pascual-Leone, 2014; Im-Bolter, Johnson, Ling, & Pascual-Leone, 2015). In contrast, mental or executive attention is mobilized to investigate and solve affective and cognitive problems in situations, perhaps brought up by prior arousal and orienting.

These two sorts of attention are complementary and may apply in coordination. However, when they apply in contradiction, effortful attentional inhibition is needed. An example of this sort of task is the antisaccade, an established attentional inhibition measure in the cognitive literature (e.g., Howard et al., 2014; Miyake, Friedman, Emerson, Witzki, & Howerter, 2000). In one version of this computer-based task, a fixation point (+) is presented in the center of the screen followed by a blank screen (50 ms). Then there is a brief appearance of a black square on one side of the screen, which elicits a strong, instinctive orienting reaction (or saccade) toward it. As instructed, participants must willfully inhibit this saccade to solve the task (Agostino, Johnson, & Pascual-Leone, 2010; Howard et al., 2014). Indeed, signaled by the square, the participant must immediately look to the opposite side of the screen to discriminate and respond to a target stimulus. Thus, alerting attention is automatically activated by the black square, which activates an orienting reaction to look at it. However, solution of the task prescribes looking in the opposite direction to find the fast-disappearing target and respond. Such cancellation of an automatic orienting reaction demands use of effortful/mental attentional inhibition that we discuss in the next chapter. Without this gaze control, the required response would not occur better than chance. This task illustrates situations when automatic attention is irrelevant or distracting and must be inhibited, by mental attention, to succeed. Problem-solving tasks, and other (cognitive or affective) misleading tasks, often present such competitive or conflict formulas. Automatic attention and mental/executive attention function complementarily within facilitating situations, but they compete in misleading situations.

Automatic Personal/Perceptual Attention Can Be Simple or Complex

Automatic attention is the normative way of functioning in familiar, facilitating situations. Recall that a situation is facilitating if it does not elicit schemes inconvenient to cognitive goals of the current executive (i.e., the dominant executive set of schemes) or the current affective tendency (affective goals or motives). In these situations, no misleading cues are present, and salient cues induce appropriate responses or interpretations. Effortless (and often subconscious) performance in these situations is enabled by schemes (perceptual, motor, affective, figurative, operative, and executive) already coordinated by associative learning (*C, LC, LCLM,* see chapter 5). In facilitating situations, because schemes are self-propelling (Piaget's assimilation function—spreading of activation in the brain circuits), performance occurs easily, being overdetermined by all activated and compatible schemes. Mental effort is not needed: subjective consciousness may occur (the "core self" of Damasio, 2012), but self-consciousness proper is often not there (see chapter 3).

When experiential-and-conceptual content is well practiced and overlearned, an often subconscious automatic mode of performance occurs because misleading schemes are no longer elicited (i.e., difficulties were ironed out by prior learning). In this case, both personal schemes (which combine within themselves affects and cognitions—often called feelings or emotions) and purely cognitive schemes (sensorial or representational perception, mentation, planning, or action) are together, activated or synthesized in the "flow" of subjective, experiential activity. This processing occurs along a continuum from unconsciousness to more or less explicit consciousness, but usually without self-consciousness. The partly or fully compatible schemes involved do not compete but combine to overdetermine performance, producing effortless attention that is automatically personal-perceptual-cognitive.

Cognitive complexity demands resources such as mental/executive attention to open to consciousness schemes that are not spontaneously salient or dominant in the inner or outer situation. Freud or Jung's construct of an *active unconscious* may now be interpreted (as Piaget and others did) as a *field of activated schemes* of any sort, within the person's *repertoire of schemes* (i.e., long-term memory), together with the hidden resource operators and principles that regulate the schemes' function. Effortful mental attention and suitable executive schemes often are needed to contain or control manifestations of unconscious self-propelling schemes, which haunt consciousness (i.e., shadow schemes).

Intelligent-perception, personal (i.e., affective and cognitive) insights, and reflective intellective or intellectual (high-cognitive) thinking are tools in this control and clarification of the dynamic unconscious. Thus, automatic-effortless attention (often driven by unconscious schemes and hidden operators) and mental-effortful attention (often driven by conscious executives) are in dialectical opposition. Together they regulate schemes' access to consciousness, as Claparede's "consciousness law" anticipated. This common idea of dialectically contradictory processes regulating access to consciousness owes much to Freud's and Jung's work (if deconstructed).

Freud distinguished primary from secondary modes of processing (e.g., Rapaport, 1960). A sort of liberalized Freudian primary processing (automatically self-propelling schemes) recently has been described as Type 1 or System 1 thinking (Evans & Stanovich, 2013; Kahneman, 2011). The dialectical counterpoint of Type 1 thinking is Type 2 thinking, produced by executive-driven mental-attentional intellective processes (Freud's secondary processes). When, through experience and mental growth, automatic (primary) and initially effortful (secondary) processes evolve to be coordinated harmoniously within given situations, a new high point of mental processing is reached: easy and spontaneously well-adjusted performances. When that happens, intellective/intellectual personal (affective and cognitive) processes can occur quasi-automatically, because they are often overlearned (our *LCLM* learning). Thus, complex and potentially misleading activities (motor or otherwise) can become easy and well controlled with insightful practice. This is the mode of goal-aligned activities and experiences called *flow* (fluent performance) by Csikszentmihalyi (Bruya, 2010; Csikszentmihalyi & Nakamura, 2010). This flow experience within the domain of action, although different, has interesting convergences with the sort of state often described as *contemplative meditation*.

Complex automatic attentional processing, such as flow, may lead to personal activities embedded in feeling (perhaps unconscious affects, calm emotions). This may bring up into performance an important bipolar dimension of feeling that has been called impression versus expression (Cassirer, 1944/1966, 1940/1996; Zubiri, 1999, 1966/2001). *Impression* is the state of being affectively or emotionally touched by feelings spontaneously induced by the encountered situation, object, or event. Any impressive experience is as such, with its own distinct characteristics of feeling that correspond to the particular reality (objective and subjective, cognitive and affective) encountered. Zubiri (1999), the Spanish philosopher, called this state *impression of reality*. He considered it a foundation (a Reality-based aspect) of intelligence; for this reason, he characterized intelligence as *feeling-sentient intelligence* (Zubiri's "inteligencia sentiente"). For him, this impression function was essential: "By impression, therefore, and not by understanding, is how a human is at first open to himself and to other things ... an essence that is an opening" (Zubiri, 1966/2001, pp. 195–196; JPL translation).

This impression function is dialectically intertwined with *expression* (Cassirer, 1929/1957, 1944/1966, 1940/1996), an open representation or manifestation in direct acts, art, poetry, communication in flow, and so on, of a person's feelings and thoughts. Expression can be signalic or symbolic. It very often is iconic (epistemically reflective—epireflective), that is, configurally congruent with aspects of the Reality encountered and its inner-feeling states. Expression, as a mode of communication or artistic creation, takes preeminence in history of art as *expressionism* (forms, colors, relational compositions that impressively induce ideas and feelings, which artists unconsciously or consciously "want" to express). Notice again that impression and expression use iconicity, the function of semiotic representation that Peirce described (see chapter 4) and which we have explained as produced by the neo-Gestaltist internal field-factor (*F*-operator), together with the principle of Schemes' Overdetermination of Performance (*F-SOP*, see chapter 1 and below). Spontaneity and openness of impression/expression dispositions often are found in spontaneous low-cognition personal processing, as happens with babies and young children. It also can be found, however, within (often personal) high cognition, in some sensitive or creative adults—artists, or people caring for others, often in spiritually developed individuals.

In the remainder of this chapter, we examine brain resources and generic processes that can cause automatic attention in the organism, fostering coordination with mental attentional processes, until flow and contemplative meditation become possible. We begin with differences in the cortical pathways monitoring attention. We then revisit the neo-Gestaltist field factor and discuss the space-structuring causal factor, the time-structuring factor (also involved in object identity), affects and personal emotions, and learning resource-factors, all of which together can explain emergence of flow and meditative contemplation.

Automatic Attention in Facilitating Situations versus Mental Attention in Misleading Situations: Their Differential Expression in Brain Processes

As mentioned, facilitating situations are those in which all schemes activated are relevant or neutral (not inconvenient or misleading) for the task at hand. Misleading situations, in contrast, activate schemes distracting for, or contradictory with, the task at hand.[2] In misleading situations, because competing schemes must be suppressed and controlled, mental/executive attention is necessary. In facilitating situations, automatic attention (with automatized schemes) may suffice for task solution, although mental attention also is used.

Lower lateral (or ventral-lateral) cortex suffices when the task is facilitating and all activated schemes are relevant for the task, which may allow automatic processing (particularly with the right hemisphere). Middle (dorsolateral and often prefrontal) cortex is involved in misleading situations, particularly if they are complex. A phylogenic study of brain evolution (e.g., Striedter, 2005) shows that dorsal pulvinar (in the thalamus) and lateral prefrontal cortex are more developed in humans than in other primates. This is consistent with the importance in humans of both low-lateral prefrontal areas (complex automatic-attentional processes) and dorsolateral prefrontal areas (complex effortful attentional processes, often in misleading situations, a site of truly novel performances, Striedter's "unconventional" performances, p. 329ff.).

There is now recognition (e.g., Christoff & Gabrieli, 2000) of different prefrontal areas activated according to the complexity of tasks at hand. (1) Facilitating situations that demand more or less complex, automatic attentional processing engage low-lateral prefrontal regions. (2) Misleading situations, which involve complex and effortful attention to produce truly novel performances, engage the dorsolateral (middle) prefrontal cortex. (3) More complex and temporally deep mentation, often engaging the future, involves the anterior region of prefrontal lobe (frontopolar or Brodmann area 10). This graded arrangement of prefrontal regions for processing progressively more complex cognitive tasks illustrates two important ideas. First, automatic-versus-mental attentional processes are distinct and complementary. Second, there is in the brain a functional local-gradation of sites related to both degree of interference found in the task at hand (having more or fewer distracting or misleading schemes) and the spatial-temporal (or computational) complexity of the task. In every case, the more affect and personal/emotion schemes are elicited by the task/situation, the more the lateral prefrontal activity is accompanied or preceded by activity in medial and ventral cortical regions. Further, the more automatized or habitual is the required process (and the less it demands truly novel performances), the more this processing will engage the right hemisphere—the site where, we believe, automatized schemes are preferentially stored (Arsalidou et al., 2018; Pascual-Leone, 1989).

Schemes' Overdetermination of Performance (*SOP*) and the Brain's Internal-Field Factor (or *F*-Operator Resource) as Causes in Automatic Attention and Automatisms

To explain automatic attention in perception and mentation, we must assume existence of self-propelling schemes that apply together to codetermine actual performances, including representations. Again, we call this co-application the principle of Schemes' Overdetermination of Performance (*SOP*). This dynamic resolution is indexed in the brain by spreading of neuronal activation over interconnected neurons, an activation that converges to form an actual and momentary neuronal *final common path,* a variable-path outcome first discovered with motor neurons but occurring with all dynamic brain-activity resolutions.[3] First formulated for human motivation and performance by Freud, the *SOP* principle is consistent with current neuroscience. We describe *SOP* as a causal-organismic construct using five rules. These rules (here with schemes) resemble the interrelations found among people who form dynamic groups of opinion in meetings, politics, or society.

(**SOP1**) Schemes activated within a situation, when they are semantic-pragmatically compatible to codetermine performance, coordinate into functional *compatibility clusters* because of the neurons' dynamic grouping. This is Edelman's (1987, p. 169) "group selection."

(**SOP2**) Every compatibility cluster has a different activation weight (assimilation propensity or strength) that combines/adds the assimilation strength of its members in the current context situation. This is their *total assimilation strength.*

(**SOP3**) Compatibility clusters that are mutually contradictory in their application compete to codetermine performance by applying their schemes and excluding schemes of other clusters.

(**SOP4**) Actual performance emerges by progressive application of different clusters of schemes to the task or situation (via application of, assimilation by, all compatible schemes activated in the situation). Priority order of their application results from the relative magnitude of clusters' total assimilation strength or propensity: stronger clusters will apply first, pre-empting application of other contradictory clusters—group competition. Subsequently, other applying clusters adapt their expression (they accommodate) to be compatible with the scheme clusters that have already applied to inform ("inject" form into) performance. As a result of this repeated process of selection-competition-adaptation, the to-be-produced final performance emerges with many diverse characteristics contributed by the successively applying clusters. Let us call *characteristic set of constraints* (or degrees of semantic-pragmatic "freedom") the variable total array of defining compatible characteristics that together will make up the actual concrete performance being produced (these characteristics should change with the *SOP*-selected scheme clusters, or with their total assimilation propensity). Thus, progressive application of different compatible clusters to complete performance adds semantic-pragmatic constraints (subtracts degrees of freedom) to the possible performance—until the performance (action, representation, or mentation) is fully actualized. Actual performance is overdetermined by all compatible characteristics available in schemes currently active in the person's unconscious or conscious mind (his or her total repertoire of schemes).

(**SOP5** or *processing-F rule*) Schemes that overdetermine performance combine by accommodating one another according to two rules. First is by minimizing (*F-min rule,* Pascual-Leone & Goodman, 1979, p. 314) the number of distinct schemes in the resulting performance (i.e., reducing redundancy among schemes applied). Second is by concurrently maximizing (*F-max rule*) the semantic-pragmatic and structural connectedness, the hierarchical organization and integration, of schemes that overdetermine this performance (including figurative representations). Less explicit versions of this *mini-max rule* were pioneered by Gestalt psychologists and neo-Gestaltists (*Pragnanz principle,* Koffka, 1935/1963; Rock, 1983; *minimum principle,* Attneave, 1959; Hochberg, 1964; *stimulus-response compatibility,* Proctor & Reeve, 1990; *field factor,* Piaget). This rule can be explained by what is called neuronal *lateral inhibition* in the cortex, which functions as the brain's *field (F) regulation operator* that automatically biases performance in this mini-max manner (conjointly with causal overdetermination, *SOP*).

Consider, in a simplified way, how this cortical lateral inhibition functionally affects the integration of scheme clusters. Activated neuronal circuits and networks (clusters) that carry and transmit relevant information also make lateral contact with inhibitory interneurons, which will tend to inhibit neighboring competitive circuits and networks. Consequently, activated neurons and circuits that fail to receive enough activatory neuronal connections (perhaps because they are not within a strong-enough compatibility cluster) will tend to lose degrees of activation and eventually be automatically inhibited. Thus, this lateral inhibition will preserve mostly neuronal circuits that carry semantic-pragmatic features rich in compatible/functional connections with other related circuits. Analogously to what happens with meaning within social groups (or within hereditable features in our genes), aspects (neuronal connections) that are not "popularly" shared (in the brain, across neuronal circuits), that is, not compatible with related meanings or features held by other brain circuits (or other people, in the case of social groups) will tend to drop out from competition; they will not contribute to define semantically the final common outcome.

A good example of this *F*-resource mini-max rule within figurative representation and perception was offered by Kanizsa (1955; Rock, 1983). Figure 6.1 shows sample Kanizsa triangles. Two equilateral triangles appear in an area with incomplete solid circles when three corner angles are added (left-hand figure). The triangles appear as ghosts, integrating and bringing closure to these incomplete parts, making gaps appear as occlusions produced by the ghost triangle. This automatic and effortless perceptual transformation that perceptually "creates" the triangles (caused by the *F* regulation operator) minimizes complexity of round parts, turning them into occluded full circles and resolves the three separate corners into another triangle partly occluded by the main ghost triangle. Thus, the number of distinct objects/aspects has been minimized (*F-min rule*) and, simultaneously, connectedness and integration (hierarchical organization of this perceptual whole) has been increased (*F-max rule*). Local neuronal lateral inhibition (cause of our *F*-factor) explains this effortless perceptual transformation that automatically creates Kanizsa triangles.

Similar automatic mini-max rules of transformation appear in motor action. A good example is found in the inferential water-level task (WLT; Morra, 2008; Pascual-Leone, 1969, 1989; Pascual-Leone & Morra, 1991) that originated with Piaget and Inhelder. In WLT the child is presented with an actual or drawn tilted bottle and asked to imagine that it is half filled with water and to draw the water line and mark the water location. Young children distort their water line, influenced by the tilt of the rectangular bottle shape. A related effect (the influence of the immediate irrelevant context on the

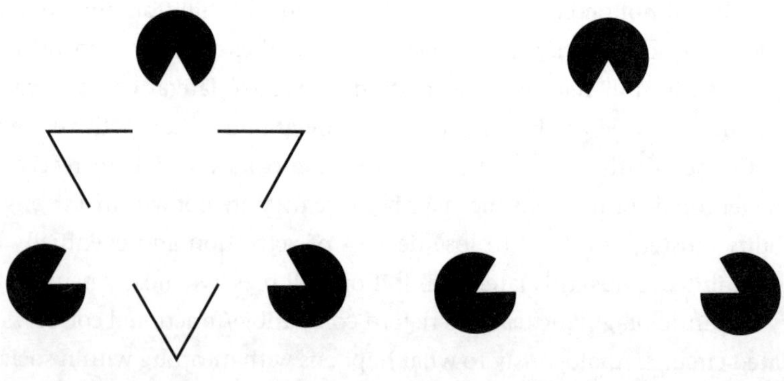

Figure 6.1
Sample Kanizsa triangles.

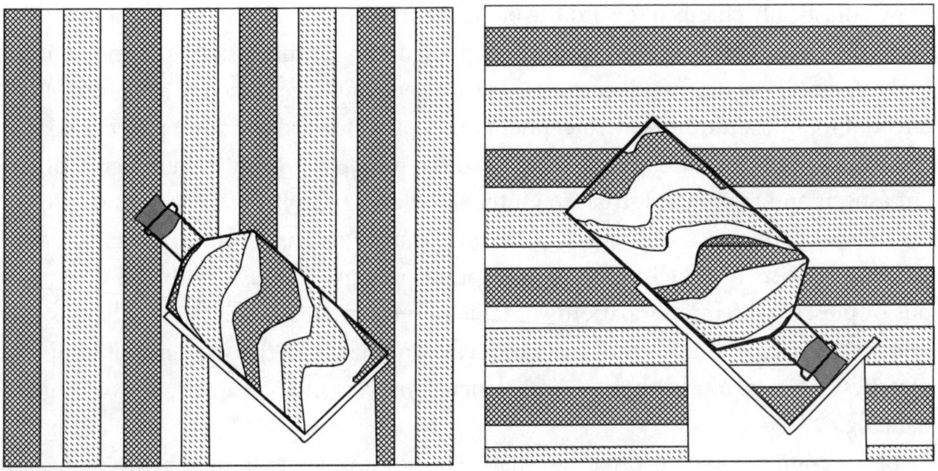

Figure 6.2
Illustration of the water-level task and copy-the-stripes task.

response) can also be seen in a purely perceptual-copying task, the copy-the-stripe task (CST) of Pascual-Leone, which can be used as noninferential task control for the WLT (Pascual-Leone, 1969, 1989; Pascual-Leone & Morra, 1991). Figure 6.2 shows sample items used in the WLT and the CST. Note that in the CST, the stripes are distorted when seen through the actual glass of the physical bottles. Thus, when older children are asked to copy the stripes as they see them through the bottle, they copy the stripes as wavy, not straight (although the background stripes are straight).

The actual physical bottles presented in either task were tilted 45 degrees to the right or left, vertical, or sideways. As mentioned, in the CST the task was always to copy as they saw them, on a paper outline of the bottle, the thick stripes participants could see through the bottle. In the WLT, which had bottles with and without a striped background, participants had to imagine the bottle half-filled with water and draw the water line on a bottle outline, placing a mark for the water location. WLT is inferential, a problem-solving task, whereas CST is a simple copying task. In both, however, similar illusory effects occur. In WLT there was a significant tendency in children and in field-dependent adults (Pascual-Leone, 1989; Witkin & Goodenough, 1981) to draw the water line tilted, making it more or less parallel to the bottom, the walls, or corners of the bottle. These drawn water line patterns covaried with the bottle's orientation.

In items where WLT had the striped screen behind (as illustrated in figure 6.2), the water line drawn by most participants was also influenced by the stripes. A related, even stronger effect occurred in the CST's copied line (Pascual-Leone, 1989). CST lines drawn drastically changed toward the background stripes' (vertical or horizontal) dominant direction, with the wavy patterns being ignored. In this task there is no conceptual inference, the illusory distortion is clearly perceptual, yet (like the inferential WLT effect) it follows a mini-max *F* rule: lines are distorted toward the dominant contextual features of bottle or background stripes, and the response minimizes number of different aspects and maximizes structural integration and "simplicity" of the whole.

Another common motor effect of the mini-max *F* rule is found in dancing. Good dancing consists of matching temporal sound patterns of music with the spontaneous or prescribed temporal patterns of dancing movements. A key process here is this mini-max *F*-factor, which effortlessly induces congruence in the sequence of temporal matches between music rhythm and dancing movements, ensuring flow and good dancing.

Let us examine now how this automatic *F*-factor intervenes in mental problem solving. Consider the classic liquid-transfer task, wine and water problem (Pascual-Leone & Johnson, 2017; the following description includes a small correction in task analysis). This is a brain teaser, popular on Internet sites and perhaps first described by W. W. Rouse Ball (1905). The qualitative findings are already well established (e.g., Case, 1975). A common version of this problem asks participants to imagine two containers, one (C1) with only water (Wa) and the other (C2) with only wine (Wi). The problem assumes (although this is not needed) that wine does not initially contain water. A spoonful (S1) of wine is taken from C2 and transferred to C1. A spoonful (S2) of the C1 mixture then is transferred to C2. *Is there now more wine in the water container* (Wi[C1]) *or more water in the wine container* (Wa[C2])*?* If unfamiliar with this problem, readers may try solving it before reading on.

There are different strategies for solving this task, both quantitative and qualitative. The basic one that we shall mention is qualitative and metasubjective, focused on the core causal process: the content carried back and forth by the spoon, which brings about change. This task is *misleading*, because a superficial look at the spoon transactions suggests that there is more wine in the water container than water in the wine container (i.e., Wi[C1] > Wa[C2]), because S1 carried only wine, whereas S2 carried both wine and water. Such incomplete analysis is the most frequent error response that we and others have found. A deeper spoon-sequence examination shows that, because the same spoon is always used, the quantity of wine that S2 carried is the quantity of water that S2 did not carry, and the water that S2 carried is the same as the wine in C1 that S2 did not carry. In other words, the wine in C1 after the second transaction (S2) is the capacity of the spoon (S) minus the wine returned in S2; and the water in C2 after the second transaction is the water carried by S2, which is the capacity of S minus the wine returned in S2. Consequently, wine remaining in C1 is quantitatively the same as water in C2 at the end of the transaction (i.e., Wi[C1] = Wa[C2]). The algebraic form of the state description of this analysis, after the two (S1, S2) spoon transfers, is

$$(\mathrm{Wi[C1]} = \mathrm{S1} - \mathrm{Wi[S2]})\ \&\ (\mathrm{Wa[C2]} = \mathrm{S2} - \mathrm{Wi[S2]})\ \&\ (\mathrm{S2} = \mathrm{S1}) \rightarrow (\mathrm{Wi[C1]} = \mathrm{Wa[C2]})$$

To understand this algebraic derivation organismically, we must imagine "from within" (i.e., metasubjectively) subjects in the mental act of problem solving. They have to represent intuitively essential task components by means of schemes. The three main constituents of this algebraic derivation are Wi[C1], Wa[C2], and S (spoon). They stand for three complex schemes that subjects must synthesize dynamically before a higher-level synthesis can coordinate them into the task solution. Notice that the algebraic formulation shows well why these three schemes are complex (having distinct coordinated constituents at different hierarchical levels). Indeed, each scheme mentioned in the equation requires an equality sign to be properly represented.

Six distinct constituents (subordinate schemes) may have to be boosted with mental activation to ensure the synthesis of task solution. These six symbolic constituents are:

$$\underline{\mathrm{S2} = \mathrm{S1}}$$

$$\underline{\mathrm{Wi[C1]}} == \underline{\mathrm{S1} - \mathrm{Wi[S2]}}$$

$$\underline{\mathrm{Wa[C2]}} == \underline{\mathrm{S2} -}\,\underline{- \mathrm{Wi[S2]}}$$

In this representation, the underlines demarcate separate constituent schemes. The markers (= and –) are repeated to indicate semantic-pragmatic connections between the schemes. The three-line sequence demarcates the three sets of coordinated schemes: constituents of the three complex schemes Wi[C1], Wa[C2], and S. We have written the scheme Wi[C1] with two constituents (marked by underlines), whereas we write scheme Wa[C2] with three constituents, because we assume that process analysis-and-synthesis of Wi[C1] took place before, and its constituents are now partly chunked. In contrast, chunking in scheme Wa[C2] should be more confusing, since in it Wi[S2] is an integral part of S2, now considered separately. Such mental demand of six symbolic schemes to be coordinated can be handled by children only when formal operations begin, at about 13 to 14 years of age, as we shall discuss in chapter 7.

The role of the brain's internal-field *F*-resource in this problem and its solution can be seen in our algebraic analysis. Indeed, repetition of Wi[S2] within two different contexts (i.e., S2 – Wi[S2] and S1 – Wi[S2]) shows that the same entity Wi[S2] takes two distinct meanings in the analysis: *water not carried* and *wine carried*. This state of affairs strongly contravenes the *F*-operator's simplicity constraint or mini-max *F*-rule (i.e., a minimum principle here describable as "when one scheme, one meaning (and vice-versa)," which corresponds to an instance of stimulus-response compatibility, in which the scheme is seen as stimulus and its meaning as response). Notice that this organismic constraint to assign to every scheme just one meaning was formulated in chapter 5 as postulate of schemes **Sch4**. The failure of the wine-and-water problem to maintain this particular constraint induces a major misleading factor (a brain bias against retaining both meanings of Wi[S2]), which is the functional core of this problem. This misleadingness imposes a need to use mental, effortful attentional capacity to maintain two semantically distinct schemes for the same entity, thus clarifying the issue for the subject who can reach the solution synthesizing six schemes. This task's solution demands the effortful mental-attentional processes that we discuss in chapter 7. Otherwise, during problem solving in misleading situations (such as this task), automatic attention (and its factors, such as *F*) would lead to error. This need to inhibit misleading automatic factors may be a major reason why effortful mental-attentional resources evolved, as Janet (1889) may have implied.

Learning Factors and Automatic Inhibition Involved in Successful Automatic Attention and Action

As Dormashev (2010) indicates in the second epigraph, attention (whether automatic or mental) carries activity of reflective, when not self-reflective, focusing. Attention highlights or activates schemes suitable for the current experience, available in the person's repertoire. These schemes tend to be adapted to the present circumstances by the participant's executive choice and by both *F*-operator and *SOP* principle, among other processes. Thus, attention, including automatic/effortless attention, expresses the working of the mind (i.e., mentation) and only indirectly informs about the experienced environment, as the first epigraph claims.

Automatic attention and automatic action are mediated by the repertoire of schemes. This repertoire was generated through personal life experiences by *C* learning and *LC* learning. *LC* learning includes *LCLM* learning. These often-complex schemes are part of the repertoire of mental habits that creates our "know-how." Know-how is knowledge that may encompass automatized high-cognition (e.g., *LCLM*), available for use without the need of problem solving and mental attention. These automatized complex schemes (schemas) are the foundation of spontaneous life experience and knowing. They are the data of our *intuition* (based largely, but not entirely, on unconscious automatized thinking). In intuition, affective and personal/emotive schemes are as important as pure cognition.

Evolution has created two different resources to prevent automatisms from interfering with adaptive action or mentation. The simplest one is automatic attentional inhibition, which complements and partly controls excess of automatic attentional activation. The second resource is effortful attentional inhibition/interruption (our *I*-operator), a component of mental/executive attention. We discuss mental/executive attention in chapters 7, 10, and 11.

Thus, we distinguish between *effortful* and *automatic inhibition* (Howard et al., 2014; Johnson, Im-Bolter, & Pascual-Leone, 2003; Pascual-Leone, 1984). Other researchers have made a similar distinction (e.g., Munakata et al., 2011). The implicit "choices" of the brain are driven by the most highly activated (dominant) cluster(s) of compatible schemes that applies. However, when this solution is not congruent with the currently dominant executive schemes, inhibition of this solution may take place, automatically or effortfully, or both. Automatic inhibition spontaneously deactivates mental operations or schemes found outside the zone of controlled effortful attention (i.e., outside the field or focus of centration); this outside-of-attention set of schemes lacks clarity of consciousness but may still have a *horizon* (or possibility) of meaning, as Husserl (1973) indicated.

Such automatic inhibition reduces interference of distractors and increases mental/executive focus on task-relevant schemes (Arsalidou, Pascual-Leone, Johnson, Morris, & Taylor, 2013; Pascual-Leone, 1984). Schemes within the focus or field of mental attention (Piaget's centration) emerge as dominant for determining performance, using help from automatic inhibition. Automatic inhibition is released when alternative (task-irrelevant) competing schemes making incompatible courses of action are simultaneously activated. In misleading situations, like those typical of inhibition tasks, correct performance demands that task-relevant schemes be hyperactivated by means of effortful mental attention, while task-irrelevant schemes are inhibited (Pascual-Leone, 1984). Such intentional suppression of task-incompatible mental operations cannot just be the result of automatic inhibition (a global No-Go move) but demands use of more specific and effortful inhibition. Further, attentional inhibition (automatic or effortful) requires the prior existence of an automatic *criterion of relevance*, to help recognize schemes as relevant (to be boosted) or not relevant (to be inhibited). This issue has not been well investigated. A plausible criterion of relevance could be learned and automatized if there existed in the psychological brain/organism *a decision rule* such as the following (Pascual-Leone, 1984): schemes that have repeatedly been highly activated during a performance, serving particular specific goals (affective and cognitive), should be automatically treated by the organism as relevant for that performance. These implicit positive "ratings" would be proportional to the schemes' degree of activation during the performance. Schemes not so activated would be marked as irrelevant—and perhaps inhibited when these goals are pursued. One place in the brain where this sort of *affective-relevance* calculation could take place (in a context sensitive manner) is the amygdala, one of the most connected brain sites for contextualized affects and emotions (Pessoa, 2013; Sander, Grafman, & Zalla, 2003). Indeed, all cognitive goals are initially promoted by affective goals (Pascual-Leone, Pascual-Leone, & Arsalidou, 2015).

Mental attention (effortful activation and effortful inhibition) increases with age along developmental stages until 15 or 16 years of age (see chapter 7; Pascual-Leone & Johnson, 2005, 2011). In contrast, automatic personal/perceptual attention (automatic attention and automatic inhibition), perhaps related to the brain's default network (Arsalidou & Pascual-Leone, 2016; Arsalidou et al., 2013), reaches maturity at a younger age. Task-based research indicates that automatic inhibition reaches adult-like levels by 5 years of age (Ford, Keating, & Patel, 2004; Lechuga, Moreno, Pelegrina, Gomez-Ariza, & Bajo, 2006). At least two other findings force a distinction between automatic and effortful inhibition: (1) the neural network that underlies automatic inhibition differs from that of effortful inhibition (Cockburn & Frank, 2011; Lechuga et al., 2006) and (2) intellectually precocious children outperform their same-aged control peers on tasks requiring effortful inhibition but not on tasks of automatic inhibition (Johnson et al., 2003). This dissociation was found also in older adults (Collette, Germain, Hogge, & van der Linden, 2009) and in schizophrenics (Huddy et al., 2009).

Affective and Personal/Emotion Factors That Influence Automatic Attention and Action

Affects, cognition, and motivation are often intertwined in automatic performances and automatic attention. They must be differentiated to reach a deeper understanding of automatic processes, which may be much more complex than innate or acquired reflex reactions. Affective schemes differ radically from cognitive schemes, but both sorts of schemes carry out evaluations of experience. Evaluations done by cognitive schemes are *truth values*, whereas evaluations done by affective schemes are *vital values* (i.e., life-informing, biological, or psychosocial), positive or negative. Truth is an alternative sort of value that claims congruence or incongruence (match or mismatch, true or false) between ideas or representations entertained and their assumed counterparts within encountered reality. Thus, cognition assigns truth values to experience. Affective processes assign vital values, evaluating the importance of particular experiences for one's (past or present) life and living, and all is done by specific, positive or negative, affect systems—love, mastery-seeking, guilt, joy, fear, and so forth, which have their distinct innate determinations (e.g., Panksepp & Biven, 2012; Pascual-Leone, 1991b). Thus, feelings may carry values of affect that color subjective/objective situations with vital meaning relevant to one's biological and psychosocial life (Panksepp & Biven, 2012; Pascual-Leone et al., 2015). These feelings often are automatic and effortless, and they may come from unconscious habits (e.g., the person's character structure).

Truth and vital values are compatible, and they can co-exist inside the same schemes. Schemes that contain both of them are often called emotions or personal schemes, because persons are individual human beings who normally experience both of these values. Thus, emotions are hybrid schemes that conjointly carry both cognitive/truth and affective/vital values. However, affect and cognition are not well distinguished in the literature, and the term "emotion" often is taken as synonymous with "affect" (even though the latter only carries vital values). It is well recognized that pure cognition and affects (or emotions) have distinct complementary sites in the brain.

Another issue concerns the organismic functional distinction between affect/emotion versus motivation. Affect and motivation are related in that they involve values, automatic in some situations, and often are accompanied by automatic attention; nonetheless they differ markedly (Pascual-Leone & Johnson, 2004). *Motivation* has three conjoint aspects that often drive automatic attention. First is the *affect motive*, an implicit conative tendency or conation (i.e., a purpose-seeking, often unconscious, quasi-volitional tendency/disposition to do a particular something). Affective motives (conations) convert conscious or unconscious affective goals into conscious or unconscious cognitive goals. Second, the *strength* or energy (magnitude of activation) of this conative tendency is high when the person is motivated. Third, well-learned, purely cognitive schemes also have a strong tendency to apply, because schemes are self-propelling (this is Piaget's assimilation tendency), and so, they have a strong *intrinsic propensity* to apply, causing competition (if they are incompatible schemes) or overdetermination (with compatible schemes).

Cognitive goals are dispositions to do something known or believed congruent with the person's affective goals. *Affective goals* are dispositions toward the future to seek certain vital outcomes or consequences (e.g., escape with fear, approach with love, attack with anger). Coordination of cognition, affect, emotion, and motivation usually is automatic, until a misleading situation emerges that induces performances incompatible with current executive plans. This is apparent in the personal domains of life experience. Personal and interpersonal domains are not covered in this book. However, we briefly discuss at the end of this chapter some positive outcomes resulting from collaboration between automatic attention and mental-attention processes when they are compatible.

Automatic Spatial and Temporal Resource Factors in Learning

Leibniz may have been the first to formulate *space* as the product of coordinated *relations of coexistence* between or among things or states. He defined *time* as resulting from *movements of change,* such as dynamic sequencing or duration of states/things. For Leibniz (1966; Alexander, 1976) both space and time were purely relational. Time or space by itself (i.e., without any reference to relations among states or things) was nothing and did not exist. This idea was taken up in modified form by Kant (1929/1965; Alexander, 1976), who claimed that "space and time are conditions of the objective reality of all our sense experience, given in intuition" (Jaspers, 1962, p. 13; Kant, 1929/1965). For Kant, space and time are the automatic forms of sensory organization that impose order onto external or internal experiences, and they produce emergence of *intuition* (which in Kant means actual experience of complex configural states represented as objects or things). This intuition serves as input to mental understanding and reason. This way of forming intuitions was, for Kant, a priori (i.e., not learned but innate to the brain). Neuroscience has confirmed in modified form this major insight of Kant's. The so-called dorsal representational network (see chapter 11), connecting among others occipital to parietal lobes, can be interpreted as neurological substratum of a liberalized Kantian space—functionally contextualized and relational, but innate in its original substratum (this innate substratum we call organismic *S*-factor or hidden regulation operator *S*). In continuous dialectical interaction with this system, the ventral representational network, which connects occipital with temporal lobes, is the neural substratum of Kant's liberalized time as representation (our hidden regulation operator *T*).

By virtue of these two resource systems, the brain can coordinate effortlessly (for the sake of perception and intuition—imaginal representation) relations of coexistence and relations of duration (dynamic sequencing) that exist among objects or configural patterns or states. These automatic coordinations propitiate perceptual or configural emergence of reality-based representational totalities. Using these durations (integrated sequences synthesized by the *T*-operator) helps to make salient the individual identities of objects (their relational, semantic-pragmatic characteristics beyond simple here-and-now perception), which allows distal-object identities to be learned. Leibniz (1966; Alexander, 1976) emphasized that space and time are organizers ("un ordre," Leibniz, 1966, p. 125) important for differentiating *things* (e.g., proximal objects), as well as their individuality and *identity*—this is their distal object. However, he added, "it is instead by way of the things that one must distinguish a site or a time from another" (Leibniz, 1966, p. 196, JPL's translation). In other words, space and time are purely relational, and the concrete content (*C*) of things is necessary to experience space (*S*) or time (*T*). On the other hand, spontaneous observation of temporal duration (or sequences) is needed to learn the true identity of things. The brain's ventral pathway usually is described as serving only to structure objects' identity, perhaps because the identity of things (for Leibniz, 1966, p. 196, their "*principle* of internal *distinction*") is more concretely salient than the temporal sequences themselves.

In our view, the combined coordination of multiple effortless resources, for example, the operator *F* (field factor of simplicity), the operators of learning (*C, LC, LM, LCLM, LA,* etc.), the operators for space (*S* factor) and for time (*T* factor), automatic attentional inhibition, and so forth yield basic mechanisms for more or less automatic performance in ordinary, facilitating situations. Furthermore, when activated schemes are not compatible, competition can be solved in facilitating situations with automatic attentional inhibition. However, when the situation is misleading, the schemes' competition requires use of effortful mental/executive attention, as discussed in chapter 7.

Strong Automatic Attention and Action Aided by Mental Attention: Mental Flow and Meditation/Contemplation

The peak successful use of automatic attention occurs when it is functionally coordinated with mental attention, to foster efficient intuitive thinking and learning. This is done by using various brain resources: content learning (*C*), automatic structural/relational learning (*LC*), executive-driven mental-attentional capacities (*M* and *I*), mental executive-driven structural learning (*LM*), affective learning (*LA*), the internal-field factor of simplicity (*F*), and so on. There are moments in activities and experiences when automatic processing and mental processing are smoothly intertwined, and the activity unfolds easily in effortless self-propelling ("autotelic") ways. In this case, as our third epigraph indicates, "the goal of each action is the following action. *Action and awareness merge*" (Dormashev, 2010, p. 313). This fully conscious but not self-conscious engagement with activity, in which the self is empathically fused (happily synthesized) with the well-adjusted activity, is what Csikszentmihalyi has called a *flow experience*, "because so many of the persons describing it used the analogy of being effortlessly carried by a current—of being in a flow" (Csikszentmihalyi & Nakamura, 2010, p. 181). The flow experience is very familiar to high performers in creative activities in the arts or science (e.g., immersive acts of problem solving) or during successful, protracted and absorbing, sport performances (such as running, tennis, or soccer).

Intense concentration of attention, blending both the automatic and the mental kind, is critical for the flow experience. As Dormashev (2010) pointed out, "any activity can become autotelic [happily self-propelling] and the decisive role in this transformation is attention ... attentional activity that provides for a cognitive restructuring of the problematic situation" (p. 307). Notice that the term *activity* as used by Dormashev (and by us) refers to Leontiev's key concept of activity: "We apply this term only to processes that, by implementing a certain relationship of a human to the world, answer a corresponding specific need" (quoted in Dormashev, 2010, p. 309). Such concept of "activity" is related to the Western concepts of agency or praxis. The experience of flow occurs when intention (motivation), attention (automatic and mental—executive driven but very well practiced), and actual performance get fused into the happy feeling of momentary achievement. Mental attention and thinking alone cannot reach this level of reborn ultimate spontaneity. We say reborn, because it occurs spontaneously in happy and unconcerned young children. However, it often disappears until, with a long labor of insightful practice, the older child's or adult's intention, attention, and actual performance become truly fused. This is an anticipation of moments when the *spirit* (not the religious one, but the brain's synthesis of a harmonious, psychologically integrated, and organismic "functional totality") emerges in us to make us wiser humans.

Is there a way to propitiate emergence of this organismic spiritual balance? One such way is the practice of *contemplative activities* like meditation, yoga, or tai chi (Kim, Pascual-Leone, Johnson, & Tamim, 2016). All meditation/contemplation techniques (**MED**) can be formulated (Pascual-Leone, 2000a) as having in common six conditions or aspects that together constitute the practice: (**MED1**) they involve a *ritual* (routine) to be pursued. This includes a prescription to maximize willful and mentally energized, open participation in the contemplative ritual activity, without paying attention to (quietly ignoring) distracting thoughts, images, feelings, noises, and so forth that are not part of the ritual, and to do so without actively rejecting (indifferently accepting) distractors. (**MED2**) This ritual has an *object or target* that the meditator maintains and attends to. (**MED3**) There is a *practicing person* (the meditator's mentally calm and vigilant self-consciousness and self-agency) quietly contemplating the object and his- or herself, the subject, as the meditator pursues the ritual. This self-conscious practicing subject makes sure that the prescribed practice is regular (e.g., from 20 minutes to an hour, once or twice a day), because regularity is fundamental. (**MED4**) There is a *belief system or philosophical foundation* that highly motivates the practice and explains or justifies the method. This belief may or not be explicit within the ritual, but the practitioner tacitly keeps it in mind as the reason (organizer and motivator) for the practice. (**MED5**) The person should have a very *high motivation* for this regular practice. This practice requires mental (and playful) self-discipline and persistence over time. (**MED6**) Although quietly ignoring distractors and persisting in the meditation practice, the person chooses to practice in situations where distractors can be minimized.

Across various schools, examples of rituals (**MED1**) and object (**MED2**), include the meditator's repetition of a mantra (e.g., in transcendental meditation), attention to prescribed breathing or body postures (e.g., in yoga), attention to ritualized movements (tai chi), seeking and contemplating thoughtless awareness from which self/nonself distinctions end up vanishing (e.g., Zen Buddhism), and so on. Examples of belief systems (**MED4**) are the Vedic metaphysics and practices simplified and adapted by Maharishi, yoga philosophy, Zen Buddhist or Taoist philosophy, or related neuropsychological and neuroscientific contemporary theories about brain, attention, and consciousness. These and other belief systems (philosophical or scientific) are often combined and variously adapted by practitioners, but a belief system is always present in any successful practice.

Austin (2009) distinguishes in Zen meditation two different, but complementary, kinds of practice that have distinct expressions in the brain. In *concentrative*, or actively energizing, *meditation* the object of meditation tends to be more concrete and focused on. This type of meditation engages particularly the dorsal brain pathway of the attentional system, more "top-down" and voluntary. In contrast, in *receptive*, or passively enactive, *meditation* the object of meditation tends to be more global or changing (e.g., attending receptively to anything in the situation) and less focused upon. Receptive meditation engages in particular the ventral pathway of attention, more "bottom-up" and involuntary. This distinction is applicable to all sorts of contemplation/meditation. Both kinds, concentrative and receptive, may be intertwined and seem to involve both mental and automatic/spontaneous attention, but at the beginning of practice mental attention is likely to be stronger in concentrative meditation, and automatic attention tends to be more dominant in receptive meditation. With persistent practice, and possibly thanks to resource-operators mentioned in this chapter, meditation/contemplation of both kinds becomes progressively more flowing and automatic, although mental attention and a willful dedication may remain necessary in both. Some reflection suggests that meditation/contemplation and the experience of flow require spontaneous coordination of many dialectically interacting processes: automatic versus mental attention, concentrative versus receptive attending, cognitive versus affective/personal processing, spatial versus temporal coordinations, and so forth.

Conclusions

We have discussed automatic attention (effortless, often perceptual and personal) and emphasized that it can be simple or complex. Complex automatic activities in facilitating situations may bring special mental (and behavioral) states of high adaptation, as found in behavioral and subjective "flow" or states of meditation/contemplation. These complex states of automatic attention usually are achieved with harmonious coordination of mental (effortful) attention and much assiduous practice (constructivist learning). Brain processes distinguish between these modes of attention.

We have also discussed different causal-organismic factors that contribute to complex automatic attention, such as the organismic principle of Schemes' Overdetermination of Performance (*SOP*) and the internal field (*F*) operator. The great adaptability of automatic attention is due to these *F-SOP* organismic constraints. The role of learning (*L*-operators), affective (*A*), personal (*B*), spatial (*S*), and temporal (*T*) organismic operators were also reviewed in relation to automatic attention. Flow and meditation/contemplation emerge when these organismic factors and mental attention combine, functionally harmonized in a stable manner, resulting from practice.

Automatic attention and mental attention are contradictory in many situations, but jointly complementary in others. They are always in continuous, dialectical interaction, tending to supplement, cancel, or replace one another. This is apparent in the tendency to find an inverse relation of activation between the brain areas that deal with mental/executive attention (such as dorsal and lateral cortical networks) versus those that index automatic/personal attention (cortical medial and ventral networks; see chapters 10 and 11). These two sorts of processes emerge separately in evolution; automatic attention appears first, and mental/executive attention later (e.g., Striedter, 2005). Mental/executive attention may appear in evolution to cope with misleading situations, in which automatic processes interfering with intended action can dominate and must be inhibited. Yet, if these two forms of attention are suitably combined and become well coordinated, performance turns efficient and flow appears.

7 Mental Attention, Intelligence, and Consciousness

We analyze the mental-attention (mental effort) mode of processing and summarize developmental data to illustrate its quantitative maturational growth (the *M*-measurement system). We discuss mental attention as a major constituent of intelligence, justifying the distinction between developmental intelligence (sensorimotor and/or symbolic) versus individual-difference (ordinary psychometric) intelligence. Consciousness may be a striking consequence of applying mental attention. Innate "hardware" mechanisms (maturational constraints, hidden operators, and principles) and various sorts of learning cause emergence of consciousness via *M*-processing and *LM* learning. Consciousness is a flow of dynamic representational syntheses that use schemes as units—driven by executives and *F-SOP* resources, which automatically combine into a "stream of consciousness."

> In general, *attention* is a *tending of the ego toward an intentional object,* toward the unity which "appears" continually in the change of the modes of givenness.
>
> —Husserl, 1973, p. 80

> Thought [a product of mental attention] is a relationship with oneself and with the world, as well as a relationship with the other; hence it is established in the three dimensions at the same time.
>
> —Merleau-Ponty, 1968, p. 145

> Separate causal factors operate through development to affect somewhat separately cognitive processes and produce individual differences.... Further evidence is needed to elucidate the nature of these factors.
>
> —Noll & Horn, 1998, p. 280

We discussed automatic attentional processes in the previous chapter, along with some organismic factors that in normal people may cause automatic behavior. Mental attention applies on schemes to control and enhance the working mind's stream of consciousness, the working mind's momentary internal complexity (i.e., the field of activated schemes in memory; Greenberg & Pascual-Leone, 1995, 2001). Mental attention enhances this dynamic internal complexity by increasing activation of some task-relevant schemes, coordinating them into objects of knowledge, procedures, adjunct information, or acts, and thereby creating spatial-temporal or verbal fields of consciousness or centrations (see epigraph by Husserl). This focal centration on schemes refers to outer or inner reality (see epigraph by Merleau-Ponty). Such intentional attention is characteristic of humans and other animals, as ethologists and behaviorists know well.

In the current psychological research literature, executive function (EF) refers to brain processes that (among other tasks) mobilize, control, and allocate mental attention. As Buss and Spencer (2014) define it, EF is both an active inhibitory function (related to our central attentional interrupt or *I*-operator) and a working memory function (related to but distinct from our activation-boosting *M*-operator); the two together highlight within mental centration task-relevant information. To these two constructs researchers add control processes (related to our executive schemes) that regulate allocation, re-allocation, and control of the two resources. Note that, in our Theory of Constructive Operators (TCO), executive control processes are (in contrast to *M, I,* and other resources) sets of more-or-less complex high-level operative schemes, which we call *executives* (or *E-operator*). The TCO and EF theories address some of the same control functions of organismic reality, yet they are two very different sorts of theory. They may complement one another, as the TCO and Cowan (2016) and Engle (2002, 2018) theories of working memory (WM) in fact do.

Unlike EF theory, the TCO provides a metasubjective analysis, that is, analysis of tasks' strategies studied from within the subject's own processing. This demands use of units of processing, the schemes, which explicate the vague but important notion of "information." For this reason, the "E" that stands for "executive" in EF theories should not be confused with our symbol *E*. Our *E* explicitly stands for a general executive function and, within situations, an activated *set of executive schemes* monitoring the psychological (brain-and-body) organism as a working functional totality. Also, our theory enables a method of task/process analysis using *constructive operators* (i.e., subjective operators—*schemes*—and hidden-resource *operators*) to clearly differentiate between automatic-perceptual attention versus effortful-mental attention. Such distinction is not clear in EF theories. Nor can EF theories analyze with clarity differences between the neo-Gestaltist field factor (our *F*-operator—lateral inhibition in the brain) versus automatic attentional inhibition—our automatic *I*-operator or attentional interruption. Note that we differentiate between automatic versus effortful attentional inhibition: the latter is part of mental attention.

Our model of mental attention encompasses operators *E, M, I,* and *F,* which can apply to boost activation of explicit semantic and action units, the schemes. Driven by *E,* however, any hidden operator could serve as a scheme booster. Hidden operators serve to control, adapt, or create schemes, allowing the TCO to quantify the individuals' capacity of mental attention, also assessing the tasks' mental demand (see chapters 8 and 9). EF theories cannot do so. Finally, theories of EF fail to explain how and why mental-attentional capacity increases in a graded manner until adolescence, even when such theories accept that effortful/endogenous attention is carried by genes and grows developmentally (e.g., Cowan, 2016; Cowan, Ricker, Clark, Hinrichs, & Glass, 2015; Hansell et al., 2015). EF theories need more clarity in explaining how this developmental growth affects intelligence and consciousness. Such issues can be clarified if our TCO approach is coordinated with current theories.

The significance of mental attention and its products (such as symbolic processing and problem solving) is apparent when we compare the intelligence of humans and other animals. Humans can confront reality/Reality in its complexity by going beyond immediate perception. But other animals are driven by appearances, proximal objects, or stimuli (e.g., Rock, 1983). As Zubiri, philosopher of the "*sentient intelligence*" (Zubiri, 1999), put it, "intelligence… is the capacity a man has to confront things as realities, not just as stimuli" (Zubiri, 1966/2001, p. 194, translation by JPL). By *realities* Zubiri meant what psychologists have called distal objects, that is, experiential as well as semiotic referents interpreted by learned schemas or complexes of coordinated schemes. These complexes emerge as functional invariants from practical activities. By *experiential* we mean *concretely real* and not conceptual, that is, what Husserl (1973) called "prepredicative" and Merleau-Ponty (1968) called "felt flesh" (embodied) experience.

Acquisition of complex distal objects (and their corresponding operative processes) occurs by coordinating schemes to extract functional invariants. This involves observation and learning of functionally related characteristic aspects (e.g., reliable features, cognitive distinctiveness) that express aspects of distal objects and related operative schemes. These organized collections of functionally necessary aspects make up what Zubiri (1999, 1966/2001) called the functional *essence* of experienced reality (in our view relative to agency and praxis). Complex realities (distal objects or distal processes) cannot be noticed or internalized until these essential (probabilistically invariant) data and interrelations are synthesized and internalized into functional totalities (organized schemes or schemas). To carry out this internalization a person usually must keep many aspects in mind simultaneously.

The minimal number of aspects needed to internalize a scheme, and the number of schemes required to solve a task, express their order of effective complexity. Such essential constructive complexity is estimated via task analysis (see chapters 8 and 9), which quantifies the tasks' mental-attentional *M*-demand. Intelligence (supported by mental attention and the already acquired knowledge schemes) is the discovery and internalization of *essential realities* of tasks (e.g., distal complex objects, distal processes), which is done by going beyond the mere stimuli or proximal objects (i.e., the information given) into the actual, manifest or hidden, processes of agency and praxis.

Going beyond the Information Given

In earlier chapters, we suggested how misleading factors affect complexity of a task (e.g., its executive and *M*- and *I*-demand). We emphasize the difference between "mere stimuli" and "realities" (distal objects, distal processes) by referring again to the wine and water problem (see chapter 6). In this problem, participants are asked to imagine two containers, C1 (with only water Wa) and C2 (with only wine Wi). A spoonful (S1) of wine from C2 is transferred to C1.Then a spoonful (S2) of liquid from C1 is transferred to C2. The question is whether now there is more wine in the water container (Wi[C1]) or more water in the wine container (Wa[C2]). As shown in chapter 6, a logical strategy for solving the problem, represented algebraically, is as follows:

$\underline{\mathrm{S{=}S2{=}S1}}$

$\underline{\mathrm{Wi[C1]}} == \underline{\mathrm{S1{-}Wi[S2]}}$

$\underline{\mathrm{Wa[C2]}} == \underline{\mathrm{S2}\ -}\,\underline{-\mathrm{Wi[S2]}}$

➔ Wi[C1]==Wa[C2]

The underlines demarcate separate constituent schemes. The markers (= and –) are repeated to emphasize separation of the six schemes. The three-line sequence prior to the arrow (➔) demarcates three sets of coordinated schemes, each set constituting a complex scheme, that is, S (scheme of the actual spoon), or Wi[C1], or Wa[C2]. Scheme Wi[C1] has two constituents (marked by underlines), whereas scheme Wa[C2] has three. This is because we assume that the subject's synthesis of Wi[C1] took place earlier, and its constituents are now partly chunked.

A "mere stimulus" level of analysis is to consider S2 to be a spoonful transfer of liquid from C1 to C2 and think that because S1 carried only wine but S2 carried both wine and water, there is more wine in C1 (Wi[C1]) than water in C2 (Wa[C2]). This view ignores the relational complexity of the broader situation. The deeper "reality" is that S2 contains both wine (Wi[S2]) and water (Wa[S2]) and that Wa[S2] (the water carried back to C2) has the same liquid amount as the wine left behind in C1 (because the water in S2 is precisely equal to the wine not being returned in S2); thus Wi[C1]=Wa[C2].

The arrow in the algebraic representation stands for the principle Schemes' Overdetermination of Performance *(SOP)*.

Thus, six essential distinct schemes (shown in the algebraic task analysis) must be attended to in order to solve the problem, making this task not accessible before early adolescence (Pascual-Leone, 1970; Pascual-Leone & Johnson, 2005). This sort of task-analytical model predicts age-bound, developmental substages described by Piaget and Case (Case, 1998; Pascual-Leone, Johnson, & Agostino, 2010). Such substages (levels of *M*-capacity) are basically invariant across very diverse populations and across types of tasks, suggesting that mental attention is a maturational capacity. Cowan and others have indeed concluded, with their own experiments, that this construct (which they call WM) is maturational (Cowan, 2016; Cowan et al., 2015; Hansell et al., 2015). Let us examine some evidence from developmental simulation and neuroscience suggesting the need for independent (some maturational) mental attentional or WM brain processes.

Need for an Independent Scheme-Activation Boosting Function in Humans

Associative learning, unless aided by an independent mental attention (a *separate* activation-boosting function), cannot explain emergence or growth of cognitive processes. Early research on neural networks, when properly interpreted, showed such limitation (Elman, 1993; Mars, Sallet, Rushworth, & Yeung, 2011; Munakata & Stedron, 2001). Elman's neural modeling of working memory suggested that WM (from our perspective, mental-attention activation), in interaction with associative learning mechanisms, is essential to cope with misleading aspects ("noise") in language learning (Elman, 1993; Elman et al., 1996). He attempted to design a network that could learn by itself grammatical properties of English sentences, using sentences of various lengths or degrees of complexity. This network had to predict what word would come next at each point in a sentence. Sample sentences included "Boy chases dog," "Mary walks," "Boy who dogs chase feeds cat," and "Girl who chases dogs hits cats." The network failed to learn the task (could not learn positional constraints that regulate sequencing of words indexing grammatical categories).

Notice that long sentences in this task should initially be misleading, because they induce grammatically invalid associative-sequence links between words (or within the learning program's hidden-units space), and so do not help to abstract basic grammatical rules. To see why long sentences are misleading, consider the meaning of a sentence such as, "Boy who dogs chase feeds cat." Its meaning is clarified by analyzing the sentence using informal operator-logic, based on our method of task analysis. The words *boy, cat,* and *dogs* stand for *objects* (*figurative schemes* symbolized by lowercase words plus a postfixed asterisk *). *Feeds* and *chase* are *procedures or operatives* (symbolized by capitals) that we place to the left of the objects on which they apply. Finally, we symbolize *adjunct information* relative to a particular object or procedure by affixing to the object this information enclosed in square brackets (i.e., [...]). In the sentence just given the relative clause (i.e., "who dogs chase") is adjunct information about the boy, which we can represent as boy*[...]. Using this notation, we informally represent the experiential meaning of the sentence as follows:

"Boy who dogs chase feeds cat" →
FEEDS(boy*[*WHO*: CHASE(dogs*,boy*)], cat*) **(f1)**

In this formula, *WHO*: stands for the pronoun *who* referring to the belief-expectancy that this boy was/is chased by dogs. The first term (or logical argument) of the operatives FEED and CHASE stands for the subject of the action, and the second term stands for the object. The corresponding verbal sentence is before the arrow (→). A comparison of the sentence and its meaning-formula shows that the sentence is misleading because it fails to make explicit the sort of scheme (operative, figurative, expectancy) each word stands for and does not give the hierarchical level of words in the sentence—explicit in our formula. The more a word is nested within parentheses or brackets, the more concrete or lower it is in the flexible hierarchy. All this is, of course, given in the grammatical structure, which is only implicit and was not trained into the learning-program network.

When the network failed to learn the task after its exposure to sentences of varying complexity, Elman proceeded to grade the input. First, he presented very short sentences, and when the network had learned to cope with them, he progressively introduced more complex ones. This allowed the network to learn the task, responding correctly to the whole set. Elman reasoned that this solution, by initially simplifying the input, may mimic nature (children's ontogenetic development), which initially endows the child with a small WM capacity. A small WM capacity should allow the child to receive and process only short sentences. In this manner, children initially would be constrained to learn simple inputs (much less misleading, because their word order relates to grammatical categories). After this level of language competence had been acquired (and the child's WM capacity had grown with maturation), longer sentences could be processed.

To simulate and test this idea, Elman added a WM parameter to the network. "Young" developing networks were able to read patterns in the sentence only over three or four contiguous words, because their WM was erased (units in the network's memory were reset to random values) at that point. As the network "aged" and acquired more experience, the number of words that could be retained within WM was progressively increased. This was in contrast to the initial "old" network, in which WM (number of context units) had no limit. Both types of network ("young" and "old") were exposed to sentences of varying complexity from the beginning of training. Elman found that only the developing "young" networks, with WM that started small and increased progressively, were able to master all sentences. The reason for the failure of "old" networks should be apparent by looking at formula **f1**. The "old" network attempted to extract invariant sequence patterns of words across sentences of any length, but complex sentences, because they contain implicit nesting of simpler sentences, mislead any learning process based in abstracting invariant ordinal relations between or across words.

Elman et al. (1996) discussed neural network experiments by Shrager and Johnson (1996) and by Rebotier and Elman (1996). These experiments independently reinforce conclusions drawn from Elman's earlier research. They show that *abstract logical structures* (i.e., complex relational structures defining various possible categories of relations between two input sources such as A and B) could not emerge in the network (**N**) unless two conditions were met.[1] (**N1**) Input to the network units (or to the neurons in the brain) must enter the network in a suitably organized fashion, in the sense that (relative to this input) the total population of units (neurons) must be organized in a nested hierarchy of areas. In this way, input first arrives at the first-level areas of units (i.e., semantically most concrete). It is then transmitted to the second-level areas, then to the third-level, and so forth (each progressively more abstract). The input activation is always transmitted along with the processing that earlier areas in the hierarchy have produced. (**N2**) To enable learning of complex relational invariants based on input sources A and B, units (neurons) must have been primed by means of at least one source of unit activation that is *independent* from both the input source and the associative-learning network of units in question. Shrager and Johnson called this independent priming source of activation a "trophic factor" that arrives in waves. Elman et al. (1996) referred to it as "a neurotrophic dynamic (whether produced by a natural wave of trophic factor or by some other endogenous or exogenous phenomenon)" (p. 140).

Well before this neural-network experimental demonstration of the need for a neurotrophic dynamic factor, and even before the concept of working memory had been introduced in the literature, we had recognized in our cognitive-developmental theory the need for such a factor (Pascual-Leone, 1969, 1970, 1987, 1995, 1996b, 2000b; Pascual-Leone & Johnson, 1991, 1999, 2017; Pascual-Leone & Smith, 1969). We call this neurotrophic dynamic factor(s) mental or effortful attention, and the *M*-operator is fundamental here. We define *M* as a developmentally growing endogenous mental-attentional capacity that can simultaneously boost activation of a limited number of schemes. This number (i.e., the power of *M*-capacity) increases with chronological age and serves as bootstrap for enabling transitions to new developmental stages (Piagetian or neo-Piagetian), provided enough learning opportunities are available. Misleading situations mark occasions when subjects cannot solve the task unless they effortfully use their *M*-capacity, if the task's mental (*M*-) demand is not greater than the subjects' available *M*-power. This is the reason why stable stages of development are reliably found only in misleading situations (Pascual-Leone, 1987, 1996b). Misleading situations are always relative to the person's repertoire of prior knowledge (and his or her *F-SOP* factor [see table 7.1] that induces misleadingness in the current situation).

Essential Function of Mental Attention and Its Plausible Evolutionary Emergence

With the theory of schemes alone, one cannot explain *general* organismic constraints (those applying across all kinds of schemes or situations) like "central" WM capacity limits, "central" inhibitory mechanisms, structural versus content learning, "central" (executive driven) resolutions of scheme competition in the network, nor emergence of truly novel performances via unplanned dynamic syntheses. We discussed in the previous chapter general organismic ("central") processes that intervene to cause automatic attention. We now discuss in more detail processes that intervene to produce mental attention.

It is often believed that prefrontal lobes control task-relevant processes elsewhere in the cortex (e.g., posterior areas) that are not automatic and whose performance must be synthesized. Prefrontal lobes are the brain's control area, indexing mental attention (including executive schemes). However, their control power is limited. Ruchkin, Grafman, Cameron, and Berndt (2003) expressed this idea with the metaphor of a restricted number of attentional "pointers." A better formulation consists in assuming that the brain, in addition to a *repertoire of schemes* (neural circuits and networks that carry "information"), possesses the set of general-purpose functional resources we call *hidden* or silent *operators* (Pascual-Leone, 1987, 1995; Pascual-Leone & Johnson, 1991, 2005, 2011, 2017; Pascual-Leone, Johnson, Baskind, Dworsky, & Severtston, 2000). These hidden operators would be produced by the brain's maturation of certain anatomical structures and of local functioning patterns in neuronal connections with special neurotransmitters (e.g., dopamine). We call each of these functional neuroanatomical utilities operators (**OP**) for two reasons. (**OP1**) They are functional mechanisms of brain "hardware" that actively operate on schemes (e.g., boost, inhibit, functionally coordinate, and create them). (**OP2**) They are defined as molar procedures whose computational and anatomical details are left unspecified (being defined only by releasing conditions and their effects on schemes and performance). Psychologically speaking, the brain operators apply on schemes to constrain them to synthesize truly novel performances, or be inhibited, or change and produce new schemes, and so on. These operators are *hidden* because they lack explicit content-specific markers in performance and experience, markers that schemes clearly have (such as perceptual, motor, representational, and other aspects of experience). Instead, these organismic operators express purely relational functional constraints of the brain, which cause surprising patterns or "anomalies" (exhibited under specific empirical circumstances) that a pure theory of schemes alone cannot explain (such as general-developmental cognitive stages, holistic/Gestaltist effects, or even automatic or associative versus effortful learning).

Operators express and formulate (as explicit functional constructs within a psychological theory) key organismic constraints (specific regulations) that the brain's cortical architecture imposes on psychological processes and behavior. Table 7.1 summarizes eleven hidden operators that we currently consider. These eleven (often compound) categories of hidden operators are ordered in table 7.1 according to their likely evolutionary emergence, as we critically speculate.

As this table shows, operators intervening in automatic attention, discussed in the previous chapter, appear first in the phylogenetic ladder. The first two operators to appear should be *A* (pure *affect*—appetitive or aversive dispositions and their biological regulations) and *C* (*content* learning and substantive-content schemes); most animals would have these two, intertwined or distinct. The third operator to appear phylogenetically may be *F*. As mentioned before, this is the tendency of the brain to effortlessly integrate performances following a simplicity (mini-max *F*) rule, which (in conjunction with *SOP* principle) causes the active schemes to overdetermine a minimal-complexity (mental or overt) performance. The right-hand column of this table gives, for humans, some regions of the brain from which the operators in question may primordially come. Next, in our evolutionary ladder, we find strong associative learning, that is, our *LC* (*logical-content*, or content-based structural) learning, possibly available in birds and mammals (see also chapters 5 and 6).

Then, at least in mammals, the *S*-operator (causing effortless relations-of-coexistence—such as *spatial* relations) and the *T*-operator (causing effortless *temporal/fluent* sequences) appear. These two operators, very important for automatic attention, expand considerably the power of associative (*LC*) learning. This *S*-and-*T* expansion may potentiate coordinations of affect (*A*-operator) with associative learning to enable frequent emergence of affective-and-cognitive, psychosocial, sociocultural schemes. We call the complex affective-learning psychosocial processes the *B*-operator to emphasize that these

Table 7.1
TCO's hidden operators listed in order of their likely evolutionary emergence

Operator	Description	Main Brain Region
A	Set of **affective** processes that intervene in motivation and attentive arousal.	Brainstem, hypothalamus, extended amygdala, limbic system
C	Both the process of **content** learning and the schemes derived from associative content learning.	Thalamus; Brodmann primary and secondary areas
F (*SOP*)	The **field** operator, which acts as a binding mechanism in the brain and brings closure to mental representations in a neo-Gestaltist manner. It often functions intertwined with the principle of Schemes' Overdetermination of Performance (SOP)	All areas
LC	The process of automatized **logical-structural** learning derived from *C*-**learning** through overpractice	Right hemisphere (RH)
T	**Temporarily** and effortlessly collates **sequences** of figurative schemes, thus facilitating the coordination that constitutes distal objects	Hippocampal complex, occipito-temporal cortex
S	Effortlessly coordinates relations of coexistence among activated schemes, during operative activity (praxis). It, thereby, facilitates emergence of **spatial** schemes or schemas	Hippocampal complex, occipito-parietal cortex
LA, B, LB	**Psychosocial and self-schemas** (*B*). **Logical-structural** learning primed by strong **affects** (*LA*), or by the personal **being** preferences—including emotions (*LB*).	Limbic system, orbito- and medial prefrontal, inferotemporal, medial parietal cortex
I	The attentional **interrupt**, which corresponds to the power of central active inhibition of unwanted schemes activated in the situation.	Prefrontal, RH-medial cortex, dorsolateral cortex, basal ganglia, thalamus
M	**Mental** attentional capacity of the individual.	Prefrontal, lateral, and dorsolateral cortex; basal ganglia; thalamus
LM	**Logical-structural learning** caused by the effortful use of **mental** attentional capacity	Left hemisphere tertiary areas, polymodal
E	**Executive** schemes in the person's repertoire, for the task at hand	Prefrontal, lateral, dorsolateral, and frontopolar areas

Adapted from Pascual-Leone, J., & Johnson, J. (2005). A dialectical constructivist view of developmental intelligence. In O. Wilhelm & R. W. Engle (Eds.), *Handbook of understanding and measuring intelligence* (p. 181). Sage. Copyright 2005 by Sage.

processes (which possibly have innate/instinctual roots) causally determine the social being of animals.

An unwelcome consequence of this increasing power of associative learning, however, is that it becomes progressively harder for animals to control overlearned habits and automatisms (both attentional and action automatisms) when they are unsuitable for the task at hand. To reduce unwanted consequences of such overlearned functional "armour" of habits, evolution has endowed high mammals with the power of a "central" attentional inhibition or *interruption* (*I*-operator—Pascual-Leone 1984; Howard, Johnson, & Pascual-Leone, 2014) to enable fast and flexible inhibition of habits (i.e., overlearned automatized schemes). Inhibition alone is not enough, however. To boost low-activated but task-relevant schemes, otherwise suppressed by stronger task-irrelevant schemes, evolution had to create a mental *scheme-activation booster* controlled by affective processes and executives. This is the *M-capacity* operator (Pascual-Leone, 1970; Pascual-Leone & Johnson, 2005, 2011), a key maturational component of working memory that, in our view, tacitly underlies Cowan's (2016), Engle's (2002, 2018), Case's (1998), Halford et al.'s (Halford, Wilson, Andrews, & Phillips, 2014), and Demetriou's (Demetriou & Spanoudis, 2018) concepts of working memory.

Because the neurotrophic resource *M*-capacity appears empirically with a developmental stagewise pattern of growth, in normal children it *relates the highest task complexity* a child can handle *to his or her chronological age*. This is most noticeable when the child first solves a problem without specialized training. As we parametrically increase the number of aspects (essential notes) to be entertained simultaneously for coping with a task, the age of children capable of passing this task increases predictably. This pattern indexes endogenous growth of *M*-capacity in an age-bound manner till adolescence (e.g., Case, 1998; Johnson, Fabian, & Pascual-Leone, 1989; Pascual-Leone, 1970, 1987; Pascual-Leone & Baillargeon, 1994).

Our data and theoretical analyses indicate that *M*-capacity already is manifest in human infants during the second month of age, driven/activated by affect (*A*-operator) and motivation. This maturational graded-growth of *M* limits relational complexity of sensorimotor tasks a child can cope with (see chapter 3). At about 18 months, with a *sensorimotor* (*Me*) *M*-capacity of five, babies are capable of subjective consciousness. This is Damasio's (2012) core consciousness, which allows the child to pass the mirror test of self-recognition (to notice in the mirror a red mark on his or her own face—Gallup's [1982] test). This test is also passed by other animals high in the evolutionary ladder (e.g., primates, dolphins, and perhaps also elephants or parrots). According to our task analysis (chapter 3), this test demands from the subject a capacity to keep in mind simultaneously no fewer than five sensorimotor schemes (Pascual-Leone, 2000a,

2006). At about 26 months the child can simultaneously attend mentally to six sensorimotor schemes. Finally, at 35 months of age (3 years) children can simultaneously activate seven sensorimotor schemes, which enables truly mental symbolic processing (mentation with symbols of symbols). Mental-attentional complexity in primates' problem-solving behavior (in chimpanzees, bonobos, gorillas, and orangutans) suggests that they can match the capacity of 26-month-old humans, and at times that of 35-month-olds (Pascual-Leone, 2006). Beyond this age, humans appear to be alone across species in their developmental growth of mental (*M*-) capacity, which keeps growing after 3 years up to the ages of 15 or 16 years in healthy people. Note that the *M*-capacity growth in infants up to 3 years (sensorimotor *M* or *Me* capacity) is graded in a much smaller amounts than for older children (with symbolic *M* or *Mk* capacity), as explained in detail in chapter 3.

As table 7.1 shows, *LM*- and *E*-operators follow on the emergence of the *M*-operator. The compound operator *LM* emerges when *M*-capacity is used to synthesize new performances, leading to effortful (*M*-driven) logical-structural learning, which we call *LM* learning. *LM* learning in turn becomes automatized into *LCLM* complex schemes, which no longer need *M*-capacity to be activated (this is a complex operational *chunk* of information). Such power to form complex logical-relational *LM*-schemes permits creation of operative/procedural schemes that stand for generic plans, controls, and task organizers. These are *executive* schemes. We call *E-operator*—"the executive"—the set of highly activated and relevant executive schemes in a given task or situation. Executive schemes are essential to plan analytically into the future, something that also requires use of the other hidden operators. Finally, when affect (*A*) and cognitive learning processes (*C, L, LC, T, S, I, M, LM, E*) become coordinated in the psychosocial and social-group domains (*LA, LB*), they produce *personal/emotion schemes*, whose repertoire we call *B*-operator (a complex affective-and-cognitive, existential, and social schemes' coordination, with human being affect-related beliefs and biases—the individual person schemas). These *B*-schemes, more or less complex hybrid (affective and cognitive) schemas, are learned during psychosocial, or self-existential and sociocultural experiences.

Figure 1.1 complements table 7.1, showing symbolically that the hidden operators can influence performance by regulating activation or bringing change to the schemes of the person. In fact, figure 1.1 symbolically stands for what we call the metasubject (the brain's "psychological organism" as represented in a gist manner by our theory, the TCO). This metasubject includes as a functional part an operative self (William James' I-self) with its conative (unconscious self-agency) propensities and its conscious Will that may drive decision making. So defined, the metasubject is functionally close to what Demetriou (Demetriou & Spanoudis, 2018) called *cognizance*—the partly conscious self-agency and intentional control processes of the subject.

Model of Endogenous Mental Attention

Mental attention (<*E, M, I, F*>, as explained below) is endogenous, active, effortful, voluntary, executive attention. Often called working memory, this is better seen as a key aspect of the working mind. *M* (power or capacity) is an operator constituent of mental attention, which stands for the limited "mental energy" or "cortical activation tone" of task-relevant processes in the cortex if the person is vigilant. In some sense, Janet (1889), James (1892/1961), Spearman (1927), Freud (Rapaport, 1960), Luria (1973), and many others have had intuitions of this endogenous, limited-energy resource that we call *M*. We call this form of attention *mental*, and not executive attention, as it is often called, for three reasons mentioned in previous chapters. (1) Mental attention can be recognized in children after 2 months of age (see chapter 3), but executive schemes cannot be found until after 12 months. (2) It is mobilized/driven not just by executives but also by affective and emotional schemes. (3) It is used to boost activation of all sorts of schemes, not just executives.

Mental attention can join forces with automatic attention within facilitating situations, as in "flow" situations mentioned in chapter 6. In misleading situations, however, mental attention is in competition with automatic attention. Misleading situations induce strong, prepotent responses (strongly activated by often unconscious shadow schemes) contradicting what the person wants to do. Salient cues exist in such situations that lead to error, and these dynamic error-factor schemes cannot be ignored, because they are intertwined with task-relevant cues or schemes. Examples of such misleading contexts can be found shopping in supermarkets or pharmacies. Products often are placed inside excessively large packaging to create the illusion of containing more product; this is a misleading factor when appraising the relative cost.

By their very nature, problem-solving tasks involve misleading situations (as the wine and water problem illustrates). The need to suppress misleading schemes (i.e., unwanted activated neural circuits) forces use of mental attention to inhibit the "bad" schemes and concurrently boost task-relevant ones. Mental capacity is estimated by the number of separate task-essential schemes that can be activated simultaneously by mental attention to solve a task: we call such measure *mental (M-) power*. *M*-power increases with age up to adolescence. This growth of *M*-power enables transition across developmental stages (Pascual-Leone, 1970; Pascual-Leone & Johnson, 2005, 2011; Pascual-Leone et al., 2010).

We symbolize mental attention (often abbreviated as *Matt*) as a *searchlight model*, constituted by a dynamic system of four brain operators in coordination: Matt=<*E, M, I, F*>, key operators that we described before. The *field factor F* (related to lateral inhibition in layer 4 of the cortex; e.g., Edelman, 1987; see chapters 10 and 11) and the inhibition factor *I* help to produce closure in performance by inhibiting schemes that are outside the "beam of attention" and not relevant.

The beam of attention is symbolized in figure 7.1. Operators *E, M,* and *I* are expressed in the prefrontal lobe, and their coordinated application to action schemes produces the *M-centration* (mental centration). This is the inner ellipse. Notice that in figure 7.1 we call H the repertoire of habitual, well-learned schemes (part of long-term memory) found within *M*-centration and call H′ the set of schemes in long-term memory but outside *M*-centration. The outer ellipse symbolizes the field of more or less activated schemes (*field of activation*) in the repertoire or long-term memory. The middle ellipse symbolizes the field of all hyperactivated schemes, which corresponds to the full attention (i.e., working memory). When hyperactivation of schemes is done exclusively with *M*-capacity, these *M*-centrated schemes constitute the *M-centration* or *focus of attention,* different from, but included into, full attention or working memory. Some schemes in full attentional centration may be hyperactivated by factors other than *M*-capacity boosting. Their high activation may be caused by associative content learning (*C*), logical-content learning (*LC*), effortful logical-structural learning (*LM*; notice that *LM*-schemes are effortful because, unlike *LC,* they still require use of *M*-boosting to fully apply), perceptual salience (*F, C*), affect (*A*), and so forth. Task-irrelevant hyperactivated schemes from outside *M*-centration tend to be automatically (perhaps also effortfully, I_e) interrupted/inhibited to some degree with each new act of *M*-centration (Howard et al., 2014; Im-Bolter, Johnson, Ling, & Pascual-Leone, 2015; Morra, 2000; Romero Escobar, 2006). This is a distinct variant of attentional inhibition (*I*), the automatic inhibition/interruption (I_{au}), different from effortful inhibition (I_e).

Note that we equate working memory or full attention with the field of hyperactivation (i.e., middle ellipse) which includes schemes being boosted by *M*-capacity (*M*-centration—inner ellipse) as well as schemes boosted instead by other organismic factors (e.g., hidden operators *C, LC, A, F*). In the case of misleading situations, highly activated misleading schemes are in full attention (working memory) and so executive processes (*E*-operator) must inhibit (*I*-operator) them to cope with the task. As a result, hyperactivated schemes placed outside *M*-centration should become inhibited, eliminating the middle ellipse and reducing full attention to *M*-centration (i.e., the inner ellipse; Pascual-Leone, 1984; Pascual-Leone & Johnson, 2005). This account of misleading situations is consistent with Engle, Tuholski, Laughlin, and Conway (1999),

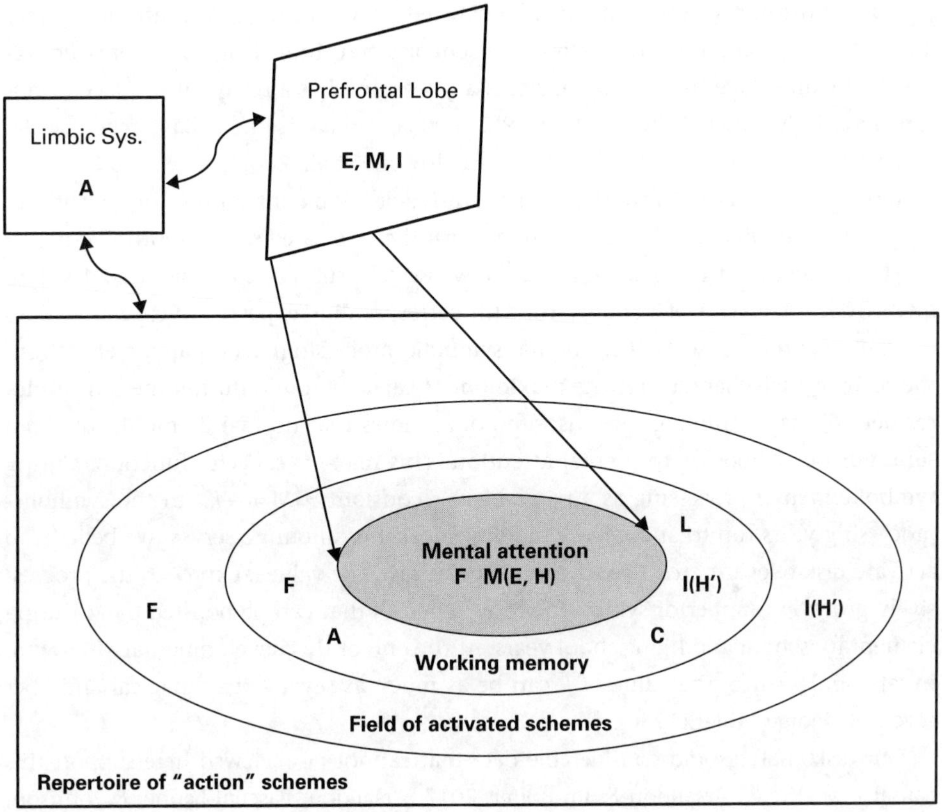

Figure 7.1
The TCO model of endogenous mental attention. (Adapted from Pascual-Leone, J., & Baillargeon, R. [1994]. Developmental measurement of mental attention. *International Journal of Behavioral Development, 17*[1], 169. Copyright 1994 by the International Society for the Study of Behavioral Development.)

among others. This is in part why misleading situations make better tasks for estimating *M*-demand via task analysis (Pascual-Leone et al., 2000). Notice that in our general theory (TCO), and in the model of figure 7.1, Matt is driven by the metasubject, which includes high-level personal schemes and developmentally evolving operative-self schemes (*self.1, self2.1, self2.2, self2.3, self2.4, self2.5, self2.6, self2.7*—see chapter 3).

Maturational Growth of *M*-Capacity in Humans

Mental attention increases during normal child development from early months of life to adolescence. This increment indexes the *power* of *M-capacity* (*Mp*), or maximum number of distinct schemes that can be boosted by *M* within an *M*-centration. Every maturationally achieved *M*-capacity level enables transition from one constructive-developmental stage to the next, if necessary learning has already taken place (Johnson et al., 1989; Pascual-Leone, 1970, 1987, 2000a; Pascual-Leone & Baillargeon, 1994; Pascual-Leone & Johnson, 1999, 2005; Pascual-Leone et al., 2000).

Table 3.1 gives Piaget's substages of cognitive-development during the sensorimotor period from birth to 35 months, along with the associated sensorimotor *Me*-stages; we do not discuss it further. Table 7.2 shows Piaget's substages of cognitive development from three years of age onward, with corresponding expected *M*-capacity values available for the *Mk*-scale (i.e., mental-symbolic processing). The parameter *e* (from the scale *Me*; see chapter 3) is the maximum *M*-capacity found during the sensorimotor period, that is, number of sensorimotor schemes that normal 26-month-olds can simultaneously boost with mental attention. This max-*Me* capacity functions during symbolic mental processing as a parameter or constant *e* ($M = e + k$). In the symbolic-processing years (up to and beyond adolescence) this amount *e* serves, we believe, to activate task-relevant *executive* schemes. In contrast, the value *k* expresses the progressively growing number of symbolic/action schemes that can be boosted by *M* during elementary school and high school years. At the end of this developmental *M*-growth, in late adolescence, the value of *k* can be as much as seven—the "magical" number seven of George Miller (1956).

Our task analyses and considerable data that cannot be reviewed here support this causal model (e.g., Arsalidou & Im-Bolter, 2017; Arsalidou, Pascual-Leone, & Johnson, 2010; Burtis, 1982; Johnson et al., 1989; Morra, 2000; Pascual-Leone & Baillargeon, 1994; Pascual-Leone & Johnson, 2005, 2011, 2017; Pascual-Leone et al., 2010). Participants

Table 7.2
Predicted *M*-capacity values as a function of age and their correspondence to Piagetian substage sequence

M-capacity ($e+k$)	Piagetian substage	Normative chronological age
$e+1$	Low preoperations	3–4 years
$e+2$	High preoperations	5–6 years
$e+3$	Low concrete operations	7–8 years
$e+4$	High concrete operations	9–10 years
$e+5$	Transition to formal operations	11–12 years
$e+6$	Low formal operations	13–14 years
$e+7$	High formal operations	15 years to adult

often do not mobilize all the *M*-capacity they have available (Pascual-Leone, 1970; Pascual-Leone & Johnson, 2011). There is a lower bound or *functional M*-capacity often used by subjects, which is equal to four or five *k*-units in adults, and there is an upper bound or *M-reserve* (the structural *M*-capacity, Pascual-Leone, 1970), which is their max-*Mk*. This *M*-reserve is often used only under very high mental activation. It is equal to seven in adolescents (15 to 16 years) and adults, decreasing in old age (Arsalidou & Im-Bolter, 2017; Arsalidou et al., 2010; Arsalidou, Pascual-Leone, Johnson, Morris, & Taylor, 2013; Jedrzkiewicz, 1983; Pascual-Leone, 1970, 1983). Our estimates come from many tasks and much data, some of which we briefly summarize below. These estimates of *Mk* are probabilistic, idealized averages or estimates for normal children, which can be modified by causal variables such as alertness, vigilance, fatigue, state of health, and so on.

M, whose behavioral measure is *M*-power, is a hidden construct, hidden because one must vary conjointly participants' expected *M*-power and the tasks' *M*-demand to obtain measured estimates of the participants' *M*-capacity (i.e., *Mp*) or the tasks' *M*-demand (*Md*). This is a purely relational invariant found in the data. The need to conjointly vary the two estimated organismic variables, *M*-power and *M*-demand, suggests that *M*, as empirical construct, exhibits what Bohm (1980) called an "implicate order." In our case, this is a complex, purely relational system of constraints (i.e., resistances to the subject from the task, aspects to be considered concurrently) binding together brain processing complexity and task complexity. *M* is the sort of latent variable that Bohm and many others call a *hidden variable.* This key characteristic appears in *M* as a purely relational invariant expressing trade-off between participants' *Mp* and tasks' *Md* (items are passed only when *Mp* is equal to or greater than *Md*). We have repeatedly tested empirically this *Mp/Md trade-off* and its relationship to participants' cognitive style (e.g., Arsalidou et al., 2010; Arsalidou, Pawliw-Levac, Sadeghi, & Pascual-Leone, 2018; Raymond Baillargeon, Pascual-Leone, & Roncadin, 1998; Pascual-Leone, 1970, 1978, 1989; Pascual-Leone & Baillargeon, 1994; Pascual-Leone & Johnson, 2011). Tasks assessing *Mp/Md* trade-off tend to show increase in performance level at the ages when *M*-operator capacity increases developmentally in normal children, that is, with Piaget's substages and Case's stages.

Staircase Model: Another Comparison of *M*-Capacity Levels with Developmental Stages

Piaget knew that an organismic general process, different from learning, was at work in cognitive development. He ambiguously called this general process "regulations" (Pascual-Leone, 2012a; see chapter 1). He also knew that a growing mobility and scope of decentration (flexibly changing the content/schemes within mental centration—focusing and re-focusing on different chosen schemes) were a major indicator of cognitive growth. This is what we now call executive-function factors such as updating, shifting, and so on. For Piaget, regulations were the cognitive processes that cause mental decentration. Reflective abstraction was his unexplicated mechanism of emergence, via constructivist learning, of schemes/schemas characteristic in developmental stages (i.e., sensorimotor, preoperational, concrete operational, formal operational) and substages within them (Piaget, 1975/1985).

Figure 7.2 (Pascual-Leone & Johnson, 2005; Pascual-Leone et al., 2010) presents the *staircase model* of stages put forward by Robbie Case (1985, 1992, 1998), modified by us to demonstrate correspondence between Piaget's substages (Case's stages) and *M*-operator levels. The initiative to change the causal significance from Piaget's developmental stages to his substages was taken first by Pascual-Leone, after inferring that

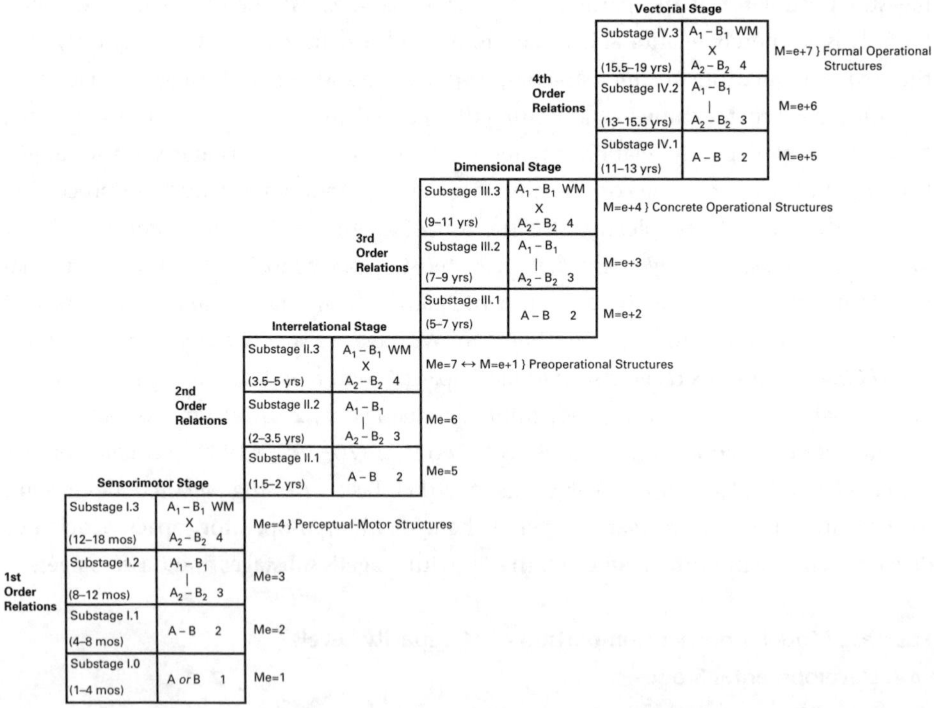

Figure 7.2

Case's staircase model of developmental stages integrated with the TCO model of *M*-capacity growth. (From Pascual-Leone, J., & Johnson, J. [2005]. A dialectical constructivist view of developmental intelligence. In O. Wilhelm & R. W. Engle [Eds.], *Handbook of understanding and measuring intelligence* [p. 184]. Sage. Copyright 2005 by Sage.)

mental-attentional capacity (the *M*-operator) was a determinant of stages. He did this empirically, via task analyses of Piaget and others' data (Pascual-Leone, 1969, 1970; Pascual-Leone & Smith, 1969; Stewart & Pascual-Leone, 1992).

In the neo-Piagetian staircase model (see figure 7.2), substages are given along with the empirically confirmed estimates of *M*-capacity (on the right side of the staircase blocks). Notice that in our view stable stages appear only within misleading situations: only these situations clearly exhibit *M*-demand constraints (because few facilitating schemes are available). Because familiarity or practice may transform (via learning) misleading situations into facilitating ones, any developmental stage could in principle be changed, with practice, in its local expression. That is, maturational development and learning maintain with each other a dialectical relation of complementarity: both are jointly needed, but each tends to obscure the clear manifestation of the other (which may justify the use of cross-sectional research in investigating developmental stages).

Children grow cognitively by moving from one developmental stage to another: from sensorimotor to pre-operational (Case's interrelational), to concrete-operational (Case's dimensional), and formal-operational (Case's vectorial). As they pass through these stages, they shift their focus of interest from one level of thinking/mentation to the next, although these levels are complementary and cumulative. As they do, children's concern moves from the here-and-now present (this is sensorial perception/action, sensorimotor); to the present inferred via some analysis, synthesis, and interpretation (this is intelligent perception/action, preoperations); to the empirically inferred future (this is intellection/action, concrete operations); to the possible future, inferred not from experience but from rational analysis of noncontradictory possibilities (this is intellectual processing or reason, formal operations). In table 5.3 we named these levels of "phenomenological" processing, giving the corresponding developmental stages and basic brain areas.

Within each Piagetian stage there are substages, each of which uses as bootstraps the maturational growth of mental attention, our Matt system of hidden operators. In it, at least constituents *M* and *I* grow maturationally in a concurrent manner with age, up to adolescence. This growth impacts on a child's ability to learn. A mental-processing *M*-unit is a quantum of *M*-capacity needed to maximally activate ("hyperactivate") learned, but not automatized, schemes.

As discussed in chapter 3, it is reasonable to assume that units of *M*-capacity are acquired cumulatively with maturation, possibly developing at a constant rate (so that time taken to acquire a given amount could be used as an indirect estimate of this amount). Initially, during the first two years of life, transition from one *M*-level to the next occurs about every four (first year) or eight months (second year). However, stage transitions after 3 years of age occur only every twenty-four months. This change in rate of stage transition suggests that sensorimotor *M*-units (the *Me* scale) may be about three to six times smaller than the symbolic *M*-units (Mk scale). Thus, sensorimotor (*Me*-) units and symbolic (*Mk*-) units cannot be combined or interchanged, because the mental (*k*) units are much larger.

We assess *M*-units behaviorally by counting, within suitably misleading tasks, the maximum number of distinct schemes that *M* would have to boost simultaneously to solve the task; thus, the need for task analysis "from within" (e.g., Pascual-Leone, 1970, 2013; Pascual-Leone & Baillargeon, 1994; Pascual-Leone et al., 2012; Pascual-Leone & Johnson, 1991, 1999, 2005, 2011, 2017; Pascual-Leone et al., 2010). In figure 7.2, we follow Case's (1992) model to indicate for each Piagetian stage (I, II, III, IV) and substage (0, 1, 2, 3) the average age at which the substage in question normally occurs. We symbolize, using repeatedly the letters A and B with a developmentally changing concrete meaning, the emergence of different sorts of integrative functional structures (schemes or schemas) at the end of each stage. This succession illustrates graphically reflective abstraction (Piaget, 1975/1985)—the levels of abstraction that Karmiloff-Smith tacitly adapted within cognitive science under the name of *representational redescription* model (Elman et al., 1996; Karmiloff-Smith, 1992). Such renaming is misleading, however. It ignores that these levels are found equally well (as Piaget asserted, e.g., Piaget & Garcia, 1983) in operative procedures and in figurative representations, when proper task analyses are done.

The theory-estimated complexity (*M*-demand) of tasks or items, and the person's quantitative levels of *M*-capacity (which *M*-power, *Mp,* assesses) are general constructs, in principle applicable across all content-domains (based on organismic processing assumptions). This is in line with both Vygotsky's concept of internalization (R. Miller, 2011; Kozulin, 1990) and with a decentralized, situated construal of Piaget's reflective abstraction process (Pascual-Leone, 1995, 1996b; Pascual-Leone et al., 2012; Pascual-Leone & Johnson, 1999, 2011; Piaget, 1975/1985). Content (*C*) learning and relational or logical-structural (*L*) learning are epistemically reflective because they internalize experiential/ecological resistances to agency/praxis that the person has encountered. Both forms of learning express the contingencies or "causal texture" of the actual environment (Pascual-Leone & Irwin, 1998; Pascual-Leone & Johnson, 2005, 2011; Tolman & Brunswik, 1935).

Developmental data supporting our claims on mental attention were presented among others by Pascual-Leone (1970, 1978, 1987, 1995, 1996b, 2000a, 2000b); Pascual-Leone and Sparkman (1980); Pascual-Leone and Johnson (2005, 2011); Pascual-Leone et al. (2010); Johnson et al. (1989); Arsalidou et al. (2013), and more. Complementary data and discussions of this mental-attention model, in particular about *M*- and *I*-resources, appear in Arsalidou et al. (2010); Howard et al. (2014); Im-Bolter, Johnson, and Pascual-Leone (2006); Johnson, Im-Bolter, and Pascual-Leone (2003); Pascual-Leone and Baillargeon (1994); Pascual-Leone and Goodman (1979); Pascual-Leone and Morra (1991). Chapters of this book, in particular chapters 3, 8, and 9, give task analyses illustrating how mental-attentional demand of tasks can be estimated across content domains. Morra, Gobbo, Marini, and Sheese (2008) and Troadec and Martinot (2003) each offer a chapter with good overviews of our *M*-theory and TCO.

Is *M*-Capacity a Maturational General Resource of Developmental Intelligence?

Data supporting the construct validity of *M* as a measure of developmental intelligence are many. We will limit ourselves in this section to summarize three types of evidence.

1 Relationship between *M*-Task Performance and Standardized Ability Score

The discussion that follows is adapted from Pascual-Leone and Johnson (2017). Pascual-Leone, Johnson, and Calvo (2004; Calvo, 2004) sought to demonstrate the general and content-free characteristics of mental-attentional capacity, by treating it as a constructivist-developmental measure causally related to Gf (fluid intelligence, the functional core of *g*, which may have as key organismic causal determinant a developmental growth of *M* and *I*). We studied 1,148 grade-four children (9 to 10 years old), from twenty-one Toronto-area schools. Children completed the figural intersections task (FIT) measure of *M*-capacity (Arsalidou & Im-Bolter, 2017; Pascual-Leone & Baillargeon, 1994; Pascual-Leone & Johnson, 2005, 2011). FIT items ask one to find the area of intersection of a variable number of geometric shapes (from two to eight). The *M*-capacity estimate is the maximum number of shapes the child can reliably intersect in his or her item responses, theoretically interpreted as maximal number of task-relevant schemes a subject can hold in mind together while solving items (FIT *M*k-score). The percentile scores on the school-administered *Canadian Cognitive Abilities Test (CCAT,* 1998)—a standardized measure of cognitive ability with verbal, nonverbal, and quantitative subscales—were available for 1,052 of the children. Average FIT score for the sample was 3.92 ($SD = 1.34$), matching the theoretical prediction that 9- to 10-year-olds have an *M*-capacity of four symbolic schemes simultaneously coordinated (Pascual-Leone, 1970; Pascual-Leone & Johnson, 2005). Mean percentile score on CCAT ($M = 54.96$, $SD = 26.22$) was as expected from age norms.

FIT correlated strongly with total CCAT score, $r(1050) = .59$, $p < .0001$. We also computed mean test scores for each of the twenty-one schools and ran correlations on the school means. Again FIT was highly related to CCAT, $r(19) = .73$, $p < .01$. We partialed out the percentage of ESL (English as second language) students at each school and average family income of the census tracts (neighborhoods) where the schools were located (obtained from 2001 Census of Canada data). FIT remained strongly associated with CCAT, partial $r(17) = .74$, $p < .01$. We can infer that variance due to *M* is strong and likely distinct from the variance as a result of sociocultural learning/experience.

To examine more closely the pattern of covariation between FIT and CCAT, we grouped participants in terms of deciles in CCAT percentile score and computed the mean FIT and CCAT score for children falling into each CCAT decile group. Figure 7.3 shows the covariation of mean FIT and CCAT. Up to the 81st to 90th percentile group, mean FIT score increases linearly with CCAT score. However, in the highest percentile group (91 to 100), mean FIT score abruptly increases, breaking the linear pattern. Within age, linear growth of FIT scores with the CCAT percentiles is likely to be due to constructivist learning: children's available *M* and growth in sociocultural experience and school learning, moderated by individual differences in *M*-capacity. However, the spurt in FIT scores after the 90th percentile is unlikely to be due to sociocultural and

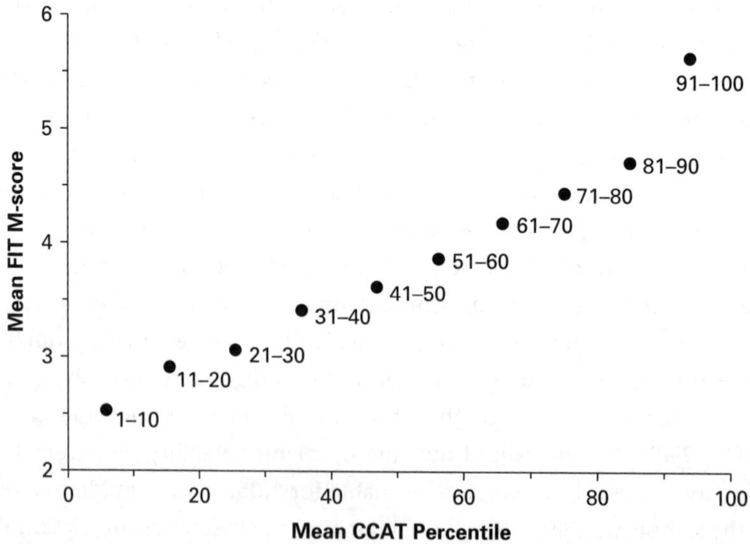

Figure 7.3
Pascual-Leone et al. (2004) sample divided into ten groups according to CCAT percentile score (1–10, 11–20, etc.). For each group, mean CCAT is plotted against mean FIT *M*-score. (From Pascual-Leone, J., & Johnson, J. [2017]. Organismic-causal models "from within" clarify developmental change and stages. In N. Budwig, E. Turiel, & P. Zelazo [Eds.], *New perspectives on human development* [p. 77]. Cambridge University Press. Copyright 2017 by Cambridge University Press.)

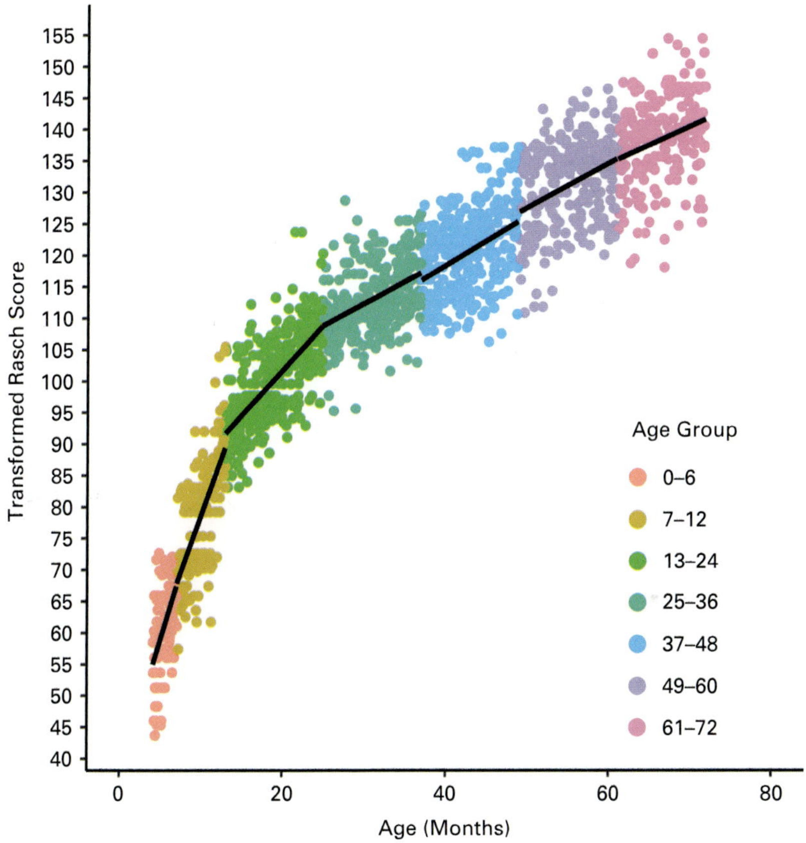

Plate 1

Association between child age and transformed cognitive-development Rasch score on Dimensional Inventory for Child Development Assessment. (From de Mendonça Filho, E. J. [2017]. "Evidências de Validade Relacionadas à Estrutura Interna da Escala Cognitiva do Inventário Dimensional de Avaliação do Desenvolvimento Infantil" [Unpublished master's thesis]. Universidade Federal do Rio Grande do Sul, Porto Alegre, Brazil. Reprinted with permission.)

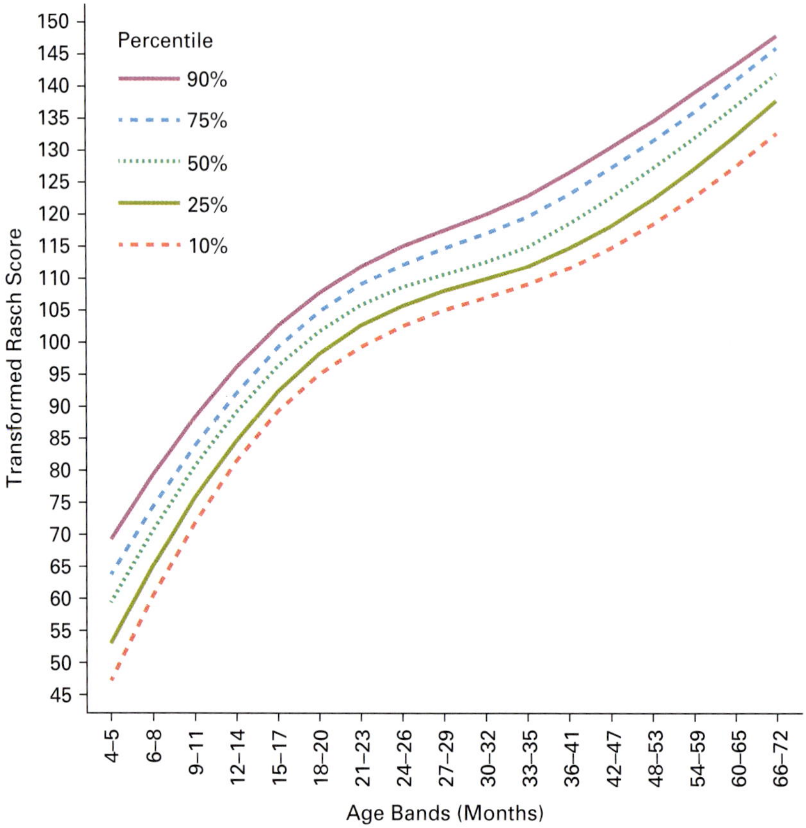

Plate 2

Normative cognitive-developmental curves for various percentile levels on the Dimensional Inventory for Child Development Assessment. (Adapted from Silva, M. A., de Mendonça Filho, E. J. & Bandeira, D. R. [in press]. *Inventário Dimensional de Avaliação do Desenvolvimento Infantil* [Dimensional Inventory for Child Development Assessment]. São Paulo, Brazil: Vetor Editora. With permission of the authors and the Publisher.)

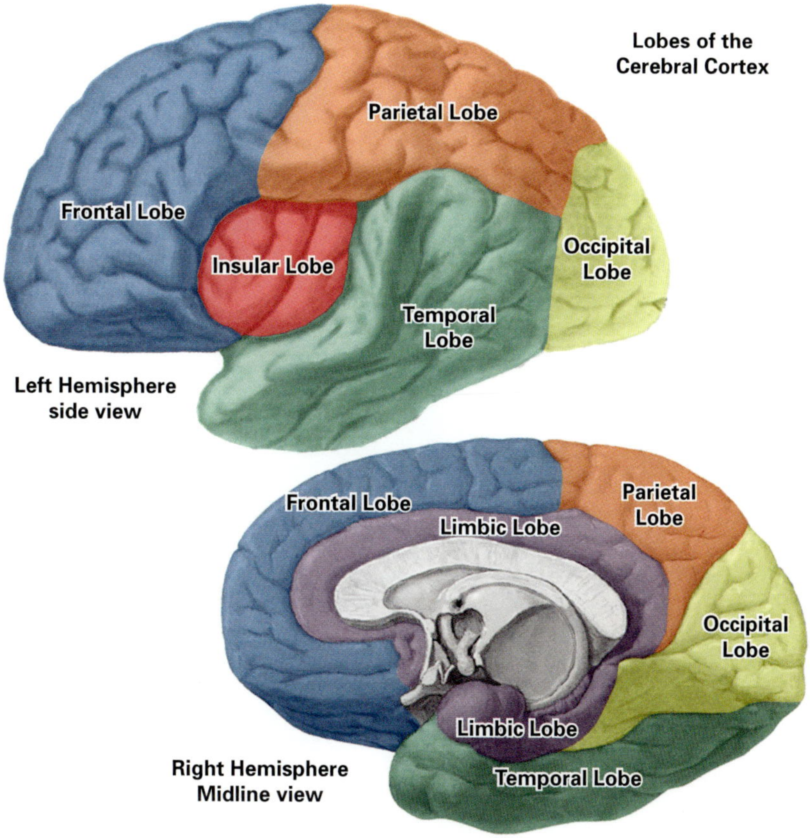

Plate 3

Lobes of human cerebral cortex: Lateral left hemisphere (*top*) and medial right hemisphere (*bottom*). In left hemisphere parietal and temporal lobes are pulled apart to exhibit the insula lobe. (From BrainInfo [1991–present], National Primate Research Center, University of Washington, http://www.braininfo.org. CC BY.)

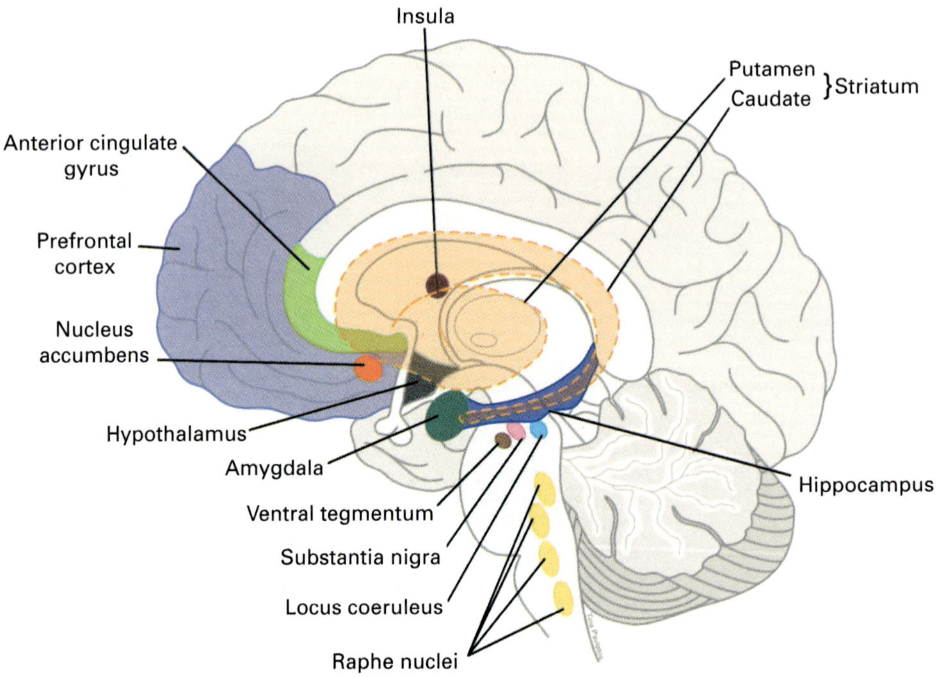

Plate 4

Inner (medial) side of the right hemisphere, exhibiting location of various important limbic or subcortical centers. (Illustration by Tina Pavlatos.)

school learning or experience, because it does not follow the age-bound linear growth characteristic of learning factors. Performance at the highest level in general ability is likely to express mostly a higher maturational *M*-capacity for children in the last CCAT percentile, perhaps a sign of true cognitive giftedness.

There is a way to appraise whether the variance boosting FIT score in the 91st to 100th CCAT-percentile corresponds mostly to sociocultural learning versus maturational *M*-capacity. Because the average 9-to-10-year-old has (in our well-supported developmental model) an *M*-capacity that can coordinate easily no more than four symbolic schemes, FIT items with more than five figures should be hard for this sample. Success on hard items may rely on learned perceptual strategies, such as guessing that the correct response is likely to fall in an area with high density of intersecting lines. If learning enhances performance on hard items, we thus may expect these items to exhibit higher correlation with CCAT in the very high percentiles. However, if learning is not an important factor in FIT performance for the highest percentile CCAT subsample, we should expect higher correlation between CCAT and the easier FIT items that children's *M*-capacity can handle. We, therefore, split the FIT test into two subtests: FIT-easy items (with two to five overlapping shapes) and FIT-hard items (with six to eight shapes). Children of this age should be able to solve the FIT-easy items using only *M*-capacity. In FIT-hard items, however, they must use both *M* and learned perceptual strategies.

We then correlated scores of the two types of FIT items with the CCAT score. For the total sample, the correlation was [$r(1050) = .57$] for FIT-easy versus [$r(1050) = .53$] for FIT-hard—a statistically significant ($p < .05$) difference. We then examined correlations for just the group in the 91st-plus CCAT-percentile group. Because of reduced variance, correlations were lower, but they maintained the same pattern: [$r(93) = .26$, $p < .05$] with FIT-easy and [$r(93) = .19$, $p > .05$] with FIT-hard. This finding adds credence to our claim that the abnormally high FIT score in the highest CCAT-percentile group was due to the factor *M*, a general-purpose maturational (not learning) constituent of Spearman's *g* (fluid intelligence, to be precise).

2 Individual-Difference Intelligence versus Developmental Intelligence

The question of whether these are causally two distinct forms of intelligence, raised by Pascual-Leone and Johnson (2005), is relevant here, because main organismic-causal determinants of ID-intelligence (the individual's intelligence variation) may be distinctly different from, albeit complementary with, developmental intelligence (the age-bound, maturational, normal growth in intelligence). Given the universality of Piagetian and neo-Piagetian stages/cycles, the component of general intelligence corresponding to developmental intelligence may be largely due to the portion of intelligence variance known to be maturational. Janet, James, Spearman, Freud, Luria, and other researchers called mental "energy" this maturational cause of intelligence (as Pascual-Leone, 1970, assumed for *M*-capacity). Thus, if maturational growth of mental attention exists, two distinct causal sources of general intelligence should exist that are independent.

General intelligence, as measured in individuals, is a complex functional probabilistic-invariant found across cognitive content domains. It subsumes both ID-intelligence (individuals' intelligence variation mainly due to learning) and developmental intelligence (maturational, age-bound variation). Factor scores of the *g* factor, often expressed by Gf (fluid general intelligence), are good individual evaluations of this compounded general intelligence. It was always assumed without proof that ID-intelligence is the same as developmental intelligence studied by Piaget and many others, but is it causally so? The discussion of CSMS and FIT data below is adapted with permission from Pascual-Leone (2019).

Figure 7.4, provided by Dr. Michael Shayer and based on research by Adey and Shayer (1994; see also Demetriou & Spanoudis, 2018; Shayer, Demetriou, & Pervez, 1988), presents these two causally distinct aspects of intelligence: one is the mean average score in Piagetian intelligence performance within every given degree of competence, exhibiting the ID variation (assessed by means of all different percentile-level samples from 12,000 children). Second is the age-bound mean developmental growth in children of Shayer's excellent Piaget tasks' point scale. The two aspects appear respectively as ordinate and abscissa of figure 7.4. The range in ID variation in intelligence may in part be innate, but it often should be due to variation in prior intellective or intellectual learning experiences, as figure 7.3 already suggested.

Shayer's CSMS survey data (1975–1978) on 12,000 students from 9 to 16 years of age (Adey & Shayer, 1994; Shayer et al. 1988) used a carefully designed stage-level scaling similar to our *M*-measures, but using only Piagetian Reasoning tasks—a unified scale based on three major Piagetian tasks (spatial relations, volume and weight conservations, and pendulum). The three tasks were given on three different occasions and covered every substage/level of development from Piaget's early concrete to mature formal operations. Point scores like those of *M*-measures were assigned to the obtained Piaget's substages in all tasks and then averaged to generate individual scores. A striking feature of Shayer's data, previously known, is the wide variability of the children's competence level in Piagetian tasks at any given age (range of variation in the ordinate of figure 7.4). In this figure the curves express, across age groups, overall performance levels for each of seven different percentile samples (graded from low to high competence levels, from the average).

All percentile samples exhibit similar patterns in their age-bound developmental growth. This growth is estimated from the substage-levels passed. The curves maintain

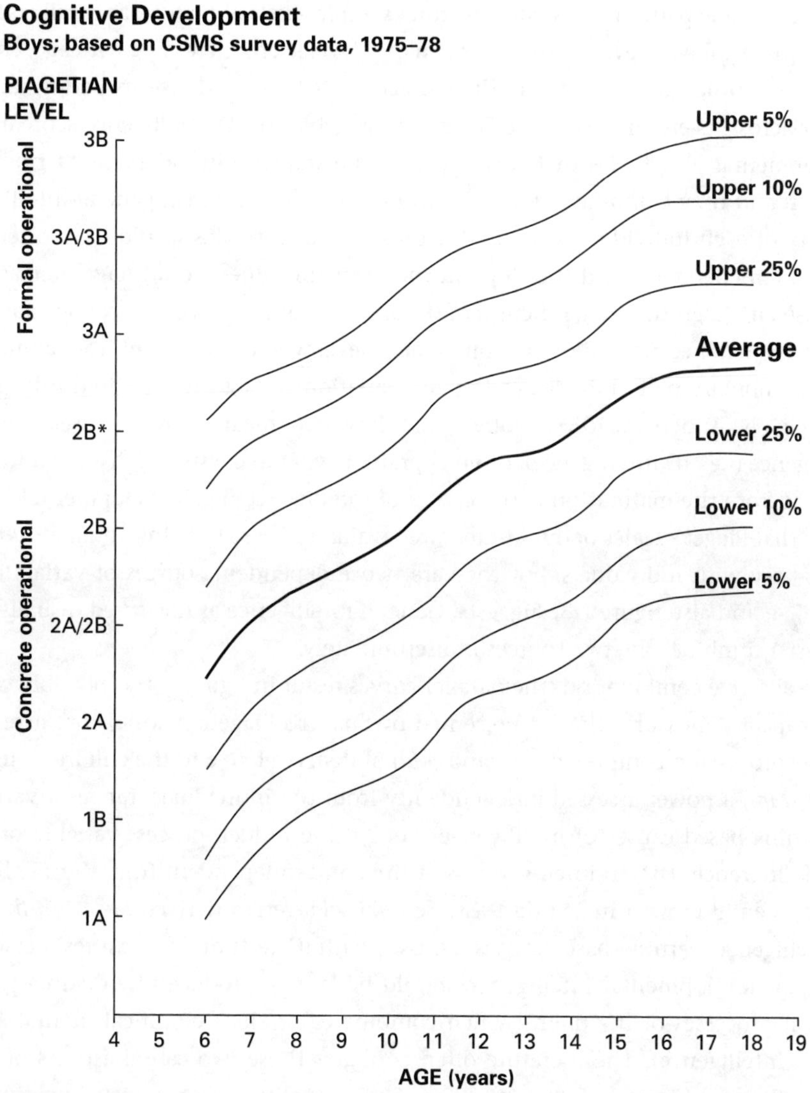

Figure 7.4
Adey and Shayer's (1994) Piagetian tasks' substage point-scale scores, within each level-of-competence sample ($N = 12{,}000$). Note that percentage curves are based on performance on Piagetian tasks. (Reprinted from *Intelligence, 72,* J. Pascual-Leone, Growing Minds Have a Maturing Mental Attention: A Review of Demetriou and Spandoudis [2018], p. 63. Copyright 2018, with permission from Elsevier.)

the same unique pattern across all percentile-sample levels of performance, all collinear and nearly parallel. In every curve, performance level keeps increasing with age, equally across percentile samples. This finding suggests that the variance in performance is similar across levels and ages. The key causal variables of ID intelligence score and of developmental growth (*M* or Matt) appear to be statistically independent from one another, and they combine (additively) to produce the empirical general-intelligence capacity of each individual. Note that if these causal variables would have interacted (been interdependent) and the slope in the ordinate values would have increased or decreased in magnitude as a function of the abscissa values, percentile curves may have spread as a fan, because children's competence level would then result from a multiplicative combination of individual-difference variation and age, relative to this Piagetian scale (e.g., N. Anderson, 1981). Notice what these data mean: learning-based causes of intelligence (i.e., ID intelligence), often appraised by IQ-like tests, are distinct and independent from the maturation-based causes of intelligence (i.e., developmental intelligence) that Piaget's scales or our *M*-measures evaluate. These learning and maturational causes vary with individuals, but they are two independent sources of variability, as figure 7.4, and also figure 7.3, suggests. General intelligence as measured (usually with I.Q. tests) combines the two sources indiscriminately.

The additive combination (the parallel-curves result in figure 7.4) is possibly caused by *M*-capacity measures (here represented by Shayer's Piagetian scale) that assess the average processing-complexity demand (*Md*) of items, relative to the children's mental power (*Mp*), a power assessed independently from the individuals' range of variation in learning-based causes of intelligence. Thus *if* the hidden-process variable of individual differences (ID) in intelligence is distinct and independent from the age-bound developmental growth in *M* (Matt assessed via a Piagetian derived *M*-scale), *then* the ID intelligence learning-based causes (assessed with IQ tests or Gf measures) versus the causes of developmental intelligence should be largely produced by distinctly independent processes of the brain (which combine to produce empirical mental age or general intelligence). The literature often conflates these two causal aspects of intelligence (Pascual-Leone & Johnson, 2005). Such conclusion contradicts experimental researchers and psychometric developmentalists like Demetriou and Spanoudis (2018), who would explain developmental intelligence by appealing only to constructivist learning. Notice further that this causal separation of subjects' learning-based response variability (the ordinate) and the maturational developmental growth in *M*-capacity was postulated by Pascual-Leone in 1970 ("Thus it seems that any attempt to save the general-stages construct must account *separately* for the general structural invariants and for the response variability," p. 304).

Figure 7.5 shows a similar pattern of relative independence between variability in *M*-competence levels and children's cumulative *M*-score (*k* +, i.e., *k* or more) as a function of age. This figure was produced by Dr. Shayer, using data on the FIT from our Developmental Processes Lab at York University. These FIT data were collected over several years from 2,684 participants of various ages. FIT is a key *M*-measure, and the figure shows cumulative *M*-scores for age groups 7 to 14, and 17-and-over (i.e., young adults). For example, the curve for *M*-level 4+ shows proportion of participants at each age level obtaining a score of four or above. Note that subsamples defined by cumulative *M*-score levels are epistemologically equivalent to the percentile-passing-score samples of figure 7.4. Figure 7.5 shows that when we exclude the extreme *M*-competence levels two (well within the competence of all age samples) and seven (beyond the capacity of most participants), all other cumulative *M*-levels (three, four, five, and six) show curves that approximate parallel lines, suggesting that growth of the *M* variable with age is relatively independent from *M*-score variability within and across age groups—as found already in figure 7.4 for Shayer's Piagetian-derived *"M" scoring*.

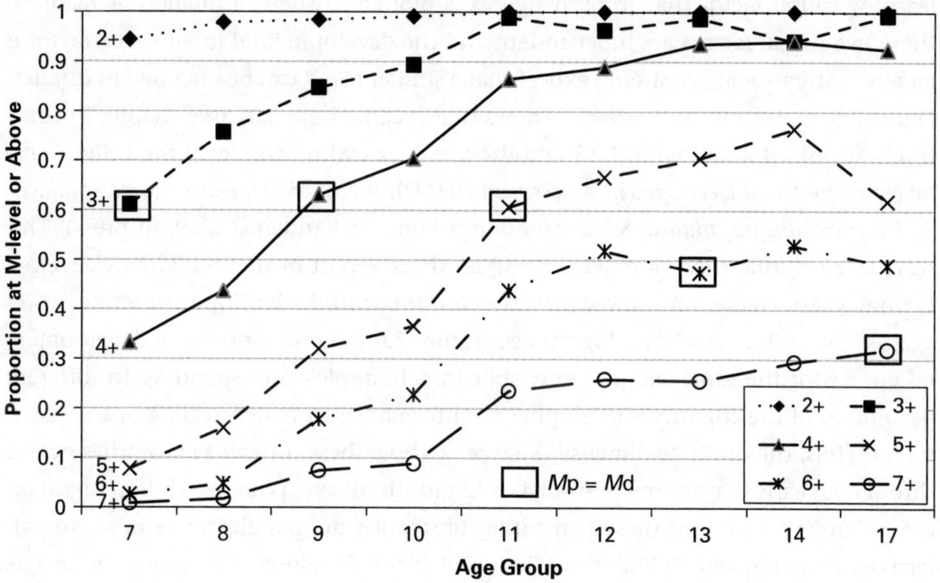

Figure 7.5
Proportion of age samples performing at different *M*-levels or above on figural intersections task (N=2,684). (Reprinted from *Intelligence, 72*, J. Pascual-Leone, Growing Minds Have a Maturing Mental Attention: A Review of Demetriou and Spandoudis [2018], p. 64. Copyright 2018, with permission from Elsevier.)

Notice further that data points in figure 7.5 marked by a small square are those in which children's theoretical *Mp* and items' *Md* are matched (*Mp*=*Md*). In our theory, when *Mp* and *Md* are equal, the subject's task difficulty should be the same, across these items, irrespective of absolute values of *Mp* and *Md*; it is their value difference that regulates task difficulty in *M*-measures (Arsalidou et al., 2010; Pascual-Leone, 1970; Pascual-Leone & Johnson, 2011). Thus square-marked points should be aligned horizontally as they are in figure 7.5, because at these points *Mp*=*Md,* provided that the *M*-measure constitutes, or at least approximates, an interval-scale within a measurement structure (Krantz, Luce, Suppes, & Tversky, 1971; Pascual-Leone & Baillargeon, 1994; Pascual-Leone & Johnson, 2011). The square-marked values are horizontally aligned from 7 to 11 years of age. Beyond this age they adopt a softly descending line, possibly because in late adolescence cognitive motivation decreases, perhaps because of hormonal changes. Such functional change in late adolescence has been found in brain waves (Uhlhaas et al., 2009).

As mentioned, the developmental growth scores plotted on the abscissa of figures 7.4 and 7.5 are likely expressing a maturation causal factor, independent from the learning causal factor that in both figures is plotted in their ordinate dimension. If these two causal factors are independent and the developmental intelligence factor is largely maturational, we should expect that a similar result can be obtained in infancy. Figure 7.6, derived from figure 3.3, shows that it can. These data (see chapter 3) come from analysis of a sample of 1,336 mothers in Brazil who completed the *Dimensional Inventory for Child Development Assessment* (IDADI; *Inventário Dimensional de Avaliação do Desenvolvimento Infantil;* Silva, Mendonça Filho, & Bandeira, 2019, in press). This scale assesses infant development (using mothers' report of their children's developmental level). Figure 3.3 showed a scatterplot of cognitive-development Rasch scores against the children's chronological age. Figure 7.6 presents normative developmental curves for the same sample, now split in subsamples corresponding to different percentiles of the cognitive-development performance score (a rescaling of the Rasch score). Thus, the ordinate dimension expresses how these different percentiles evolve throughout early development, from 4 to 72 months of age. What we wish to highlight is that growth curves of these percentile subsamples are parallel and similar to patterns exhibited by older children in figure 7.4. Thus, developmental growth with age, throughout infancy and childhood, is largely caused by maturation, in relative independence from the learned experience expressed in the figures' ordinate dimension.

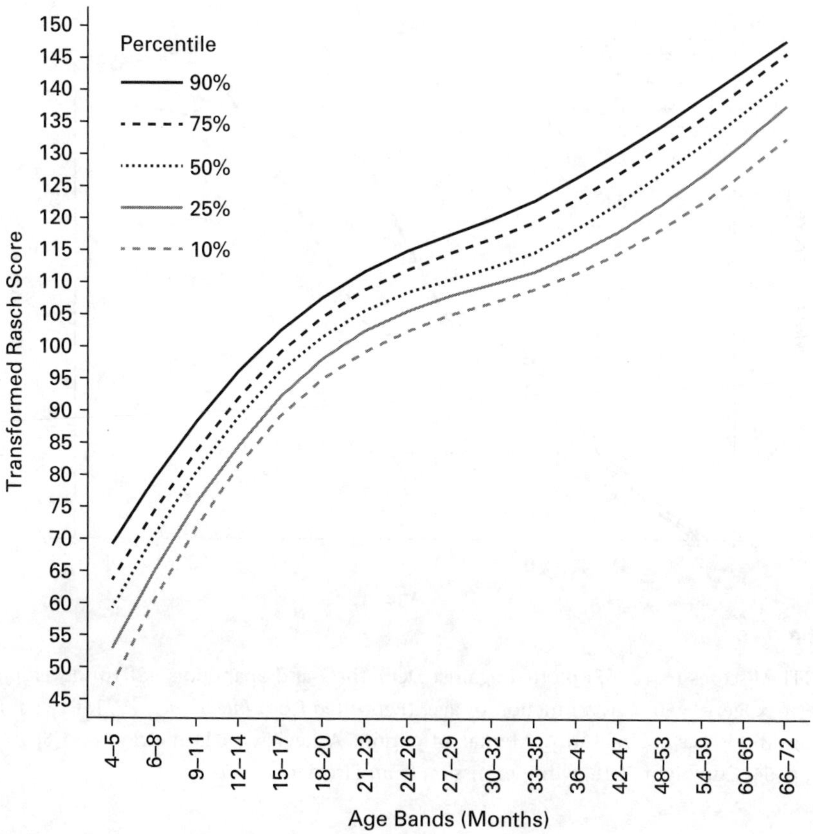

Figure 7.6
Normative cognitive-developmental curves for various percentile levels on the Dimensional Inventory for Child Development Assessment. (Adapted from Silva, M. A., de Mendonça Filho, E. J. & Bandeira, D. R. [in press]. *Inventário Dimensional de Avaliação do Desenvolvimento Infantil* [Dimensional Inventory for Child Development Assessment]. São Paulo, Brazil: Vetor Editora. With permission of the authors and the Publisher. See color plate 2.)

3 Correspondence of *M* (Matt) with the Developmental *g* Factor

This section is adapted with permission from Pascual-Leone (2019). Demetriou and Spanoudis (2018) found that the ID general-intelligence factor *g* did not capture all developmental variance in their data. This led them to adopt a *developmental g* score, based on a scoring method proposed by Tucker-Drob (2009). They obtained individual factor scores for *g* and multiplied them by the chronological age of individual children. Because this developmental *g* score is semantically/epistemologically congruent with our definition of *M*-power, we asked Andreas Demetriou to send his data, which

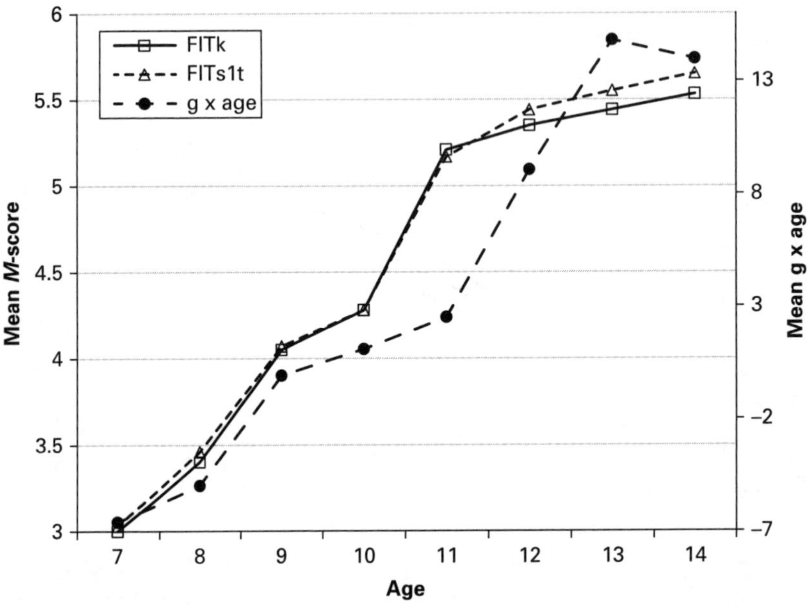

Figure 7.7
Mean FIT *M*-scores (*N*=2567) plotted against Demetriou and Spanoudis' (2018) developmental g scores (g×age; *N*=506), as a function of age. (Reprinted from *Intelligence, 72,* J. Pascual-Leone, Growing Minds Have a Maturing Mental Attention: A Review of Demetriou and Spandoudis [2018], p. 64. Copyright 2018, with permission from Elsevier.)

we used to plot figure 7.7. It compares Demetriou's developmental *g* scores (*N*=506) with developmental scores from our FIT. FIT data came from multiple studies and ages varied from 7 to 14 years (*N*=2,567). Demetriou and Spanoudis' developmental *g* score (*g times age,* i.e., g×age) is very close indeed to the two complementary FIT *M*-scores plotted in figure 7.7: FIT k-score (i.e., item class with highest number of relevant shapes that the person can conjointly intersect) and FIT s1t-score (i.e., the total number of items passed rescaled to the theoretical *M*-distribution).

The only point at which the curves diverge is at 11 years of age, where g×age falls below FIT *M*-score. According to our mental attention (Matt) theory, growth of *Mp* occurs every two years after 3 years of age (Arsalidou et al., 2010; Pascual-Leone & Baillargeon, 1994; Pascual-Leone & Johnson, 2011). Thus, all contiguous odd-even years after three theoretically have the same *M*-capacity (e.g., 7 and 8, 9 and 10, 11 and 12, 13 and 14). FIT is a single visuospatial task designed to measure *M*. In contrast, Demetriou's *g* factor score is calculated over a diverse multiplicity of tasks, across content domains, which often have considerable executive and knowledge demands. Thus, we could expect that in the first year of every *M*-level stage (i.e., in the odd years of participants' age) performance increments with age in FIT could already be manifested, whereas in the *g* factor omnibus task one may have to wait until the even years to observe effects of developmental *M*-growth in the data. This predicted contrast explains the anomalous departure of FIT versus g×age curve at age eleven. As expected, FIT's upsurge of *M* occurs between 10 and 11 years, whereas sharp increase in g×age occurs between 11 and 12 years, but, in these two *M*-capacity performances (FIT at 11 years and g×age at 12 years), the respective values are very close; thus, FIT *M*-measure is very close to the g×age developmental-general-intelligence variable, confirming the construct validity of *M* as a measurement of developmental intelligence.

M-Measures Are Not Just Working Memory Measures

M is the brain resource or specific regulation (hidden operator) that allows simultaneous holding in mind of *k* task-relevant symbolic schemes ($M=e+k$) within misleading tasks. By task design, *M* is measured in terms of Matt (Matt=<*E, M, I, F*>), parametrically increasing the *M*-demand across item classes (homogeneous scales) while holding constant across classes the demands of *E, I,* and *F.* This *M*-measurement is a new form of behavioral measurement, modeled from-within a person's mental process (Pascual-Leone & Johnson, 2017) and indexing the trade-off between the subject's *M*-power and the task/item's *M*-demand. This is a "from-within"/organismic form of *fundamental measurement* (Krantz et al., 1971).

Fundamental measurement occurs when numbers are assigned to objects/entities so that (1) real relations among objects can be mapped to relations among numbers and (2) this measure does not require prior measurement of other quantities. *Additive conjoint measurement* is a kind of fundamental measurement in which two distinct variables (in our case, the child's mental capacity, estimated from age, versus the task's mental demand, estimated from task analysis) combine to produce a single fundamental measurement scale. Our estimation of the two variables is based on modeling from-within the subjects' mental processes (metasubjective task analysis, MTA). With this MTA method the *M*-demand of tasks or test items are estimated using empirical developmental-experimental results. Then a scale of measurement is constructed as indicated below.

When the *M*-measurement scale (*M*-scale) has been empirically validated with results such as those illustrated below, using developmental *M*-scales created in different content-domains (e.g., visuospatial, verbal, visual, attentional memory), we are in a position to define average *M*-capacity of an age group. Our estimate of a group or individual child's current *M*-capacity (expressing his or her *M*-power or energy of attention) is the highest *M*-demand that the child or age group can master to solve *M*-task classes (each with a different *M*-demand) of items. Once empirically validated, the task analyses on items in each class (a homogeneous scale because its items all have the same *M*-demand) warrant the value estimates for the task's *M*-demand.

Thus, an *M-measure* is characterized by six points (Pascual-Leone et al., 2000). **(Mm1)** *M*-measures have classes of items, each presenting one level of *M*-demand, obtained via developmentally verified MTA. **(Mm2)** All possible levels of *M*-capacity are represented in the complete *M*-scale. **(Mm3)** Other aspects that can affect performance (such as executive task/item demand, prior learning or familiarity, perceptual field *F*-factors, whether facilitating or misleading) are held constant across classes of items. **(Mm4)** Items within the total scale may be randomly ordered or ordered from easier to difficult across classes, and item executive processes involved are basically constant across items and classes, so executive procedures can be learned in easy items and transferred to harder *M*-demand items without difficulty. **(Mm5)** Because *M*-measurement characteristics are structural and not content bound, we can obtain equivalent *M*-measures across very different content-domains, and these *domain-different M-scales* exhibit predictable same (or very similar) average quantitative scores, validated across ages (preserving predicted relations in *M*-growth with increasing age) and across populations (e.g., Arsalidou & Im-Bolter, 2017; Pascual-Leone et al., 2000). Some illustrative data are given below. Such predicted scale/score invariance is not mathematically possible without having an interval scale of measurement. Yaremko, Harari, Harrison, and Lynn (1982) define an interval scale as one in which the units are equal (constant) throughout, but which lack an absolute zero point. Our units are equal on the basis of metasubjective task analysis and the *M*-theory, and they empirically behave as such. **(Mm6)** In these measures the person's *M*-score (his or her *M*-capacity) is the *M*-demand of the highest item class that he or she can pass.

Notice that this empirically found invariance of *M*-scales across different content-domains and populations (and across human development) gives validity to our task-analytical method—metasubjective task analysis (MTA). It does because MTA is used to establish score values for the two conjoint variables (i.e., *M*-capacity versus *M*-demand) for the "from within"/organismic fundamental measurement procedure (we discuss the MTA method in detail in chapters 8 and 9). In conclusion, *M*-tasks assess trade-off between mental demand of task items and the mental capacity of a child. In contrast, working memory (WM) tasks lack distinct item classes of theoretically estimated mental demand, and test developers use only their refined common sense to decide on the mental demand of the few items, with validity often insufficient because of a lack in process-analytical theory. Other aspects or factors intervening on the WM item (e.g., executive demand, learning, perceptual field factors) often are not constant across items or not controlled. WM tasks often yield ordinal or perhaps metric ordinal scales, but not interval scales.

Invariance of *M*-Scores across Content Domains and Populations

Our average estimates of attentional *M*-capacity exhibit the characteristic (uncommon for a psychological measure) of having values reasonably (probabilistically) invariant across content domains and across populations, whenever task analysis and task construction are properly done. This quantitative invariance shows our *M*-measures to have the power of interval scales of (metasubjective) measurement (Pascual-Leone & Johnson, 2011, 2017). We briefly illustrate these important characteristics of *M* measurement using data from different studies in our lab.

Figure 7.8 shows the mean *M*-scores of twenty-six mainstream and twenty-six cognitively gifted 9- to 11-year-old children (gifted children identified by their school board) on three of our measures (Johnson, Pascual-Leone, Im-Bolter, & Verrilli, 2004; Pascual-Leone & Johnson, 2005, 2011). Mainstream peers were schoolmates who did not meet criteria for gifted programming. We originally hypothesized that gifted children (identified on the basis of IQ and ability test scores), who possess a superior executive repertoire relative to mainstream children, may not necessarily have superior *M* (Johnson et al., 2003; Pascual-Leone et al., 2000).

M tasks were the figural intersections test (FIT), the mental attention memory (MAM) task, and the direction-following task (DFT). As described above, FIT asks children to find the one area of common intersection of two to eight overlapping geometric shapes. The MAM (Agostino, Johnson, & Pascual-Leone, 2010; Im-Bolter et al., 2006) is a novel WM task with low executive demand, and (as task analyses suggest) variable degree of misleadingness in its items, making it an *M*-measure that minimally taxes executive repertoire. Children read aloud sets of consonants arranged in a circular pattern and then recall the consonants under differing degrees of interference (i.e., simple free recall, dialing each letter recalled on a rotary phone, or responding to a Stroop item before voicing each recalled letter). *M*-score was the average number of consonants correctly recalled across items in each of the three conditions. The DFT involves verbal directions of increasing complexity that children must carry out (Cunning, 2003; Pascual-Leone & Johnson, 2011). These directions require placement of tokens (which vary in size, color, and shape) on locations that vary in size and color. *M*-demand of each complexity level (estimated via task analysis; Pascual-Leone & Johnson, 2011)

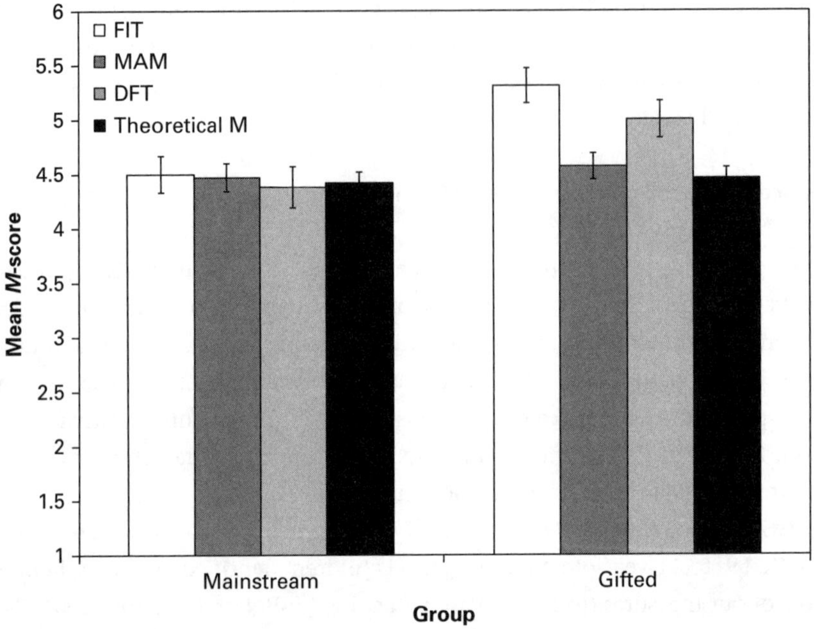

Figure 7.8
Mean *M*-scores on the figural intersections task (FIT), mental attention memory (MAM) task, and direction-following task (DFT) for age-matched gifted and mainstream children; bars show standard errors. (Adapted from Pascual-Leone, J., & Johnson, J. [2005]. A dialectical constructivist view of developmental intelligence. In O. Wilhelm & R. W. Engle [Eds.], *Handbook of understanding and measuring intelligence* [p. 196]. Sage. Copyright 2005 by Sage.)

varies with the number of objects or attributes specified in the instructions. *M*-score is the *M*-demand of the highest complexity level that the child can reliably enact.

Figure 7.8 shows mean *M*-scores, as a function of task and group, as well as expected theoretical mean *M*-score based on the children's ages. There is considerable group-mean invariance of *M*-scores across the content domains for the three *M*-tasks (visuospatial, verbal, and linguistic) in the mainstream sample. A mixed analysis of variance indicated that gifted children scored higher than mainstream children only on the FIT and DFT (which have greater executive demand than MAM), consistent with the common idea of giftedness. In other research we have demonstrated that FIT scores can be used to predict giftedness in children (Johnson et al., 2003; Pascual-Leone, Johnson, Calvo, & Verrilli, 2005; Pascual-Leone & Johnson, 2017). As described above (see figure 7.3), children with performance at the highest level of general cognitive ability (using the Canadian Cognitive Abilities Test, CCAT) tended to have a very high *M*-capacity for their age. Research is needed to investigate whether gifted children (as identified by school boards) have superior *M*-power independently from their executive know-how, which tends to be high (Johnson et al., 2003; Pascual-Leone et al., 2000).

Figure 7.9 shows (with the same samples) mean proportion pass on FIT and DFT, with items grouped by their predicted *M*-demands (*M*-complexity of the item classes; Johnson et al., 2004; Pascual-Leone & Johnson, 2011). Consistent with data in figure 7.8, gifted children perform better than mainstream children. However, within each sample, for each *M*-demand class of items, the proportion correct levels for FIT and DFT are almost identical.

Figure 7.10 shows the mean *M*-scores on FIT, DFT, and MAM of children with specific language impairment (7- to 12-year-olds) versus children with normal language. The

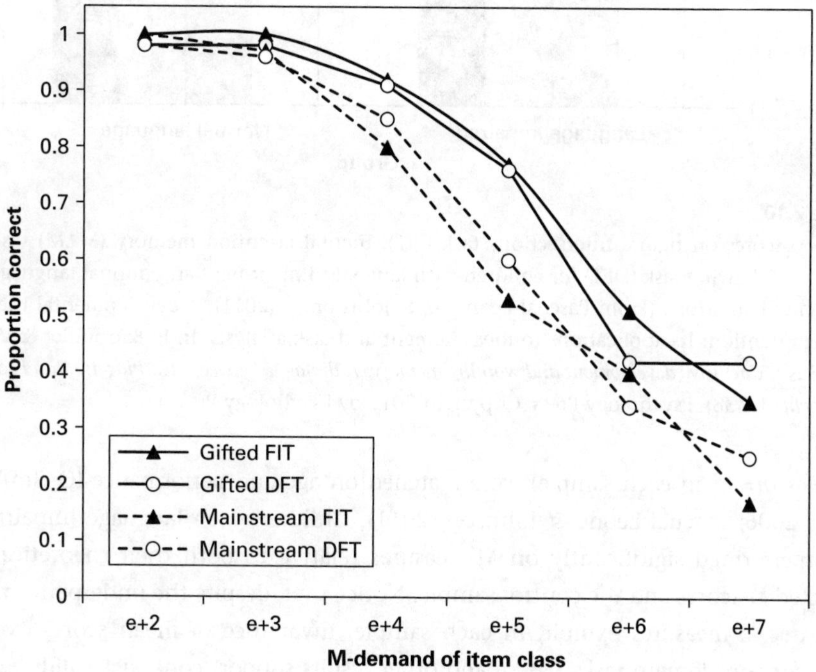

Figure 7.9
Mean proportion of items passed on figural intersections task (FIT) and direction-following task (DFT), as a function of items' *M*-demand. Samples are age-matched gifted and mainstream children. (From Pascual-Leone, J., & Johnson, J. [2011]. A developmental theory of mental attention: Its applications to measurement and task analysis. In P. Barrouillet & V. Gaillard [Eds.], *Cognitive development and working memory: A dialogue between neo-Piagetian and cognitive approaches* [p. 37]. Psychology Press. Copyright 2011 by Psychology Press.)

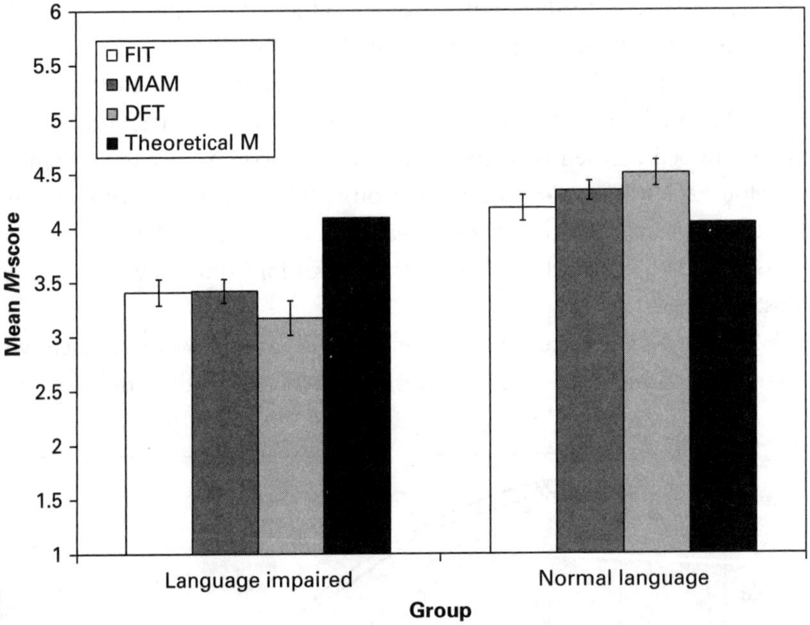

Figure 7.10

Mean *M*-scores on figural intersections task (FIT), mental attention memory (MAM) task, and direction-following task (DFT) for children with language impairment and normal language; bars show standard errors. (From Pascual-Leone, J., & Johnson, J. [2011]. A developmental theory of mental attention: Its applications to measurement and task analysis. In P. Barrouillet & V. Gaillard [Eds.], *Cognitive development and working memory: A dialogue between neo-Piagetian and cognitive approaches* [p. 38]. Psychology Press. Copyright 2011 by Psychology Press.)

samples ($n=45$ in each sample) were matched on age and performance-IQ (Im-Bolter et al., 2006; Pascual-Leone & Johnson, 2011). Children with language impairments underperformed significantly on *M*-measures relative to both their theoretical age-expected *M*-scores and the control sample. Notice that, despite the underperformance, the three *M*-measures exhibit, in each sample, invariance of mean scores over the three content-domain tasks. These and other results support construct validity of our *M*-measures and show that the scale of measurement behaves as an interval scale.

Upper Bound of *M*-Capacity in Adults

It often is claimed that the upper bound of WM in adults is about 4 (Cowan, 2005). The upper bound level of WM capacity is important, because we (e.g., Pascual-Leone, 1970) and others (Case, 1998; Cowan, 2005, 2016; Hansell et al., 2015) claim that endogenous attention is maturational during both the sensorimotor and symbolic-processing periods. After the sensorimotor period we claim that *M*-capacity has an upper bound of 7 (this is the adults' structural reserve of *M* or max *M*) but it also has a lower bound of 4 or 5 (adults' frequent functional *M*-level). Thus, we expect probabilistic oscillation of adults' measured *M*-capacity, often between the values 4 and 7 (Arsalidou et al., 2010, 2013; Pascual-Leone 1970).

Figure 7.11 shows as example the mean *M*-score obtained by young adult samples in five studies, using FIT, DFT, and MAM (Pascual-Leone & Johnson, 2011). The average *M*-score across tasks and samples is 6, consistent with our prediction (see also the 17+ sample in figure 7.5). Figure 7.12 shows data from four studies using the DFT with adults in Canada and Italy (Pascual-Leone & Johnson, 2011). Sample Canada-3 received DFT items in random order. Other samples received them in order of increasing complexity,

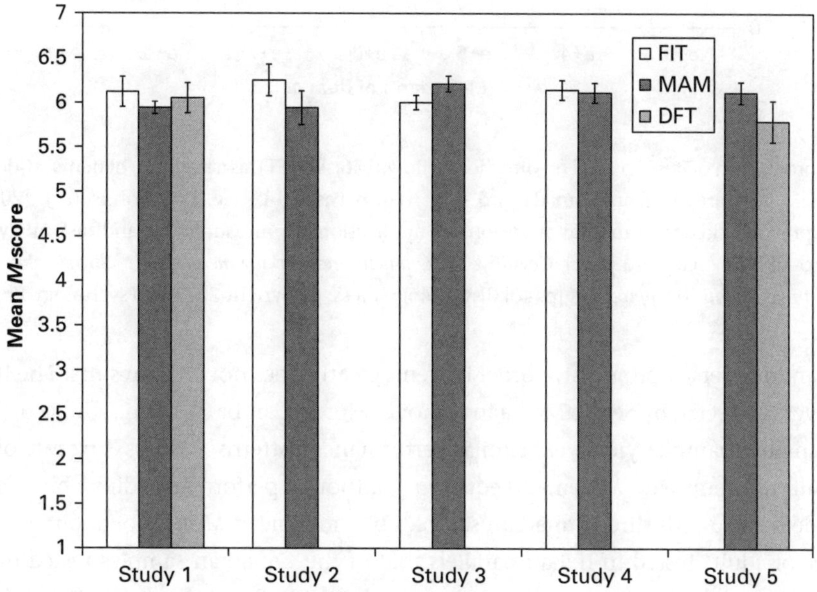

Figure 7.11
Mean *M*-scores on figural intersections task (FIT), mental attention memory (MAM) task, and direction-following task (DFT) from five studies with university-student samples; bars show standard errors. (From Pascual-Leone, J., & Johnson, J. [2011]. A developmental theory of mental attention: Its applications to measurement and task analysis. In P. Barrouillet & V. Gaillard [Eds.], *Cognitive development and working memory: A dialogue between neo-Piagetian and cognitive approaches* [p. 39]. Psychology Press. Copyright 2011 by Psychology Press.)

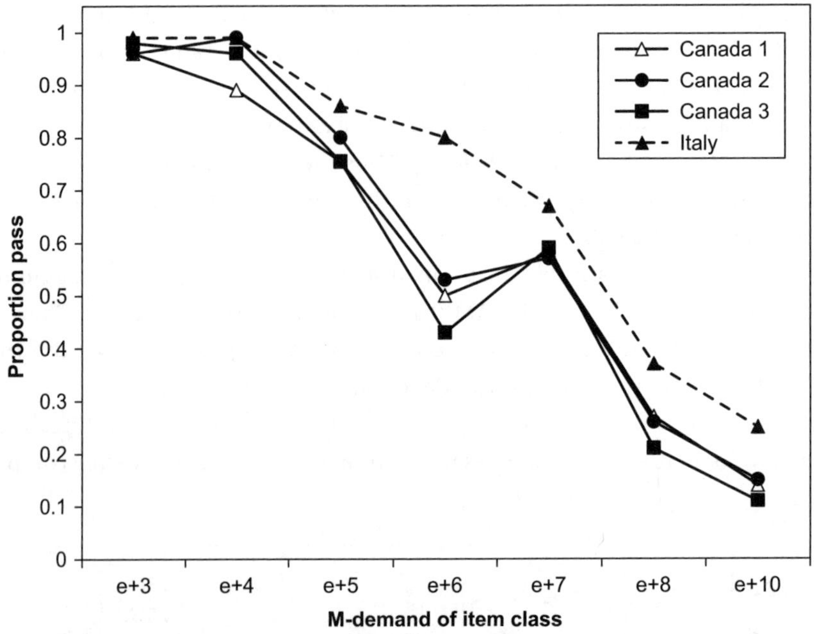

Figure 7.12
Mean proportion of items passed on direction-following task (DFT), as a function of items' *M*-demand for four adult samples from Canada and Italy. (From Pascual-Leone, J., & Johnson, J. [2011]. A developmental theory of mental attention: Its applications to measurement and task analysis. In P. Barrouillet & V. Gaillard [Eds.], *Cognitive development and working memory: A dialogue between neo-Piagetian and cognitive approaches* [p. 40]. Psychology Press. Copyright 2011 by Psychology Press.)

and random-versus-progressive order of items clearly does not affect results. The Italian data were collected by Sergio Morra in Genova (Morra, Camba, Calvini, & Bracco, 2013). All Canadian samples yield very similar performance patterns, with passing rate of 60% in adults for items with *M*-demand equal to 7 (although performance dips a bit for items of *M*-demand 6). All three Canadian samples obtain a mean *M*-score of about 6. Performance of adults tested in Italian parallels that of the Canadian samples tested in English, with an important exception: Italians maintain high performance on items with demand of 6, and their mean *M*-score is 7. Morra et al. (2013) suggested that this higher performance of the Italian sample is due to facilitating linguistic factors introduced by the Italian translation of the DFT items. Morra's linguistic interpretation, and his revised task analysis, were investigated and supported by research in our lab (Giuliano, Johnson, & Pascual-Leone, 2015; Johnson, Tsvetkov Kristen, & Pascual-Leone, 2014).

We return to this research in the next section, because it illustrates how situational/contextual factors can affect mental attentional (*M*) demand of tasks.

In all, the results support our prediction of an upper bound in *M* (often globally interpreted as working memory) equal to 7 symbolic schemes in adults. This result contradicts the majority view about adult WM capacity (a max-WM equal to 4 or 5).

Dynamic Context-Sensitive Character of Tasks' *M*-Demand

The current literature on WM research recognizes that a task's WM-demand (its *M*-demand when properly measured) is dynamically variable with contextual or organismic circumstances. Yet it is often not explicit why or how this is so. The key organismic factor in this regard may be task familiarity—familiarity with the repertoire of schemes that participants need in the task, including degree of automatization (habituation, chunking). Estimation of the *M*-units required in a task should count only the essential schemes that are distinct, that is, not already overlearned and chunked in their coordination (not *LC*-structured). Therefore, prior repeated experience with the task can change (decrease via schemes' learning and *LCLM* coordination) the task's actual *M*-demand. Part of the current problem for quantifying reliably WM-demand of tasks may be related to neglect of this important structural/relational factor of prior learning. In chapter 5 we discussed learning from a constructivist (metasubjective) perspective, including the role of associative learning/chunking (*LC* learning); we do not discuss it further.

We address here instead another distinct factor, often neglected in the literature. We refer to the internal field factor, or *F*-operator, known as holistic or Gestaltist factor, stimulus-response (S-R) compatibility, minimum principle, simplicity principle, and so on (see chapters 1 and 6). We illustrate the function of this contextual *F*-operator factor as modulator of *M-power/M-demand* trade-off, using three examples.

1 Color-Matching Task (CMT)

Arsalidou et al. (2010) showed the role of this organismic *F*-factor interacting with both learning and mental (*M*-) processing. We constructed two tasks very similar in informational and executive-demand characteristics but radically different in the role (facilitating vs. misleading) played by the *F*-factor. Using suitable samples of children, we showed the *F* factor affecting task difficulty.

The CMT paradigm adapts an updating design known as 1-back task (e.g., Owen, McMillan, Laird, & Bullmore, 2005). In this timed paradigm, participants see (one item at a time) a set of figures (either a clown or a bunch of balloons, depending on task version). For each item they must indicate whether relevant features (or schemes) of the current figure match those of the immediately preceding figure. We varied number of features/schemes (concretely, relevant colors) needed to determine a match. Thus, while keeping executive demand constant (i.e., number and kind of operations needed), the *M*-demand was varied across classes of items by adjusting number of relevant colors. Number of relevant colors (varied from one to six) determined *M*-demand (basically, WM complexity level) of each item. Participants had to indicate whether the current (i.e., target) item had the same set of relevant colors as the previous (i.e., criterion) item, irrespective of color location. Blue and green were colors to be ignored (irrelevant), as were colors in the clown's face. Presented in different item blocks were classes of items that varied in difficulty level in terms of *M*-demand. These levels spanned the expected *M*-capacity of our participants (children from 7 to 14 years, and adults).

To examine contextual influences, we contrasted a task with items having a set of balloons (CMT-balloon) with one whose items had the single integral figure of a clown (CMT-clown). CMT-balloon is facilitating, because task relevant features (i.e., colors) of the stimuli are salient, being part of segregated shapes (balloons) with relatively constant form and size, shapes best identified by their color. The bundle of distinct balloons was not salient as an integral whole. In contrast, CMT-clown is a misleading task: it contains in each item an attractive integral whole, the clown itself, very salient but irrelevant to the task—a "frame" in which relevant colors are embedded. Further, the costume parts (e.g., shoes, hat, patch) are attractive distractors, variable in shape and size, contiguous, and task irrelevant, which make the search for colors in the holistic Gestalt of the clown mental-attentionally demanding (because of *F*-misleadingness, because the holistic-clown Gestalt and the irrelevant distractors must be overcome and inhibited). These contextual differences should influence *M*-demand at every item level of the two CMT versions. We estimated this demand using our method of metasubjective task analysis.

Let us now examine some summary results from Arsalidou et al. (2010). One hundred forty-nine participants completed the CMT, covering age groups that represented five levels of *M*-growth: 7 to 8 years, 9 to 10 years, 11 to 12 years, 13 to 14 years, and adults (18 and older). The CMT *M*-score corresponds to the *M*-demand of the highest level of items passed. Metasubjective task analyses of these tasks, presented in chapter 9, showed that *M*-demand of each class of items is equal to the number of relevant colors to be matched plus two. This added value of two corresponds to needed operative and parameter schemes.

Arsalidou et al. (2010, table 1) found that performance for each age group on CMT-clown was high whenever the estimated *M*-demand (*Md*) of the item's class was less than or equal to the predicted *M*-capacity (*Mp*) of the age group in question. That is to say: For as long as $Mp \geq Md$, participants exhibited a very high probability of passing the item; when $Mp < Md$, participants had very low probability of succeeding. Furthermore, Arsalidou et al. (2010, table 3) used an equivalence test, that is, a test that directly examines whether means are statistically indistinguishable (Wellek, 2003), to compare CMT-clown mean *M*-score with the expected theoretical *Mp* for the age group. The mean values in all ages were close to the theoretically predicted, and in three of them (for samples 7 to 8 years, 9 to 10 years, and young adults of 18 years or more) equivalence vis-à-vis the theoretical prediction was found. Qualitatively and quantitatively our predicted *Mp/Md* trade-off was confirmed for CMT-clown.

In contrast, this was not so for the facilitating task CMT-balloon, which tended to show subjects passing item classes with *Md* one unit beyond the predicted *Mp*. In addition, Wellek test for the balloon task showed equivalence only for 13 to 14 years and adults. To conclude, a suitably chosen context-misleadingness (a negative *F*-factor), as found in CMT-clown, facilitates fitness to predictions for developmental *M*-theory. In contrast, congruently with the theory, a facilitating context makes estimation of *M*-capacity unstable (Pascual-Leone et al., 2000). In chapter 10, we briefly summarize two fMRI studies that used the CMT-clown. Results with adults show activity in the dorsal attention network that increases in a graded manner with increasing demand of CMT-clown items. This is congruent with the developmental *M*-power (participants) versus *M*-demand (tasks) trade-off, which we have discussed in this chapter.

2 Linguistic "Italian" DFT Effect

The DFT (Pascual-Leone & Johnson, 2011) is an excellent verbal/linguistic measure of *M*-capacity (see figures 7.8–7.10). It is composed of a series of increasingly complex verbal commands to move shapes on the lower part of a task board to spaces on the upper part of the board (e.g., "Place a green circle on a yellow space," "Place a red square and a white circle on a small yellow space"). Task analysis was used to estimate the *M*-demand of each complexity level in the DFT. Research with English-speaking participants has supported DFT's validity as a measure of *M*-capacity (Agostino et al., 2010; Baliousssis, Johnson, & Pascual-Leone, 2012; Im-Bolter et al., 2006; Pascual-Leone & Johnson, 2011).

Morra et al. (2013) translated the DFT into Italian and found that performance of adult and child Italian samples paralleled that of our English Canadian subjects (see figure 7.12). However, the Italian version appeared to carry one unit less of *M*-demand. In Italian, the adjective often is placed after the noun (equivalent to "Place a circle green on a space yellow"). Morra et al. reasoned that Italian word order may create a *recency effect* (an *F*-factor effect) that would facilitate DFT performance, because the final word (e.g., "yellow") may remain sufficiently active and not require *M*-boosting. Note that in English the final word of a DFT instruction always is "space," which is not informative. Johnson et al. (2014) tested this hypothesis, using an English DFT version with "Italian" phrasing (e.g., "Place a green circle on a space that's yellow"). English-speaking adults were randomly assigned to complete the standard DFT or a DFT that used English instructions with "Italian" word order. They also received the FIT *M*-measure and counting span—a commonly used WM task (e.g., Conway et al., 2005). They found that mean DFT *M*-score was higher in the "Italian" condition, exceeding DFT *M*-score in the standard condition (between groups) as well as FIT *M*-score (within-subjects). As predicted, performance on the "Italian" version exhibited the recency effect but only at higher *M*-demand levels. Both DFT versions correlated significantly with FIT and counting span.

Results support Morra et al.'s (2013) findings: Italian word order facilitates DFT performance and appears to reduce *M*-demand of the task by one unit. This effect is seen only at the *M*-demand levels that approach the expected *M*-capacity of the age sample, as one would expect if performance were expression of an *Mp*/*Md* trade-off. Giuliano et al. (2015) have replicated these findings by testing adult Spanish-English bilingual participants with both Spanish and English DFT versions (Spanish has word order similar to Italian). Thus, word order and other linguistic aspects that may affect *M*-demand should be considered when task-analyzing or translating verbal *M*-measures (or WM tasks).

3 Simon Effect and the Joint Simon Effect

There are other ways in which characteristics of the situation (particularly the context and the participant's affective/emotional mood) can change task performance. Indeed, because *F*-factor (whether facilitating or misleading) applies within the focus of attention (i.e., *M*-centration), if the content (the activated schemes) of this attentional focus changes, so could the task's *M*-demand. Motives also could change with the task construal. Consider as examples the Simon task and its derivative, the joint Simon task.

In the standard Simon task, the child sits before a monitor and has two response buttons (Left vs. Right). Stimuli appear one at a time, on the left or right side of screen. The task is to press the right button when a given stimulus (e.g., a frog) appears; and press the left button when another stimulus (e.g., a butterfly) appears. The joint Simon

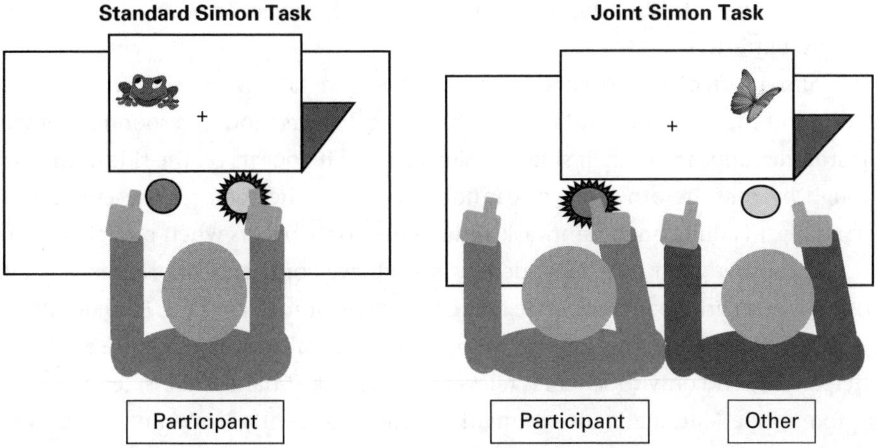

Figure 7.13
Diagram of Simon and joint Simon tasks.

task (e.g., Saby, Bouquet, & Marshall, 2014) is a variation in which two participants sit side by side and play concurrently, each responding to a different stimulus (e.g., either frog or butterfly, not both). Figure 7.13 illustrates both tasks. The *Simon effect* is the tendency of participants to respond faster, and with fewer errors, when the stimulus appears on the same side of the screen (right or left) as the appropriate response button. This is an S-R compatibility effect produced by the *F*-factor. It can facilitate fast and accurate performance on spatially compatible trials, but becomes a misleading factor on spatially incompatible trials. A metasubjective analysis of the Simon task is presented in chapter 9. At this point, we just remark that players know (via task instructions) that they must respond to butterfly, for example, by pressing the left button and respond to frog by pressing the right button. Similar rules are followed in the joint Simon task, but now one player responds only to the frog and the other only to the butterfly.

To respond correctly, participants must synthesize a response by suitably coordinating two symbolic schemes, which require the boosting power of *M*-capacity: (1) a PRESS-the-button scheme that already has chunked to it the operative actions of LOCATE the stimulus that appears and IDENTIFY it (as being a frog or butterfly) and (2) an operative parameter to PRESS, which evaluates the response button to choose by appraising (in Simon task) identity of the present stimulus, OR (in joint Simon) deciding whether this is the stimulus one should respond to. In the joint Simon task, if subjects construe the other player as participating in the same game (so that they experience the other's play as part of their own activity), parameter (2) is replaced by one reminding players of their own player-to-stimulus assignment (i.e., who, subject or other, should respond to frog vs. butterfly).

The Simon effect also occurs in the joint Simon task when the other player is friendly, and the subject identifies with this other. Players tend to respond faster when their stimulus appears on their side, but slower when it appears on the side of the other, although by rule the other player cannot respond to the subject's present stimulus. Remarkably, in adults, such joint-task Simon effect only occurs when participants have with one another good interdependence and achieve cordial collaboration. The effect is absent when participants compete or are hostile (Hommel, Colzato, & va der Wildenberg, 2009). This surprising finding becomes theoretically clear when we realize that participants attend only to schemes relevant to the task at hand, that is, tend automatically to exclude (automatic attentional inhibition) presently irrelevant schemes they are not attending to. When the other player is experienced emotionally as a cordial collaborator, both the schemes of the other and the subject's own schemes for the task are construed as task relevant and so are automatically included in the task's field of centration. Thus, as a result, the schemes are attended to, and interact dynamically with one another (prompted by the *F*-factor), which prompts the Simon effect. In contrast, if the other is hostile or unfriendly, and thus is not construed as being part of the subject's own task, schemes of the other are not task relevant and not *M*-centrated and would be automatically inhibited, preventing a joint Simon effect.

How Mental Attention Is Expressed in Tests of Intelligence

William Stern (1914/1977), contemporary of Binet and inventor of the IQ concept, defined intelligence as the person's skill in successfully coping with novel problem situations. This coping synthesizes or constructs performance by composing previously acquired processes (schemes) in novel ways. Thus, Burt, according to Stankov, defined intelligence as "the 'integrative function of the mind' that encompasses processes at *all* levels" (Stankov, 2002, p. 25). Other definitions have been offered, often less apt, but all assume that intelligence increases with age in normal persons, up to adulthood. They all assume that this growth with age is due to learning and maturation (in that order), and they do not differentiate between individual-differences (ID) intelligence and developmental intelligence, as we do. In spite of its beginning (in Binet's and Spearman's work), current conceptions of intelligence often suffer from lack of integration with cognitive-developmental research. Ever since Reuchlin (1962, 1964), researchers have recognized the need to integrate constructivist developmental (e.g., Piagetian or neo-Piagetian) theory with ID conceptions of intelligence (e.g., Case, Demetriou, Platsidou, & Kazi, 2001;

Demetriou & Spanoudis, 2018; Pascual-Leone, 1969, 1970; Pascual-Leone & Goodman, 1979; Pascual-Leone & Johnson, 2005, 2011; Pascual-Leone et al., 2012). The hope is that developmental theory and its empirical methods can help to create theory-based measures that can be used at all ages. This goal has not yet been met.

A good enough organismic-causal theory must explain these two sorts of measures: intelligence task/tests mostly rooted in individual differences other than age (our ID causal organismic factors) and those mostly rooted in developmental age (due to developmental growth in mental attention within relatively misleading problem-solving situations). Although these two aspects of intelligence were traditionally considered to have the same organismic causes, we suggest that they may in part have different causes: those of *individual-difference intelligence* (indexed by cognitive tests scaled with reference to across-age *reference groups*, or "populations") and those of *developmental intelligence* (which evaluates from a within-the-subject perspective, effective complexity of tasks/tests, by *using the age when ordinary children can first cope*, without special training, with the task/item in question). Both causal sources of intelligence empirically combine to form general intelligence, as usually measured. Some form of endogenous mental attention was pioneered under various names by many classics (e.g., Janet, James, Freud, Binet, Spearman, Jung, Luria). For them it was a mental energy or volitional effort used during acts of judgment or problem solving (perhaps as part of a self that includes voluntary/mental attention and preconscious or conscious plans).

In contrast to developmental intelligence, ID intelligence stems mostly from the study of adults using the correlational method and factor analysis. Factor analysis is a descriptive, abstracting tool, a way of condensing relational structures found in the data across tasks (e.g., Deary, 2000; Reuchlin, 1962; Spearman, 1927). Other approaches, like development and neuroscience, anchor and clarify constructively these intelligence causal factors (e.g., Arsalidou & Pascual-Leone, 2016; Horn, 1998; Horn & Noll, 1994; Reuchlin, 1964).

In table 7.3 we summarize correlations of the FIT and DFT *M*-tasks with various ID intelligence or IQ measures. Of these cognitive ID tasks, Otis-Lennon, OLSAT (Otis-Lennon School Ability Test), and WASI IQ (Wechsler Abbreviated Scale of Intelligence) are recognized general ability measures. Ravens Matrices is the known standard for fluid intelligence measurement, and Matrix Analogies (Naglieri, 1985) is a closely related test. The CCAT (Canadian Cognitive Abilities Test) and CAT-3 (Canadian Achievement Test) provide excellent assessments of ability and school achievement, respectively. All correlations are significantly different from zero, and most are $r = .50$ and above. *M*-measures also correlate with standard complex WM tasks. For instance, correlations from our lab for 7-to-12-year-olds: backward digit span with FIT, $r(45) = .54$, $p < .05$; sentence span with FIT,

Table 7.3

Correlations of FIT and DFT *M*-scores with standardized ability measures

Ability Test	Figural Intersections Task (FIT)		Direction-Following Task (DFT)		Sample
CCAT percentile		**.59****			N=1000, 9 years
CAT-3 percentile		**.51****			N=1000, 9 years
CCAT percentile		**.52****		**.60****	N=156, FIT & DFT at 7 years; CCAT at 9 years
CAT-3 percentile		**.53****		**.58****	N=156, FIT & DFT at 7 years; CAT-3 at 9 years
WASI IQ		**.45****		**.50****	N=156, 7 years
OL mental age	.71**	**.40****			N=161, 6–12 years
OLSAT raw score	.37**	**.39****			N=80, 9–13 years
Ravens total score	.40**	**.36***			N=80, 9–13 years
Matrix Analogies	.55**	**.36****	.45**	**.28****	N=175, 7–12 years

Note: CCAT=Canadian Abilities Test; CAT-3=Canadian Achievement Test; WASI=Wechsler Abbreviated Scale of Intelligence; OL=Otis Lennon; OLSAT=Otis-Lennon School Ability Test. **Bolded *r's*** are without age variance; non-bolded *r*'s are with age variance.

**$p<.001$. *$p<.01$.

$r(173)=.40$, $p<.01$, and with DFT $r(173)=.52$, $p<.01$; running span with FIT, $r(104)=.40$, $p<.01$, and DFT $r(104)=.50$, $p<.01$; counting span with FIT, $r(104)=.48$, $p<.01$.

We (Arsalidou & Pascual-Leone, 2016; Arsalidou et al., 2010; Pascual-Leone & Goodman, 1979; Pascual-Leone & Johnson, 2005, 2011) have a process-causal interpretation of organismic factors traditionally seen to underlie intelligence (e.g., Wilhelm & Engle, 2005). Clearly (see the last epigraph) there are multiple distinct factors (in our theory, different schemes and multiple hidden-resource operators) that help to configure *mental strategy formulas* for organismic functioning. Using task analysis, we can characterize formulas for different tasks and types of paradigms, such as those that assess fluid intelligence (Gf). These formulas describe demands in cognitive-processing strategy raised by tasks (or types of tasks/tests) or situations, when subjects pursue intended goals. Individual-differences in intelligence correspond to the quality and degree of availability in individuals of the resources and schemes required by these formulas.

From such theoretical perspective, psychometric factors can be explained semantically and process-causally. For instance, Gf (Horn & Hofer, 1992; Horn & Noll, 1994), the *fluid intelligence* factor, would have as main causal-organismic determinants the resource-operators and executive schemes of mental attention (Matt=<*E, M, I, F*>), together with learned effortful schemes (*LM* learning) that may be automatized (*LCLM* learning) in part (Pascual-Leone, 1995; Pascual-Leone & Goodman, 1979; Pascual-Leone & Johnson, 2005). In addition, tasks indexing Gf present a mental strategy formula that involves *intellective convergent* problems (problems with an explicit unique solution) in the context of misleading situations, usually tapping the participant's experiential (nonconceptual) domain—Piaget's infra-logical or our mereological processes. In contrast, Gc, *crystallized intelligence* (or acculturation knowledge) may have as a key determinant a good conceptual or logical knowledge (overlearned or automatized schemas from past effortful, analytical thinking—complex schemes' coordinations that we call *LCLM* structures). This knowledge is well exhibited in tasks that use mental attention within *facilitating* situations (see chapter 6).

In general, the nine psychometric second-order factors of Horn (1998), and of other psychometricians, are descriptively congenial to our approach. For instance, visual processing factor (Gv) may have as key resource determinants the *S*- and *T*-operators, in addition to the operators of mental attention (*E, M, I, F*) and schemes needed for coping with misleading situations. In contrast, the auditory processing factor (Ga), as well as language, may have the *T*-operator as a dominant constituent, often within facilitating tasks involving the auditory domain.

Our theory of constructive operators also can contribute to four other research enterprises: (1) developmental theories of neo-Piagetians (e.g., Case, 1998; Case et al., 2001; Demetriou et al., 2018; Demetriou & Spanoudis, 2018), (2) theories of WM (e.g., Cowan, 2005, 2016; Engle, 2002, 2018), (3) issues involving the working mind such as Matt and *M*-capacity measurement, briefly discussed in this book, and (4) helping to understand metasubjectively developmental changes that occur with aging, that is, the "mechanics" of normal human intelligence change during the later years (Baltes, Lindenberger, & Staudinger, 2006; Lövdén & Lindenberger, 2005; Pascual-Leone, 1983, 1990a, 1990b).

Mental Attention and Consciousness: Relation, Similarities, and Differences

What is consciousness? As suggested in chapter 3, it is a special representational state. According to Damasio (2012), "consciousness is a *state of mind in which there is* [conceptual and experiential] *knowledge of one's own existence and of the existence of surroundings*" (p. 167). Neuroscientific formulations of consciousness are offered by Damasio (2012), Dehaene (2014), and many others, but this vast literature is not reviewed here. To comprehend consciousness, we need a clear understanding of *representation* (Pascual-Leone, 1976a, 1976b, 1983, 1990b, 2000a; Pascual-Leone & Johnson, 1999; Pascual-Leone &

Irwin, 1998). As Bickhard (2015a) and others have emphasized, representation is not a passive *presentation* of what is given in experience. Bickhard (2015b) says that "interactive representation is anticipatory, and the anticipations of interactive potentialities may be true or may be false" (p. 64). A similar descriptive conception of representation, cast in very different language, was offered by Husserl (e.g., 1970, 1973). If we concretize this abstract formulation causally and use our own language (Pascual-Leone & Johnson, 1999), representation appears to be a subject's own creative dynamic synthesis of more or less conceptualized experience (issues, states of affairs). This dynamic synthesis emerges by way of competitive coordination (overdetermination) among clusters of highly activated compatible schemes of all sorts, present in the subject's field of activation.

Bickhard emphasized four key characteristics of representations. (**RE1**) They are *emergent*, that is, they result from (perhaps conscious) creative dynamic syntheses of distinct activated schemes. (**RE2**) All representations point to the future with tacit or explicit anticipations of plausible outcomes, and consequences that express both semiotic intentionality and a cognitive/affective "horizon" (i.e., expectancies that aim to some future or possible goals or outcomes, evoking other related schemes; Husserl, 1973). (**RE3**) Representations may carry a truth value (a true-versus-false evaluation criterion for anticipations, expectancies, and outcomes), so errors (false evaluations) can be recognized and changed by the psychological organism, whether consciously or not. (**RE4**) The truth value of a representation is knowable or accessible to the subject's metasubject and often can become conscious. Thus, a representation should not be defined from an observer's perspective, but from within the subject's own processes as we have defined these terms (i.e., a metasubjective perspective).

To complete our characterization of a symbolic or interactive representation, we must add two other key conditions. (**RE5**) Although Bickhard does not address this important issue, representations (and consciousness) can also involve operative schemes, which express transformations, and so representations can also carry what may be called deliberate, willful or conative, *intentions* (i.e., conscious or unconscious goals to do such or such). In this case, new figurative states and truth values appear when figurative schemes apply to characterize the newly transformed state-of-affairs produced by application of operative schemes. (**RE6**) Affective schemes (pure affects) or affective-and-cognitive schemes, such as emotions or other personal (psycho-cultural) schemes, may not carry truth values by themselves but carry instead vital (evolutionary, bio- and sociocultural) values. Both semantic-pragmatic facets (i.e., deliberate intentions and vital values) could be expressed, tacitly or consciously, in any representation.

Ordinary consciousness is the aptitude of a metasubject (the subject's inner organization, both operative and figurative self) to have a distinct representation of some constituents of its own thinking (perception, cognition), feeling (affects and emotions), or willing (deliberate intentions). As consciousness emerges, self-schemes differentiate that produce the *ego* (which for us is the *outer* self or *persona*—as Jung called it), the inner *self* (operative and figurative), the *Will*, and self-conscious reflective representations. Demetriou's (Demetriou et al. 2018; Demetriou & Spanoudis, 2018) construct of "cognizance" resembles considerably this sort of developmentally advanced metasubject; a metasubject that is representation driven, often conscious, and with a conative self-agency (an *operative self*) that can produce a Will function.

Ordinary subjective experience may not be a true representational experience. Ordinary low-level perception may not qualify as a deep-enough consciousness, because experiences in low-level perception are not distinct, not organized in distinct sorts or levels of knowing; they are holistic. Therefore Husserl, Piaget, and others talked in this case of "presentation" (presentational processes), which they contrasted with "re-presentation" (see Pascual-Leone & Johnson, 1999). From this developmental perspective we have emphasized (Pascual-Leone, 2000a; chapter 3), independently but converging with Damasio's (1999, 2012) neuroscience perspective, that consciousness emerges in the child, and in the brain, in a sequence of constructivist stages: from a level of sentience that in humans Damasio calls *protoself* to various intermediate levels that culminate in a personal, social-and-historical self-consciousness that Damasio calls *autobiographical self*. The agent who does this representation (the knower or knowing self) first appears in child development at the end of infancy (Legerstee, 1998). At about 35 months a reflective/conceptual self-consciousness has begun its developmental run (in chapter 3 we called it *self2.1*—for us the precursor stage of Damasio's autobiographical self). Consciousness of any real-life situation involves many distinct, different schemes, most of which are not automatized and cannot by themselves be hyperactivated. "Hardware" exists in the brain, monitored by executive schemes, that brings about this hyperactivation. In our view, a major "hardware" mechanism is endogenous *mental attention*.

III TCO, Task Analysis, and Neuroscience

8 Process Analysis and Mental Task Analysis: Foundations

We outline a method of mental-process task analysis, describing its epistemological foundations. This metasubjective task analysis (MTA) models mental qualitative processes that individuals would have to use when solving tasks according to given strategies. This is a useful way to find the effective processing complexity of tasks. The analyses are constructed using schemes, concretely interpreted within the task, and the subjects' brain resources or hidden operators, as formulated by our theory. These general principles of mental-process task analysis could be applied (using suitable organismic-causal general models or theories) to any sort of task. Our focus is to model macro-processes in problem-solving, learning, and developmental tasks.

> Whenever a *system of conditions* is given that can be realized in different contents, there we can hold to the form of the system itself as an *invariant*, undisturbed by the differences of the contexts, and develop its laws deductively.
>
> —Cassirer, 1953, p. 40

> I have searched in vain for a system that allows one to systematically identify and catalog the elements of a cognitive task that ratchet up its complexity. Such a system would have practical applications as well.
>
> —Gottfredson, 2016, p. 216

Unlike physical scientists, psychologists find it difficult to distinguish between two sorts of equally necessary (and complementary) theories: descriptive-empirical (meta-empiricist) theories and their models or statements versus causal-organismic (meta-subjective) theories and their models/statements. Epistemologically, this is the difference between an empirical modeling done from an observer's perspective (from "outside" the subject-matter or the organism—a *visible process*, as Merleau-Ponty, 1968, would call it) versus a theoretical modeling done from the perspective of "within" the subject matter or functional organism itself (the *invisible process* of Merleau-Ponty). Only the latter can lead to causal-organismic modeling. Here, *"cause"* should be understood not as a single classic cause, but as an organismic one, that is, a constellation of organismic causal determinants, simple or complex, that together overdetermine the empirical manifestations. In every case both approaches are needed, because a descriptive-empirical or meta-empiricist moment is functionally essential (Pascual-Leone & Johnson, 2017).

Congruence between causal-organismic and descriptive-empirical models is the ultimate criterion of a theory's worth. Perhaps for this reason, in human science a valid observer's description of processes often is seen as a causal explanation and not just the description that it is. For instance, it is often expected in developmental research that fine-grained sequential delineations of descriptive steps of children's development are causal accounts of this development. This is not so, however. Proper causal-organismic theories and models must have distinct organismic-causal accounts that are metasubjective models from within the processing subject. We call them metasubjective because they hypothesize causal determinants as working from within the organism and distinct from the descriptive-empirical aspects they attempt to explain (Pascual-Leone & Johnson, 2005, 2017). For example, well-known developmental sequences in math tasks, expressing what children understand and solve at different age levels (developmental stages), are not causally explained just by appealing to working memory or to task difficulty. To causally explain them, we need metasubjective ("from within") models of the subjective processes intuitively explaining the task solution. Such models may include constructs like schemes and working memory or mental/executive attention (e.g., Matt or *M*-capacity) and so forth, processes that can be qualitatively appraised and measured. All these organismic causal factors must be empirically distinct from the task's descriptive process (e.g., from the schemes' application sequence) they help to explain.

Organismic-causal (metasubjective) models can appraise complexity within stipulated acts of mental (sensorimotor or symbolic) processing. In psychology to do so one must adopt a processing complexity viewpoint analogous to that of Gell-Mann (1994). For Gell-Mann, the complexity of processing involves effective (as opposed to crude) complexity. He defines *crude complexity* as follows: "The length of the shortest message that will describe a system, a given level of coarse graining, to someone at a distance, employing language, knowledge, and understanding that both parties share (and know they share) beforehand" (Gell-Mann, 1994, p. 34). This definition is related to the complexity of algorithmic-information content, often called *algorithmic randomness*. Algorithmic-information content is largest for random strings. However, randomness is not what we mean by complexity in rational analysis (or psychology). Instead Gell-Mann (1994, p. 50) defines *effective complexity* with this remark: "In fact it is just the nonrandom aspects of a system or a string that contribute to its effective complexity, which can be roughly characterized as the length of a concise description of the *regularities* of that system or string."

Crude complexity refers to the length of a concise description of the whole system (or string) that includes its random features and not regularities alone. For Gell-Mann (1994, p. 50), effective complexity is "related to the description of regularities of a system by a complex adaptive system [such as a person] that is observing it." Like Piaget, Gell-Mann understands by regularities the relevant probabilistic (often functional) invariants that emerge in the context of repeatable situations (Ullmo, 1958, 1967) relevant for an adaptive system, such as human producers or observers. *Invariant* meaningful aspects are preserved over repetitions even when incidental random aspects vary. In effective complexity, one appraises the power (perhaps estimated by the number of distinct characteristics essential or relevant for the task, i.e., task regularities) of a set of invariants. This concept of probabilistic invariants includes *coordinating relations*—all essential constituents that produce meaning and relevance of a message or agency/praxis. Piaget and developmental constructivists often call such experiential (often nonverbal) invariants schemes or schemas, which are relative to given situations, to agency/praxis, and to human intelligence.

Mathematicians and logicians regard math or logic as being generic explicit abstractions of this sort of invariants, which (relative to a given activity) express general principles of rational procedure with "ideographic symbols and in a form which exhibits the connection of these principles one with another" (to use words of C. I. Lewis, 1960, p. 1). A major implication of these definitions is that effective complexity is not only objective but also the basis for *metasubjective modeling,* that is, a formulation or formula from within the process itself, as synthesized by a human analyst. An implicit or explicit process/task analysis is involved in such modeling.

General causal (e.g., physical, chemical, organismic) theories in science can be applied to concrete situations only when using specific theoretical models that incorporate the task constraints (the resistances of the Real). Local models in classic physical science (e.g., those that instantiate calculus in particular situations) illustrate this sort of implicit task analysis. The equivalent in human science is *process task analysis,* often implicit and unexplicated. Process task analysis modeling tends to be qualitative and descriptive in human science. It becomes explicit and organismically explanatory when the subject's own task activity is modeled from within, particularly if key semantic-pragmatic, spatial, and often temporal constraints of the task are retained. We call this method *metasubjective task analysis* (MTA). Such analysis models functional infrastructure of a given strategy relative to a task.

Effective complexity is relative to the user's goals and context of use, the user's own criteria, and the intended agency/praxis. MTA helps formulating the task processing and effective complexity. We have informally illustrated this method in previous chapters and introduced some notation. In this chapter, we explain and justify it.

Construction of MTA models, with their rational reconstructions of task-relevant processes within the subject, assumes (at least implicitly) an *organismic model* of the subject-agent and his or her goals, together with a model of the task's *situation*. With these initial conditions in mind, analysts infer (doing what Peirce called abductions) the processes and effective complexity of a performance strategy that could in principle solve the task when implemented. This is *specific operative modeling*, from within the person, of processes that may cause the performance. Such rational reconstructions can be called *mereological models*, because they usually intend implicit or explicit experiential thinking (often nonverbal, semantic-pragmatic, spatial and temporal), actions, mental acts, or operations dealing with mental/physical objects[1]; this is not verbal-propositional thought or utterances. Note that situations can be seen as objects that the subject's agency/praxis attempts to control, understand, or modify. In MTA, we assume processes and logical effective complexity of a given strategy, highlighting the manifold psychological sequence involved. When explicitly done, this task analysis uses graphic or symbolic representations of distal objects or entities and represents how mental operations (operative schemes) apply to change relevant objects and anticipate or cause emergence of a performance.

Analysis of Conservation of Substance

Some tasks, such as Piaget's conservation tasks or adult versions of conservation such as the wine and water problem (see chapters 6 and 7), can help to illustrate similarities between analysis-of-figures in classic geometry (Hintikka & Remes, 1974) and analysis-of-schemes done in MTA. Because metasubjective analysis models a task solution from the perspective of a given strategy, schemes (information-bearing processes for task solution) are important. A common problem of task analyses is demarcation of schemes that inform the task. Piaget (Piaget & Morf, 1958, p. 86) defined a scheme as "an organized set of reactions that can be transferred from one situation [a task] to another by the assimilation of the second to the first." As explained in chapter 5, this functional construct can be defined as the (brain) coordination of three components: (1) a *functional component (fc)* that formulates gist and goals of the scheme, often providing expectancies that anticipate what could result from its application; (2) a *releasing component (rc)* that carries conditions under which this scheme can apply; and (3) an *effecting component (ec)* that produces effects that follow from the scheme's application and can also recognize when the scheme should terminate this application (this is a terminal or stop subcomponent, important in recursive schemes). Expectancies and terminal subcomponents show that schemes are minimal but very active functional systems that can recursively apply (schemes of schemes of schemes) and form functional/adaptive hierarchies.

When a scheme is proposed in MTA as component of a strategy, the assumption is made that this scheme is sufficiently learned or overlearned to function as a unit in its activation (activation by input, other schemes, or hidden operators). The plausibility of such an assumption should be evaluated rationally by means of a method of psychogenetic reconstructions (**Rec**). This method has four analytical phases:

(**Rec1**) The first phase is a *concrete scheme definition* (instantiation) as a functional semantic-pragmatic system (i.e., *fc*, *rc*, and *ec*). (**Rec2**) The second phase is the schemes' *ecological evaluation* to back the assumption that it is already in the subject's repertoire (i.e., long-term memory). Does the subject's usual expectable concrete environment offer sufficiently frequent opportunities for practicing and thus learning such scheme? (**Rec3**) The third phase involves *motivational/operative evaluation* to assess whether the subject's activities, motives, and interests may have led to learning this scheme in the past. (**Rec4**) The fourth phase, *psychogenetic evaluation,* appraises whether children of this age or younger, from the same population, in view of their known actual performance in suitable tasks (or their known age-bound mental attentional and learning capacities), could have acquired the scheme(s) in question. This fourth phase explicates Piaget's tacit psychogenetic assumption: if younger or peer-age children can be said to have acquired a certain scheme, then older (or peer-age) children of the same sociocultural background also should have it. A scheme that has passed three of these four evaluations is an acceptable scheme for metasubjective task analysis.

Pascual-Leone, Goodman, Ammon, and Subelman (1978) illustrated schemes, their dimensionality, and their codetermination of a performance by using Piaget's conservation of substance task. It may be useful to begin here with such a well-known task. In one of Piaget's versions, children were given two balls of Plasticine (**A** and **B**) and asked to make believe they were food or candy (see figure 2.1 and the upper half of figure 8.1 for a schematic). One ball was red and the other blue. Children were asked whether the two balls had the same amount. If they said "no," changes were made until the children accepted the balls as equivalent (e.g., if the balls were food to be eaten, they would be equally filling). Thus **B = A** represents a higher relational scheme stating that the balls as food amounts were practically equivalent. The tester then transformed one of the balls into a sausage, and children were asked whether now the ball and sausage had the same amount. Young children often would claim (even though nothing had been added or removed) that now they did not have the same amount, thinking that

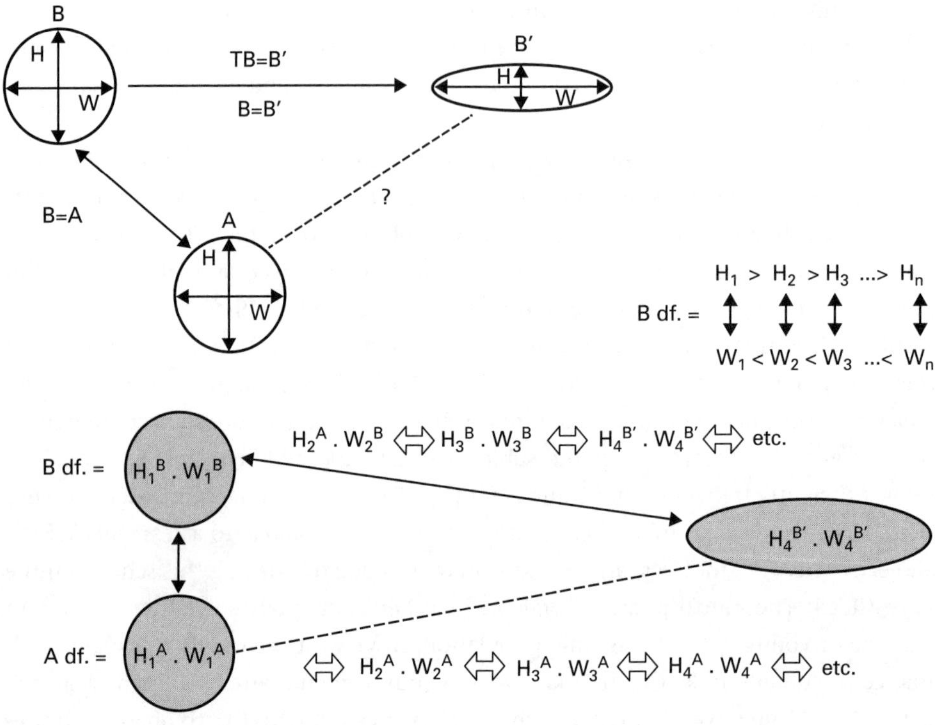

Figure 8.1
Piaget's structural model of conservation.

the sausage (**B′**) had more because it was longer (some children may claim that **A** had more because it is thicker).

A and **B** as object-schemes can be defined as purely experiential coordinations of three components, *fc, rc,* and *ec*. Assuming an experiential, nonlinguistic knowledge, these object schemes could be described as follows: **A** could be [*fc*: a ball of certain size presumed to be food, brought to the game by the tester, and so forth; *rc*: a round patch of red color with features of a manipulable substance and perhaps other features; and *ec*: if held it will not be heavy, will be soft and yielding]. Likewise, for the scheme **B=A** [*fc*: if the two objects were food and we each would eat one of them, each of us would have had as much food to eat as the other; *rc*: both have same size, same shape, although one is red and other blue; *ec*: the two balls are the same in amount]. Notice that other symbols in this subjective analysis, which stand for schemes, could be similarly formulated. For the sake of clarity, task analysts should be prepared to formulate explicitly any scheme used in the analysis if it raises any doubt in their or others' mind.

We call **T** the actual transformation of **B** into **B′** (the ball **B** transformed into sausage **B′**). This is a transformation in which no matter was added, lost, or taken away; for us, **T** is an *identity transformation* relative to amount. We call **TBB′** the child's concrete memory-scheme for this change (a complex experiential scheme that expresses a temporally ordered set of events—an experienced sequence, which we call a *fluent* or *temporal scheme*). Such clear memory is important, because data on children show that 5- or 6-year-olds already recognize the transformation **TBB′** as preserving the original amount (preschoolers know experientially that this particular T was, in our terms, an identity transformation).[2] As a result, **B** is seen by the child as having the same amount as **B′** (we represent this idea by scheme **B=B′**). The mental-attentional demand (*M*-demand) for the identity task that uses only one ball, **B**, and transforms it, is just two symbolic schemes: **TB** and **B′** to be held in mind simultaneously, via mental attention (*M*-demand $= e+2$); this is accessible to 5- or 6-year-olds.

Notice that both conservation of equivalence and conservation of identity involve verbal questions about future or past events (e.g., transformations of shape do not affect amount of substance), and so the tasks are not sensorimotor but symbolic. Executive processes (boosted by the "e" of *Mk*-demand $= e+2$) must be used. In symbolic tasks executive schemes are activated, we believe, by the sensorimotor *Me*-capacity, which in turn is driven by affects and personal schemes. Because 7-year-olds already have an experiential scheme of substance related to identity transformations, we can interpret the upper part of figure 8.1 as symbolically representing the working problem-solving mind (*M*-centration and dynamic synthesis) of a child facing the conservation of equivalence task. It represents the moment when **A** has already been compared to **B**, leading the child to conclude **B=A**, and **B** has been transformed into **B′**, leading the child to infer spontaneously the scheme **TB=B′**. At this point, object **B** has already disappeared, perceptually transformed into **B′**.

Such mental representation of the child is not just perceptual but symbolic. If we use, for brevity, algebraic terms that children do not have, their intuitive problem can be described by the formula: **A=B & TB=B′→A ? B′** (i.e., Does **A** have equal amount as **B′**?). This algebraic transcription of the upper diagram in figure 8.1 symbolically exhibits the problem-solving conflict of children. They know that **A=B** and **B=B′**, but what can they do with this information? The theory of schemes alone (and this is the only theory that Piaget and other constructivist-learning developmentalists have) could not explain how a 7- or 8-year-old solves this problem, unless we unrealistically assume that these children possess a scheme for the logical rule of transitivity (i.e., if A=B & B=B′→A=B′). However, such logical rule could not be innate because of its abstractness and cannot be learned without prior experience. Piaget had a different, mistaken, constructivist-learning explanation of how children solve this problem. We shall discuss his explanation later on, to illustrate the structuralist method of analysis that Piaget used (not unlike that of current constructivist-learning theoreticians).

Although preschoolers lack logical structures of transitivity, children after 7 years of age can answer this question correctly, when they have constructed the mental representation of the problem symbolized in figure 8.1. They solve the task, we claim, by using a neo-Gestaltist "field" factor heuristic (of *F*-operator). This factor induces the inference that when two of the relations connecting the three objects (the two balls and the sausage) are equivalence relations, then the third relation that connects A and **B′** "should" (likely will) also be an equivalence. This "should" inference (possibly an instance of Peirce's abduction) coordinates, using minimal effective complexity, the active schemes within this mental set. Such is the brain dynamics that Berthoz (2012) has called "simplexity." Indeed, if the child's representation of this problem were called the stimulus and the answer to the question called the response, then this "simplexity" (classically known as Gestaltist "simplicity" principle, but also known as S-R compatibility, Proctor & Reeve, 1990) applies. It induces the inference or response that **A=B′**. In our theory, this principle expresses the *F*-factor, which is related to cortical lateral inhibition in the brain, as discussed by Edelman (1987). Only after this correct response has been heuristically (not logically!) evoked, the child may (with much practice) acquire the constructivist structural system or psycho-logical *grouping* (Piaget's groupement) that Piaget proposed for solving this and other problems. Constructivist learning does occur, as Piaget believed, but it follows and cannot precede the solution of this problem (see chapter 5). Notice that this sort of *F*-facilitated "simplicity" of reasoning also appears in a common implicit inference: "the friends of my friends are my friends," often conceptualized in psychology by means of balance and social congruence theories (e.g., Zajonc, 1960). Piaget interpreted this sort of inference using his "equilibration" principle (Pascual-Leone, 2012a), a dialectical and abductive, but not a formal-logical, way of solving problems.

How to Do Metasubjective Task Analysis (MTA)

An illustration of objective task analysis (tacitly, also of subjective analysis) is available in Polya's (1973) heuristic method for mathematical reasoning. In part inspired by Polya's methods, MTA investigates heuristically how to solve a problem or task by first finding a strategy (the idea of a plan), which is the first necessary step of our method. In the words of Polya (1973), this is done as follows:

> [W]e shall distinguish four phases of the work: First we have to *understand* the problem; we have to see clearly what is required. Second, we have to see how the various items are connected, how the unknown is linked to the data, in order to obtain the idea of the solution, to make a *plan*. Third we *carry out* our plan. Fourth, we *look back* at the completed solution, we review and discuss it.... It is generally useless to carry out details without having seen the main connection, or having made a sort of *plan*. Many mistakes can be avoided if, carrying out his plan, the student *checks each step*. Some of the best effects may be lost if the student fails to reexamine and to *reconsider* the completed solution.... The student should also be able to point out the principal parts of the problem, the unknown, the data, the conditions.... *What is the unknown? What are the data? What is the condition?... Is it possible to satisfy the condition?* (pp. 5–7)

Then Polya (1973, p. 7) recommends the student/analyst to represent ("*draw a figure*" of) the data and unknown and design a suitable notation.

Whereas Polya's method suggests a heuristic for conducting an objective/subjective logical analysis of effective complexity and how to solve the problem in question, MTA methods attempt to formulate models of how psychological organismic schemes and hidden operators of participants could actually achieve what Polya talks about. The following questions may be raised in MTA to translate Polya's own questions: What is the task? What does the subject have to do or solve (this is the "unknown")? Which are the schemes elicited or needed by the task situation? Which schemes in the subject's repertoire are related or necessary for the task solution (this corresponds to the "data" of Polya)? How do these effective-complexity schemes relate to one another according to the chosen organismic general model (e.g., the TCO)? How are schemes affected and changed in their functioning by organismic hidden operators/principles elicited locally within the task situation and ongoing plan (this corresponds to Polya's "condition")? Can schemes activated by the situation, or retrieved from memory by an executive strategy, suffice to solve the task if aided by elicited hidden operators/principles (this is Polya's question: "Is it possible to satisfy the condition?")?

Notice that the "condition" of Polya tacitly (or explicitly) includes schemes and hidden operators/principles not directly elicited by the situation but contributed by subjects' willfully using their cognitive resources and following the executive strategy. This is related to what Hintikka and Remes (1974) called "auxiliary constructions" in classic geometry. Notice further that in MTA one also has to recognize automatically elicited schemes (i. e., shadow schemes) and strategies that are irrelevant or misleading for the task at hand. One may have to task-analyze and understand them as well, because control of these automatic irrelevant schemes (and their related automatic hidden-operator applications) can impose constraints on the strategy's effective complexity.

Because human organisms tend to follow a mini-max principle (minimal complexity that maximizes adaptation) we seek to represent only the essential task processes (the minimal effective complexity that could solve the task). Thus, an important first principle of metasubjective task analysis (**MTA1**) is that organisms that process information and perform *take as much error as they can afford*. That is to say: for as long as the intended results are adequate enough for the goals pursued, the task's performance may not be improved; consequently, task analysis should adopt the simplest and least precise executive strategy that still meets the intended goals. An explicit formulation of these supposed goals (the tacit or explicit strategy/plan) of the subject is important, and these goals should be assumed to be as simple as they can be to still satisfy strategy and task constraints. Secondly (**MTA2**), because task analysis always and only bears on a given, chosen strategy for a given task, one should take up first the *most likely strategy for the intended population of subjects*. Then, if other strategies appear that we should investigate, other MTAs on the new strategies can be done.

Given a strategy, analysts can infer or derive the predicted or postdicted performance by coordinating four sources of constraint in subjects' performance. (**Source 1**) First is the *structure of the situation* in which the performance is produced, including objects, instructions, expectations, emotions, habits or automatisms, and other structural elements, which could affect performance within the situation. In this way an objective/subjective model of the situation can be formulated. (**Source 2**) Second is a *general* (explicit as a functional totality) *model* of the subject's psychological organism, that is, a modeling "from within" of the processes that enable subjects to cope with the task, given the chosen strategy. This involves what we may call a *metasubject*—an idealized working model of the subjects' general processing formulated "from within" to explain how they cope with task problems. Our metasubjective models tend to have three sorts of processes: schemes, hidden operators, and hidden principles. The latter express the most general regulations or "functional axioms" of the psychological organism (or metasubject). The three most important hidden principles for metasubjective task analysis, because they must be assumed by the analyst, are the principle of Schemes' Overdetermination of Performance (*SOP*), the principle of Equilibration (*EQU*), and the principle of Schemes Inhibition and Decay (*SID*) (Pascual-Leone et al., 1978; Pascual-Leone & Goodman, 1979).

The *SOP* principle says that performances (perceptions, imaginal representations, actions, mental operations) are coproduced at any moment by the currently dominant set of most highly activated schemes within the subject's internal field of activation (in long-term memory). The *EQU* principle says that the psychological organism (or metasubject) tacitly pursues, and changes toward, three concurrent tacit goals: (a) maximize internal consistency among its activated schemes; (b) maximize adaptation (functional payoff) in its dealing with the environment; and (c) minimize internal complexity within and among its coactivated schemes. The *SID* principle says that activated schemes are subject to various sorts of inhibition: (a) *automatic inhibition* (our I_{au}-operator), caused by interference from other dominant, currently applied, contradictory, and competing schemes; (b) *effortful inhibition* (i.e., mental-attention driven, our I_e-operator) willfully applied to competing/contradictory ongoing schemes; (c) automatic *habituation* of the schemes in question (e.g., scheme attenuation via habit-making, most apparent in perception, also attributed by us to the I_{au}-operator); and (d) *decay* propensity of schemes' activation (forgetting, attributed to the I_{au}-operator) resulting from time of inactivity or repeated inhibition because of other competing or contradictory, shadow, or task-irrelevant schemes. Schemes and hidden operators, along with these few organismic principles, formulate, explain, and model effective complexity of a given task, analyzed from the perspective of the metasubject (Pascual-Leone & Goodman, 1979; Pascual-Leone & Johnson, 2005).

(**Source 3**) The third source of constraint in subjects' performance are the *plausible motives* (affective goals) *and motivational factors*, that **Source 1** could elicit from **Source 2**, in terms of potential actions or reactions. (**Source 4**) The final source is a detailed observer's description of *the performance to be explained* (which corresponds to Polya's "unknown" in a predictive task analysis). The performance to be produced must be inferred if it is unknown. Some sort of objective-subjective problem solving, equivalent to Polya's, must be achieved before our subjective-metasubjective analysis can be done on it.

Two different sorts of task analysis have different requirements on the strength of **Source 2** (i.e., the theoretical model of the metasubject). In *postdictive task analysis* there is an already completed performance to be task-analyzed. This task analysis is a generic metasubjective operative model of what the subjects may have attended to or done (within their mind or body actions) to produce the performance. In this sort of analysis our working model of the subject could be quite weak, because performance is known. A key outcome of this analysis often is differentiation and strengthening (via abduction) of our general working model of a metasubject. In contrast, in *predictive task analysis*, where performance is not known, the subject's working model (**Source 2**) has to be well differentiated, and the context (**Source 1**), motivation (**Source 3**), and the coordinating task analysis itself must be carefully done, if we are to infer the performance (**Source 4**) with plausibility.

Dialectical Steps in Metasubjective Task Analysis

There are three equally important nested sorts of task analysis that constitute different and dialectically complementary moments; progress in any of them aids and benefits from completion of the others. Although analysts could work concurrently with all these moments and change from one to another as convenient, they all must be completed (they constitute a dialectical triplet). These three nested analytical moments are called *objective*, *subjective* (subjective 1 & 2), and *metasubjective*, or organismic-process, analysis (Pascual-Leone & Johnson, 1991, 2011, 2017).

Objective Analysis

Objective analysis is a method that psychology shares with all sciences. Here, one describes objects, procedures, and relations known about the actual situation, task, subject, and performance. This method is a common practice familiar to scientists. It can be described as some form of liberalized descriptive or phenomenological analysis coupled to a rational semantic-pragmatic analysis of relevant causal (or organismic-causal) processes, which fully contextualize the situations. This constructive analysis uses results from empirical science when possible. The analyst may begin by asking a few guiding questions: what do we have to model? That is, what is the intended (internal or external) performance? What does the subject have to do in this task (i.e., what is the strategy or procedure chosen)? What aspects of the empirical situation could affect (facilitate or hinder) the subject's task—either intended internal process or external performance? What is the effective complexity (or functional infrastructure) of the task? That is, which are the essential or necessary skills (in concrete acts and percepts) needed to produce the performance?

This is the first essential moment in MTA. One does not describe at this stage natural organismic units of processing (such as schemes) nor refer to the brain constraints and maturational capacities (the brain's "functional hardware" of hidden operators and principles) that produce the internal/external performance.

Subjective-1 Analysis

Subjective-1 is a more advanced form of analysis. It includes functional-*structuralist* (structural constructivist) modeling, which Piaget's psycho-logic pioneered. Piaget's functional schemes/structures such as (psycho-logical) groupings or the INRC group (of four transformations, i.e., Identity, Negation, Reciprocal, and Correlative; see Beth & Piaget, 1961; Pascual-Leone, Escobar, & Johnson, 2012) illustrate well his form of structuralist modeling. A more recent but analogous form of neo-Piagetian structuralism is the functionalist analysis of relational complexity suggested by Halford and associates (Andrews & Halford, 2011; Halford, Cowan, & Andrews, 2007; Halford, Wilson, Andrews, & Phillips, 2014). Such relational analyses, as Piaget's own, go beyond (while including) objective analysis. They make assumptions about logical/functional structures of the subject's task-solving processes and adopt a perspective that is both constructivist and tacitly "from within" the process. They attempt to capture coordinated task-essential aspects (dimensions of variation) that subjects must master and use in their minds (or externally) to produce the performance.

Both Piaget's psycho-logic and Halford's relational complexity theory state that meaning occurs when a link is formed between or among schemes by way of interrelations (a view that we, Case, Demetriou, Cowan, and many others also share). Chapter 1 discusses Halford's relational complexity in more detail and provides examples, However, Halford's relational complexity or Piaget's psycho-logic are not the only task-relevant sorts of effective complexity. Tasks' mental demand may not increase with relations when they are overlearned. As well, mental demand can accrue also with other not interrelated but relevant relations or pieces of knowledge, such as those that must be kept in mind for later use in a task.

Subjective-1 analysis adds to objective analysis the modeling of processes in terms of organismic units (such as schemes/schemas or structures). Thus, after we have intuitively grasped useful process-structural relations in the task, a first question to ask should be, "Which are the schemes involved in this task?" Analysts then must instantiate these relational structures as schemes of a concrete functional totality that could model the performance (such as Piaget's concrete operative/operational structures, Case's concrete central conceptual structures, or Halford's concrete relations of various complexities). In this manner, we can obtain distinct, independent but coordinated, functional constituents of the task processing. Every constituent should be formulated as a scheme/schema with its three functional components: releasing, effective, and functional.

To achieve such performance decomposition, one adopts a within-the-subject perspective and analyzes what processual information a subject must attend to and coordinate to solve the task. This is the *first heuristic prescription* of MTA. The analyst must express this information deconstructed into organismic units (to repeat: schemes implicitly or explicitly defined in terms of their three components *fc, rc,* and *ec*). The set of necessary (relevant) units for the task solution is part of the effective complexity in the task. These are essential constituents, because without them the intended outcome cannot be attained. *Essential* constituents are not absolute but relative to motives and goals of subjects, relative to the needed task and strategy adopted.

In the analysis we should retain all influential, and not only the essential, schemes for the task and strategy adopted. This is the *second heuristic prescription* of MTA. We should also describe some haunting schemes that are not infrastructural, are nonrelevant, and often are unconscious, because they are in fact like a shadow that can affect performance (we call them shadow schemes). Shadow schemes are important in modeling MTA, whether they strongly facilitate the task-intended performance or strongly interfere with it. Eminent neuroscientists, who do not do any explicit metasubjective analyses (and in theorizing do not use well-defined scheme units) recognize nonetheless the great importance of often unconscious processes as codeterminers of task performance (e.g., Dehaene, 2014). A major problem of structuralist analysis, including relational-complexity analysis, is that these methods fail to analyze relevant shadow/unconscious schemes, and because scheme units are not well defined in these methods, process-analytical contradictions caused by shadow schemes (a frequent cause of misleadingness) often are not detected. To illustrate this important point, we shall now discuss critically the conservation of equivalence model that Piaget developed. We shall present Piaget's as the only example, but similar criticisms can be addressed to many contemporary learning developmentalists and structuralists.

Piaget's Final Structuralist Model of Conservation of Equivalence of Substance The lower half of figure 8.1 presents Piaget's structural model of equivalence conservation as he formulated it, although the figure is ours. The formulas in this part of the figure are our representation of Piaget's own theoretical model. Figure 8.1 represents the mental state of a 7- or 8-year-old child who, after been exposed to the whole task (i.e., the comparison between **A** and **B** to conclude equality, and identity transformation of **B** into **B′**), is confronted with the conservation-problem question: "If I eat this ball of candy (**A**) and you eat this sausage of candy (**B′**), do you think that we will have eaten as much (same amount of) candy?" Young children may reply that they would eat less, because **B′** is much thinner. Alternatively, they may say they would eat more, because **B′** is much longer than **A**.

These responses are induced by two factors: (1) the *F*-factor of simplicity (e.g., because **A** looks to have more in the thickness dimension, it must have more amount) and (2) automatized past-learning experience (things that have more surface tend to have more substance, e.g., an orange has more substance than a clementine). These shadow schemes (unconscious but compelling, often unsuitable habits from past learning) interfere with the problem solution. How then could children ever come to solve a conservation problem by themselves? Piaget thought that this solution comes via constructivist learning: experiential construction of a relational structure that Piaget called a psycho-logical "groupement," that is, the grouping of vicariances (Beth & Piaget, 1961). The term vicarious has an uncommon meaning in Piaget that fits the etymology (from Latin, *vicarious*: substitute). This grouping structure is best explained with the help of three other examples briefly described in table 8.1.

The first example comes from Piaget. Consider the concept (logical class) of European. One way of understanding "European" is as a logical class: a collection or set of people who share an invariant relational characteristic—being distinct national-types within Europe. So European can be described using different alternative facets, all equal or equivalent: for example, the set of Spaniards and of non-Spaniards is equal to the set of French and non-French, which is equal to the set of Polish and non-Polish. Each of these partitions of the collective "persons from Europe" constitutes a qualitative aspect or facet of the European concept, which quantitatively (number of European members) is equal to other facets or partitions of the same collective; this shared invariant corresponds to the people from Europe. Such logical relational invariant (qualitative and the quantitative equality of all vicariances/partitions for this collective) is what Piaget's logic defines as the concept "European." Leaving aside the practical relevance of this definition, this is an instance of Piaget's *grouping of vicariances*: the system of dichotomous partitions of a large set, or collective, such that all partitions are equal in number of members; and all partitions, qualitatively sorted in kinds as facets, are instances of the same invariant concept—"European," in this case.

Another example comes from Frege (1980), who seems to have thought that a number can be characterized as a relational invariant (quantitative identity) defined on a large interrelated collection of numerical units. From this perspective, the number seven would emerge, under various arithmetic operations, as the invariant generated by $3+4 \Leftrightarrow 1+6 \Leftrightarrow 10-3 \Leftrightarrow 21 \div 3 \Leftrightarrow \ldots$ and so on. This relational structure is another example of Piaget's grouping of vicariances, and each of the equal (identity) terms is a facet of the number seven.

Consider a last example. Driving a standard-gear car requires coordinating the foot pedals to get the car moving. This coordination as a relational structure also corresponds to Piaget's grouping of vicariances. If we call Accel_n the graded act of pressing

Table 8.1
Examples of grouping of vicariances

1. **Piaget's logical class** of "European" is the conceptual invariant common to alternative partitions of the set: [Spaniards & Non-Spaniards ⇔ Polish & Non-Polish ⇔ French & Non-French ⇔ ... etc.]
2. **Frege's notion of number as relational object/symbol.** For instance, "7" is the quantitative invariant common to an infinite family of arithmetic operations: [$3+4 \Leftrightarrow 1+6 \Leftrightarrow 21/3 \Leftrightarrow \ldots$ etc.]
3. **Pedal coordination in driving a manual-shift car**: [$\text{Accel}_1.\text{Clutch}_1 \Leftrightarrow \text{Accel}_2.\text{Clutch}_2 \Leftrightarrow \text{Accel}_3.\text{Clutch}_3 \Leftrightarrow$ etc.]

(or relaxing) the accelerator pedal (this is a graded series 1, 2, 3, … n); and call $Clutch_n$ the graded relaxing (or pressing) of pressure on the clutch pedal, it is evident that a suitable coordination of these two motor series is necessary to ensure that the car can start moving smoothly, without too much noise or stalling. At each graded moment (1, 2, 3, etc.), actions in the respective series must be adjusted to constitute a proper facet or moment: one where smooth car movement—a coordinated dual series—occurs. This coordination can be interpreted as a grouping of vicariances of the form shown in table 8.1.

These distinct concrete examples of grouping of vicariances should make Piaget's logical model clear. Piaget thought that one such grouping, acquired by coordinating schemes during prior experience with suitable objects (constructivist learning), could explain why children of at least 7 years can solve a conservation of equivalence problem. Our symbolic explanation of Piaget's vicariances model for conservation is represented in the lower part of figure 8.1.

In this figure, the letters *H* and *W* stand, respectively, for the height and width measures of the perceived objects (ball, sausage). Each object, for instance ball **B**, emerges in cognition as a functional invariant constructed by coordination of a grouping of vicariances—not unlike the number seven or the concept of European, just mentioned. Only now the facets are constituted by coordinated graded values, in inverse relation, of height (H) and width (W) of the object. These two series represent the distal object perceptually presented. As H decreases (>) because ball **B** is being rolled, the corresponding W increases (<). All these facets (H × W) are equivalently indexing the distal object, indexing the set of state-configurations the object could adopt under any transformation that preserves amount of substance as an invariant. The transition from one facet to another is any identity transformation (where no matter is added or taken away).

A similar logical grouping of facets could of course be attributed to **A**, as the lower part of figure 8.1 shows. Because the particular facet corresponding to **B′** (i.e., $H_4^{B'} \times W_4^{B'}$) can also be found as a facet in the vicariances grouping of **A** (the H × W pair 4), the child (Piaget would say) can recognize that **A** and **B′** belong to the same distal object (i.e., the same concrete grouping of vicariances) and therefore **A = B′**. However, this elegant explanation of Piaget's model fails, because past learning could not possibly have led to the emergence of such grouping of vicariances for conservation. The reason is that there is no prior objective empirical invariant here, which could induce and guide empirical coordination of the object facets.

This is in sharp contrast with the three examples of vicariances given before. Piaget's grouping for the concept of Europeans does have an independent empirical invariant (i.e., the nationals inhabiting geographic Europe), which can guide relational abstraction of this grouping. There is also an independent empirical invariant for the number seven. By counting the numerical units, or counting objects, we can generate the number seven prior to its emergence as invariant of its grouping of vicariances. The same happens with the pedal-coordination grouping. In contrast, in the case of equivalence conservation there is no prior independent empirical invariant: the equality of substance/matter (as a quantitative invariant) would have to be hypothesized ("invented") to be confirmed after experimentation.

Thus, this conservation grouping is totally abstract and could not have been inferred in prior experience, unless heuristic methods of a different sort lead the subject to guess an equivalence. In conclusion, conservation invariants in Piagetian tasks are not empirically grounded, and the subject must hypothesize them (Peirce's abduction) before they can be empirically discovered and reflectively abstracted; the invariant of conservations is a rationalist fact and is not an empiricist fact (Pascual-Leone & Sparkman, 1980). This learning-paradox problem of Piaget's structuralist models often is shared by other structuralisms.

The abduction process needed to solve the problem is presented in formula **f1**.

$$\{\ldots M[\text{COM.AMOUNT}(A,B'),\ TB=B',\ B=A]\ \ldots\} \rightarrow [B'=A]_{F,LC} \qquad \textbf{(f1)}$$

In this mental attention model, when children have reached the problem representation indicated, they can coordinate three schemes: (1) an operative scheme that says COMPARE(A, B′); (2) a figurative relational scheme that says TB = B′, where T preserves identity of amount, that is, B = B′; and (3) another figurative scheme that says B = A. However, as mentioned above, there is a problem. Some organismic factor must exist that enables, without a logical rule of transitivity, this global heuristic inference: Because two terms (B = A, TB = B′)[3] have an equality in them, the third term (B′?A) of this formula should also have an equality (B′ = A). Indeed, simplicity *F*-factor should induce such inference, according to our TCO model. If this is the solution, it has also demanded use of hidden operators, such as the *F* and *M* (factors of mental attention) to boost activation of the three essential schemes and synthesize mental representation of figure 8.1. But we are now ahead of our discussion; we introduce hidden operators in task analysis in our third and final moment of task analysis.

Limitations of Subjective-1 Modeling A good subjective task analysis begins by examining relevant objects and procedures, with their coordinating relations. These entities are modeled into schemes (or related constructs) that could have been created by the subject during his or her past history (via constructivist learning, e.g., *C*- or *L*-operators; see chapter 5). Before a scheme is posited for introduction in task analysis, a rational reconstruction must be done to appraise whether all posited schemes could have been acquired by the subjects in their expectable past, given their personal history (as we have just shown, Piaget's schema/structure for conservation—his vicariances model—cannot have been acquired).

At the end of this subjective-1 analysis, researchers should have formulated (and, at best, represented symbolically with suitably compact logical notation) all task-relevant schemes/schemas, and all the shadow (unconscious compelling habit) schemes, that may have an effect (positive or negative) on performance. Clear definition and coordination of these schemes may still be missing, however; and relevant processes may have been modeled top-down holistically as functional-structural totalities for solving tasks (illustrated with Piaget's vicariances model of equivalence conservation). Many contemporary task analyses are done as simple flowcharts, diagrams, theoretical figures, or simple enumeration of relevant processes, without detailed explanatory organismic models. This is generally inadequate. Neither true parallel processing of schemes nor models of real-time unfolding (i.e., step-by-step natural sequencing of task processes) is available in this sort of task model. Also omitted is the relevant "functional hardware" the organism needs to generate performance, as our analysis of Piaget's model shows.

However, there are intermediate methods. The task analyses used by Case and his students (Case, 1992, 1998; Case, Okamoto, Henderson, & McKeough, 1993), for instance, go beyond subjective-1 analysis in that they occasionally include real-time reconstructions of some task-solving processes. They have executive processes and also organismic hardware constraints such as those of working memory (mental/executive attention).

Subjective-2 Analysis

Also called *ultrasubjective analysis*, subjective-2 task analysis goes beyond structuralism and associative relational learning, because its focus is on the real-time deconstructing of tasks and total-task structures of the subject. Subjective-2 analysis deconstructs the task in terms of schemes (whether task-compatible or incompatible) that may be simultaneously coactivated or sequentially coactivated to constitute structures that can together describe a task-relevant functional totality. This temporally sequenced form of analysis coordinates relevant and shadow schemes to explicate task processing and performance. It often begins with a simple question: can the process-task analysis be obtained in steps of (perhaps competing) scheme applications?

Adding this finer level of discrete schemes to the analysis and modeling real-time process force analysts to use a clear and explicit formulation of schemes. This creates the need to differentiate among four distinct and necessary sorts of schemes, which may be intentionally conscious (during their manifestation) or unconscious (e.g., shadow schemes). For this unfolding of sequential process, we distinguish between figurative schemes that stand for logical "objects" (logical "arguments") and operative schemes. The latter are processor operations that apply on figuratives. Notice that figurative and operative schemes must be regulated in their intertwining to produce the intended processing change. This is done by temporally structured schemes that we call *fluents*. Fluents can function either as figuratives ("objects") or operatives (functors), depending upon the task circumstances. In our task analyses we call these fluent schemes parameters, when they stipulate (boundary conditions) how to achieve successful application of specific operatives on the given figuratives. We use the term fluent (Pascual-Leone et al., 1978), because this sort of operative-or-figurative scheme expresses or regulates the changing flow of processing and involves expectancies. A simple example of a fluent is the scheme TB = B′ of our model of conservation (see figure 8.1 and accompanying text). In task analysis, parameters often are relational fluent schemes.

Subjective-2 task analysis also requires executive schemes, needed to plan and coordinate into the future action schemes (figuratives, operatives, and parameters) to produce performance. Finally, planned anticipations and preparations are driven by affects/emotions (affective goals, motives) and social/personal schemes with their values, biases, and preferences. These affective and personal schemes in fact regulate and produce complex, often intrinsic, motivation (Pascual-Leone & Johnson, 2004; Pascual-Leone, Pascual-Leone, & Arsalidou, 2015)—the sort needed for cognitive problem solving and cognitive development. These schemes, although important, are assumed, but not directly represented in the usual sorts of cognitive task analysis. However, they can be made the direct focus of analysis if necessary (Antonio Pascual-Leone, Greenberg, & Pascual-Leone, 2009, 2014).

Attention to the interactions between essential/infrastructural schemes and shadow schemes leads analysts to discover error factors or facilitating factors in situations. These usually are caused either by shadow schemes or sensorial-perceptual task features that influence cognitive/perceptual scheme representations. Misleading schemes may force adoption of a more effortful (*M*-demanding) strategy, for example, one that segments the originally intended schemes into relevant subschemes that must be *M*-boosted separately, to bypass the misleading aspects (by opening alternative strategies). In contrast, facilitating schemes could induce suitable cue-generating moves in the solution process that may spare the need for *M*-boosting some task-relevant schemes (e.g., making more salient some cue that activates relevant schemes).

Metasubjective Analysis (MTA)

An MTA can be dimensional (an analysis of task-relevant executive processes), a real-time mental-process reconstruction (which we call an *M*-construction), or a synthesis of both methods. In any case, it is a generic summary representation of real-time organismic processes responsible for a subject's performance. A step in this analysis need not correspond to a single moment or unit of time but may represent a time segment unfoldable into a sequence. In this respect, the steps of a metasubjective analysis are like block units of a detailed flowchart, but better defined and including hidden operators that regulate schemes.

Objective and subjective (1 & 2) methods of analysis do not suffice. As our discussion of these methods and of equivalence conservation have shown, we are forced to refer again and again to hidden operators such as the *F*-, *M*-, *C*-, and *L*-factors, as well as hidden principles (e.g., *SOP, SID, EQU*). These hidden processes are alternative general-theoretical constructs that enable task completion, and they should be represented and carefully studied in MTA.

At this point, if we aim toward a predictive task analysis, it is necessary to adopt an explicitly general metasubject's model of organismic processes that is detailed enough. Analysts may already have from subjective analysis a list of scheme types and their characteristics. As this subjective analysis progressed, they may also have recorded various distinct schemes that appeared necessary for the task, each explicitly defined (perhaps with their functional, releasing, and effecting components). Now analysts need a list of all hidden operators of the organismic theory that could be relevant for the current task analysis. It could be used as a checklist to decide the hidden operators needed for the chosen strategy. Table 7.1 offers one such a list.

In MTA, and after having demarcated relevant schemes for a given strategy in the task, analysts should take this list and ask themselves, as a heuristic, how the various operators and key principles (e.g., overdetermination, inhibition/decay, equilibration, and reflective abstraction) may intervene in the chosen task strategy. To illustrate this checklist heuristic, consider again Piaget's conservation of equivalence task. We have discussed two models for conservation of equivalence. One expresses Piaget's constructivist-learning strategy, which could not have been learned in time and is therefore contradictory. The other presents our own maturational-attention (i.e., *M*-operator) problem-solving or synthesis strategy, which may be correct. Let us follow the list of hidden operators in table 7.1.

A-operator, affective schemes and motives (affective goals), and with it the *LA* (logical-structural learning of affective/personal processes), and *B*-operator (the self-defining, personal or sociocultural schemes of a particular human being) are usually mediated by "good" or "bad" shadow schemes, more or less automatized for cognitive psychosocial tasks. To maximize a child's motivation, these complex affective/personal schemes must modulate cordial, affectively warm interactions between the child and experimenter. In many symbolic problem-solving tasks, like this one, participants often are not intrinsically motivated to solve the problem, but they are indirectly motivated because they like the experimenter and want to answer his or her questions.

C-operator appears in perceptual processes that lead to a representation of proximal/distal objects, such as **A** and **B** or **B′**, often intervening tacitly but decisively. The *F*-operator is the factor imposing a simplicity (a mini-max *F*) bias in perception and cognition. As mentioned, dynamic synthesis of the conclusion **A = B′** is mediated by *F*. To summarize, we illustrate this dynamic synthesis using a logical representation of the equivalence conservation task:

$$\underline{\mathrm{COM{:}A,B'}}\ (\underline{\mathrm{A[A{=}B]}},\ \underline{\mathrm{B'}[\mathit{TB{=}B'}]},\ \#\{\mathrm{MIN{:}Rel.Diff}\}_{F,\,C}) \rightarrow \mathrm{A{=}B'} \qquad \textbf{(f2)}$$

This logical formula is an alternative representation of formula **f1** that carries the solution to the problem. The parentheses (…) in **f2** demarcate schemes on which the operative COM applies. A[A = B] is the distal object **A**; the square bracket enclosure […] signifies that this is a cognitive-intellective (symbolic) distal-object scheme and has as part of its meaning the prior relational conclusion that A = B. The first step in this task was precisely intended to induce in the participant this intellective representation of A as a distal object, via automatization based on mental attention (*LCLM*-schemes; see chapter 5 and below). A similar meaning applies to B′[TB = B′], except that here the identity scheme of TB = B′ has the structure of a fluent, the sequence of events that transformed **B** into **B′**.

This structuring would not be possible without effortless participation of two hidden operators and the principle of overdetermination. The first operator contributing to the TB = B′ synthesis is the *T*-operator (number 5 in table 7.1), discussed in chapter 6. The *T*-operator is needed to learn relational sequences that make objects meaningful. When a concrete transformation **T** (e.g., rolling the ball on the table) is actually applied on **B** to progressively turn it into **B′** (a sausage), the *T*-operator in the child's working mind internalizes these sequentially related elements as a relational structure, without need of mental attention. These evolving coactivated schemes are unified into a temporal structure by *T* with the help of *F*-operator and *SOP*, which induce **B** and **B′** to combine as two facets of the same distal object.

The role of mental/executive attention is shown in formula **f2** by underlining of three schemes (one operative and the two figuratives), to signify that *M* is boosting them, which causes the mental operation. *M*-boosting is necessary because the schemes are symbolic and not directly cued (in contrast, **A** and **B′** are perceptually given and nonsymbolic object-schemes—radically different schemes from their symbolic counterparts). The fourth scheme, a parameter (#) does not need to be boosted with *M* because *F* is boosting it; which in the formula's notation is indicated by $\{\ldots\}_{F,\,LC}$. In our method of notation, the braces enclosure indicates that schemes inside do not need to be boosted by the *M*-operator. The subscripts at the end of the enclosure (*F, LC*) show that *F*-operator and *LC*-operator are boosting the enclosed scheme. In **f2**, this fluent scheme is a parameter (indicated by prefix #) that constrains the operative scheme to **MIN**imize **Diff**erences among **Rel**ations interconnecting the various object-schemes (this is the mini-max *F*-tendency described in chapter 6). Such constraint, imposed by *F* (principle of simplicity) is likely to be already automatized (*LC*-learned).

If the four schemes mentioned (three of them boosted by *M*) are highly activated together in the mind (i.e., the focus of attention or centration), *F-SOP* can impose a dynamic synthesis that appear after the arrow (➔ symbolizes *SOP*). This is the conclusion scheme A = B′. Formula **f2** illustrates emergence of a scheme solving the conservation problem. The *I*-operator, promoting mental-attentional (active, effortful) interruption or inhibition, intervenes in the task but only implicitly. It is not represented in the formula, because this would require modeling misleading shadow schemes. To illustrate the competition of the task-relevant schemes in **f2** versus the task's interfering shadow strategy, we now present in formula **f3** the alternative misleading scheme, likely to be automatized in perception:

$\underline{\text{COM:A}}^{L1}$, $(\{\text{A}\}_{L1}$, $\underline{\text{B}}'$, $\#\{\text{MAX:Per.Diff}\}_{F,\,LC})$ ➔ A > B′ **(f3)**

In this formula subjects try to compare **A** with **B′**, but here the two objects are purely perceptual (proximal objects). Now operative scheme COM already carries (because of perceptual *LC*-learning comparing objects) the reference to **A**, its first proximal object to compare. Thus, during the actual comparison this term ($\{\text{A}\}_{L1}$) is boosted (chunked) by *L1* with COM. However, **B′** has to be explored to appraise its shape to compare amount with **A**. In perception, our cognitive system automatically would minimize complexity by grouping features into objects that MAXimize Differences among these objects. This is the meaning of parameter $\#\{\text{MAX:Per.Diff}\}_{F,\,LC}$. Such perceptual mechanism, facilitated by *F* and by *LC*, is a parameter (#) inducing incorrect judgment. Thus, **B′** may be perceived as having less matter than **A**, because **B′** is much thinner, that is, **f3** infers that A > B′.[4]

Because conclusions of the automatic scheme of **f3** contradict the correct emergent scheme of **f2**, an internal activation of mental attention (Matt = <*E, M, I, F*>) must intervene in **f2**, boosting relevant schemes with *M* and inhibiting **f3** with the *I*-operator. All this is monitored by executive schemes (*E*-operator). Such a competitive misleading situation can be expressed in our MTA method as a *strategy-relational formula* that shows how various strategies elicited by the task relate:

$(*E, M, I, LM* [formula **f2**]) $\underline{\text{V}}$ ^(*F, LC* [formula **f3**]) **(f4)**

Formula **f4** shows that **f2** and **f3** are in competition prior to the correct response. Strategy **f3**, which is wrong (prefixed ^), will be activated first, because *F* and *LC*, two fast automatic processes, are boosting it. To overcome this initial dominance of **f3**, the correct (prefixed $) strategy **f2** needs executive schemes (*E*-operator) to effortfully inhibit **f3** with *I*-operator and allow **f2** to unfold. Because both strategies share lower-level schemes, an initial dominant activation of **f3** should decrease (competitive interference) probability of subsequent activation and application of **f2**, unless Matt is mobilized by *E*. We call such a state of affairs a misleading situation. We represent this misleadingness with $\underline{\vee}$, a logical connective of mutual incompatibility.

People differ in the relative weight (propensity, probability of activation) their own hidden operators have in their working mind. *Field-dependent* people's propensity is to have very high weight for *F* and *LC*, with a lower weight for *E*, *M*, and *LM*, whereas the contrary is true for *field-independent* people. Research is congruent with such interpretation (Hederich-Martinez, & Camargo-Uribe, 2016; Pascual-Leone, 1989; Witkin, Dyk, Faterson, Goodenough, & Karp, 1962; Witkin & Goodenough, 1981). We should expect, and it happens, that in tasks exhibiting misleading competition as in formula **f4**, such as conservation tasks or the water and wine problem, field-independent people will do much better, provided that they have enough Matt to synthesize the correct solution (formula **f2**).

Judgment Tasks versus Action/Operative Tasks, and Knowledge Metadomains

Conservation is an example of a *judgment task*. Such tasks involve collecting information regarding schemes from the recent past about relevant events and circumstances—relevant to reach some problem-solving judgment conclusion. Judgment tasks tend to be representational (involving figurative schemes, whether perceptual or relational-conceptual), which often require recognizing/synthesizing relations of actual coexistence among relevant schemes. Note that abstracting undiscovered relations of coexistence from structural totalities (e.g., in space, in rational abstraction, in music) is characteristic of the brain's dorsal pathway activity (Pascual-Leone & Johnson, 2017; see chapter 11), which we formulate as *S*-operator, an operator that facilitates coordinated simultaneous integration of several distinct processes (see chapter 6 and table 7.1).

A common and very different kind of task could be called *temporal-meaning operative integration* (episodic) *task*, because in it one's attention is applied to a sequence or series of specific actions or steps, in proper order, to appraise objects/situations or achieve a given concrete result. Many learning tasks, such as learning to drive, experimental (motor or matching) learning tasks, many language tasks, or music, illustrate in part this sort of action/operative task, which often uses temporally structured (fluent) schemes. Abstraction and coordination of temporal sequences of events, or steps to appraise objects or situations, are essential in these tasks, which often involve use of the brain's temporal lobes (e.g., hippocampus and related areas, lateral temporal lobes). This sort of temporal-meaning structuring of experience often is done using automatic attention and our *T*-operator (see chapter 6 and table 7.1).

The distinction just made is important in task analysis, because the required modes of processing are different in one sort or another, engaging preferentially different hidden operators (as we suggested, among others, *S*- vs. *T*-operators). To this end, it is important to distinguish between metastrategies of task solution and metadomains where task analyses can be made. *Metastrategies* (or *operative modes of processing*) are functional categories of strategies that differ in operative viewpoint or perspective—adopted because of task characteristics, or subject bias, or bias of the analyst. The distinction between judgment tasks and temporal-meaning operative tasks is an example of metastrategies imposed by task characteristics.

An example of metastrategies adopted because of subjects' individual differences bias is the field-dependence-independence distinction just made. An example of analysts' bias is found in two alternative ways of doing science or problem-solving: by adopting an observer's perspective from outside the subject (i.e., a meta-empiricist perspective) or adopting a perspective from within the subject's own thinking process, which we have called metasubjective perspective. The latter informs our methods of task analysis (Pascual-Leone, 2013, 2019; Pascual-Leone & Johnson, 2017; Pascual-Leone et al., 2015; see previous chapters). There are other types of metastrategies.

Metadomains (or content-abstraction modes of processing) are another important functional category relevant for task analysis. Expanding on categories initiated by Piaget (who contrasted infralogical versus logical versus linguistic), we differentiate among four functionally ordered metadomains. They are ordered, because as knowledge advances the earlier metadomains "force" emergence of the more symbolically abstracted metadomains that follow. This is the (developmental) weakly ordered set of categories that follow:

Metadomains:= infralogical ➔ mereological ➔ logological ➔ linguistic **(f5)**

The *infralogical category* was formulated by Piaget: it refers to purely experiential objects (i.e., transformations and relations felt prior to logical or conceptual characterization) and to sensorimotor *prelogical* (prepredicative) experiencing. The *mereological category* refers to the experiential construction of objects and their coordination or interrelations (e.g., transformations). Such coordination creates, among others, collective classes (collections of actual infralogical objects or entities). Formally formulated by Lesniewski as an alternative logic based on concrete-activity (Mieville &Vernant, 1996), *mereologic*, if psychologically interpreted, becomes a mode of psycho-logical modeling in the infralogical domain, using conceptual (logological) tools for qualitative modeling of activity (praxis, goals, operations, object relations) in the world. This modeling produces mereological entities.

The *logological category* was called logic by Piaget (in the sense of *psycho-logic*; Pascual-Leone et al., 2012). It clarifies from within a child's processing developmental progress of conceptual psycho-logic. Psycho-logic expresses the most general relations in experiential activity. From this perspective, Ferdinand Gonseth (1936/1974) defined logic as the physics of any object whatsoever. For instance, for babies not older than about 6 months, the mother they seek, smile to, and embrace, is an infralogical (emotion arousing and purely experiential) entity. For this baby, mother becomes a mereological entity when the baby is about 10 to 12 months. At this time, mother is an individuated and specific object-person (a subject), who comes to feed and play with the child, who is on call, and at times disappears. Perhaps at 18 to 35 months, with the use of symbolic function and the parents' linguistic communication, logological schemes finally appear in the baby (who then can classify, relate, and organize generically his or her own experiences). When the child's language emerges as an explicit communication tool, its elements are sign representations of mereological and logological entities (including relations) extracted from actual experience.

In formula **f5** the arrow (➔) symbolizes an overdetermination (*SOP*). Metadomains to the left of the arrow force emergence (i.e., are semantic-pragmatic origin, semiotic foundation, and referent) of all metadomains to the arrow's right. Conversely, metadomains to the right progressively abstract, make compact, and refine the effective complexity of the metadomains to the left, which thus become explicit and communicable. Understanding all these functional categories of processing is useful when doing task analysis. MTA often is modeled using linguistic and logological skills of the analyst but usually represents mereological or infralogical schemes of the subject. We give examples below.

In task analysis, it is useful to investigate various metadomains, using distinct methods (i.e., objective, subjective-1, subjective-2, and metasubjective). These analyses can be done in any order or done concurrently (perhaps using various pages to separate methods and metadomains). The analysis is completed when the pages become coordinated and mutually congruent.

Symbolic Notation in Metasubjective Task Analysis (MTA)

Task analysis can be made verbally if it is not too complicated. However, a symbolic notation and representation of the analysis by psycho-logical process formulas helps reasoning and makes results explicit and compact, facilitating awareness of deficiencies and helping in process modeling. In this section, we summarize some notational conventions but give a more detailed list of conventions in the appendix at the end of the book. As with all heuristics, notation and rules should be changed when desirable. The best MTA representations are always intuitive, simple, and clear. Usually, compactness is best. It is useful to omit from MTA formulas aspects or features of little current importance and retain what is essential. Whenever formulas are contrasted (as in formula **f4**) only differential characteristics of the formulas being compared should be given.

One distinct advantage of using standard notation and rules is that, once known to practitioners, they need not be repeated. However, we are repeating much notation in this book to facilitate learning. Readers of previous chapters are familiar with basic conventions (see table 3.2), which we now summarize again. Operative schemes are written in CAPITALS, and figuratives written in lowercase with an optional asterisk (*) prefixed or postfixed. Parameters (adjunct information to regulate application of operative to figuratives) are written with a number-sign prefix (#); and executives, if we should need to highlight them, can be represented in capitals with a prefixed superscripted E (^{E}OPERATIVE). Temporally structured schemes (fluents) are represented in italics (*OPERATIVE*) or with a pre-superscripted *fl-* (e.g., $^{fl\text{-}}$OPERATIVE, $^{fl\text{-}}$figurative). Information that belongs to the meaning of a main symbolic scheme is written inside square brackets to the right of this scheme, for example, *figurative[…], or OPERATIVE[…]. This meaning may be expressed in ordinary language by a relative clause. Schemes that in the formula must be boosted by *M* usually carry an underline (OP (#scheme, *scheme … scheme)) to indicate this boosting; three schemes are being *M*-boosted in this example. Operatives control (and apply on) schemes located to their immediate right inside parentheses; the parentheses mark the scope of application for the operative. Application/performance of operators in a formula, with its line sequence of operations, goes from the rightmost to the leftmost operator, in that order. Usually rightmost operations are more concrete, and the leftmost ones apply on their results. When a formula has several steps, the order of application goes from the first to last step (i.e., from right to left in the formula). Result of a formula's application is written after an arrow (➔), which stands for Schemes' Overdetermination of Performance (*SOP*). Consult the appendix at the end of the book for more details.

Metasubjective Analysis and Analysis of Figures in Geometry

People uninterested in relating this analysis to classic geometry could, on first reading, skip this section. Metasubjective analysis is always intuitive in Kant's sense, that is, it deals with objects and processes—concrete or abstract—in a truly experiential-representational way, adopting the perspective of the subject and often using infralogical or mereological modes. The method of analysis in classic geometry, or in Newton and Galileo (Hintikka & Remes, 1974), is intuitive in this Kantian-sense of representational concreteness. It is also constructivist, within our definition, as Newton's (1974) writing on his experimental philosophy shows. As described by Hintikka and Remes (1974), two dialectically complementary ways (backward/bottom-up and forward/top-down) of analysis exist that in combination produce the "analysis of figures" (geometrical analysis) of ancient Greek philosophers such as Pappus. Pappus' work was recognized by Polya as an influence. Polya is a twentieth-century innovator in heuristic methods for problem solving in mathematics, and his book *How to Solve It* became very popular (Polya, 1973).

In contrast to Polya's, our metasubjective method is process-organismic and qualitative (dealing with schemes), not fully objective or quantitative. Whereas in the Greek geometry (or analysis of figures) the objects given or created as auxiliary constructions represent actual external figures, lines, or objects, in our task analyses these (given or constructed) objects are schemes, regulated by hidden operators or principles. The contrast is striking with regard to *auxiliary constructions*, which in analysis of figures are drawn to become new objects on which geometry proofs can be based. In the case of MTA, invoked auxiliary constructions are organismic schemes and hidden operators posited as activated within the subjects (when they confront a problem-solving situation). The theory of geometry is essential in analysis of figures, whereas an explicit organismic theory of schemes and operators is needed in our analysis. The two complementary (dialectically intertwined) ways for these analyses follow.

The first way or sub-method is the *backward* (abductive, *ascendent*, bottom-up) way, which begins from the given (problem to be explained), in our case the predicted or postdicted performance, and then proceeds to infer, via abduction, causal determinants (in our case organismic causal processes such as needed schemes or operators). The second, *forward* (*descendent*, deductive, top-down) way of synthesis begins with principles, laws, and theorems (in our case, objective task analysis of the situation and a general operative model of the subject/metasubject), to derive in various steps the results to be explained.

A description of Newton's research analysis, provided by Hintikka and Remes, shows clearly these concrete and dialectical intertwining backward-and-forward ways. Hintikka and Remes (1974, p. 110) wrote:

> The Newtonian method may perhaps be schematized as follows:
>
> (i) An analysis of certain situations into its ingredients and factors →
>
> (ii) An examination into the interdependencies between these factors →

(iii) A generalization of the relationships so discovered to all similar situations →

(iv) Deductive applications of these general laws to explain and to predict other situations.

It is easy to see that steps (i) and (ii) belong to the backward/ascendent submethod, whereas steps (iii) and (iv) belong to the forward/descendent submethod. Further, step (iii) very often constitutes an instance of abduction, as Peirce formulated this form of inference. An abduction is an "explanation" of results by postulating a theoretical hypothesis as a fact under the presupposition of a general law (here we are paraphrasing Apel, 1995, p. 40; see also Johansen, 1993). This inference may be true or false. It is a hypothesis that requires experimental/formal verification.

The descendent, or forward, way derives the intended performance from the subject's organismic operative model and the particulars of the dynamic situation as elucidated by the ascendent/backward way (which for us includes consideration of shadow schemes, affective goals, and intended cognitive goals of the subject). Eventually the analyst can confirm that the ascendent procedure and the descendent procedure are congruent, which heuristically makes acceptable as a model the metasubjective analysis in question.

To illustrate this intertwined ascendent (often based on abduction) and descendent (often based on deduction) method, let us again use the conservation task of Piaget summarized in figure 8.1. After an operation on balls **A** and **B** to determine their equality in mass, and the transformation (**T**) of ball **B** into **B′**, the child is asked whether there is more amount (more to eat if the pieces of candy were to be eaten) in **A** or **B′**. These two balls are the only task-relevant objects present to the subject when this question is raised. To properly address the issue, the child must somewhat proceed as Polya's method suggested, but referring to relevant schemes and other organismic processes: "What is the unknown?" (i.e., What is the problem issue?); "What are the data?" (i.e., What is it that we know?); "What do we have to do in this task?" A here-and-now simplistic analysis may lead the child to conclude that **A** has more (or less) mass than **B′** because the latter appears perceptually to be much thinner (or much longer) and so could have less (or more) mass. These are *F*-operator's effects when the task has not been analytically explored, and the longer length (or greater thinness) of **B′** has not been noticed.

To correct this error of judgment, caused by the "bad" scheme represented in formula **f3**, participants must go beyond the information given (i.e., ball **A** and "sausage" **B′**) and use a mentally ascendent way from the schemes immediately given (the ball and "sausage") and the problem issue (which has more **A** or **B′**?), to the events that have preceded the current situation. Participants may then remember the no-longer-visible ball **B** transformed into **B′** by hand rolling; a transformation that reminds children, who can solve the problem, of transformations in which nothing is added or taken away, and so quantity is preserved—identity transformations, as we adults call them. With the help of this auxiliary construction (recollected inference), the child may conclude that **B** under an identity transformation (i.e., **TB**) has the same amount as **B′** (**TB = B′**). Another ascendent recollection is the initial comparison between **A** and **B**, which leads the child to conclude (a second auxiliary construction) that the object-schemes involved are equal in amount (**A = B**). With these assumptions in place, **TB = B′** and **A = B**, participants can represent, in a descendent analysis manner, the task as the mental-relational representation of the whole problem illustrated in figure 8.1 above (i.e., [A = B & TB = B′ → A?B′]).

This strategy representation raises the issue, within a descendent analysis, of whether these findings can yield the correct problem solution. For a solution, we need to appeal to other auxiliary constructions (i.e., schemes or hidden operators and principles) possibly existent in the participant. As we pointed out before, the auxiliary elements needed cannot be a learned logical principle of transitivity (if A = B and B = B′, then A = B′). Children have not had life opportunities to learn this logical transitivity rule. It could, however, instead be an innate/maturational *F*-operator (principle of simplicity in the brain), as already discussed. This example and Piaget's own model for conservation illustrate a key function of the descendent analysis: to appraise whether a given auxiliary construction or element in a solution model is likely to be available to participants as scheme or as a dynamic synthesis (a momentary emergent representation).

Let us show that this sort of analysis can be applied to geometry. An example of geometric analysis using auxiliary constructions (instantiated schemes or scheme relations) is given in figure 8.2, modified from Polya (1973, p. 216). As this figure illustrates, the participant is given a triangle BCA and asked to prove the theorem that in any triangle the sum of the three angles is equal to two right angles. With this problem and the present triangle, the unknown is the sum of the three angles a, b, and c. Using an ascending/abductive way of inference to go from the triangle to the final construction, which already appears in figure 8.2, a participant may think of physically bringing together the three angles (this would be an executive strategy or plan). This goal may lead him or her to recall a theorem (an instantiated complex relational scheme available in the repertoire—a mental auxiliary construction previously learned). Such theorem is that alternate angles constituted by a straight line and two parallel straight lines are equal. The participant then proceeds (this is a physical auxiliary construction) to draw the parallel line passing by point A illustrated by figure 8.2. They may then notice that this parallel line creates not only an angle alternate to b (b′—marked with double curved line) but also another similar angle c′ that is alternate to c (another complex, now synthesized, relational scheme). Using a descendent/deductive way of analysis (which evaluates all relations found in the total construction) he or she then notices that b′+c′+a are equal to two right angles, because the three together encompass 180 degrees. This proves the theorem.

If this mathematical proof were to be re-presented as a metasubjective process-model of the participant's mind as he or she solved the problem, we should also consider the hidden general-purpose organismic resources, that is, hidden operators (*M, I, E, F*,…) and principles (e.g., *SOP*) that intervene to make possible this solution discovery. Key to the discovery is the moment when participants have the idea of drawing the parallel line on the point A—a psychological (metasubjective) process that Polya (1973, p. 216) may not have understood, because he introduces figure 8.2 by saying: "Fig 23 [i.e., our figure 8.2] which is an inalienable mental property of most of us, needs little explanation." Let us represent this moment when the participant thinks of implementing the executive plan of physically producing the sum of the three angles (a, b, and c) by drawing a parallel line at A. This moment of the analysis is represented by formula **f6**.

M [SUM:a,b,c (#{b=b′ ← 2parallels +1oblique line},
DRAW^{L1}({*parallel}$_{L1?}$, #TO:BC))] ➔ Figure 8.2 **(f6)**

Translating into English, this formula says that the scheme operative seeking the physical sum of angles a, b, and c (i.e., SUM:a,b,c—note that here : symbolizes that SUM is an operative that applies on a, b, and c) has a parameter # stating that when two parallel lines are crossed by one oblique line the two appearing alternative angles, b and b′, are equal. Operative SUM and its parameter # are monitoring the DRAWing of a line parallel to the line BC (we represent the adverb "*to*" as an operative TO that applies on the line BC to create its parallel). The result of this operation induced by SOP and *F* (here symbolized by ➔) is represented in figure 8.2. Note that all these schemes are complex and symbolic. Their meaning is mediated by intuitive/experiential schemes that function as signs (not necessarily verbal) of previously found complex scheme

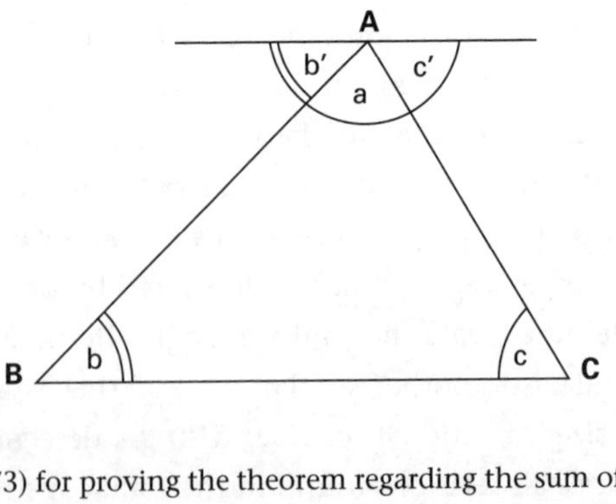

Figure 8.2
Diagram (after Polya, 1973) for proving the theorem regarding the sum of the angles of a triangle.

relations. When participants are practiced in drawing lines, the DRAW operative may perhaps be already *L*-structured (chunked) with the scheme for drawing the line (the superscripted *L1* and $\{\ldots\}_{L1?}$ indicate that this is uncertain). If we count the underlined schemes (there are six—an *M*-demand of $e+6$), our developmental prediction would be that this problem is accessible at the age of 13 to 14 years—unless *L-structuring* has lowered the *M*-demand to $e+5$, accessible to 11- to 12-year-olds.

Final Comments on the Two Modes of Metasubjective Task Analysis

The ascendent/abductive and descendent/deductive mental ways of an effective task analysis are dialectically complementary and in continuous interaction. Together they help to synthesize problem solutions, as we have just illustrated. Our method of task analysis is, in this regard, analogous to an analysis of classic geometry. This is particularly the case when participants must synthesize complex figures or aspects (in our case, schemes at progressively higher levels) to reach the complexity needed for a total proving/processing formula. *Executive MTA* or *executive M-dimensional analysis* is our name for a sort of procedure that uses ascendent/abductive less often than descendent/deductive ways. We illustrated this sort of analysis above with the conservation task and the proof of the sum of angles in a triangle.

Another complementary procedure exists that is mostly step-by-step in an ascendent-abductive manner and can be made as concrete as desirable, by deconstructing and segmenting the often descendent/deductive macrosteps of any executive MTA. This would be a *sequential MTA* or *step M-construction*. Such step-MTA is a sequence, each step representing a moment of process analysis concretely represented. Unlike executive MTA, this sequential MTA may not often require changing levels of processing (decentration or shifting of mental attention during processing of schemes) and may limit itself to recentration (updating of mental-attentional application) when the task does not demand changing levels of abstraction. Sequential step-MTA is initially more intuitive and can be easily unfolded to any degree of refined concreteness by deconstructing steps and unfolding further their sequence. The next chapter uses these foundations to present in a more focused manner our methods of MTA followed by concrete illustrations in various domains.

9 Process and Mental Task Analysis: Methods and Examples

We illustrate specific methods of metasubjective task analysis (MTA). Two distinct complementary methods are outlined: the sequential, or step, MTA (an ascendent, abductive way) and the executive, or dimensional, MTA (a descendent or deductive way). To delineate these methods, we give a list of eight guidepoints. For further illustration, four very different (and age-distinct) examples of task process are analyzed in detail. We then show how these two ascending/abductive versus descendent/deductive methods are intertwined in practice and can be combined in theory. Examples are emergence of jealousy in infants and children's performance on a motor task that has moments of automatic processing and of mental-attentional processing.

The ascendent/abductive and descendent/deductive methods of an effective task analysis are dialectically complementary and in continuous interaction. Together they help to synthesize a problem solution (or to explain performance in postdictive analyses). These two methods can be used together as convenient. In a step *M-construction* (i.e., sequential MTA) the ascendent/abductive method may be more frequent, whereas in the *M-dimensional* (i.e., executive) MTA the descendent/deductive method is more common.

Sequential MTA (Step *M*-Construction) versus Executive MTA (*M*-Dimensional Analysis)

To see how these two methods of task analysis relate, we examine them in the context of one of our color-matching tasks (CMT, Arsalidou, Pascual-Leone, & Johnson, 2010; Arsalidou, Pascual-Leone, Johnson, Morris, & Taylor, 2013; Pascual-Leone & Johnson, 2011; see chapter 7). These are modified 1-back tasks, in which the stimuli are patterned figures with different colors (either a clown or a set of balloons; we focus here on the clown task). Colors of successive figures must be compared to determine whether the set of current-figure colors (i.e., *target colors*) is the same or different from that of the previous figure (i.e., *criterion colors*). Irrelevant features to be ignored are colors blue and green, location of colors in the figure, repetition of colors in the same figure, and colors in the clown's face. The two tasks differ in the form of the stimulus figure: CMT-balloon has as figure a bundle of balloons of various colors; CMT-clown has a clown with a varicolored costume. The balloon task presents a facilitating situation: colors are the balloons' most salient feature, because basic balloons are distinguishable only by their colors in everyday life. In contrast, the clown task is more difficult because it presents a misleading situation: the salient aspect of the stimulus is the charming clown itself with its expressive face (which must be ignored). As a result, relevant colors in the clown figure are more difficult to notice, hindered by the salience of the clown's body and costume, where relevant colors are embedded. CMT-clown was designed as an *M*-measure and has its classes of items differing only in the items' *M*-demand. To ensure that they had the executive and action schemes needed for the task, children were tested first with CMT-balloon and then with CMT-clown.

The two CMT versions (balloon and clown) are computerized tasks (Arsalidou et al., 2010, 2013). They have a brief pretraining in which relevant strategy and needed executive and action schemes are suggested and practiced. Figures are presented one at a time, and for each of them participants indicate, by pressing one of two buttons, whether the current set of relevant target colors is the same as those (criterion set) in the previous figure. There also are baseline items (containing only irrelevant colors, blue and green) that do not require a response. Items are presented in pseudo-random blocks of classes (a class of items is indexed by the number of relevant colors found). *M*-score in the CMT is the *M*-demand of the highest item class passed reliably.

The executive *M*-dimensional analysis of the clown task has the following generic form, where the suspension points stand for omitted schemes (and *M*[…] is also omitted):

$$\text{RESET (RECUR (… MATCH\&PRESS (SCAN\&IDEN}^{L1}\text{ (\{…\}}_{L1}\text{<… ccn>, tci))))} \qquad \textbf{(f1)}$$

In this formula we see four different main operative schemes (symbolized in capitals) with their operations separated by four sets of parentheses. Conventions of *M*-dimensional analysis have two functional orders jointly represented: (1) the *order of actual temporal occurrence* when the formula is enacted, which goes from right to left (every rightmost operation is applied or instantiated first), and (2) the *order of executive dominance or control* that goes from left to right. The leftmost operator is always the task-general one, which controls operatives on its right.

Formula **f1** is represented again in a completed manner in **f2**, which unfolds its four main temporal (right to left) operations into a step sequence:

(1) $M[\underline{\text{SCAN\&IDEN}}^{L1}\ (\{\#\text{IGN: f, irrC, locC, repC}\}_{L1} < \underline{\text{cc1}}, \underline{\text{cc2}}, \ldots, \underline{\text{ccn}}>, \underline{\text{tci}})]$

(2) $M[\underline{\text{MATCH\&PRESS}}\ (< \underline{\text{cc1}}, \underline{\text{cc2}}, \ldots, \underline{\text{ccn}}>, \underline{\text{tci}})]$

(3) $M[\underline{\text{RECUR}}^{L2}\ (\#\underline{\text{set.t:tci}}' \leftarrow \{\text{tci}\}_{L2})]$

(4) $M[\underline{\text{RESET}}^{L3}(\#\underline{\text{set.c}}' \leftarrow \{\text{set.t}\}_{L3})]$ **(f2)**

Step (1) shows the rightmost (temporally first) operation of **f1**, and steps (2), (3), and (4) are the other operations that complete the task's executive procedure. Added to the step formulas in this model are the key schemes of all sorts needed for the executive strategy to be carried out (or outlined) in a suitably concrete manner. Notice that the functional organization found in **f1** is respected in **f2**, albeit using an alternative representation (i.e., the same temporal occurrence and control dominance orders are assumed). Representation in generic formulas like **f1** and **f2** is metasubjective (modeling process sequences from within) and can be unfolded into finer and finer steps until a truly concrete *M*-construction (a step-MTA) has been produced when necessary. Indeed, the step sequence in **f2** is not a true *M*-construction, because each represented step is in fact a generic compacting of a number of truly concrete steps that are implied but not explicitly represented. We shall comment on this point below.

What we present in this task analysis is the sort of action schemes (operatives, figuratives, expectancies, parameters) that may instantiate the executive strategy. However, executive schemes themselves are not represented (nor are the truly concrete action schemes, unless unfolded into an *M*-construction). Executives are not included, because we assume that they are not boosted with the *M*-capacity used for symbolic action processing (i.e., the *Mk* scale of *M*-power). Instead, they are boosted with the *M*-capacity for sensorimotor processes (the *Me* scale of *M*-power; Arsalidou et al., 2010; Pascual-Leone & Johnson, 2005, 2011).

In the first step of **f2**, a participant must scan, one by one, and identify (SCAN&IDEN) the relevant target colors found in the current clown item. As the participant scans and identifies, he or she must ignore (#IGN) the face colors (f), irrelevant colors (irrC), location of colors (locC), and repeated colors (repC). These aspects and the embedding context that is the clown itself are features making the task misleading. We assume that this injunction to ignore is already chunked with the operative SCAN&IDEN. We symbolize such chunking by placing these #IGN schemes inside braces and subscripting the letter *L1* to the second brace, while placing a superscripted *L1* on SCAN&IDEN. This signifies that the latter is the operative-scheme portion of a chunk (a relational scheme) that controls the former. The *L*-boosting process, symbolized by *L1, L2,* and *L3* in formula **f2**, corresponds to multiple schemes that are so highly associated (structured together) that the chunk needs only one unit of *M*-energy to be hyperactivated, because the *L*-operator (*LC* or *LM* learning) is doing it already.

Step (1) in this model shows a moment when the participant, having just scanned the current set of target colors (set.t), now is keeping one of them (tci) in mind along with the total set of criterion colors (set.c, i.e., <cc1, cc2, …, ccn>) from the previous item. Note that a true *M*-construction analysis would have to present, one by one, all steps where each of the criterion colors was scanned, identified, and stored in the *M*-centration during following steps for subsequent use. Then other steps would have scanned the current target colors in front of the subject and retained one of them, tci, in *M*-centration for matching with criterion colors. Step (1) of our *M*-dimensional analysis shows the schemes kept in mind at such precise moment, with all previous steps of the possible *M*-construction being omitted, because they are less complex, although implied by our compacted step (1). In this step the participant must boost with *M*-capacity each of the criterion colors separately, because they are no longer present. Next he or she will have to match them, one by one, with each of the target colors (tci). This total set of criterion colors is symbolized in step (1) as <cc1, cc2, …, ccn>.

In step (2) participants, pursuing analysis begun in step (1), must match each and every target color (tci) with the total set of criterion colors (checking whether the color in question is among the criterion colors). To do this they keep in mind (boosted with *M*-capacity) the operative scheme MATCH&PRESS, the current target color tci, and the *n* criterion-color schemes of step (1). Again, this *M*-dimensional step collapses a number of truly concrete *M*-construction steps that we imply here, in which each of the target colors is matched with the whole set of criterion colors, and then the appropriate button (match or nonmatch) is pressed to respond. Every target color has in turn to be so compared with the set of criterion colors until one mismatch appears. These *M*-construction steps are implied in our step (2).

Step (3) shows the RECUR operation. RECUR applies within the processing of each item; it changes *M*-centration within the set of target colors from one target color already matched (tci) to another target color (tci′) not yet matched (#set.t: tci′←tci). Notice that RECUR is chunked with tci (superscripted/subscripted L2). Finally step (4) shows the RESET operation. RESET applies whenever a new item is introduced, to ensure that the set of target colors (set.t) in the just finished item is retained with its function changed (←), to become the new criterion-colors set (set.c′). That is, the previous target set becomes the new criterion set at the end of the item (#set.c′←set.t). Notice that set.t is chunked with RESET (superscripted L3). Again, steps (3) and (4) are compacting within each finer M-construction steps that would be needed to model in real time every act required.

In formula **f2** the underlined schemes are those that in each step must have their activation boosted by using *M*-capacity. They must be boosted, because they are neither salient nor subordinate parts of an already hyperactivated scheme/schema (or chunk). Counting these underlined schemes for each step we find that in steps (1) and (2) the number of schemes to be boosted by *M* is equal to $2+n$, where *n* is the number of criterion colors in the trial item. For example, in step (1) participants must use *M*-capacity to boost the operative SCAN&IDEN, each of the colors cc1 to cc*n* of the criterion set, and the target-set color (tci), which they next try to match against the criterion set. Because steps (3) and (4) have lower *M*-demand (which is equal to 2), participants should be able to solve a given CMT-clown item, if they have the necessary schemes, when their *M*-power (*Mp*) is equal to $2+n$. Because classes of items differ in the number of relevant colors presented, the value of *n* varies with the item class from 1 to 6 so that corresponding *M*-demand varies from 3 (i.e., $2+1$) to 8 (i.e., $2+6$). Notice further that assumptions of the *M*-demand in items are predicated on the positing of chunking (*L*-structures) and perceptual or cognitive facilitation factors (e.g., due to *F*-operator's simplicity principles), plus other hidden operators/principles that can increase or decrease the task's effective complexity. This is why it is important to differentiate analytically between facilitating versus misleading situations. Once we have examined in the formula these facilitating or misleading factors (checking potential influence of every hidden operator and principle in the model adopted, which in our work is TCO) we may end up with a range of possible *M*-demand values for the task, each contingent on alternative plausible assumptions.

Thus executive *M*-dimensional analysis allows a first approximate estimation of the task's *M*-demand, less credible than that of the sequential *M*-construction. In the dimensional case, *M*-demand (as illustrated by **f1**) corresponds to the sum of various independent (not already *L*-structured) operatives involved, plus all essential data schemes on which these operations apply. This first approximative estimate may be revised when we deconstruct that formula, as done in **f2**, to obtain a finer sequential analysis of what is to be done and how. *M*-demand of this sequential *M*-construction will always be the *M*-demand of the step that has highest *M*-demand in the total sequence. Comparing **f1** with **f2**, it becomes clear that although the set of *n* criterion colors (ccn) and the current target color (tci) must be *M*-boosted, only one of the operatives needs *M*-boosting at any one time.

These two procedures of analysis are different, top-down versus bottom-up, although both methods address the same processes. The descendent strategy estimates task *M*-demand by counting the essential operative constraints needed for the total executive process, whereas in the ascending strategy (called *M*-construction) *M*-demand is estimated in terms of the most *M*-demanding step within the sequence. The two sorts of *M*-estimates are semantic-pragmatically equivalent, but the second is likely to be more precise.

Some Heuristics of MTA: An Incomplete Heuristics Checklist

We list several points that together may help to guide (G) new analysts. The order of points is somewhat arbitrary.

(**G1**) To do a task analysis, we must have already an intuitive model of the psychological organism (metasubject), which allows one to formulate causal operative models of the metasubject's functioning in the task. Second, we need an intuitive objective (figurative and operative) model of the situation, including the problem-to-be-solved (or performance to be produced) as well as facts, conditions, and unknowns that help in formulating the problem/performance. Third, an observer's (operative and figurative) objective model of the task to be performed is needed, with an explicit choice of the executive strategy to be followed.

To formulate this strategy, it is useful to begin by considering the task as an organized totality, from the initial (a-) to the final (-b) situation in which the task can be solved. One keeps deconstructing or differentiating these a-b transitions (often working backward from plausible or intended final solutions) and creating nested functional hierarchies of progressively finer a-b transitions that allow us to mentally actualize and eventually solve the strategy. These subjective a-b transitions are not conceptual but rather are intuitive mentation, modeled as actual mental or physical acts (i.e., infralogical processes modeled mereologically). Finally, the strategy to be chosen must be basic—as simple as it can be—and must take/accept as much error, or variation in performance, as the subject could afford given the chosen task. Thus, from the perspective of the analyst, the executive demand of the strategy adopted has to be minimized. Consequently, we do not try to enumerate all possible schemes of the final performance, but rather only the functional-infrastructure ones (schemes that belong to the minimal effective complexity) that must be activated (perhaps boosted by *M*) to cause organismically the intended performance.

(**G2**) The purpose of metasubjective task analysis is to produce a causal (via overdetermination), specific (operative and figurative) from-within model of the subject's intended task performance, given the chosen strategy (the simplest possible, with minimal executive demand, because subjects like to minimize complexity).

(**G3**) The intuitive causal model of the metasubject (psychological organism) is constituted in our TCO by a repertoire of schemes of all sorts, which are plausible for the type of subject, in terms of his or her expectable personal history. The subject is assumed to be motivated, highly active (schemes follow Piaget's assimilation principle—a propensity to apply under minimal elicitation if not suppressed by some other scheme), and driven by motives and expectancies. Organismic hidden operators and principles must be assumed explicitly, to explain how schemes can be boosted or inhibited in their activation, or dynamically synthesized to produce performance. The problem-solving performance is always overdetermined, that is, multidetermined by hidden brain factors that are variable in potency (weight) relative to both individual subjects and the situation (i.e., are context sensitive). These factors together combine to overdetermine an outcome that conforms to the *F*-mini-max rule: an outcome that minimizes effective complexity while maximizing adaptation to the situation within the goals and strategy pursued.

Learning takes place through reflective abstraction. Schemes emerge as constructive abstractions from task-relevant functional invariants in the situation—task-directed (intentional) actions and expectancies released by schemes previously developed by internalizing empirical invariants (e.g., abstracted via *LC, LM*). This recursive constructivist learning leads to functional hierarchies (schemes of schemes of schemes) defined at various levels of abstraction, which together, and often by steps, can internalize recurrent sequences of performance patterns. This sort of learning can occur within the task itself and be modeled in task analyses.

Performance is either automatic (attentionally effortless) or is constrained by mental-attentional demands and executive demands of the sequence in question (i.e., the needed a-b connections to be carried out). To be automatic, the metasubject must already have in its repertoire suitable schemes facilitated (directly cued) by the situation. Effortful performance, in contrast, requires sufficient mental-attentional capacity (particularly *E, M,* and *I* capacities) to cope with these demands. To minimize mental-attentional demand, subjects tend to break tasks into simpler steps. Consequently, task analysts should carry out a similar sequencing into simple (e.g., *M*-accessible) steps.

(G4) Every metasubjective analysis presupposes a strategy adopted by the subject, which the analyst should make explicit as soon as possible. Each new strategy requires a different MTA. Because subjects follow a principle of simplicity and take as much error as they can afford, we should assume the simplest possible initial strategy and apply the theory resources and principles (in the TCO hidden operators, such as *M, I, E, LC, LM, F,* and hidden principles) to see how a useful process model could evolve in solving the task or explaining subjects' errors. To demarcate a strategy, it is important to consider factors (schemes elicited, biases or dispositions such as those induced by

F-factor and others) that could distract or mislead problem solvers away from simple, correct solutions. Searching for and recognizing these misleading/distracting factors in the problem situation is a golden thread for establishing appropriate strategies, because correct strategies must avoid or control misleading and distracting factors (e.g., automatized shadow schemes, biases). Thus, good schemes chosen for the winning strategy are those that can bypass or control misleading factors when necessary.

(G5) It is fundamental to decide whether the task and the schemes used in it are sensorimotor or symbolic. Sensorimotor (e.g., signalic) schemes are concrete and mediate no meaning beyond themselves; in addition, they are directly activated by affects—affectively immediate. In contrast, symbolic schemes mediate meaning beyond themselves (e.g., the hat on mother's head means for Piaget's 12-month-old daughter that her mother is about to leave the house), and they are directly activated not by affects but by executive schemes and cognitive-perceptual cues. Symbolic but not signalic schemes are executive-mediated and, as symbols, they are not affectively immediate. Clear differentiation between symbolic versus signalic (sensorimotor) schemes is fundamental in task analysis. As discussed in chapters 3, 4, and 7, two different scales of measurement for *M*-capacity, *Me* (signalic) versus *Mk* (symbolic), must be distinguished. These two scales cannot be mixed. In facilitating situations, subjects can use sensorimotor-scale processing. However, in complex misleading situations they must use symbolic-scale processing.

(G6) Although the distinction between sensorimotor and symbolic schemes must be maintained, actual *M*-demand of a scheme within each category is unrelated to the amount of information it carries. All well-learned schemes have in some sense similar *M*-demand (which counts as one *M*-unit), because when well constituted, schemes become structurally unitized or chunked by logical-relational (*L*-) learning. Once constituted as distinct schemes, each needs only one unit of *M*-boosting, and the organism allocates the mental energy needed. To accept a scheme as a suitable unit in task analysis, we must estimate the plausibility of such scheme being already well learned within the subject's repertoire. This depends on the expectable personal learning history of the subjects in question.

(G7) As mentioned, a series of questions should be explored to clarify task-analytical issues: What is the task, and what does the subject have to do to solve it? What is the structure of the situation(s) serving to define the task? What is the unknown? What is the strategy to be analyzed? Which are the relevant schemes in the subject's repertoire elicited by the situation? Which affective/personal schemes (such as motive or shadow schemes) could facilitate or hinder the intended performance? Which are the compelling task-irrelevant (shadow) schemes elicited by the situation? What relevant schemes can be chunked to reduce the mental load of the problem? How do these schemes relate to one another in the context of the task? How can hidden operators or principles (from the chosen organismic theory) affect and change activation or application of schemes to the task? Do they suffice with their dynamic interaction to cause the intended performance? Which other schemes or hidden operators may be needed?

(G8) A plausible analytical sequence for approaching metasubjective analysis (concurrently pursuing its four key moments, i.e., objective, subjective-1, subjective-2, metasubjective) consists in adopting six complementary analytical perspectives. First is *informal ascendent/abductive analysis* (predominant in sequential *M*-constructions) to evoke, from the intended task results, the essential schemes that subjects may have in their repertoire or construct during the task process. Second, *informal descendent-deductive analysis* (predominant in executive *M*-dimensional analyses) helps to clarify how various constructs (e.g., schemes, hidden operators, principles) of the adopted organismic general model can induce the subject to infer or evoke schemes (or biases) necessary to solve the task. In this regard it is mandatory to ask oneself how each construct of the chosen organismic model may be mobilized by the task solution process. Third is a more *formal descendent-deductive analysis* (with explicit symbolic notation and suitable completeness). Fourth is a more *formal ascendent-abductive analysis*, with all its explicit steps toward the task solution. This sequential *M*-construction should be consistent (and complementary) with the executive *M*-dimensional analysis obtained in the prior analysis, therefore helping it to be corrected for the final MTA model. Fifth, coordination and integration of the first four perspectives produces the final causal (operative and figurative) process model. When various distinct competing strategies have been elicited by the task (often a correct effortful strategy vs. an automatized misleading one), the functional relations holding between or among these strategies must be investigated theoretically (schemes and hidden operators involved) and empirically (feature/cues in the situation eliciting one or another scheme) to determine functional interrelations. The analysis of complex and misleading tasks may be culminated with a strategy-relational MTA (see formula **f4** in chapter 8 for an example of this strategy-relational analysis).

Notice that these various perspectives should be explored and coordinated, although the order of exploration can vary. With practice, this often can be done concurrently and implicitly, without requiring explicit formulation of each perspective. However, explicit formulation may be needed to resolve contradictions or confusions.

Illustrative Examples of MTA in Chosen Ages and Domains

Example 1: Attachment in the Sensorimotor Period

Consider first an example of task analysis in the sensorimotor period: developmental emergence of the attachment motive. A current definition of *attachment* would say that it is a personal, emotional bonding to a primary caregiver or protector-companion. It is indexed when the child uses the primary caregiver as secure base from which to explore unfamiliar situations and as a haven of safety and a source of comfort. Attachment begins in infants at 8 to 9 months, and it peaks at 12 to 18 months. It "seems to emerge in every society around the age of one to one-and-a-half years" (Oerter, 2000, p. 65). Nobody has explained why it peaks at 12 to 18 months. What seems clear from animal research is that it is founded on an innate complex disposition of *affective-social beings* (i.e., *affect*-driven beings, affective-plus-cognitive schemes). We express with a *B*- (personal/social beings) operator the intertwining of affects and cognitions, the coordinated affective plus cognitive functional totality of schemes/schemas that enable psychosocial and personal interactions. According to our task analysis (Pascual-Leone & Johnson, 2004), the following essential schemes constitute this attachment disposition:

- **mother*OB***: Mother is the first *internalized social object*—a complex synthesized personal inner-object. This is an instance of internalized (*LCLM*-learned) complex scheme, which we call an *OB* to signify a distal (in this case, personal) object.
- **mother***per:* Mother's *OB-percept.* The internalized representation of the mother's coordinated percepts, which index her actual presence in situations.
- ***Self1*** (Damasio's, 2012, core self): The baby's *first basic awareness of Self vis-à-vis an immediate* Other (mother, father, caretaker).
- **#context**: Concrete and specific cognitive/perceptual/affective representation of the actual here-and-now *situation.*
- **BE-WITH**: Personal/social operative scheme expressing *need to BE WITH*—protected, loved by the intimate protector-companion.

$$\underline{\textit{BE-WITH}}_{L2?}\ (\#\underline{\text{context}},\ \underline{\textit{Self1}}^{L2?},\ \underline{\textit{MotherOB}}^{L1,AB}\ \{\text{mother}\textit{per}\}_{L1,AB}) \rightarrow \textit{Attachment-representation} \qquad \textbf{(f3)}$$

Again, operatives are written in capitals (i.e., *BE-WITH*). Schemes that involve expectancies and are temporally structured in themselves (fluents in our terminology), we write in italics. Three of the schemes in **f3** are written in italics for this reason: *BE-WITH*, *Self1*, and *MotherOB*. These personal schemes emerge within social interactions, and they present as a feature their having fluent (temporally structured and often causal) interrelations. The superscript *AB* stands for the affective (A) and personal (B) operators, which boost and chunk the mother schemes. The core-self scheme *Self1* may be (?) boosting, and be chunked with, the personal scheme *BE-WITH,* as indicated with the superscripted and subscripted *L2*? Overdetermination (**➔**) by these schemes explains emergence of attachment at about 12 months whenever affect (AB) is low, causing the *Self1* to have weak bonding with mother and *BE-WITH* (in which case four schemes have to be boosted by *M*; they are underlined in formula **f3**). However, when affect (AB) is high, causing a strong bonding of *Self1* with *BE-WITH,* only three sensorimotor schemes would have to be boosted by *M* (which should reduce the age of attachment acquisition to 8 or 9 months, see table 3.1). Notice that all the schemes in this task are sensorimotor (e.g., signalic), because they are concrete and mediate no meaning beyond themselves; also, they are directly boosted by affect and do not need executive-driven mediation. The scale of *M*-capacity used is the *Me* scale discussed in chapter 3.

Example 2: Anticipation of Mother Leaving

Another spontaneous, sensorimotor behavior is Piaget's example of Jacqueline's crying when mother puts on a hat—a cognitive expectancy about separation anxiety anticipating mother's leaving. The child had learned to connect the hat on Mom's head with the absent-mother who does not appear when called. Piaget (1963) writes, "[F]rom 0;11(15) Jacqueline cries as soon as her mother puts her hat on. This is not due to fear or anxiety as before, but due to the certainty of the departure" (p. 250). Jacqueline seems to know that a hat (or coat) on her mother signals that mom will disappear and not return when called.

According to our dialectical constructivist interpretation, this behavior expresses the learning (and dialectical competition) of two *WANT-MOM* schemes that constitute two competing alternatives (a dialectical pair) of the *Mother-OB* scheme. Namely, (1) a *Good-MotherOB* scheme, which will make the mom appear on call (Present-Mom) when the mom is not wearing a hat and (2) a *Bad-MotherOB* scheme, manifest when mother wears a hat, who does not appear (Absent-Mom) when called. Constructive abstraction of these two schemes requires the possibility of keeping in the sensorimotor *M*-centration the essential constituent schemes that make up the particular functional facet of the complex scheme *Good-MotherOB* (formula **f4**), and of the contradictory complex scheme of *Bad-MotherOB* (formula **f5**), that our task is addressing.

WANT.MOM1 (*#Mom-no-hat*, CALL (*Mom[Present.Mom]*)) **➔**
CALLING.*Mom[→Present.Mom]* **(f4)**

WANT.MOM2 (*#Hat-on-Mom*, CALL (*Mom[Absent.Mom]*)) **➔**
CALLING.*Mom[→Absent.Mom]* **(f5)**

The two schemes in question have the same mental (*Me*) demand, so they will be formed concurrently, and in dialectical interaction, one is the reciprocal (reverse) of the other. As the number of underlined schemes shows, *Me*-demand of both formulas is four sensorimotor schemes, which makes their constructive learning (in the *Me*-scale) accessible to 12-month-olds. The schemes are indeed sensorimotor for reasons given in the previous example.

The two formulas are complementary and similar. The experience that initiates the functional facets of the two *MotherOB* complex schemes is a desire or longing to be with mother, which we express with schemes *WANT.MOM1* and *WANT.MOM2*. These and other schemes in the formulas are written in italics because they represent fluent schemes (i.e., temporally structured schemes that have emerged through daily interaction with mother across different situations and moments). As these formulas show, in **f4** the child expects that mother will appear (*Present.Mom*) when called, and at this moment the mother *should not* have a hat on (*Mom-no-hat*). In contrast, in **f5**, calling the mom will bring no response (*Absent.Mom*) when mother has the hat on. Thus, Jacqueline cried seeing her mom wearing the hat.

Example 3: Object Permanence

The object permanence task (Barth & Call, 2006; Pascual-Leone & Johnson, 1991; Piaget, 1954) is paradigmatic of Piaget's constructivist epistemological idea (new at the time) that children's learning occurs by internalizing functional invariants abstracted from experience. For him distal (cognitive, not just perceptual) objects are represented by complex schemes (functional-relational systems of structures probabilistically invariant—more or less constant—across space and time, which relate to other schemes). This explains object permanence, as well as (later on) various sorts of conservation properties. We present here and in chapter 3 versions of this family of tasks to illustrate our method (Benson, 1989; Benson & Pascual-Leone, 1997; Pascual-Leone & Johnson, 1991). We cannot review the abundant literature. A recent study of this sort of tasks, using apes and children, is provided by Barth and Call (2006). Complementing, we hope, their empirical work, we briefly outline here a theoretical model of metasubjective processes for this family of tasks, thus explaining differences in age of acquisition by children, which vary with task characteristics.

In this family of tasks there is always a desired object (e.g., a toy) that disappears behind a screen (it may be one screen, **a**, or several of them, **a**, **b**, **c**, ... all aligned in proximity). The paradigm can be complexified in different ways: for example, by having more than one displacement of the toy behind screens, or making the toy displacements visible (child can see the toy) versus invisible (toy is place inside a container, hand or cup; and the container, with the toy hidden inside, brought behind the screens), or by varying the distance (adjunct or nonadjunct) between screens where the toy is being hidden, or by varying the waiting time (long/short delay, nondelay) until the child/ape is allowed to reach for a screen as he or she looks for the object. In all cases, the subject (child or ape) who desires the object is given an opportunity to reach to a screen, and remove it, seeking the desired object (which in apes is food).

Single Visible Displacement The simplest version of this paradigm is a single visible displacement and a single screen. After some suitable motivating introduction, the tester presents the child with one screen and places the toy, visibly displaced with the hand, behind it. The tester then encourages the child to find the toy. The child can remember that the toy disappeared behind the screen (this is acquired at about 4 months; Renée Baillargeon, 1987). Thus infants would have schemes *screen[toy behind]* and *#toy[FOUNDbehind screen]*. Consequently, the child may bring the hand to the screen (*screen [toy behind]*) to remove it (GET.TOY:.*REMOVE.screen*) and obtain the toy. With these schemes (sensorimotor because their activation is affectively immediate, not deferred or executive driven, and there are no misleading factors) our task-solving formula is:

$$\underline{\text{GET.TOY}}^{L2,A}\ (\textit{screen[toy behind]}_{L1}, \{\textit{REMOVE.screen}\}_{L2}\ (\{\textit{\#toy[FOUNDbehind screen]}\}_{L1}, \underline{\text{toy}}^{L1,A})) \rightarrow \text{toy} \qquad \textbf{(f6)}$$

Notice that this formula has only two schemes underlined (i.e., requiring *M*-boosting), because *screen[toy behind]* is already *L*-boosted, along with the parameter (#) scheme, by the scheme toy^{L1}. All these schemes are also *immediately* boosted by affects (*A*-operator), and so no executive scheme is needed to mediate them: the *Me*-scale of *M*-capacity has to be used, which predicts (see table 3.1) success at about 4 or 5 months—a bit after the age when the child can use his or her gaze to show knowledge of where the toy is hidden (Renée Baillargeon, 1987).

More difficult is when the child must choose which of two or more screens (**a**, **b**, **c** ...) the toy was hidden behind. After the toy is placed behind a screen, to solve the problem the child needs a process formula such as the following:

$$\underline{\text{GET.TOY}}^{L2,A}\ (\{\textit{screen[toy behind]}\}_{L1}, \{\text{SELECT \& }\textit{REMOVE.screen}\ \}_{L2}\ (\textit{\#toy[FOUNDbehind } \mathbf{a}]_{L1}, \underline{\text{toy}}^{L1,A}, \underline{\text{a}}) \rightarrow \text{toy} \qquad \textbf{(f7)}$$

In this formula, *#toy[FOUNDbehind **a**]* is the expectancy scheme that stems from the experience of the toy disappearing behind **a**. This scheme serves as parameter to the operative SELECT that chooses which screen (**a**, **b**, **c**) to REMOVE to find the toy. This is clearly an affectively immediate task driven by perception and without executive mediation—a sensorimotor task. The *Me*-scale (see table 3.1) shows that three distinct schemes can be simultaneously *M*-boosted when the infant is about 8 months old, not earlier.

The well-known A-not-B error discovered by Piaget has a formula similar to **(f7)** when the child in fact commits the **a**-response error (directly driven by affect, as analyzed in formula **f6**). This error consists in the following: when the infant is presented with two or three screens **a, b, c,** and the desired object (e.g., toy) is hidden by the tester behind **a**, the child, seeing the tester's action, selects **a** (an active SELECT operation may not be needed because the response is facilitated by a perceptual and learned habit to do the **a**-response), then the child REMOVES **a** to get the toy. The tester repeats the same procedure two or three times, and the child gives the same response (which induces in the child an **a**-response habit). Next, without warning, the tester hides the toy behind screen **b** or **c**, thus creating a dialectical contradiction between the **a**-habit and the recent-scheme memory of where the toy is hidden. Reliably, infants younger than 12 months or so commit the error of persisting in responding to **a** (because of the learned expectancy that toy is behind **a**), even though the child just saw the toy being hidden behind **b** or **c**. This is the A-not-B response. A more complex version of the same task appears using multiple visible displacements, in which errors can be called A-not-B/C/D responses.

Multiple Visible Displacements with Multiple Screens To solve the multiple visible-displacement task, when the child has seen the toy go behind **a** (one or more times) and gotten the toy by removing **a**, the child must overcome the **a**-response habit. He or she can do so if enough task-relevant and compatible schemes opposing the **a**-response have been hyperactivated by mental attention to force suppression of the **a**-response habit (via overdetermination against this habit). Four different schemes are needed: (1) the memory of the screen (**c**) where the hand with the visible toy last disappeared (*toy-at-hand-TO-c*); (2) the learned intuitive rule that the object will be found where it disappeared last (*toy-FOUNDbehind-LAST-screen*); (3) the remembered-fact that toy was no longer in the hand when it reappeared after being hidden behind **c** (No-toy-at-hand-after-c); (4) The main operative scheme described above (GET.TOY:.*SELECT & REMOVE.screen*). Thus, the solution formula could be this operative model:

$$\underline{\text{GET.TOY}}^{L2}\ (\#\underline{\textit{toy-FOUNDbehind-LAST-screen}},\ \{\text{SELECT } \&\ \textit{REMOVE.screen}\}_{L2}\ (\#\{\underline{\textit{toy-at-hand-TO-c}}\}^{L1,Sit} \leftarrow \{\text{No-toy-at-hand-after-}\mathbf{c}\}_{L1,\ Sit}),\ \underline{\mathbf{c}}) \rightarrow \text{c-RESPONSE} \qquad \textbf{(f8)}$$

Let us analyze this formula showing a dialectical interaction between the ascendent (bottom-up, abductive) method and the descendent (top-down, deductive) method. The child wants to GET.TOY but when the tester has completed the hand-with-toy displacements, the toy is no longer in the tester's hand (scheme No-toy-at-hand-after-c). This fact-scheme and the child's desire to get the toy, make children recall (←) that hand and toy were last behind **c** (*toy-at-hand-TO-c*). This scheme is temporally structured (a fluent), because it describes the movement of hand-toy going behind **c** (we emphasize this fluent idea by writing the scheme in italics, and its operative sub-scheme of going *TO*, behind the screen, in capitals). The child intuitively thinks the toy is likely to be behind screen **c**, because the actual situation (*Sit*) suggests so. The memory (←) that *toy-at-hand-TO-c* together with scheme No-toy-at-hand-after-**c** should raise the inference that toy is to be FOUND behind **c**. This thinking causes the child to aim for **c** (**c**-RESPONSE), to *SELECT* and *REMOVE* this screen and find the toy. It should be obvious that this is an affectively immediate task using sensorimotor schemes. The *Me*-demand is four schemes, a capacity that emerges at about 12 months.

Multiple Invisible Displacements with Multiple Screens The task is virtually the same, but now the toy is not visible, because it is always hidden by the tester's hand (or in Barth & Call's, 2006, research with apes, the toy is hidden inside a cup being displaced). In Piaget's method, the tester exclaims, *"The toy! the toy!"* as the toy, hidden in the tester's hand, leaves a given screen to enter behind another screen, and so on till the last screen.

GET.TOY^{L2} (#*λtoy!*[*toy behind screen*] ({SELECT & *REMOVE.screen*}$_{L2}$
(#*toy-FOUNDbehind-LAST-screen*,
(#{*toy-at-hand-TO-***c**}^{L1} ← {No-toy-at-hand-after-**c**}$_{L1,Sit}$), **c**))) ➔ *c-RESPONSE* **(f9)**

For this reason, as shown in **f9**, children of 12 or 18 months understand the utterance (*λ*—lambda is our symbol for linguistic schemes) *λ:The toy! The toy!* as meaning that the toy is being carried behind the screen when the utterance is voiced (scheme #*λtoy!*[*toy behind screen*]). This scheme results from the inference that if utterance *λtoy!* were voiced prior to the hand entering behind a screen, then the toy would be inside the hand entering the screen. This parameter (#) is boosted by *M*, because it involves a novel inference essential for task solution. When the child sees that the toy is not inside the hand emerging from **c** (scheme No-toy-at-hand-after-**c**), he or she experiences a contradiction with the prior inference that the hand with the toy went TO **c** (an inference based on the schemes #*λtoy!*[*toy behind screen*] and *toy-at-hand-TO*-**c**). However, to be able to SELECT the suitable screen, the child needs to be guided by the empirical concept-rule abstracted after 12 months of age: the toy is always found where it disappears last (*toy-FOUNDbehind-LAST-screen*). This scheme, an expectancy-parameter (#), is available after 12 months, because its essential constituent schemes are at most three or four units: *FOUNDbehind*, *LAST*, *screen, and *toy; but notice that schemes *FOUNDbehind* and *toy may be chunked (*L1* subscript/superscript) because of the child's prior playing experience. As the underlines of formula **f9** suggest, the sensorimotor *Me*-demand of the total task is five, accessible at about 18 months of age. The important categorical scheme *toy-FOUNDbehind-LAST-screen* should be available in the child's repertoire at this point.

Example 4: Simon Task

The Simon task, discussed in chapter 7 (see Figure 7.13), illustrates a symbolic processing task. In the standard Simon task, the child sits before a monitor and has two response buttons (left vs. right). Stimuli appear one at a time, on the left or right side of the screen. The task is to press the right button when a given stimulus (e.g., the frog) appears and press the left button when another stimulus (e.g., the butterfly) appears. The Simon effect is the strong tendency (present in 4-year-olds) to press the right-side button when a stimulus appears on the screen's right, and press left when stimulus appears on left, irrespective of the stimulus. This tendency (induced by the subject's *F*-operator) slows responses to any *incongruent* trial—one in which the stimulus appears on the opposite side to where the response must be made. The task analysis formula for the standard Simon, in the case of an incongruent trial, is:

$$\underline{\text{PRESS}}^{L1}(\#\textit{RESPbutton-left}_{L2}, \{\text{IDEN}(\text{LOCATE\# right}\}_{L1}, \underline{\text{butterfly}[\textit{RESP:left}]}^{L2})) \rightarrow \text{RESPONSE [correct button press]} \qquad \textbf{(f10)}$$

The task-process sequence begins at the right side of the formula before the arrow (as always in *M*-dimensional analysis): the stimulus butterfly appears located at the right side of the screen. The subject perceives this location using automatic/perceptual attention (LOCATE# right) and then cognitively identifies it (IDEN) as one to be responded to with the left button (butterfly[*RESP:left*]). This recognition may already be chunked (*L2*) with the parameter (#) *RESPbutton-left*, which stipulates the button, right or left, to press. A dynamic overdetermination (➔) by these schemes brings the child to the moment of PRESS, producing the correct RESPONSE. Notice that cognitive identification (IDEN) is needed because the task is misleading, because of the *F*-factor. With congruent trials, in contrast, fast processes of automatic/perceptual attention and S-R compatibility can suffice to produce the performance. Mental attention is needed to overcome these automatic processes in the incongruent trials.

The cognitive/conceptual identification and the necessary verbal instructions show this task to be symbolic. The mental (*Mk*-scale) task demand is $e+2$ (i.e., two symbolic schemes plus sensorimotor *Me*-capacity, used to activate executive schemes not directly represented in formula **f10**). The task should be accessible to 5- or 6-year-olds. The

Simon effect persists into adulthood, yet adults are not affected by the *M*-demand of the task, because it is well within their capacity. However, they still experience misleadingness induced by S-R compatibility.

Example 5: Fraction Word Problems

The problems that follow and their task analyses stem from the doctoral research work (in progress) of our York University graduate student, Cheryl Lee. Consider the following word problem: "Carla's pencil case contains 12 things, including 3 pens. The pens are what fraction of all the things in Carla's pencil case?" We assume that participants recognize that the question requires computing the *part of a whole*. Formula **(f11)** symbolizes the schemes that must be activated during task solution.

$$\underline{\text{PARTWHOLE}}^{L1}\ (\{\textit{DIV.equal.shares}\}_{L1}\ (\#\underline{\text{num–denom}},\ {*}\underline{\text{pens}[\textit{PART}]},\ {*}\underline{\text{things}[\textit{WHOLE}]}))\ \rightarrow \text{SOLUTION} \qquad \textbf{(f11)}$$

PARTWHOLE refers to the conceptual and procedural understanding of this type of problem. As we have mentioned, superscript *L1* stands for a logical (*L* learning) structure or schema—an overlearned connection between schemes. Here it conveys that participants know the part-whole concept as linked to a scheme for division into equal shares. The subscript *L1* outside the braces indicates that operative *DIV.equal.shares* is *L*-boosted by PARTWHOLE, because of prior learning.

The next three schemes are the parameter #num–denom and figuratives *pens[*PART*] and *things[*WHOLE*]. The parameter indicates that PARTWHOLE{*DIV.equal.shares*} is applied by forming the numerator and denominator of a fraction. The parameter indicates that *pens[*PART*] is numerator over the denominator *things[*WHOLE*]. The predicted *M*-demand of this strategy is four symbolic units (as the number of underlined schemes in **f11** indicates). Such *M*-demand is accessible to children of 9 years or older.

More challenging is multiplication of two proper fractions, as exemplified by the following problem: "Kathy has 4/5 of a bag of peanuts [fraction 1: $a_1/\ b_1$]. Ryan has 2/3 as many peanuts as Kathy has [fraction 2: $a_2/\ b_2$]. What fraction of a bag of peanuts does Ryan have?" The full formula for solution of this problem is:

$$\underline{\text{PARTWHOLE}}^{L2}\ \{\textit{DIV.equal.shares}\}_{L2}\ (\#\underline{\text{num–denom}},\ \underline{\text{SCALAR}}^{L1}\ (\{\text{MULT}_2\}_{L1}\ (\#\underline{\text{mult-tab},\ {*}b_1,\ {*}b_2) \rightarrow {*}\text{denom}},\ \{\text{MULT}_1\}_{L1}\ (\#\underline{\text{mult-tab},\ {*}a_1,\ {*}a_2) \rightarrow {*}\text{num}}))\ \rightarrow \text{SOLUTION} \qquad \textbf{(f12)}$$

Formula **f12** is an example of *M*-dimensional analysis (an executive MTA). Here the thin arrow (→) indicates the result of the locally preceding operation. The strategy modeled can be unfolded in several steps that proceed from right to left in the formula.

We show this in formulas **f13** to **f15** below. When the following steps are put together, we have an instance of a sequential *M*-construction analysis (a step MTA).

$$\underline{\text{PARTWHOLE}},\ \underline{\text{SCALAR}}^{L1}\ (\{\text{MULT}_1\}_{L1}\ (\#\underline{\text{mult-tab}},\ {}^*\underline{a}_1,{}^*\ \underline{a}_2)) \rightarrow {}^*\text{num} \qquad \textbf{(f13)}$$

Step 1 (**f13**) requires participants to hold in mind five symbolic units: (1) the general operative scheme for a part-whole problem; (2) because the problem requires rescaling one quantity by the value of another, the operative stipulates a scalar type of multiplication; (3) the scheme of the multiplication table; (4) the numerator of first fraction; and (5) the numerator of second fraction. In formula **f13**, starting from right to left, the figurative schemes $^*a_2$ and $^*a_1$ refer to numerators of the two fractions. The scheme #mult-tab is the parameter indicating that the multiplication table (or relevant number facts) must be held in mind. $\{\text{MULT}_1\}_{L1}$ is an operative of multiplication. It carries number 1 subscripted because it is the first multiplication to be calculated. The subscripted *L1* outside the braces indicates that this scheme is not being boosted by the *M*-operator but by SCALAR^{L1} to which it is chunked; this corresponds to the conceptualization of the problem as requiring scalar multiplication in the service of the part-whole solution. PARTWHOLE is the overall concept (general operative) of the problem. The solid arrow (➔) indicates the product of this dynamic synthesis of schemes (which apply together via overdetermination) yielding the actual numerical result (*num) of the operation. Predicted *M*-demand of this step is five symbolic schemes.

$$\underline{\text{PARTWHOLE,}}\ {}^*\underline{\text{num}},\ \underline{\text{SCALAR}}^{\ L1}\ (\{\text{MULT}_2\}_{L1}\ (\#\underline{\text{mult-tab}},{}^*\underline{b}_1,\ {}^*\underline{b}_2)) \rightarrow {}^*\text{denom} \qquad \textbf{(f14)}$$

Step 2 (**f14**) represents multiplication of the denominators (b_1 and b_2), while also holding in mind the result of the previous step (*num). The solid arrow points to the operation's actual numerical result (i.e.,*denom). Predicted *M*-demand of this step is six symbolic schemes.

$$\underline{\text{PARTWHOLE}}^{L2}\ (\{\textit{DIV.equal shares}\}_{L2}\ (\#\underline{\text{num–denom}},\ {}^*\underline{\text{denom}},\ {}^*\underline{\text{num}}))$$
$$\rightarrow {}^*\text{num/denom} \qquad \textbf{(f15)}$$

Step 3 (**f15**) mirrors the task analysis for the simpler problem discussed above (**f12**). The parameter #num–denom indicates that PARTWHOLE^{L2} $\{\textit{DIV.equal shares}\}_{L2}$ is to be applied to the previously obtained numerator (*num) and denominator (*denom). The fraction *num/denom is the obtained result. Predicted *M*-demand of this step is $e+4$.

Each step has a different *M*-demand, and the total *M*-demand of the task is given by the step with highest demand. In this example, the *M*-demand of the task is six mental schemes, derived from Step 2. Thus, this task should be accessible to 13-year-olds and older. The *M*-demand could be decreased, however, if we assume the problem solver has had relevant experiences that further chunked these schemes. For example, if the participant has memorized and automatized the multiplication table (i.e., relevant number facts), then scheme #mult-tab may not have to be *M*-boosted, and so *M*-demand would decrease by one unit. In our task analyses we usually assume that problems will be fairly novel to participants, and afterward calculate the decrease in *M*-demand that may result from further relevant experience. Additional steps to solve a problem do not necessarily add to *M*-demand. However, the need to plan and coordinate multiple steps does increase executive demand of a problem (and may require better prior experience). Agostino, Johnson, and Pascual-Leone (2010) found that children performed worse on multistep multiplication word problems than they did on one-step problems matched for *M*-demand.

Let us now compare the *M*-dimensional analysis as illustrated in formula **f12** with the *M*-construction analysis as given in combined formulas **f13**, **f14**, and **f15**. As mentioned before, *M*-dimensional analysis exhibits more clearly the overall executive-processing structure of the task and shows its key figuratives, operatives, and parameters. However, the *M*-construction analysis shows better how the executive process is implemented in real time and makes more explicit assumptions locally made to justify the *M*-demand of the task. Given that the two methods, albeit complementary, are different—the one mostly top-down (descendent) and the other bottom-up (ascendent)—why should they yield the same *M*-demand estimates (as they usually do)? The answer to this important question was explicated earlier in the chapter. Because in both cases the same strategy and effective complexity is being modeled, we must expect, if the analyses make no error, that their respective *M*-demand estimates will be the same. Thus, results of one method can serve to check and complement the other method.

Methodological Synthesis of Sequential MTA (Step *M*-Constructions) and Executive MTA (*M*-Dimensional Analysis)

Step MTA is particularly useful in sequential tasks, such as tasks involving human performance and motor activities. Its complementary alternative, dimensional/executive MTA, is more useful in problem-solving or inferential mental tasks. However, these two sorts of analysis can be synthesized into a common methodology. We illustrate again such synthesis with two very different examples: one dealing with babies' emotions, on the emergence of the feeling of jealousy, the other dealing with speeded motor performance in school children, a child's mental strategy for coping with a hard-to-do motor performance.

Example 1: Emergence of Jealousy

Consider first the emergence of jealousy in a 20-month-old. This is a true concrete happening communicated to us by the father of the child. This child (we will call him J) had had (three months before) the event of a brother (T) born into this family of four (counting Mom and Dad). Prior to the present incident, J had always expressed positive regard toward his brother, accepting to share with him Mom and Dad. "J has been known," his father says, "to give T gifts, his favorite toys, foods, even his sucky blanket. He puts these things on T's belly. Sometimes he kisses the baby." But the day we talk about was different. Although at 17 or 18 months, his mother says, J occasionally expressed distress about his brother, in the event we are presenting J's actual motive and intention seemed much clearer. Mom was holding and breastfeeding T, and J approached trying to attract Mom's attention. Mom kindly moved J away from her. The father says, "This time he was pulling at his mom's arm. Prying her hand off the baby she was feeding. He also was using his free hand to try and hit T in this process." J attempted with some emotional distress to push his brother away. Dad went to J and moved him away. Father recognized that J had now expressed jealousy of his brother, which he had never done before. Why now?

Let us reconstruct in a sequence the experiences J went through. In step (1) J notices/feels Mom's love and attention for his brother T, and Mom is not attending to J. In step (2) J has now enough M-capacity to simultaneously attend (in his working mind) to both Mom's love for T and J's love for Mom, which elicits a negative feeling that T is taking away Mom's love from J. In step (3) the previous negative feeling has elicited in J's emotional mind an operative feeling of hostility toward T, which induces J to physically show this hostility in an overt RESPONSE. The three steps are symbolized in formula (**f16**):

(1) J-NOTICES[$\underline{\text{LOVE}}_{\text{M}}$ ($\underline{\text{Mom}}$: $\underline{\text{T}}$)] ➔ *J[feels Mom's love of T]*
(2) J-THINKS[$\underline{\text{LOVE}}_{\text{J}}$ ($\underline{\text{J}}$: $\underline{\text{Mom}}^{L1}$[$\underline{\text{LOVE}}_{\text{M}}$ ($\{\text{Mom}\}_{L1}$: $\underline{\text{T}}$)])] ➔
J[feels T is taking away Mom's LOVE from him]
(3) J-FEELS[$\underline{\text{HOSTILE}}$ (#$\underline{J^{L1}\textit{[feels T is taking away Mom's LOVE from him}}$], $\{\text{J}\}_{L1}$:$\underline{\text{T}}$)] ➔
J-RESPONSE[*shows hostility*] **(f16)**

As usual, words in italics stand for fluents; square brackets [...] placed to the right of an object/noun (or operative/verb) indicate that the scheme in question embodies cognitive/affective characteristics (or states) described inside these brackets. The symbol *L1* affixed as superscript/subscript to schemes indicates the schemes in question are *L*-structured into a chunk, and so only one of them needs to be *M*-boosted to activate the two schemes.

These formulas could be read as just complicated ways of presenting a narrative of what happened. Such interpretation would miss an important point: the narrative is here deconstructed into schemes, self-propelling constructs, and their co-activation necessitates their being boosted by *M* (because, although affectively driven, they are not perceptually salient, and contradictory negative affects—products of internalized parental "interdiction"—tends to suppress them). This is a process model, not just a narrative, and it has causal-explanatory power. It shows that limitations of *M*-capacity prevented J from being jealous before the age of 18 months. The steps of formula **f16** exhibit, respectively, an *M*-demand of $Me=3$, $Me=5$, and $Me=3$ (counting the underlined schemes in formula **f16**). These schemes express sensorimotor and not symbolic processing, because there are no misleading factors at the cognitive level, and both the affects (our *A* and *B* hidden-operators) and salient cues (*C*- and *LC*-operators) of the situation facilitate this embodied/experiential emotional processing. The demand of the total formula/model is $Me=5$—the *M*-capacity of 18-month-old or older infants (see chapter 3; Pascual-Leone & Johnson, 1991).

The three personal (i.e., affective and cognitive) schemes of J, that is, J-NOTICES[…], J-THINKS[…], or J-FEELS[…] stand for J's emotional states, whose meaning is expressed inside the square brackets […]. The new emotional experience occurs in step 2 as an emotional not-verbalized THOUGHT of J, which in English may be expressed as, "J loves Mom who loves T!" This is a primal emotional conflict—a dialectical contradiction within the emotional *love-system* of the child: the frustrated love elicits its reciprocal emotion, *anger-aggression-hate* (Pascual-Leone, 1991b). This conflict immediately leads to step 3: J feels hostility toward T, which produces the overt aggressive RESPONSE (➔) we represent by scheme "*J[shows hostility*]."

This task analysis shows sequentially (*M*-step analysis) what in a dimensional/executive MTA would appeared more compactly but also more abstractly and perhaps less clearly. A dimensional/executive analysis of the same formula appears in **f16′**:

J-FEELthinksActs[HOSTILE (#J$_{L2}$[*feels T is taking away Mom's LOVE from him*], J$_{L2}$:T),
LOVE$_{J}$ (J^{L2}: Mom^{L1}[LOVE$_{M}$ ({Mom}$_{L1}$: T)])] ➔ *J-RESPONSE*[*shows hostility*] **(f16′)**

A comparison of **f16** with **f16′** shows that the three steps of **f16** have been collapsed, from right to left in the formula **f16′**, into a single "manifold step," which produces as a result the final outcome. In this formula, like in all *M*-dimensional analyses, executive control of processes goes from left to right of the formula line, but temporal occurrence of events goes from right to left of the line. With these rules, **f16** appears as an unfolded version of **16′**.

A second remark worth making for this example is that it shows how personal emotions emerge within the personal mind. This sort of personal task analysis suggests how refined analyses may be conducted in clinical (and sociocultural or psychotherapy) individual-difference studies. They are congruent with (albeit more explicit than) task analyses done within the emotionally focused therapy approach (Antonio Pascual-Leone, Greenberg, & Pascual-Leone, 2009, 2014).

Example 2: Motor Rho Task

The motor rho task is a speeded-performance task in which mental problem-solving occurs as the subject synthesizes intended motor performance. Participants need to coordinate dimensions/constraints of the action sequence to maximize speed and quality of performance; this is the effective complexity of this task. The present summary is extracted from Pascual-Leone (1987). Its data (summarized in more detail in chapter 10) were replicated in Pascual-Leone's lab. The MTA of the rho task involves three successive analytical "moments": (1) the repertoire of essential schemes needed; (2) a sequential *M*-construction analysis, which presents the dimensional/operative processes involved and their effective order; and (3) an executive *M*-dimensional analysis summarizing (1) and (2) and exhibiting its executive structure.

Originally used in human performance research, the rho task was brought to our attention by Dr. John Todor (1977, 1979) in a personal communication. It is best explained with a diagram of its apparatus, which appears in figure 9.1. The apparatus consists of a crank mounted on the top of a wooden box, with a small handle to be held with thumb and index of the stipulated hand. The crank must be rotated by the hand horizontally 360 degrees, until hitting a small bumper. At the other end of the box, two warning lights, a "ready" and a "go" light, control in every trial the initiation of hand-movement action. Onset of lights is separated by a random interval. Next to the lights there is a metallic paddle that participants must hit and knock down with the hand that moved the crank, very quickly, as they complete the crank rotation. At the end of every trial the hand will have completed a trajectory resembling the Greek letter rho. The dependent variable is speed of movement in this trajectory, measured in latencies (participants were urged to proceed as fast as possible). Participants were trained in every segment of the movement performed; their final scores were average latencies when speed practice had stabilized (i.e., at the end of thirty trials) under instructions of maximum speed.

Four consecutive latency measures were used that index clearly distinct phases in the task. Pascual-Leone predicted these differences, and Todor (1977, 1979) first demonstrated them experimentally. These were: *reaction time* (RT: time elapsed from onset of "go" light to initiation of the crank movement); *circular time* (CT: from offset of RT to about 30 degrees prior to hitting the bumper); *pause time* (PT: from offset of CT to the moment when the hand releases the handle); and *linear time* (LT: from offset of CT to time of hitting the paddle). From a developmental-constructivist perspective, RT and CT are ordinary perceptual-motor performance factors: speed of motor processing and ability to "draw" very quickly a circle with the hand on the horizontal plane, which produce fast times. In contrast, PT and LT reflect problem-solving

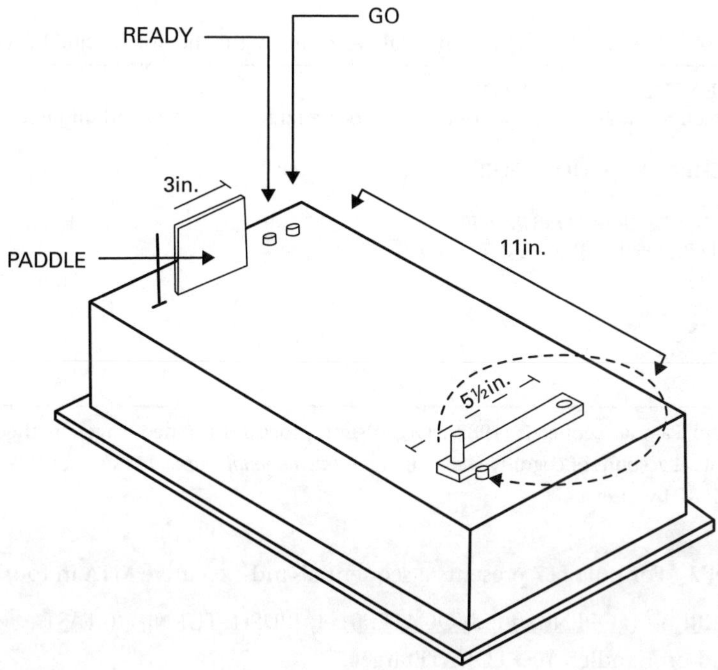

Figure 9.1
Rho task apparatus (right-hand unit). (From Pascual-Leone, J. [1987]. Organismic processes for neo-Piagetian theories: A dialectical causal account of cognitive development. *International Journal of Psychology, 22*, 559. Copyright 1987 by Elsevier.)

(concrete-operational) processing, because during them participants adjust schemes (timing of various actions) to follow the executive plan and adjust every moment to maximize speed of performance. PT and LT scores index dynamic mental-motor syntheses of mentation (mind work). A task analysis of the three "moments" or models of rho task shows its full dynamics with critical decisions taken at maximal speed. We discuss them in the natural order.

Rho Model 1 Table 9.1 shows the key repertoire of schemes required to perform the task. Identifying the schemes is first step for any task analysis. Three distinct sorts are needed: executive schemes, action schemes (both operative and figurative), and parameters (i.e., conditions, often temporally structured schemes, regulating the processing). The table evokes these schemes using intuitive terms. When the repertoire of task-required schemes is completed and refined, the other two moments in the total task are easier to complete.

Table 9.1

Three types of schemes involved in the mental processing of the rho task (PT and LT segment)

EXECUTIVE SCHEMES: PLANS FOR
"go fast," "scan for target," "open hand now," "locate target now," "go and hit target"
ACTION SCHEMES: ACTIONS FOR
GO FAST!
LOCATE TARGET PRIOR TO BUMPER!
OPEN HAND PRIOR TO BUMPER!
GO+HIT target
PARAMETERS
open hand NOW!
locate target NOW!

Adapted from Pascual-Leone, J. (1987). Organismic processes for neo-Piagetian theories: A dialectical causal account of cognitive development. *International Journal of Psychology, 22*, p. 560. Copyright 1987 by Elsevier.

Rho Model 2 Formula **f17** presents a sequential-and-executive MTA in four steps.

(1) TASKRho^{L1} (#OPENhand, #LOCATEtarget, {PUSH+TURN}$_{L1}$, (#FAST, {hand-on-handle}$_{Sit}$)) ➔ LOCATEtarget

(2) TASKRho^{L1}(#LOCATEtarget.PRIORbumper,{LOCATE}$_{L1}$(#NOW$_{1}$!, #FAST, target)) ➔ $LOCtarget[paddle]

(3) TASKRho^{L1}($LOCtarget[paddle], OPENhand^{L2} (#PRIORbumper: #OPENhand, #NOW$_{2}$!, {#FAST}$_{L1}$, {hand}$_{L2}$)) ➔ $hand[OPENED]

(4) TASKRho^{L1}({GO+HIT}$_{L1}$(#{FAST}$_{L1}$, $hand[OPENED]:$LOCtarget[paddle])) ➔ ➔ $RESPONSE **(f17)**

The task's sequence is represented. In step 1 a general operative of the task (TASKRho) is already chunked (coordinated via intratask learning) with the task's hand and handle circular movement that leads the hand toward the bumper, that is, PUSH+TURN (this chunking is symbolized by super/subscripted *L1*). In other steps TASKRho is chunked also with LOCATE (the target in space) and with GO+HIT. OPENhand has a misleading scheme (not represented in the formulas) inducing it not to open the hand until hitting the small bumper (as is done in most life activities, induced by *F* and *LC* factors); this misleading scheme is contradictory with going very FAST. Unless local cues have been fully automatized, which is not expected, TASKRho must carry two parameters (#LOCATEtarget and #OPENhand) telling it when, to optimize speed, the target/paddle must be localized and then the hand opened. PUSH+TURN has another parameter with the injunction of doing it as FAST as possible (notice that, in steps 3 and 4, FAST no longer needs to be *M*-boosted because, via intratask learning, it has been chunked with TASKRho).

In step 2 the general operative, carrying a parameter to locate the target PRIOR to hitting the bumper, implements the now-required LOCATE response. This specific operative has two parameters, #FAST and #NOW$_1$!. At the appropriate time the scheme to LOCATEtarget is called by the corresponding parameter NOW$_1$!. The figurative scheme *target, on which LOCATE applies, must be boosted with *M*. This step produces the response (➔) of location of target in true space, which will guide in step 4 movement of the hand. Notice that $LOCtarget[paddle] is prefixed with the symbol $ to mean that this scheme refers to a here-and-now real event (it is applying on a real object in space, it is actually true).

In step 3, the child proceeds to OPEN hand prior to bumper when parameter #NOW$_2$! demands it. As a result of this action (➔), the hand becomes free from the handle ($hand[OPENED]). In step 4, the hand moves toward the paddle to HIT it, producing the intended FAST and true ($) final RESPONSE.

The steps with highest *M*-demand (number of schemes underlined) are 2 and 3. This *M*-demand is five symbolic schemes (i.e., $Md=e+5$). This is the mental-attentional demand when a speeded optimal-performance strategy is used. However, children with less *M*-capacity may solve this task using less speed in their performance. Indeed, there are three possible passing strategies in the Rho task.

Passing strategy 1 is a slow strategy that uses the four key operative schemes but no parameters. With this strategy, children can perform the task without exhibiting errors, such as "forgetting to open the hand before attempting to move toward the target" or "moving toward the paddle without having located this target in space." These two errors, together or separate, can be found in 5- or 6-year-olds: the first appears as an attempt to move the hand toward the target while still holding the handle; the second as rapid hand movements that miss the target and go toward the ceiling. Using strategy 1 (with no parameters) children with a maximum *M*-capacity of $e+3$ (i.e., 7- and 8-year-olds) would be slow but succeed in the task.

Passing strategy 2 uses the three key action schemes and one parameter (one of the various parameters # indicated in formula **f17**), with an *M*-demand of $e+4$, first accessible to 9- and 10-year-olds. Thus, these two ages should exhibit significantly faster PT+LT latencies than 7- and 8-year-olds, and the 7-year-olds should not differ from 8-year-olds because they have the same maximal *M*-capacity. These predictions are supported by Todor's (1977, 1979) data and unpublished results from our lab from about 1985 (Pascual-Leone, 1987), some of which we summarize in chapter 10.

Passing strategy 3 is modeled in formula **f17**; it has four operative schemes and two key parameters: Now1! (Locate) and Now2! (Open). The *M*-demand of $e+5$ makes the strategy first accessible at 11 and 12 years of age. Thus, 11- and 12-year-olds should produce the shortest latencies. Notice that the two key parameters must be boosted with mental attention (*M*), because for both there are misleading schemes that would induce error if parameters of Now! are not highly activated. The scheme Now1! (locate target prior to bumper) is contradicted by "bad" habits induced by Gestaltist misleading factors (*F*-operator, S-R compatibility) that lead participants to forget about location of targets until after releasing the handle. Scheme Now2! (open hand prior to bumper) has a similar "bad" habit interfering with it: not to open the hand until after touching the bumper. These *F*-facilitated misleading factors interfere with fast responding, making it necessary to boost the Now-parameters with *M*.

Rho Model 3 This simpler MTA lists only operatives that constitute the executive strategy or executive plan. Some analysts may prefer in some tasks to demarcate this executive plan (rho model 3) before attempting to model fully a step *M*-construction (like rho model 2) or a full executive *M*-dimensional formula. When rho model 2 is properly done, rho model 3 often becomes a sort of a summary of steps used in the former, as formula **f18** shows:

$$\text{TASKRho}^{L1}[\{\#\text{FAST}\}_{L1},\ \mathit{(1)PUSH{+}TURN{::}(2)LOCATE{::}(3)OPEN{::}(4)GO{+}HIT}] \rightarrow \$\text{RESPONSE} \qquad \textbf{(f18)}$$

Rho Performance Data The Rho task, with its four time-scores (RT, CT, PT, LT), was designed guided by a prior metasubjective analysis similar to the one we have presented. This MTA suggested that one should minimize, via intensive training, the unavoidable misleading factors that could affect performance (particularly the Now! parameters). Participants were given practice so that the obtained time-scores were their best after much training. We therefore expected the best performance-level children could produce after learning has taken place. We now briefly summarize results of Todor (1977), which we replicated, showing the performance level that children of different ages attained under these conditions. Findings support our analyses.

Figure 9.2 shows average mean latency (PT+LT) of mental problem-solving during the rho task. Children were administered thirty trials after an initial training practice, and mean latencies reported are an average of the last five trials during testing. PT and LT are, together, the time interval used during spatial location of the target, the opening of hand, and the going to target. The stagewise age trends shown in figure 9.2 were statistically significant (Todor, 1977). The data show that 5- and 6-year-olds, who should not be able to cope with passing strategy 1, in fact produced a very long mean latency, around 325 ms. In contrast, 7- and 8-year-olds, with the *M*-capacity to cope with passing strategy 1, show a significantly faster mean latency of about 260 ms. Similarly, the two other age groups (9 and 10 years and 11 and 12 years), who should be able to handle passing strategies 2 and 3, respectively, show a mean latency of about 225 ms and 170 ms. The *M*-capacity of all these children was measured by Todor (1977) using Pascual-Leone's CSVI (compound stimuli visual information task, Pascual-Leone, 1970), and all the children had attained the *M*-capacity predicted for their age group. It is important to note that (consistent with our theoretical model, see chapter 7) odd-even contiguous age-group differences (i.e., 5 to 6, 7 to 8, 9 to 10, 11 to 12) were not statistically significant, because these children have the same *M*-capacity. Equally important, the performance attained by 11- and 12-year-olds (in PT+LT mean time) does not in fact differ from that found in adults.

The MTA-predicted time-score pattern is fully recovered in these summary data, which yield a clear, decreasing step function. Resulting performance should express our rho MTA model, after task-learning has been maximized. In chapter 10 we again discuss

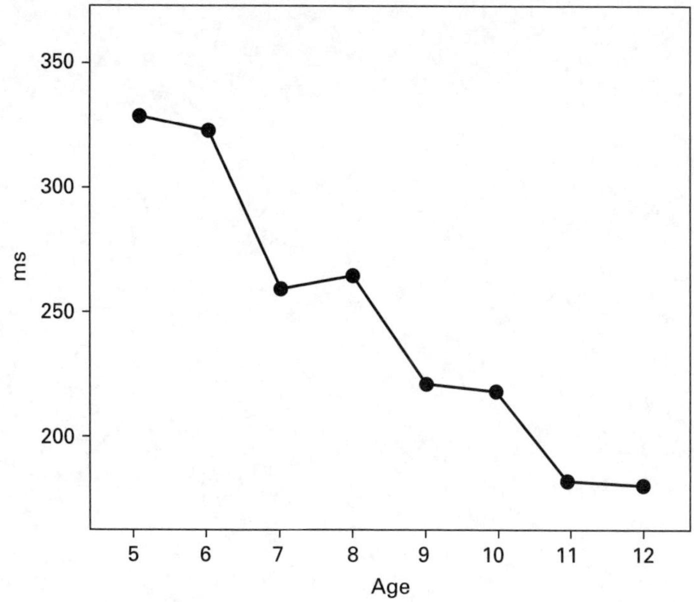

Figure 9.2
Average latency data on the mental-processing segment (PT+LT) of the rho task, as a function of chronological age. Reprinted with permission from Todor (1977, p. 226).

the rho task, and present data showing that this age-bound step function is generated by the left brain hemisphere and not by the right.

Conclusion

In this and the previous chapter we reviewed our method of metasubjective task analysis, its epistemological/ontological foundations, and notational rules. Use of congruent (iconic, epistemologically reflective) symbolic representation is important in this method, because it facilitates analysis, making easier mindful thinking on task processes. The MTA method helps to causally explain transitions between developmental stages, from infancy to adulthood, and helps to integrate cognition with affective/personal processes, and psychometric intelligence with developmental intelligence. In chapters 10 and 11, we address how it could help to integrate neuroscience with cognitive-developmental science.

10 The Working Mind Inside a Working Brain: A Neuropsychological Introduction

We sketch, as scientific speculation, a general brain-process model grounded in developmental constructivism. To this end, we reformulate developmental organismic schemes as neuropsychological units and describe how organismic (hidden) operators and principles of the TCO could be interpreted as brain regulations. Hidden operators and principles appear as general-purpose biases, or specific regulations/resources of the brain processing. We illustrate in neuropsychology the usefulness of our metasubjective task analysis (chapters 3, 8, and 9) for explaining brain semantics.

> To my mind, one must entertain the possibility that there is no "supervisory center" but a *dynamic equilibrium* constituted by processes of the type *winner takes all*... decision would then be the result of a non-supervised competition among inhibitory processes and excitatory processes.
> —Berthoz, 2003, p. 300, translated by JPL

> We have not yet figured out how the brain assigns unity to worldly objects (or to our body, which we perceive as an object).
> —Berthoz, 2012, p. 72

> [M]an, in his perception and action, his memory, speech and thinking, makes use of a highly complex system of concertedly working zones of the cerebral cortex.
> —Luria, 1973, p. 341

> From the strains of binding opposites comes harmony.
> —Heraclitus, 2001, p. 31

The idea of a neuropsychology (i.e., a psychological general theory rooted in the brain's functional characteristics and structure) is as old as psychology. Wundt, Janet, James, Pavlov, Freud, Luria, Pribram, Hebb, and many others attempted to theorize neuropsychological principles. Neuropsychology has made a great leap forward with such methods as electrophysiology, magnetic resonance imaging, positron emission tomography, and magnetoencephalography. This newly found power to observe in vivo brain processes has raised the level of sophistication of neuropsychology and neuroscience. But the research literature often fails to distinguish between neuropsychology and neuroscience. *Neuropsychology* (as pioneered by Hebb, Luria, Pribram, and others) aims to produce a psychology founded in behavior and brain processes, whereas *neuroscience* seeks a behavioral neurology—a neurology capable of explaining psychological (including behavioral) processes. In neuropsychology, the basic substratum level of analysis is that of overt behavior; in neuroscience it is neuronal processes and neuronal pathways. Because these two sciences differ considerably in levels of analysis and goals, one must formulate explicitly their respective models and then place them in informational correspondence to coordinate them. For instance, to say that the dorsolateral prefrontal lobe is a site of working memory is not meaningful in the absence of a causal model of the psychological organism (i.e., a brain-interpreted or organismic, general-causal psychological model), which can explain analytically the causes and how working memory together with other processes (such as schemes/networks) can produce performance, and how this modeling corresponds with the semiotic (semantic-pragmatic) functioning of the brain. For this purpose, a related construct such as the mental/endogenous attention discussed in this book, is more apt than working memory (see Pascual-Leone, 2019; Pascual-Leone & Johnson, 2017).

A common error is to impute psychological meaning to discrete parts of the brain organization without having an integrated neuropsychological theory (a "general organismic" model—Pascual-Leone, 1978; Pascual-Leone & Sparkman, 1980). This epistemological error has been called the *mereological fallacy*[1] of neuroscience (Bennett & Hacker, 2003). As Bennett and Hacker (2003) describe this "fallacy," psychological predicates that apply only to humans as complex functional totalities cannot intelligently be applied to their parts, or parts of their brain. "Human beings, but not the brain, can be said to be thoughtful or thoughtless; animals, but not their brains, let alone the hemispheres of their brains, can be said to see, hear, smell and taste things; people, but not their brains, can be said to make decisions or to be indecisive" (Bennett & Hacker, 2003, p. 71).

Neuropsychology or neuroscience cannot afford to fall under this fallacy, even though the brain, as an organismic functional totality, in fact generates the psychological phenomena. Local neuropsychological theories (or models) are available for descriptive areas such as working memory (usually conflated with mental/endogenous attention), long-term memory (i.e., the repertoire of organismic information-bearing schemes), learning, perception, problem-solving executive processes, and language. However, a comprehensive, truly organismic (i.e., general and causal neuropsychological) theory or model does not seem yet to exist.

As an overview of the book, particularly for those who wish to read the neuropsychological chapters (chapters 10 and 11) without having first read earlier chapters in detail, we summarize aspects of our model, re-introducing organismic schemes as neuropsychological units. Further, we show that some functional constraints of the brain organization can be interpreted neuropsychologically as "hidden operators," that is, as specific regulations (or resources or utilities) of the brain that adapt cortical functioning (scheme processes) to variable situational conditions and task demands. Using Edelman's (1987) theory of neural Darwinism, we illustrate how schemes, in their dynamic interaction with hidden operators, can help to clarify the "binding problem" (via our *F-SOP* processes). This "binding" is the aptitude of brain processes to synthesize truly novel experiences or performances as functional totalities, by relating distinct schemes—distinct assemblies or pathways active in the brain.

To have a general model in neuropsychology, one must simplify and restrict predictions to few epistemo-ontological[2] levels (which for brevity we call epilevels, e.g., maturationally ordered functional levels of brain processing). The TCO is interpretable in the brain because its constructs have plausible neuroscientific interpretations that can be investigated. To facilitate their correspondence-mapping with the brain structures (sites, circuits, networks, neuronal hubs), we first summarize some TCO constructs and then, in this and the next chapter, refer to brain sites, networks, and more to illustrate our neuroscientific empirical mapping for the TCO. After this neuropsychological summary of the general model, the next chapter attempts to show how the TCO interprets known functional networks of neuroscience and how it interprets within the subjects' agency/praxis the semiotics of key brain areas and neurotransmitters. Our TCO neuro-interpretation is schematic and incomplete because much about mapping psychology onto the functional brain remains unclear. This chapter is a theoretical essay meant to contribute something to this clarification. It is not a review of the vast neuroscience literature, and it does not justify with review references many claims we are putting forward in this and next chapter.

To maximize the clarity of this chapter and chapter 11, readers should have some understanding of the brain's topography and its functional organization. Those without training in neuroscience or neuropsychology may find it useful to refer to figures 10.1 through 10.3.

Core TCO Constructs

The TCO model is founded on Piaget's concepts of reflective abstraction and equilibration, two distinct but complementary notions. Reflective abstraction predicts that

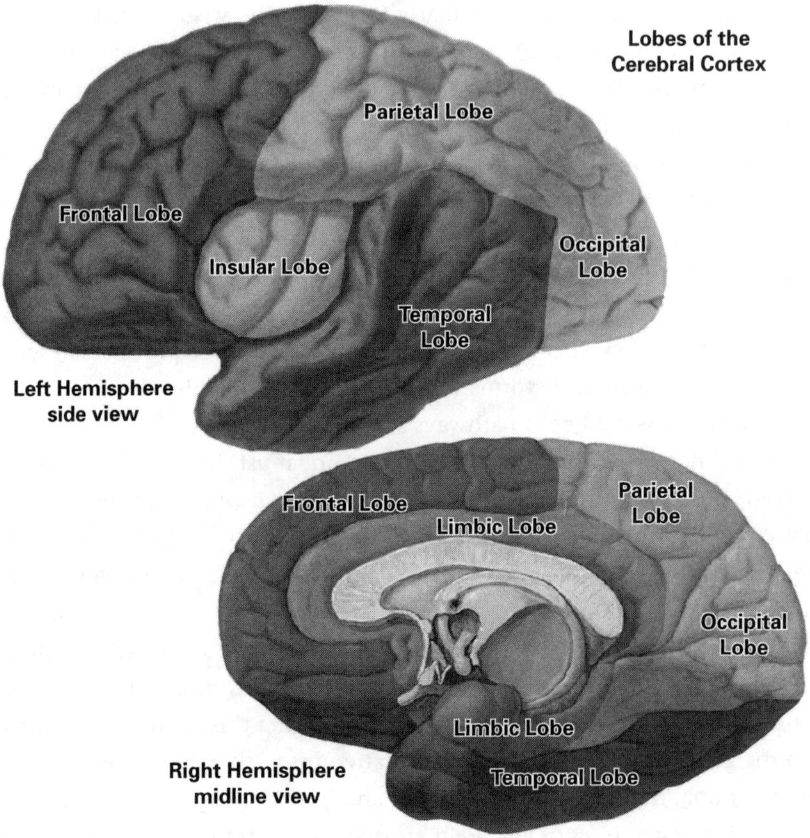

Figure 10.1
Lobes of human cerebral cortex: Lateral left hemisphere (*top*) and medial right hemisphere (*bottom*). In left hemisphere parietal and temporal lobes are pulled apart to exhibit the insula lobe. (From BrainInfo [1991–present], National Primate Research Center, University of Washington, http://www.braininfo.org. CC BY.) See color plate 3.

learning occurs by internalizing the effective complexity (see chapters 7 and 8) of functional invariants; that is, the recurrent patterns of experience, behavior, and/or brain processes that are useful as generic components to explain human agency/praxis. This constructivist form of abstraction is reflective in two distinct senses, which Piaget repeatedly emphasized. (1) The abstracted functional invariants are reflective$_1$ (as a mirror can be reflective) of some task-relevant (probabilistically invariant across repetitions) aspects of experience, or performance, or inner processing. This sense of reflective$_1$ relates to informational correspondence and to Peirce's iconicity (see chapter 4).

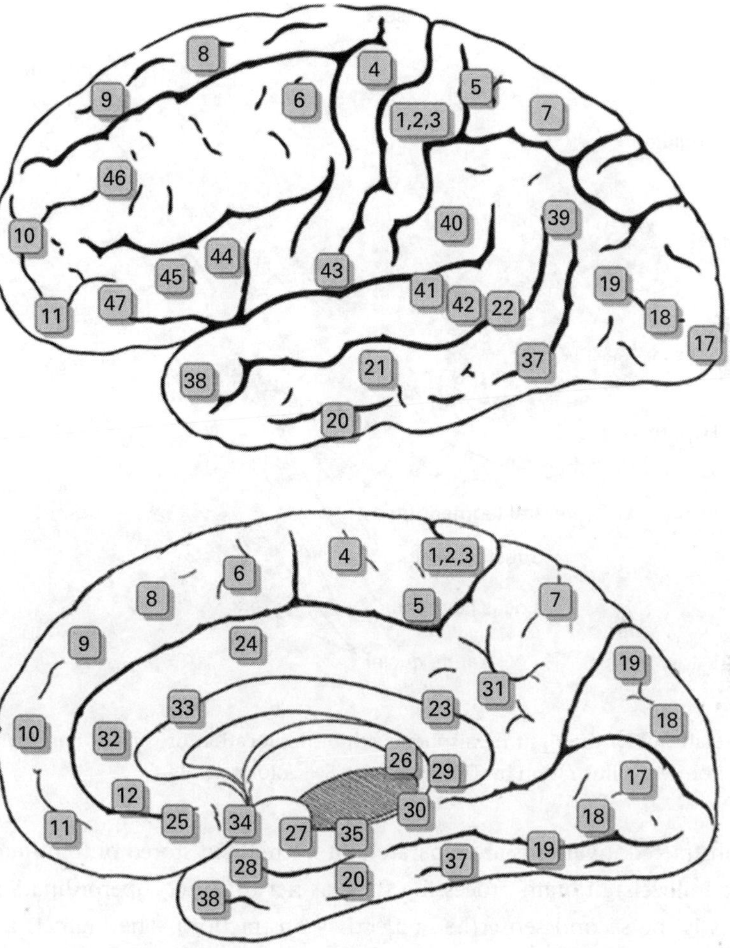

Figure 10.2
The lateral left hemisphere (*top*) and medial right hemisphere (*bottom*) exhibit the Brodmann areas of cortex microanatomy. (From Grey, H. [1918]. *Anatomy of the human body* [20th ed.]. Lea & Febiger. In the public domain. https://commons.wikimedia.org/wiki/File:Gray726-Brodman.svg #filehistory)

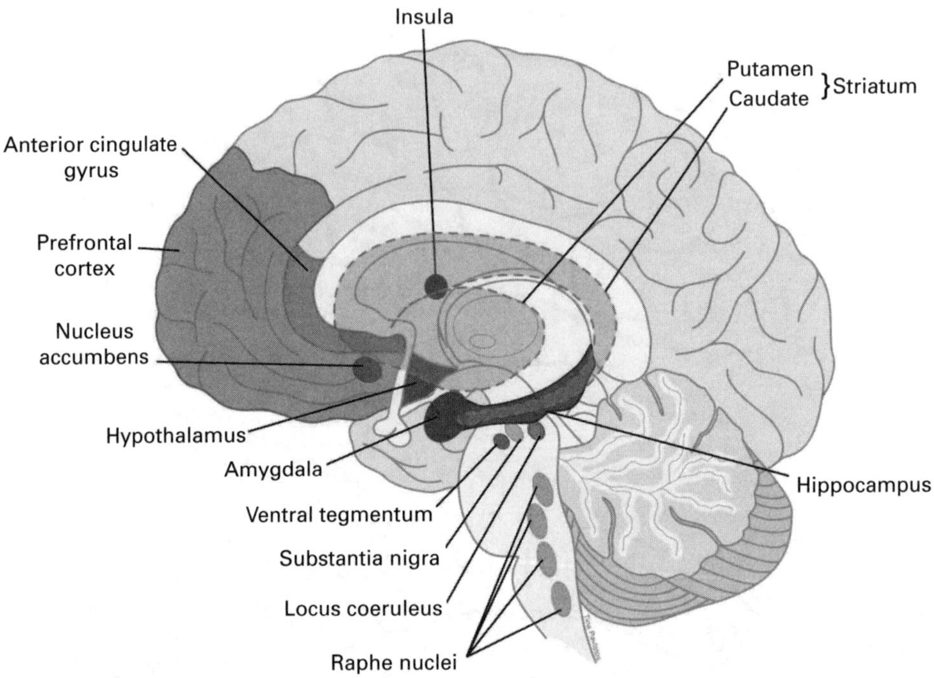

Figure 10.3
Inner (medial) side of the right hemisphere, exhibiting location of various important limbic or subcortical centers. (Illustration by Tina Pavlatos.) See color plate 4.

(2) The abstracted invariants are separately projected and stored onto different higher levels (or epilevels) of brain-processing sites, as new distinct superordinate processing units; this is the second sense (i.e., reflective$_2$ abstraction). The brain is a biological organization that produces reflective abstractions in this dual sense, being informed by life experiences that are external (situational happenings) or internal (feelings, thoughts). Implicit claims about this sort of functionally flexible hierarchical (often called heterarchical) disposition of the brain are found in classic neurology or neuropsychology (e.g., Luria, 1973) and in current neuroscience (e.g., Badre, 2008; Christoff, Keramatian, Gordon, Smith, & Mädler, 2009).

The equilibration principle (Piaget, 1975/1985) is the tendency to minimize processing complexity while concurrently maximizing scope of adaptive performance, which often uses reflective abstraction as key storage mechanism.[3] There are three distinct but dialectically complementary sorts of reflective abstraction (see table 5.2): constructive, reductive, and causal abstraction. *Constructive abstraction* is the progressive coordination and integration of empirically related probabilistic invariants (regularities in experience and reality) within a given field or domain, thus facilitating thought and praxis in this domain. This abstraction is usually descriptive and founded in perception—optimal for learning within facilitating situations. It can internalize probabilistically invariant aspects (or details of recurrent information stored in schemes by the brain) without much restriction. To do so, it uses developmentally emerging mental or endogenous attention. This is a functional system that we call Matt (i.e., *<E,M, I, F>*; see chapter 7 for an explanation), whose functioning is often called working memory (or executive function plus attentional activation/inhibition). Constructive abstraction (see Tables 5.2 and 5.3) can be naturally sorted in four (developmental) levels of complexity ranging from simple sensorial perception (i.e., the immediately given present; related to Piaget's sensorimotor); to intelligent perception (the inferred present, which has expectancies anticipating outcomes; related to Piaget's late sensorimotor and preoperations); to intellection (local concrete thinking that predicts future events—the empirical possible; related to Piaget's concrete operations); to intellectual inferences (general and explicit abstract processes, which can anticipate and analyze the formal—never experienced—possible; related to Piaget's formal operations).

We call these four sorts of levels *epilevels*, because they are epistemo-ontological, that is, marking essential functional differences in maturation (ontology) of mental attention. Such complementary constructive processes have in the brain their own distinct sites, which constitute coordinating hubs at progressively higher levels and epilevels. These in turn function as new coordinating centers of lower-level reflective abstractions. These brain sites, which topographically tend to correspond to classic neurology's primary, secondary, tertiary, and high tertiary—or quaternary—areas (Eccles, 1980; Luria, 1973), are discussed later.

With the progress of this learned abstraction, the informational selectivity within abstracted schemes progresses as well, increasing the meaning reduction (from the richness of particular actual experience to progressive reduction, until they retain only a more-or-less comprehensive gist of meaning). *Reductive abstraction* differs from constructive abstraction in that it filters the to-be-abstracted information, reducing information stored in schemes until only functionally essential or gist meaning of the relevant invariants is retained. This functional aspect expresses the brain's disposition to continuously minimize complexity (until the abstracted relational schemes are minimally complex, albeit adapted to agency/praxis). The schemes involved, nonetheless, remain encompassing and useful, maximally reflective in processing (to achieve effective complexity). Attentional inhibitory processes (our *I*-operator) are used to interrupt the abstraction of nonessential or irrelevant aspects. Reductive abstraction works together with constructive abstraction and causal abstraction to ensure that abstracted invariants are schematized into functional local systems (schemes and more complex schemas—brain circuits, networks) for viability and effective complexity. The reductive inhibition function ensures ontogenetic growth of reflective abstraction, which evolves from simple concreteness to complex abstractness: from purely experiential and simple infralogical processes (i.e., prepredicative, prior to and much more concrete than logical abstraction) to two progressively more complex complementary modes: mereological-mathematical versus logological-linguistic processes.

The *infralogical* mode is the ground from where intuitive logic is abstracted as a generic, effective-complexity descriptive representation of reality (Gonseth, 1936/1974; Husserl, 1973). The term infralogical was introduced by Piaget to signify the final referent of original experience—experiential resistances, concrete transformations, contextual characteristics, and so on, that occur prior to any logical/conceptual structuring. From this original mode of (pre)processing, two distinct and complementary constructive-versus-reductive modes emerge. In *mereological* mode the part/whole relations are abstracted analytically. Mereological abstraction leads to psycho-logical (intuitive or formal) analysis of objects and situations via internal mental models (e.g., Johnson-Laird, 2010). These models often demand a moderate amount of reductive inhibition, retaining considerable similarity with actual experience. Visuospatial, temporal, and causal processes (often globally identified as experiential or "nonverbal") are mereological processes that construct mental/internal models, which create and coordinate representational, operative-and-figurative, and often episodic, processes. The *logological mode* subtends language by providing its logic, and it needs considerable use of experiential filtering (i.e., inhibition—reductive abstraction, using more-or-less arbitrary symbols to eliminate unneeded, not-functionally invariant concrete similarities within the actual infralogical experiences).

The infralogical domain (often known also as experiential) has its sites primordially in the right hemisphere. The mereological domain uses both hemispheres. Because this domain gives detail and richness to experience, it uses the right hemisphere. Because mental attention and explicit intuitive (often abstract and nonverbal) conceptual analysis also is needed for the mereological domain's spatial-temporal and causal representations, it uses the left hemisphere. The logological mode uses mostly left hemisphere in right handers and is often accompanied by language. This distinction that we are making (perhaps new in neuroscience and founded on constructivist-developmental analysis) may clarify brain lateralization issues (Arsalidou, Pawliw-Levac, Sadeghi, & Pascual-Leone, 2018; Pascual-Leone, 1987, 1989). For instance, Casasanto (2003) claimed that left hemisphere is particularly able to practice the mode of processing he calls "verbalizability" (clearly a product of logological mode), whereas the right hemisphere is specialized in what he calls "imageability" (clearly product of infralogical or mereological modes). Stronger evidence for our mereological versus infralogical lateralization difference has been provided unwittingly by Stevens, Kahn, Wig, and Schacter (2012). Their findings are consistent with our distinction between mereological abstraction (including cognitive or conceptual form-abstraction processing, as Stevens et al., 2012, prefer to say), which is predominantly lateralized in the left hemisphere, versus infralogical concrete-content (*C, LC*) learning that we believe to be in the right hemisphere (where automatized schemes and most facilitating concrete perceptual processes may be found). Some findings about aging may be congruent with these hypotheses. It is known that with old age memory for names, faces, facts, and other information (i.e., recall of automatized-figurative schemes) becomes harder. It is also known (Alvaro Pascual-Leone et al., 2011; Cabeza, Anderson, Locantore, & McIntosh, 2002) that in older people the functional asymmetry in brain lateralization functionally decreases, and many tasks mobilize both hemispheres, perhaps a compensatory strategy for a weakened right hemisphere (Yaple, Stevens, & Arsalidou, 2019).

Causal abstractions are reflective abstractions biased toward the generation of causal ideas or models that represent actual change using plausible dynamic-organismic factors or determinants, such as temporally structured schemes (which we call fluents). These schemes, when attributed to nature, lead to inferences about external causality (Piaget & Garcia, 1983). Such determinants can bring about distinct processing results in accord with their theoretical causal characteristics. Three sorts of schemes must be coordinated to generate these models of change: (1) *figuratives,* which represent objects or situations; (2) *operatives*, which are action-producing, procedural-execution schemes; and (3) *executives*, which are a special sort of operative found (innately primed) in prefrontal lobes, to permit goal-setting, or control and planning of operative/figurative co-actions. Operatives apply on figuratives to produce changes, as prescribed explicitly by executives, and demanded by affective processes.

The brain sites for these three sorts of causal abstraction are posterior lobes (parietal, temporal, occipital) for figurative schemes, frontal lobes for operatives, and prefrontal lobes for executives. Our hypothesis (i.e., prefrontal areas—operative, executive schemes—versus posterior lobes—figurative schemes) cannot easily be appraised using only descriptive (e.g., fMRI) neuroscience. For instance, in aging there is a tendency to shift brain activity to prefrontal lobes rather than occipito-temporal on some tasks (Davis, Dennis, Daselaar, Fleck, & Cabeza, 2008). Aging does indeed lead to weakening of automatized figurative processes, but operative prefrontal processes also decay with aging; this shift of activity may be compensatory (Yaple et al., 2019). Further, because causal abstraction usually involves temporal structures, it should necessitate use of fluent (temporally structured) schemes, as well as constructive contributions from the *T*- and *S*-operators in the parietal-temporal regions. These causal-reasoning hypotheses have to be investigated.

Need for Functional Units of Analysis (Schemes) That Could Relate Psychological-System Descriptions to Brain Processes

All contexts, objects, and activities are in fact created in the brain by interpreting regularities in experience and reality (i.e., empirical probabilistic invariants). This brings up the question (Gonseth, 1936/1974; Husserl, 1973; Zubiri, 1999, 1966/2001): What then is *actual* reality? We say that, from a dialectical constructivist perspective, reality-out-there (we call it Reality) is a universe of species-specific "packages" of resistances encountered by the organism in its own agency or praxis. These "packages" are indexed by stable (often probabilistic) aspects of the input interpreted by sensory receptors, or indexed by stable relational-perceptual scheme patterning, which can be changed by recurrent effective actions yielding (probabilistically invariant) outcomes. Resistances may have dependency relations to one another (e.g., the sound of the coffee cup, as I hit it with the spoon, occurs when I see the spoon hitting the cup, if my eyes are open). Thus, Reality is populated with functional packages of interdependent resistances, elicited by agency/praxis, perhaps resistances contingent on other ones, always relative to each animal species. These packages can be interpreted, without empiricist excess, as indexing Real (i.e., present in Reality) functional invariants (Gibson, 1979; Nozick, 2001; Piaget, 1970; Ullmo, 1967) that may appear represented as relational aspects of experienced reality (i.e., the interpreted Reality).

The mammal's brain represents to itself this interpreted reality (in lowercase). And these interpretation packages relate to one another in ways that human activity exploits. Examples of these ways are what Tolman (1961, 1959) called *discriminanda* (e.g., the color or texture of oranges), *manipulanda* (e.g., the size and weight of tables, when, for instance, I enter them into rooms with small doors), and *utilitanda* (e.g., the aptitude of a chair to serve its function—the defining property of a chair that Tolman called "sit-on-ness"). Some of these resistances serve as reality supports for activity, Tolman's utilitanda, which J. J. Gibson later called affordances. Other resistances are encumbrances or obstacles to agency/praxis. We internalize (i.e., constructively learn) these packages and their interdependencies to create representations of what Tolman and Brunswik (1935) called the causal texture of the environment, which includes other people (Pascual-Leone & Johnson, 2004).

Via progressive internalization of these patterns (resistance packages) our brain learns about functional realities and causal relations. To achieve this mediation, the brain is organized by evolution to internalize informational correspondences between/among packages of resistances experienced during activities with the environment (Pascual-Leone & Johnson, 2004, 2005, 2011). These contingencies constructively abstracted by the brain, correspond to functionally created invariants of the subject's activities—as functional magnetic resonance imaging (fMRI) and other brain techniques seem to suggest (e.g., Shirer, Ryali, Rykhlevskaia, Menon, & Greicius, 2012). General processing assumptions of current dynamic-system theories seem compatible with our interpretations (Friston, 2010; Gershman, Horvitz, & Tenenbaum, 2015; Schöner, 2014).

We recognize that functional packages of reality experiences are abstracted in three distinct categories: actual targets (objects) of our activity, procedures we follow to act on the targets, and adjunct information—the conditions facilitating success when a procedure (response) is applied in relation to a target (or distal object). Relational learning permits internalization of complex experiences as packages of resistances (interrelations among resistances emerging in agency/praxis). This sort of learning is essential to acquire "objects," "procedures," and "adjunct information." The three complementary sorts of invariants are abstracted (Piaget's accommodation) into specific, locally causal and self-propelling (Piaget's assimilation) functional systems. These distinct functional systems are the *schemes* or *schemas* (complex schemes, i.e., Piaget's structures), which correspond to Hebb's (1949/1961) *assemblies* (neuronal circuits, interconnected collections) or *phase sequences* (pathways, networks) of neurons. These neuropsychological schemes distributed over the brain are learned as cofunctional and often coactivated to serve agency and praxis (Arbib, Érdi, & Szentágothai, 1998; Pascual-Leone, 1987; Pascual-Leone & Johnson, 2004, 2005, 2011).

Schemes as Processing Units in Cognitive Neuropsychology

Although discussed in detail in chapter 5, we summarize the construct of scheme as it applies to neuropsychology. Schemes are situated semantic-pragmatic causal systems that carry knowledge or know-how about relevant activities (Pascual-Leone, 1970; Pascual-Leone & Goodman, 1979; Pascual-Leone & Johnson, 1991, 2004). They can be understood as learned co-ordinations of three intertwined components: (1) a *functional system* that embodies the gist of the scheme's semiotic or action organization; (2) a set of *conditions* that release the scheme; and (3) a set of *effects* that follow when the scheme applies to experienced reality. Schemes must be internally consistent to be formed, and they are recursive: conditions and effects of complex schemes can in turn be constituted by (compacted representations of) other compatible schemes. Complex schemes, also known as schemas[4], are expressed by neural networks (Arbib et al., 1998; Fuster, 1995; Kagan, 2002; Pascual-Leone, 1995) and can be interpreted as internal models, local dynamic systems, or modules (when this construct is liberalized). Schemes often codetermine (overdetermine beyond the need) the organism's performance by applying together when compatible.

This dynamic scheme-unit construct facilitates, in cognitive development and neuropsychology, a refined process and task analysis (see chapters 8 and 9; Arsalidou, 2013; Arsalidou, Pascual-Leone, & Johnson, 2010; Arsalidou, Pascual-Leone, Johnson, Morris, & Taylor, 2013; Pascual-Leone, 1995, 2013; Pascual-Leone & Johnson, 2017). Related units are used in computation and neuroscience. For instance, Drescher (2003) developed a computing language inspired by Piaget's scheme unit, and Holroyd and Yeung (2012) used the construct of "options" in a similar manner. Arbib et al. (1998), Moscovitch, Cabeza, Winocur, and Nadel (2016), and others explicitly use "schemas" (as we do) to describe units used by the cortex to synthesize spatial-temporal information.

Schemes/schemas overdetermine performance because they are self-propelling, applying to situations (firing as collections of neurons) under minimal conditions. Every scheme has a dual reality. (1) It is a *functional* package of qualitative characteristics that can inform (i.e., inject form into) experience or action, when it applies to configure performance. (2) It has a variable quantitative parameter (an *activation weight* for its neural functional system) that conditions probability that this scheme will become activated and apply in a given context. Long-term memory rests on a manifold repertoire of schemes/schemas sorted in kinds, with different brain sites and functional characteristics: executive schemes (prefrontal lobes), operative schemes (frontal), figurative schemes (i.e., perceptual, representational, or linguistic—occipital, parietal, or temporal lobes), coordinated packages of automatized operative/motor schemes (frontal left and/or right hemisphere, basal ganglia, and cerebellum), affective/emotion schemes (limbic, paralimbic and other medial or ventral cortical areas), analytical schemes produced with mental effort (*LM*-schemes, initially in the left hemisphere), holistic figurative automatized-schemes (*LC*-schemes, usually found in the right hemisphere), and other schemes.

Schemes must be internally consistent to be formed. Because their conditions and effects can in turn be constituted by (implicit copies of) other schemes, they can naturally constitute alternative levels (and epilevels) of processing, related to what is called in the brain "hubs" or "convergence zones" (e.g., Damasio & Damasio, 1994). Schemas are organized in flexible hierarchies (often called heterarchies), whose ranking position can be reorganized to fit the context. The level of a scheme in a heterarchy can be appraised by finding the contingency relations among schemes, the nesting relations that this scheme has with other schemes. Low cognition is produced by a subrepertoire of schemes expressing low levels of relevant information. These are schemes that do not have many other schemes functionally nested under them (as often happens in sensorial perception, simple conditioning learning, simple life activities, and so forth). In contrast, high cognition occurs with schemes found at high levels and epilevels. These complex schemes can have many other schemes functionally nested under them. Neural network modeling with computers has been used to investigate these heterarchical organizations of knowledge.

The principles of scheme learning (see chapter 5) are shared by all schemes, including affective or personal ones. They differ in modalities (i.e., content of processing) and modes of processing. *Modalities* are innately prepared as distinct and irreducible content-domains of experience. They often are expressed in primary and secondary areas of the cortex (e.g., vision, hearing, somatosensory, motor), and they are related to subcortical centers (e.g., thalamus, superior and inferior colliculus, basal ganglia, amygdala, hypothalamus). *Modes* are distinct generic ways of scheme functioning that can be found across content domains (e.g., cognitive versus affective versus personal versus linguistic schemes; operative versus figurative/representational; experiential-concrete versus conceptual-abstract schemes, and others).

Functionally however schemes/schemas are best characterized as *subjective* (not hidden) *operators*, that is, operation-specific generic ways of processing in brain circuits. Alvaro Pascual-Leone and Hamilton (2001) have emphasized that brain processes are often "metamodal," because they are not restricted to concrete modalities/modes of processing (e.g., only visual processing in the occipital primary areas), but their specific functionality could cut across modalities; enabling, for instance, blind people to use the occipital primary areas as a site for synthesizing refined tactile information. Such "metamodal" processing capacity is found, we believe, in brain sites for high-level schemes/schemas, whose functionality is often applicable across modalities and is defined computationally, not concretely. Notice that these metamodal *scheme operators* are different from the *hidden operators* we discuss in this book (e.g., mental attention, learning operators, simplicity Field factor), which are also metamodal but without carrying substantive-content information—only different *regulations*.

One important pair of modes is operative (anterior in the brain) versus figurative (brain posterior). Operative processes are blueprints for actions or mental operations, constituting procedures. Figurative processes are either state descriptions or representations of objects (which may function as targets of thought or praxis), or they can carry adjunct information (e.g., relations or predicates bearing on figuratives and operatives). Another important pair of modes is experiential-automatized (found in the right hemisphere) versus problem-solving-effortful (found in the left hemisphere).

Affective versus Cognitive Schemes

A major distinction, neurologically deeper than modalities and modes, exists between cognitive and affective (including instincts and emotion) schemes. It is well recognized that affects/emotions are qualitatively-distinct evolved functional systems (Ekman & Davidson, 1994; Fredrickson, 2001; Tucker, 2001) that evaluate, in terms of life-supporting vital values ("good," "desirable," "bad," "aversive," and so forth), inner organismic states by way of feelings (Damasio, 1999, 2012; Edelman & Tononi, 2000; Panksepp & Biven, 2012). Affects/emotions inform the organism about these evaluations, preparing it for specific action (fear's tendency to escape, anger to fight, love to approach and seek contact, and so forth), as researchers have noted (Ekman & Davidson, 1994; Fredrickson, 2001; Frijda, 1987; Greenberg & Goldman, 2019).

We call these often-unconscious tendencies affective goals or motives. Affects, affective schemes (Pascual-Leone, 1991b), involve the limbic system and related brain sites. The amygdala is important in preattentive processing of situations for recognition of (positive or negative) affectively salient "stimuli," and the estimation (aided by other limbic and paralimbic centers) of a situation's affective relevance (A. Anderson & Phelps, 2001; Damasio, 1999, 2012; Pascual-Leone et al., 2015; Schaefer et al., 2002). Cognitive expression of affective goals (i.e., cognitive goals) may engage the insula (active in interoceptive affect appraisals; Arsalidou et al., 2018; Sescousse, Caldú, Segura, & Dreher, 2013), the orbitofrontal cortex (active, we believe, in exteroceptive affect appraisals; Sescousse et al., 2013), and the anterior cingulate cortex (which may convert affective into cognitive goals; Pascual-Leone, Pascual-Leone, & Arsalidou, 2015). This interpretation of the anterior cingulate gyrus is particularly clear, we think, in the higher cognitive functions (Albright, Jessell, Kandel, & Posner, 2001; Allman, Hakeem, Erwin, Nimchinsky, & Hop, 2001; Ochsner, Bunge, Gross, & Gabrieli, 2002).

Implementation of affective goals involves activation and application of cognitive schemes, which tell organisms about truth values and appraisals (meaning events and situations). This is different from purely affective schemes, which evaluate organismic states only in their organismic life importance. Cognitive schemes (predicates, relations) carry information that can be true or false about external or internal experience and may be useful to implement affective goals. When affective goals are aroused, cognitive schemes are also activated. Thus affective and cognitive schemes soon become intertwined, producing consciously experienced personal or emotional situations (Pascual-Leone, 1991b; Pascual-Leone et al., 2015; Pessoa, 2013). The resulting hybrid schemes (i.e., affective and cognitive) are the main causal determinants of emotions and of personal/personality processes; we call them *personal or emotion schemes* (Greenberg & Goldman, 2019; Greenberg & Pascual-Leone, 2001; Pascual-Leone, 1991b). Within consciousness affect is normally carried by hybrid emotions and personal feelings; this may be why so many authors (e.g., Ekman & Davidson, 1994) treat "affect" and "emotion" as synonymous. Such confusion may be acceptable subjectively for conscious states but is an error with respect to organismic causal processes (Arsalidou & Pascual-Leone, 2016; Pascual-Leone, 1991b; Pascual-Leone & Johnson, 2004; Pascual-Leone et al., 2015).

Schemes' Overdetermination of Performance (*SOP*), the *F*-Operator, and the Binding Problem

Berthoz (2003, see the first epigraph) claimed, expressing a common view, that brains have no permanent "supervisory center" or "central executive" to oversee functioning. Instead, their direction of processing results from dynamic competition among different (compatible or incompatible, activatory or inhibitory) processing streams (coactivated schemes) each carrying different strands of meaning, often expressing a distinct level of analysis, synthesized in an overdetermined but simple manner (i.e., brain follows a neo-Gestaltist simplicity/*simplexity* principle re-emphasized by Berthoz, 2012). These neuronal streams temporarily constitute compatible activation clusters (perhaps nested) that compete to inform (inject form into, constrain) performances. Although recognized by others, such view of neural networks was formulated early by Edelman's *neural Darwinism* (1987; Edelman & Mountcastle, 1978; Edelman & Tononi, 2000). According to this view, connections are formed when neurons are coactivated ("correlated") and belong to groups that can fire and influence performance together (i.e., are cofunctional). Edelman described three kinds of neuronal population processes that in their interaction produce a dynamic "map organization" (i.e., a stable dynamic pattern of cofunctional and coactivated neurons) in dynamic equilibrium. He called these neuronal processes group confinement, group selection, and group competition.

Group confinement introduces the idea that, locally in the cortex, there are functionally co-compatible neurons grouped in coordinated assemblies, generating alternative neuronal paths, using various connecting links that lead to the same output (this is Edelman's principle of network "degeneracy"). These groups of neurons grow because of neurons that join the group after repeatedly firing together (assumption first made by Hebb, 1961). Local activation/inhibition processes intervene. Using known microanatomy, Edelman (1987) described an activation/inhibition model for these neuronal groups. It has long been known (e.g., Benes, 1994; Swanson, 2003) that there is a maximum of six layers of neurons in the neocortex: counting layers from the cortex surface to the deeper end of neurons' columnar organization. Layers two and three communicate with cortical sources (i.e., groups of neurons—cortico-cortical connections) located respectively in the same or in the other hemisphere. Layer four, often called the (internal) granular layer, receives a dense input from the thalamus (the external sensorial-source brain organizer) and directs its main excitatory output to the supragranular layers (i.e., one, two, and three). In addition, layer four contains numerous inhibitory neurons (interneurons) with horizontal connections, whose activation—often by thalamic input or by high activation in the supragranular layers (i.e., two and three)—leads to inhibition of activatory neurons in layer four, thus reducing the size of the group of activatory neurons in question. This lateral inhibition ensures that local groups of cofunctional and coactivated neurons do not become too large. We believe that lateral inhibition also constitutes the mechanism streamlining mental representational, perceptual, and motor processes, making them more selectively refined, simpler, and qualitatively consistent (in their patterns of coactivation), using principles first formulated by Gestalt psychology (i.e., good form, simplicity, minimum principle, S-R compatibility, elegance, and others).

There is a dynamic inverse relation—a trade-off functional relation—between these layer-four lateral inhibition processes, which cause neuronal group contraction (with Gestalt effects), and the neuronal excitatory processes from both thalamic input and supragranular layers, which both cause expansion of the neuronal group. The trade-off between these two contradictory processes brings about dynamic balance and a stable group size. The proverbial attribution of absentmindedness to intellectuals and university professors may perhaps be related to this sort of phenomena: high activation in the supragranular layers resulting from intensive conceptual (cortico-cortical) activity that induces intense lateral inhibition on layer four and thus causes filtering (relative inhibition) of external events (which enter cortex via thalamic inputs). In our neuropsychological theory we call these cortical lateral-inhibition processes the internal field (*F*-operator) factor (see chapter 6; Pascual-Leone, 1995; Pascual-Leone & Goodman, 1979; Pascual-Leone & Johnson, 1991; Pascual-Leone & Morra, 1991). This name is meant to highlight their causal relation to Gestalt effects, to Piaget's field effects, and to S-R compatibility phenomena (Koffka, 1935/1963; Proctor & Reeve, 1990).

The second and third neuronal population processes posited by Edelman (1987) are group selection and group competition. The *group selection process* states, simplifying, that a group of interconnected neurons is recruited into high activation and becomes dominant over other local groups (i.e., achieves "specificity of group selection"), whenever afferent (i.e., input) connections to many neurons in its receptive field are more highly activated in a correlated (i.e., coactivation) manner than those of other local neuronal groups in the cortical neighborhood. This principle works in tandem with the *group competition process*, which Edelman seems to construe as the tendency of dominant neuronal groups (i.e., with neurons more often and more strongly coactivated by afferent connections) to both capture neurons from less dominant (weaker) groups and suppress their activation and expression in performance, if they are incompatible.

Independently from Edelman, our theory had formulated psychologically (Pascual-Leone & Goodman, 1979) an organismic dynamic model that can cause performance resolutions along these lines of a "winner takes all" strategy (see Albright et al., 2001; Berthoz, 2003). We believe that our dynamic process model is equivalent to the joint effects of Edelman's neuronal-group selection and group competition processes and could thus be interpreted as expressing these neuronal mechanisms at the macro-level of schemes' processing. Our "winner takes all" model is called the principle of *Schemes' Overdetermination of Performance* (*SOP*, see chapters 1 and 6). Edelman's model is defined at the microlevel of neurophysiology, whereas schemes are defined at the macrolevel of neuropsychology. Nonetheless there is correspondence: schemes express assemblies of cofunctional and often coactivated neurons—the *neuronal groups* Edelman talks about. The *SOP* principle is derivable from Piaget's principle of schemes' assimilation (i.e., self-propelling tendency of schemes). This principle asserts that schemes, once formed, have the strong propensity to apply to codetermine representations and performances, and when encountering resistances that cannot be avoided, they actively adapt by way of accommodation. Our *SOP* principle states that performance, at any time, is synthesized by the dominant (most activated) collection/cluster of compatible schemes available in the brain's field of activation at the time of responding, perhaps in competition with other incompatible activated schemes (Arsalidou et al., 2018; Pascual-Leone, 1995; Pascual-Leone & Goodman, 1979; Pascual-Leone & Johnson, 2011; Pascual-Leone & Morra, 1991). To this end, schemes incompatible with the dominant cluster automatically accommodate by relaxing (adapting) their semantic-pragmatic constraints (compatible neurons are thus recruited by the dominant cluster). Otherwise, they are locally inhibited (or inhibited via mental attention, see below). Keeping in mind the difference in levels between the *SOP* principle and Edelman's model, this convergence is nonetheless compelling.

Our model, converging with Edelman's theory, says that *SOP* often works in tandem with the *F*-operator to bring about a closure of processing to dynamically synthesize performance (mental or behavioral output of the brain). Historically, this principle is related to Tolman's (1959, 1961), Janet's (1889), and Freud's dynamic conceptions of motivation and behavior. It can be partly explained by the spreading of neuronal activation among connected neurons and by an organismic generalization of Sherrington's neural principle of a final common path (McFarland & Sibly, 1975; Sherrington, 1906), coupled with Edelman's model of lateral inhibition.

This problem of how *dynamic synthesis* of performance emerges is referred to in neuroscience as the "binding problem" (e.g., Kolb & Whishaw, 2003). This is the problem of how dominant groups of neurons, sorted into assemblies and networks, semantically expressing a multiplicity of different schemes, can together dynamically synthesize and overdetermine the obtained single performance—the integrated functional totality. Neuroscientists have reached the conclusion that to bring about dynamic syntheses in neurons of the relevant manifold groups (which semantically express dominant schemes), they must become actively synchronized in their firing (Buzsaki, 2006; Damasio, 1999, 2012; Edelman & Tononi, 2000; Koch & Crick, 1994; Llinás & Ribary, 2001; Singer, 2001). Studies on functional connectivity among brain areas also point to this conclusion (Shirer et al., 2012).

This synchronization of firing among cofunctional neurons is made easier by high activation of the neuronal assemblies, or schemes, involved. Synchronization of neurons' firing is important because when cofunctional neurons distributed over the brain begin to fire in synchrony they form a temporary (*dissipative*; Prigogine, Allen, & Herman, 1977) functional structure/schema that may express either activation of a previously learned scheme or the dynamic synthesis of an *ephemeral* (i.e., virtual and new, dynamically generated) scheme or schema, which codetermines performance and could, with repetition, be learned (Pascual-Leone, 1987; Pascual-Leone & Johnson, 1991). More than one scheme or schema can be activated simultaneously during an awakened state. These multiple separately synchronized processes, which express coactivated schemes, are kept apart by desynchronized (fast, low amplitude waves) activity of other neighboring sites of the cortex, which produces the overall effect of desynchronization in the EEG typical of awakened states (Buzsaki, 2006; Singer, 2001; Steriade, 1996).

The semiotics (semantic-pragmatics)—or psychological meaning of the brain's synchronized processes (schemes and schemas)—is carried (encoded) in various concurrent ways. We mention three intertwined and important ones: (1) the brain locations of activated assemblies or networks (interconnected collections) of neurons, locations that differentiate distinct subjective operators (schemes/schemas)—which may be metamodal in the sense of Alvaro Pascual-Leone and Hamilton (2001); (2) neurotransmitters and their neuronal activation/inhibition networks, which help to boost activation or potentiate certain sorts of processing, by biasing toward certain processing dominance (affective and/or cognitive); and (3) temporal patterns of firing within the neuronal manifold (e.g., synchronization, sequencing patterns, frequency band—theta, alpha, beta, gamma—within which the firing in question takes place; von Stein & Sarnthein, 2000). These patterns of firing, along with (1) and (2), help to make dominant some scheme clusters activated with the situation, serving to synthesize in that situation a criterion of (task) relevance. This criterion of relevance has, we hypothesize, at least one key learning aspect.

Schemes strongly activated within a situation when a performance or experience (rewarding or aversive) occurs will tend to coordinate (if compatible) into a compatibility cluster of schemes likely to condition this performance/experience. These schemes tend to interconnect and be construed by the organism as determinants of the performance/experience when they apply (related to Hebb's, 1949/1961, learning mechanism: "Neurons that fire together wire together"). With repetition, such coordination of coactivated schemes could become established and function as a criterion of relevance for schemes in this sort of situation, which would lead the cluster to increase its activation propensity in this sort of performance or experience.

Dialectical Nature, Dynamic and Competitive, of Brain Processing: Facilitating versus Misleading Situations

Neuroscientists often write as if associative learning, coupled with unexplicated maturation, were sufficient to enable development from low cognition to high cognition (e.g., Damasio, 1999; Edelman & Tononi, 2000). However, there are epistemological problems with such theoretical stand (Pascual-Leone, 1996b; see chapter 2). A simple associative-learning reductionism is internally contradictory (**CONTR**) for at least three reasons: (**CONTR1**) Schemes (or other units of knowledge) must be separately distinct and be defined at different epistemological levels (and epilevels). This enables emergence of suitable functional heterarchies (i.e., flexible hierarchies of neuronal assemblies that can change with current goals and circumstances). Without having distinctly separate and self-propelling schemes, these flexible heterarchies would not be possible. (**CONTR2**) To enable dynamic syntheses that produce schemes and superordinate schemas, brain resources other than associative learning are necessary, because reflective abstraction (Vygotsky's internalization) of superordinate schemas may not happen, unless the various nested constituent schemes were concurrently activated and synchronized in neuronal firings. Thus, special general-purpose brain resources (specific regulations), which we call hidden operators, that permit this abstraction must exist in the brain (in addition to innate schemes and associative learning coordinating schemes). (**CONTR3**) The same case can be made, even more strongly, whenever there are misleading circumstances that induce competition among sets of schemes jointly activated during processing (competing strategies). Under these conditions, strategies supported by associative learning are often faster and more concrete, but often also task-misleading. Thus, some other brain resources, contradictory with associative learning as normally understood, must be used to enable high cognition to develop. Necessity of these additional resources has been demonstrated with computer simulations using neural networks (e.g., Elman et al., 1996; Munakata & Stedron, 2001). Singer (2001) reached similar conclusions after experiments on binocular rivalry in cats.

The distinction between facilitating and misleading situations (or tasks) is important, because misleading situations (and the analysis of resources needed to cope with them) explain limitations of traditional associative learning theories used in classic and some of current neuroscience. To repeat, a situation is misleading when it elicits schemes unsuitable for the task at hand; which, when applying, lower the probability of subsequent activation of task-relevant schemes (Pascual-Leone, 1969, 1987, 1989; Pascual-Leone & Baillargeon, 1994; Pascual-Leone & Johnson, 2017). In misleading situations, a processing conflict usually emerges between two or more, implicit or explicit, strategies. One strategy (we call it strategy Y) is unsuitable but facilitated by well-learned, automatized schemes—often congruent with our *L*- and *F*-operators. The other strategy, X, is suitable but has little chance of applying, because the initially stronger Y strategy (often perceptually salient) dominates from the perspective of associative learning. By contrast, within facilitating situations only task-relevant schemes are activated, and automatized (associatively chunked) schemes suffice to solve the task. Application of automatized schemes does not then need, but may allow, mental effort (mental-attentional activation and inhibition). Acquisition of automatized schemes often depends more on the sociocultural milieu (which offers or not learning opportunities) than on maturational processes. An example of how sociocultural learning factors affect use of mental attention in cognition is found in cognitive styles—one of which is field dependence-independence (e.g., Goode, Goddard, & Pascual-Leone, 2002; Y. Li, Wu, Zhu, & O'Boyle, 2014; Pascual-Leone, 1989; Witkin & Goodenough, 1981).

Misleading situations demand use of maturational mental-attentional mechanisms driven by executive schemes. This is because in misleading situations the unsuitable Y strategy is more salient and faster (because of automatization), being mobilized sooner than the X strategy. Therefore, to enable the X strategy to determine performance, strategy Y must be interrupted by using active executive inhibition (mental-attentional interruption or *I*-operator), while concurrently strategy X is being boosted with mental attention (*M*-activation). Take the example of an *embedded-figures test* (Witkin & Goodenough, 1981), in which one must identify a simple figure embedded within a more complex drawing. Strategy Y compels the participant to notice and see only the whole drawing, which hides (has embedded into it) the relevant figure (Pascual-Leone & Goodman, 1979). This is because the neo-Gestaltist field (*F*-) factor prevents participants from seeing simultaneously the complex drawing and its embedded simple figure. Strategy X is the executive-driven mental-attentional process, making it possible to see the relevant simple figure embedded in the drawing. This process uses schemes' inhibition to break down perception of the salient whole-drawing configuration and simultaneously applies focal executive attention (*M*-activation) to schemes that can make up the relevant simple figure. Misleading situations are found in problem solving, cognitive development, complex linguistic or conceptual tasks, and emotional/interpersonal development. They exhibit cognitive-style differences in individuals. Field-independent persons usually cope well, but field-dependent people, although having a good developmental intelligence, might have difficulty (Goode et al., 2002; Johnson, Prior, & Artuso, 2000; Pascual-Leone, 1989; Wapner & Demick, 1991; Witkin & Goodenough, 1981).

To explicate this sort of issue, we need (in addition to "brain software", i.e., schemes), some *hidden operators* of the general-purpose resources and regulations of the brain. Hidden operators express organismic constraints that relate types of situations to surprising patterns or anomalies in data (such as age-bound limits in information processing due to *M*, or a simplicity resolution tendency for complex processing resulting from *F*). They are hidden (Bohm, 1980), because they lack direct manifestations (obvious functional expression) in conscious experience, something that schemes or schemas have (i.e., perceptual, motor, representational, linguistic, mental thinking—mentation). Table 7.1 summarizes twelve key hidden operators that our theory proposes (see chapter 7).

We have best methods of assessment for mental (*M*-) attentional capacity (hidden operator number nine of table 7.1). This is the key operator that, we believe, causally underlies Cowan's (2005, 2016) and Engle's (2018; Heitz, Unsworth, & Engle, 2005) concepts of working memory. *M*-capacity appears empirically as a limiting constraint on the complexity that a subject can cope with in a task at a given age (see chapter 7). As we parametrically increase the number of necessary causal factors, internal or external, to be entertained simultaneously (this is mental effective complexity of a task), the age of children that can pass the task/item increases predictably. In our model, this age differential is caused by the stagewise growth of mental attention with its effortful activation (*M*-) and inhibition (*I*-) capacity. Mental attention (Matt) results from dynamic interactions among four hidden operators that constitute the dynamic system <*E, M, I, F*>, a key component of working memory. As discussed in chapter 7, *M* is a limited attentional energy that can activate schemes by exerting effort; *I* is an attentional interrupt or effortful inhibition; *F* is the internal-field mechanism, a neo-Gestaltist minimum/simplicity principle; and *E* is the executive function as expressed by the currently dominant set of task executive schemes, which allocates *M* and/or *I*, as needed, to relevant or irrelevant schemes. *M, E,* and *I* are prefrontal functions. *F* is a result of local cortical lateral-inhibition in the brain.

Other hidden operators are also needed to explain cognitive development. There is content learning (simple perceptual or automatized conditioning), determined by the *C*-operator (see chapter 5), and logical-structural (relational) learning determined by the *L*-operator (logical, reflective abstraction). As chapter 5 discusses, *L*-operator encompasses three varieties of learning that are complementary, although occasionally in contradiction. One is automatized *C* learning (via repeated experiences), which we call *LC* learning. Another results from repeated mental acts or applications of mental attention (*M*-operations), causing *LM* learning. A third, which we call *LA* or *LB* learning, results from priming of *L* learning caused by strong affects (A) and/or by the psychosocial personal-being or emotion schemes (*B*).

These hidden operators in the brain have the role of "neurotrophic" dynamic factors (Elman et al., 1996), fostering and expanding adaptability. They may help to understand differential metamodal characteristics of some neural streams or networks (alternative modes of processing) recognized by neuroscience. These analyses are particularly important in high cognition and consciousness (Pascual-Leone, 2000a), because better neural models of consciousness are needed, although we cannot review this literature—see Dehaene (2014) and Damasio (1999, 2012) for intuitive overviews.

Levels of Abstraction in Processing Complexity

Brain sites are relevant for neuropsychology because they express/embody particular domains, modes of processing (metamodal operators), and various sorts of schemes intertwined with specific hidden operators. When these processes are coactivated, they codetermine probabilistic (overt or covert) performances overdetermined (*SOP*) by multiple sites in coordination. These sites acquire psychological significance from interactions between life experience and innate brain-site determinations (e.g., Alvaro Pascual-Leone & Hamilton, 2001; Elman et al., 1996; Stevens, Kravitz, Peng, Tessler, & Martin, 2017). Site-related pathways become coordinated and constitute context-sensitive flexible hierarchies. Some of them were described by classic neurologists as functionally nested sequences of epilevels: *primary* areas (afferent or efferent; e.g., Broadmann Areas BA 1, 2, 3, BA 4, and BA 17), *secondary* areas (coordination of local unimodal processes; e.g., BA 6, BA 5, and BA 18), *tertiary* areas (coordinations of regional multimodal areas; e.g., BA 8, BA 9, BA 7, BA 19), and finally general polymodal areas, also called *high tertiary* or *quaternary* (functionally generic, applicable across domains, and integrating the totality of regional multimodal regions; BA 47, BA 46, BA 10, BA 39, BA 40, among others).

Some, like primary cortical areas (sensorial or motor), are permanently/innately localized or anchored. Areas from higher levels of integration tend to be less permanently anchored and are more affected by life experience in their functional significance, perhaps being metamodal. The number of levels varies across domains, and it may be high. Nonetheless, it is functionally convenient to reclassify levels empirically found in different domains under the four epilevels of abstraction-complexity just mentioned, as Luria (1973) and Eccles (1980) did. They all exhibit heterarchies that have frequent reciprocal (Edelman's re-entry) connections. In this respect, *primary* are cortical areas where sensorial information arrives to (afference) or motor information (efference) leaves the cortex. They tend to be fixed sites, each defining a particular innate mode of processing or a modality content. However, if their normal content is not available, they could perhaps be used for other (metamodal) processes, as the occipital primary area in blind people (Alvaro Pascual-Leone & Hamilton, 2001). Within a given restricted content domain, *secondary areas* coordinate information (schemes) that carries different/distinct local attributes (e.g., visual categorizing, form and depth, color, complex patterns, movement) to constitute proximal (i.e., closely anchored perceptually) objects or motor aspects. *Tertiary areas* progressively coordinate schemes across distinctly diverse local-content domains (e.g., vision, somatosensory, hearing, motor), to constitute multi/polymodal distal objects (which emerge in individual praxis) or action procedures, or adjunct relational information, which serve to elaborate the situational context and express the world of actual experience (abstracting from probabilistic packages of resistances).

Finally, following Eccles (1980), we call *quaternary areas* (or high-tertiary) those that are maximally polymodal and are more generic, embodying the widest/highest integration of information (e.g., reflective abstraction of relations across high-order schemes/schemas and across lower areas). Quaternary areas also mediate special general-purpose resources (basic hidden operators of the brain such as *M, I, S, T,* and *E*). Clear examples of quaternary areas are (in operative functions, e.g., executive schemes) the prefrontal areas and (in figurative functions) the superior temporal sulcus and some sites in the occipital-temporal-parietal crossroad (gyrus angularis and gyrus supramarginalis, i.e., Brodmann areas 39 and 40, intraparietal sulcus). These parietal-temporal regions are common location for complex distal objects, model-objects expressing complex entities—like numerical structures or intuitive (mereological) organized functional totalities (e.g., cities, airports).

Developmental Stages Express Age-Bound Maximal Levels of Effective Processing, Such as Efficiency in Complex, Misleading Situations

Normal process of growth in the human cortex and related subcortical areas has various phases. First, there is a great proliferation of neurons generated by precursor cells (neuroblasts) that migrate to their genetically intended sites in the cortex. Second, once arrived, they differentiate and mature, producing dendritic arborizations. The third phase is synaptogenesis, formation of synapses interconnecting neurons. The number of synapses initially established is in excess of what finally will be used. Thus a "possibility space" (prefiguring possible connections among neurons) gets formed, in advance of experience, in the neuronal network. The density of synapses reaches a peak level in the human cortex at about 5 to 8 years of age (Benes, 2001). The fourth phase, elimination of unnecessary synapses, or pruning, reaches its peak at about adolescence. This is a dialectical/dynamic consequence of consolidation of neuronal circuits as a function of experience, perhaps genetically intended to maximize networks' effective complexity while minimizing redundancy and useless complications (a sort of mini-max rule). This consolidation takes place via Hebb's learning rule: neurons that often fire concurrently (cofunctional and coactivated) within situations tend to consolidate their links and constitute connectivity clusters, via *long-term potentiation* followed by synaptic growth (Arbib et al., 1998). Other synaptic connections that are not cofunctional and coactivated (rarely used in a cluster or network) tend to regress and be pruned. Thus, experientially consolidated circuits develop and become established within a stricter "plausibility space," to replace the original bigger (maturationally grown) "possibility-space" network. Synaptic connections are eliminated or formed to yield experientially plausible, useful and suitable, networks "according to individual needs and environmental demands" (Thatcher, 1994, p. 257). Cortical sites mature in a chronological order that mirrors their phylogenetic antiquity (i.e., paleocortical, archicortical, neocortical), with newer areas (more relationally encompassing and abstract) maturing later (Gogtay et al., 2004). Thus, our brain follows a constructivist-developmental pattern of growth.

The result of this pruning is progressive emergence of cortical networks that are both maximally adapted to effective processing of life experience and minimally complex. How this effective complexity of neuronal connections is attained becomes clearer when we examine the sort of connected network likely to serve this purpose—following current graph-theoretical interpretations of networks (Wang & Chen, 2003). The connection network constituted by a totality of neurons in the brain is expressed behaviorally by the totality of schemes within the repertoire (long-term memory), and converges developmentally toward a model of neuronal connectivity often called *small-world networks* (Pessoa, 2013; Wang & Chen, 2003; Watts & Strogatz, 1998); that is, networks with many local connections and few long-distance connections critical to constitute a coordinated functional totality.

As described in previous chapters, there are two distinct sorts of information processing in the human brain, sensorimotor (fully dominant before 3 years of age) and symbolic (available and progressively dominant at and after 3 years of age). Because brain structural resources, regulations, and so forth, which we model as hidden operators and principles, evolve in psychogenesis in a given order, only the schemes that can be constructed, given the currently available resources (e.g., mental-attentional *M*-capacity), should be able to emerge at that time. Operators *A, C, F, LC, S, T,* and *B* emerge first during early sensorimotor period, along with low-level sensorimotor mental-attention capacity (scale *Me,* see chapter 3). These resources enable associative learning (i.e., *C* and *LC* learning; see chapter 5), causing perceptual/local learning in primary and secondary or low-tertiary brain areas. Thus, various modalities/modes of concrete and situated sensorimotor schemes become organized to constitute local clusters of sensorimotor processing.

Later, after 3 years of age, symbolic mental processing emerges (see chapter 4). This is made possible by the appearance of *I,* of *high M* capacity (the *Mk scale,* see chapter 7), of *LM,* and *E.* These various processes enable complex forms of learning (e.g., high *M*-operations, *LM* of various high-levels of *M*-demand, *LCLM, LMA, LMB*). And these more complex processes cause reflective abstractions that integrate information from different cortically distal domains. In this manner, in high secondary, tertiary, and quaternary cortical areas, long distant links among many shorter links appear: very different distinct sensorimotor scheme clusters coordinated by distant links. Schemes' information storage in the brain creates small-world networks. This small-world network constitution requires the use of both *M*-power and automatic inhibitory processes, together with synthesis by schemes of performance via overdetermination. This overdetermination is caused, as we suggested in previous chapters, by *F-SOP* processes.

Our theory claims that mental attention (Matt) and *M*-power, used to construct the brain's small-world network, are primordially controlled from the left hemisphere (*M* and *LM* processes) and produced by subcortical processes. As discussed in chapter 9,

Todor (1977, 1979) and then Pascual-Leone (1987) conducted experiments on the rho task (see figure 9.1), a motor task in which the speed of children's performance is constrained by their available *M*-capacity (see figure 9.2). Results of Pascual-Leone's (1987) follow-up research strongly indicate that *M*- and *LM*-processing is preferentially controlled in the left hemisphere of normal right-handers.

In this study, we used two symmetrically equivalent versions of the rho task apparatus, one for the right hand (identical to the one used by Todor) and another for the left hand. Typically developing children ranging from 7 to 12 years were tested with both versions of the task, using this schedule: (1) first session (right hand tested first, left hand second); (2) interval of about two weeks to consolidate learning; (3) second session (left hand tested first, right hand second). In all runs the provided training, practice trials, and testing were like those originally used in Todor's simpler experiment, although our apparatus allowed computation of four distinct sequential time intervals.

Figure 10.4 shows the group-average latency data (based on three trials after training) from our follow-up on Todor's results. Data are displayed separately for the right versus left hand and the first versus the second testing session. The solid line indicates right-hand performance, known to index left hemisphere functioning because this is a practiced asymptotic performance. Figure 10.4 shows that the step function predicted by our theory (see chapter 7) and obtained by Todor (figure 9.2) can be found with much clarity in the right-hand data of both testing sessions. Further, it is apparent that overpractice provided by the second session does not change the step function, which seems insensitive to learning. In contrast, the left-hand data (which indexes right hemisphere activity) does not show the step function in either testing session. These results strongly support our claims (see chapters 5 & 6) that *M*- and *LM*-processes are mainly controlled from left hemisphere.

Some Relations between Brain Waves and Mental-Attention Growth in Development

To indirectly highlight the role of *M*-capacity in the development of brain waves, we could compare these waves at or before 3 years of age versus later on in life, because in our theory (chapters 3 and 7), the capacity of *M*-power (*Mp*) in the sensorimotor period (the *Me* scale) has smaller-capacity units than during the symbolic period (*Mk* scale). Results of Thatcher's work (1994, 2010; Thatcher, North, & Biver, 2008) on EEG coherence and phase differences are congruent with this distinction. *EEG coherence* is a measure of consistency in the waves' phase comparison, synchronic oscillation of neurons, over cortical sites (like a correlation); this yields estimates of phase synchrony or phase stability between spatially distant generators. *EEG phase difference* is a measure of wave time difference between coupled oscillators or neuronal firing waves. Thatcher

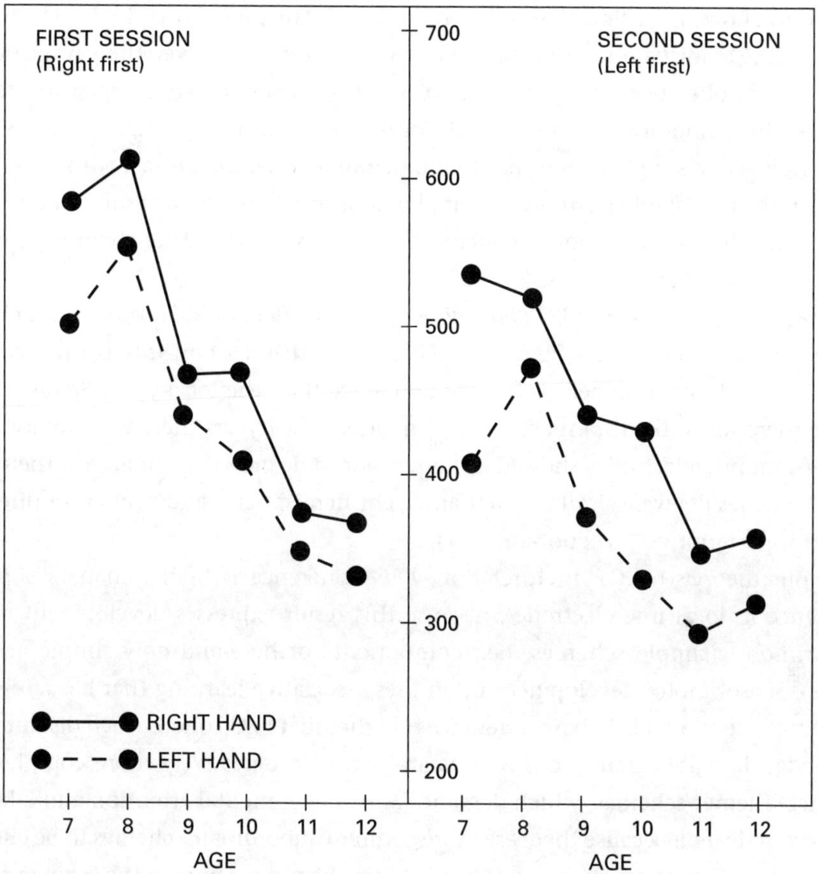

Figure 10.4
Mean mental-processing time (PT + LT in ms) in the rho task as a function of age, hand, and testing session. (From Pascual-Leone, J. [1987]. Organismic processes for neo-Piagetian theories: A dialectical causal account of cognitive development. *International Journal of Psychology, 22*, 565. Copyright 1987 by Elsevier.)

(2010) studied beta waves of 458 participants ranging from 2 months to 16.67 years of age, and theta waves (Thatcher, 1994) of 577 participants from 2 months to 26.4 years. He contrasted electrode sites differing in degree of interelectrode separation: *local* (6 cm—short distance, he also had 12 cm) versus *distant* (24 cm—long distance, he also had 18 cm). Thatcher found (figure 2 of Thatcher, 2010, p. 99) that with beta waves there is a sharp difference in data between before 3 years of age and after 3 to 4 years, when the pattern of EEG coherence and phase difference changes. Before and up to 3 years, locally connected electrodes (6 cm, short distance in the anterior-to-posterior direction) showed considerable *coherence* and maintained a sort of high plateau that peaked close to forty in the coherence measure. At 3 years, however, there was a depression of this coherence (measure close to twenty), then coherence began to increase with age in a monotonic manner (with local spurts) until 16 years. At about 8 years the coherence was higher than that found before 3 years. In contrast with these data, interconnected distant electrodes (24 cm, long-distance connection) showed under the same circumstances much lower coherence before 3 years (measure about ten), with a very low flat level after 3 years.

However, in the case of EEG *phase differences*, the effect of local versus distant electrodes reversed completely (figure 3 of Thatcher, 2010): Distant interconnected electrodes showed much higher EEG phase difference than the local ones. Before 3 years of age there was little distant activity. Then, after 3 to 4 years there was a phase-shift, and distant interelectrodes showed a sharp monotonic growth with age on their measure. These results were similar in left and right hemispheres and similar for direction (anterior-posterior vs. posterior-anterior).

In our theory's (TCO) interpretation, EEG coherence is higher than EEG phase-difference in local interelectrodes, because this result expresses development within sensorimotor (simple schemes, i.e., components of here-and-now simple agency/praxis). Sensorimotor development often uses associative learning that has a low cost in mental attention. EEG phase difference is higher that EEG coherence in distant interelectrodes, because distant connections are part of progressively more complex and abstract schemes/schemas, which demand (*Md*) greater mental attention and take longer to be activated, because there are more complex subordinate schemes to be elicited/coordinated. The longer the interelectrode path, the more complex the scheme and so the longer the neuronal transmission will take. Thus, EEG phase-difference should be found, growing with age, in distant connections much more so than in EEG coherence.

Thatcher's (1994, 1997) earlier data showed developmental patterns of change in EEG coherence within the theta band, in some 500 children from 5 to 26 years. He observed developmental cycles of growth in neuronal connections. Thatcher (1994) described them as "rapid and significant increment of EEG coherence that could be characterized as 'growth spurts'" (p. 235); that is, cycles of displaced brain activity that repeated a sequence within and across hemispheres from the left (LH) to bilateral (LRH) to the right hemisphere (RH). A similar displacement occurred, making three distinct cycles of growth spurts (Thatcher, 1997, figure 5.9). Cycle I corresponds to Piaget's sensorimotor and low pre-operations stages; cycle II corresponds to high pre-operations and concrete operations stage of Piaget; cycle III corresponds to high concrete operations and formal operations stage. These three cycles mirror closely the last three cycles (interrelations, dimensional, and vectorial stage) of Robbie Case's (1992) staircase model that we presented in chapter 7. The repeated displacement of spurts from LH to LRH to RH is congruent with our theory (TCO). Pascual-Leone (1987, 1989; Arsalidou et al., 2018) formulated a brain-lateralization hypothesis based on differences between *LM* learning versus *LC* learning (see chapter 5), as follows: "When a *LM* structure becomes automatized due to practice, with or without human mediation, this *automatization consists in the emergence, in the right hemisphere, of LC structures* [*LCLM* schemas to be more precise] *that are functional copies of their precursors,* the *LM* structures located in the left hemisphere. This process creates the dialectical 'spiral' of development emphasized by Thatcher and by so many writers, from Lenin, to Werner to Piaget" (Pascual-Leone, 1989, p. 65).

Thatcher (1994) seems to have found that there is basically a two-year growth spurt within cycles for a transition from LH to LRH, which is nested within basically a four-year intrahemispheric rotation across cycles. Perhaps not coincidentally, our empirically grounded prediction about *M*-growth level (see chapters 3 and 7) claims that *M*-capacity growth spurts in childhood occur every two years after 3 years of age (symbolic processing—*Mk* scale). This may relate to why LH schemes/schemas (via *LM* learning) take on average two years to reach automatization (*LC* learning), displacing activity to the RH as found by Thatcher.

This brain lateralization hypothesis (i.e., *LM*-schemes in LH, versus *LC*-schemes in RH) was tested using the rho task as discussed above. In substance, because *LM*-schemes need *M*-capacity to be applied (see chapter 5), when trade-off between *M*-capacity (*Mp*) versus *M*-demand (*Md*) is not in favor of *Mp*, the person cannot compute a solution using the *LM*-schemes. In that case, the person should resort to *LC*-schemes located in RH to seek a solution. Further, when the task is too simple to require *M*-computation (i.e., *Md* << *Mp*), the solution schemes will become automatized (*LC*) very quickly with some practice, displacing the processing to RH. Such hypothesis, tested by Arsalidou et al. (2018), is a dialectical "Right-Left-Right hemispheric dominance hypothesis" (table 10.1).

Quoting Piaget, but inserting neuropsychological comments in brackets, Thatcher (1994, p. 257) described the process of developmental growth as "a sort of spiral staircase in which [intra cortical connection systems] are [reorganized] each time the spiral sweeps around, forming successively higher levels of integration (Piaget, 1975, p. 123)." Thatcher (1997, figure 5.8, p. 99) shows this spiral in his data using mean percentage EEG coherence in the theta frequency band, by plotting three-dimensionally the mean percentage coherence against its first derivative and children's chronological age. These data illustrate the "sort of spiral staircase" that he mentions. Thatcher's

Table 10.1
Right-left-right hemisphere transfer model

Mp/*Md* Trade-Off	Familiarity/ Novelty of Task	Hemisphere—Strategy	Hemisphere Change with Learning/Experience
$Md \ll Mp$	Highly familiar	Right—associative heuristic to cope quickly with easy task	With experience remains familiar—*processing at Right*
$Md \leq Md$	Novel, but manageable	Left—mental attentional heuristics for problem solving	With experience becomes familiar and overlearned—*transfer to Right*
$Md \gg Mp$	Highly novel and difficult	Right—associative heuristic to cope with too difficult task	With maturation *Mp* increases, and with experience task becomes less novel—*transfer to Left*

Mp = mental-attentional power/capacity of the individual; *Md* = mental-attentional demand of the task. (Adapted from Arsalidou, M., Pawliw-Levac, M., Sadeghi, M., & Pascual-Leone, J. [2018]. Brain areas associated with numbers and calculations in children: Meta-analysis of fMRI studies. *Developmental Cognitive Neuroscience, 30*, 248, CC-BY.)

results (1994, 1997) support our model's claim that *M*-capacity has two "hidden" scales, *Me* and *Mk*. Figure 10.5 reproduces panel A of Thatcher's (1997) figure, which plots mean EEG coherence against age, using theta frequency-band output of the left frontoparietal electrodes. In this figure, there are two phases in the evolution of coherence across chronological age. In the first phase we see, from 1.5 to 5 years, a slow decrement in coherence with oscillations that have a high coherence point at 3 years. In the second phase there is a high jump in coherence at about 6 to 8 years followed by a slow decreasing trend of coherence (accompanied by spurts spaced every two years) down to 16 years of age. Interestingly, oscillations in the second phase occur about every other year. This is consistent with our *M*-growth model (see chapter 7).

Thatcher credits Case's (1985) and K. Fischer's (1980) theories, which in fact cannot explain in detail Thatcher's two-phase spiral change. However, the TCO can: *M*-growth during the sensorimotor phase occurs frequently, every few months, and this growth is in small amounts (when measured by complexity and number of schemes boosted by *M*). *M*-growth in the second phase (symbolic, after 3 years of age) happens every two years, as measured behaviorally, but it grows in much larger amounts. Thus, we expect two phases, and within each phase, the coherence should decrease as the phase proceeds (*M*-boosting causing local differentiation of new schemes). Further, because *M*-growth in the first phase occurs very often (every four to eight months) and in small increments, the rate/frequency of *M*-change with time should be greater during the sensorimotor phase than during the second, mental phase of *M*-growth. This is

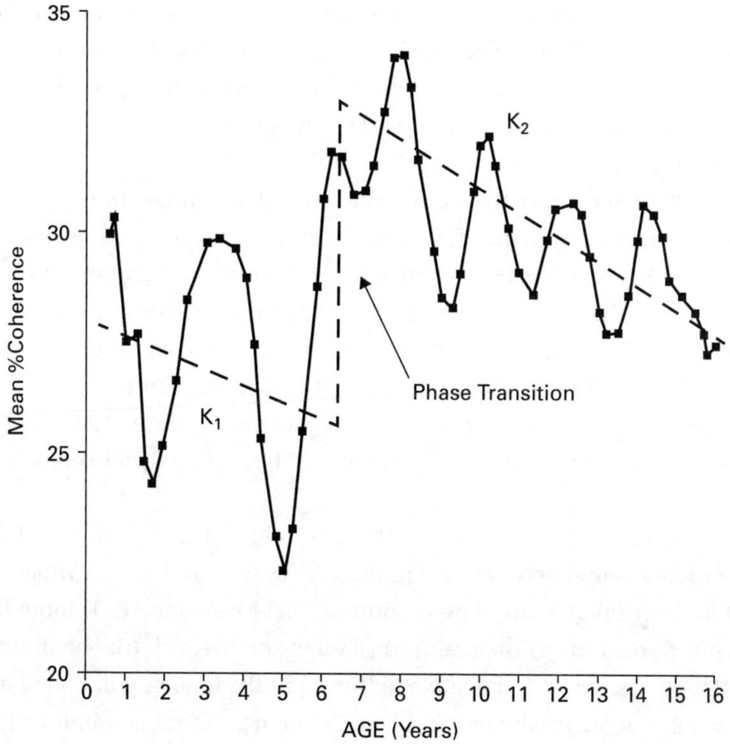

Figure 10.5
Mean percentage EEG coherence (i.e., coherence×100) in the theta frequency band from left lateral frontal-parietal regions (i.e., F7-P3) from 6 months to 16 years. Two modes of oscillation—mode 1, from birth to approximately age five, and mode 2, from approximately age seven to sixteen—are fit by regression lines K_1 and K_2. The phase transition between the two developmental states of equilibria or attractor regimes is represented by the line connecting K_1 to K_2. (Figure and caption reprinted from Thatcher, R. W. [1997]. Human frontal lobe development: A theory of cyclical cortical reorganization. In N. A. Krasnegor, G. R. Lyon, & P. S. Goldman-Rakic [Eds.], *Development of the prefrontal cortex: Evolution, neurobiology, and behavior* [p. 99, panel A]. Paul H. Brooks. Copyright 1997 by Paul H. Brooks Publishing Co., Inc. Reprinted with permission.)

consistent with Thatcher's (1997) results in panel B of his figure 5.8, where the first derivative (i.e., rate of change of coherence with the change of time) is plotted three-dimensionally along with the mean coherence of subjects and their chronological age. Thatcher shows, indeed, that changes in the first derivative and in mean coherence are more frequent during the first sensorimotor phase. The onset of the second phase at about 6 years of age (showing coherence decrementation but with decreasing spurts every two years) is also consistent with our model, because at this age there is a big increment in *M*-energy every two years. This should produce much scheme learning, creating more EEG phase differences across distant connections. Consistent with these interpretations, Case (1998, figure 15.12) has shown suggestive curves indicating considerable cross-sectional similarity between the growth with age of EEG coherence (Thatcher's results) and the growth of working memory (mean of counting span and spatial span) in Case's own data.

Thatcher (2010, pp. 100-101) wrote, "The most likely explanation [of the Thatcher findings we have just summarized] is that a common genetic factor is responsible for the regular rhythms and slopes of change in mean coherence and mean phase over a 16 years period." Thatcher's work seems to suggest a neurotrophic dynamic factor other than learning, which our theory identifies with mental attention and its *M*- and *I* operators. This neurotrophic factor emerges endogenously, in a cyclic iterative way that parallels neo-Piagetians stages/cycles (see chapter 7; Case, 1998; Demetriou & Spanoudis, 2018; Fisher, 1980; Pascual-Leone, 1970, 2019; Pascual-Leone & Johnson, 2005, 2011).

Elman et al. (1996) have discussed the possibility of connectivity relations (coexistence or sequencing) between two input sources such as A and B, which could not emerge in the network (**N**) unless two conditions have been met: **(N1)** input to the network units (e.g., neurons in the brain) must enter the network (the brain) in an organized fashion, in the sense that relative to this input the total population of units must be organized in a sequentially nested hierarchy of areas. That is, input first arrives at the first-level units (neurons), is then transmitted to the second-level areas, then to the third-level, and so forth, and the input activation is always transmitted along with the processing that earlier areas in the hierarchy have produced. **(N2)** To enable learning of complex relational invariants based on input sources A and B, units (neurons) must have been primed by means of at least one source of unit activation that is independent from both the input source and the associative-learning network of units in question. Shrager and Johnson (1996; see also Elman et al., 1996) called this independent priming source of activation a "trophic factor" that arrives in waves. Within the "spiral staircase" of cognitive development (Thatcher, 1994), the necessary condition **N1** of neural networks, described above, can be found in the brain organization; the "trophic factor" of **N2** is for us the mental-attentional boost provided by the *M*-operator, which (as we show in chapter 11) has a clear neuroscientific interpretation.

Arsalidou et al.'s (2013) research with adults also provides support for our theory of mental-attention. They found fMRI brain-activity increments congruent with the graded complexity of the *M*-measure items used. Using fMRI and a well-studied *M*-measure, the color-matching task-clown (CMT-C; Arsalidou et al., 2010), they showed that developmentally graded task items' difficulty was parametrically recovered in the degrees of brain activation as a function of item-class difficulty, within the brain's executive network (e.g., prefrontal cortex). This result was concurrently accompanied by equally graded deactivation in brain areas associated with the default mode network.

Arsalidou et al.'s (2013) results have been replicated in an fMRI study with adults, by Guevara, Arsalidou, Pascual-Leone, and Stevens (2019). They used three different *M*-measure tasks, one of which was the CMT-C used by Arsalidou et al. (2013). The three tasks replicated the 2013 results. This well-controlled replication supports and expands the prior Arsalidou et al. findings of graded activity increments in the dorsal attention and frontoparietal control networks (with graded deactivation in the default network) when items' *M*-demand is incremented in a graded manner.

Qualitative stages—and their functional structures (schemas or schemes)—are always situated (always relative to situations, experience, and contexts). However, the quantitative levels of *M*-capacity (*M*-energy invested in constructing qualitative stages) are purely organismic and general. This is new, but consistent with both Piaget's reflective abstraction (Pascual-Leone, 1995, 1996b; Pascual-Leone & Johnson, 1999) and with Vygotsky's concept of internalization (Kozulin, 1990; R. Miller, 2011), although neither of them quantified levels of effective complexity, nor did they relate levels to an explicit maturational-process of mental-attentional growth. Organismic schemes are internalized from the subject's agency/praxis (including mental activity), and they reflect encountered experiential resistances, contingencies from the environment's "causal texture" (Pascual-Leone & Irwin, 1998; Pascual-Leone & Johnson, 2005, 2011; Tolman & Brunswik, 1935).

Task Analysis as a Tool for Cognitive Neuropsychological Investigation

Our method of metasubjective task analysis (MTA), discussed in detail in chapters 8 and 9, is a useful tool in neuroscience, as we illustrate in this final section. MTA can help to clarify semiotic significance of brain areas and networks that are active when tasks are solved by participants. We discuss three tasks, of variable complexity, well recognized in neuropsychology. The first is Piaget's object permanence task in infants (Pascual-Leone & Johnson, 1991). This is known in neuroscience as the A-not-B-error task or AB task (e.g., Diamond, 2001; Stiles, 2001). We analyzed it in chapter 9 but briefly repeat this analysis in a different manner to highlight its usefulness in neuroscience.

A-Not-B Error

Infants of 8 months of age or older are presented with an attractive toy, which the tester hides behind screen A. (Piaget's original task had three closely positioned screens, but other researchers use two screens with similar results.) The infant is encouraged to find the toy, which she may do by removing screen A with her hand. After two or three of these A-trials, the same toy is now hidden behind the nearby screen B. The child is encouraged to search for the toy but persists in looking behind screen A, even though she saw the toy disappear behind B; this is the A-not-B error. Children are not able to solve the task by removing screen B without hesitation until after 12 months of age. Numerous experimental variables can affect performance, such as time elapsed between seeing the toy disappear behind B and the moment when the child can search, or the transparency/solidity of the screen (or box) used, or number of screens and distance between them. These factors act as moderator variables and simply make the task more facilitating or misleading, as our model would predict.

Diamond (2001) recognized two causal factors in the task: (1) the prepotent response tendency to go to screen A, which must be inhibited, and (2) the need to keep in mind some necessary facts, such as where the object disappeared last. Our model (Pascual-Leone & Johnson, 1991) made similar claims and clarified the nature of the schemes involved and number of schemes to be held in mind, predicting age of success in this and similar tasks. Clearly there is in this task a conflict between two mutually contradictory strategies: an error strategy and a correct-response strategy. These strategies share contextual schemes, and because the error strategy is more salient and practiced, it becomes a misleading context that the correct strategy will have to inhibit (attentionally interrupt) if it is to apply and produce the correct response.

We can represent in a model-formula each of these strategies. The notation used in these formulas was explained in chapters 3, 8, and 9. We write with capitals the names of operative (action, procedural) schemes involved in the task and write in lowercase the figurative (representation, declarative) schemes. The meaning of the schemes is indicated informally by English words, although the schemes are sensorimotor and not linguistic. The prefix # indicates the scheme in question to be a parameter or adjunct information. The A-not-B error strategy, which the child already has practiced during the initial A-trials (causing *LC* learning, i.e., automatization), can be represented as follows (where screens are **a** and **b**):

GETtoy (#*obFOUNDbehind-screen*, {*toy-behind-screen* **a**}$_{L1}$, **a**^{L1}) ➔
RESPONSE-**a** **(f1)**

GETtoy is an operative scheme for getting the object. The infant already has this scheme and instantiates/applies it here. The parameter #*obFOUND* is an expectancy scheme developed before. It asserts that objects like a toy can be found behind screens where they disappeared, and screens have to be removed to find them (this is a perceptually facilitated scheme, already available at 8 months). The third scheme is a factual figurative scheme asserting what the child has witnessed (i.e., the toy went behind **a**). The superscript *L1* of **a** indicates (as explained in chapter 9 and the book's appendix) that **a** is chunked (*L*-structured) with scheme {toy-behind-screen **a**}$_{L1}$ (this is why the subscript *L1*). Thus, **a** can boost the activation of the other scheme, which will not need additional *M*-boosting. As an abbreviation, we shall call this strategy **GETtoy>a**, which the whole formula **f1** symbolizes. We note that this strategy is facilitated by practice (*LC* learning) and by the positive affect of obtaining the toy or reward (*A*-operator). An infant who can boost with attentional capacity (i.e., sensorimotor *Me* capacity) the three underlined schemes on the left side of the arrow will have no difficulty in synthesizing the overt response of searching behind **a** (this is RESPONSE-**a**).

Competing with this error strategy is the correct strategy **GETtoy>b**, which we symbolize as follows:

GETtoy (#*obFOUNDbehindLASTscreen*, {*toy-in-hand-TO-**b***}^{L1,Sit} ←
{no-toy-at-hand-after-**b!**}$_{L1,Sit}$, **b**) ➔RESPONSE-**b** **(f2)**

Notice that the parameter #*obFOUND* is now more sophisticated: it intuitively says that the object is behind the screen where it disappeared *LAST* (this necessary parameter contradicts, and thus invalidates, the error strategy of going to **a**). The other two schemes are boosted by practice or overlearned perception (*L1*) and by situational facilitation (*Sit*) and are activated in an empirical reverse sequence (←) that violates the expectancy (!). We symbolize this expectancy-violation by an exclamation point—a tacit surprise, because if toy-in-hand went TO-**b**, it is surprising that the toy does not reappear after **b** (no-toy-at-hand-after-**b**!). This brings the infant to remember the idea (by activation of a previously learned scheme) that objects are found where they disappeared LAST (this fluent—temporally structured—scheme probably emerged as generalization of the infant's act of getting objects hidden behind a single screen—an easier scheme acquired around 8 months). Clearly the child at this moment already has a vague concept of permanence about the object. That is, the object (toy) has disappeared and is somewhere; it is not just completely gone. The awareness that a missing object is somewhere tends to appear at about 6 to 8 months of age. Piaget called this level of awareness "object permanence without object location," because the child cannot yet figure out where it may be. The next sensorimotor stage, permanence with location of the object, appears at about 12 months, thanks to the sort of inference symbolized in formula **f2**.

These two strategies **f1** and **f2** are mutually contradictory, competing for control of performance. The following *mental-strategies formula* describes this competition (see chapter 8; Pascual-Leone & Goodman, 1979):

^*A,LC* [(**f1**) GETtoy>a] **V̲** $*A,M,I* [(**f2**) GETtoy>b] **(f3)**

The two strategies are indicated inside brackets. To the left of each strategy, we have written the hidden operators that, according to theory, boost activation of each strategy. Past affective rewards (*A*) and practice (*LC*) boost the error strategy, and we also place the symbol (^) to indicate that this strategy is false—in error. Application of the correct strategy (preceded by $ to signify it is true) requires mental-attention mechanisms of sensorimotor affective processing (*A*), mental activation (*Me*), and sensorimotor attentional inhibition (*I*) of the error strategy. Special symbols indicate that the strategy is maladaptive/wrong for the task (^) or it is adaptive/correct ($), but this is an observer's evaluation—the subject often does not know the truth evaluation of a strategy—unless he or she can be symbolically reflective (i.e., can use mentation), which sensorimotor children cannot do. Finally, we place a logical symbol of mutual exclusion (V̲) to describe the functional relation between the strategies.

Because processes facilitated by previous content learning (*LC*) are activated faster than those requiring mental attention (*M*, *I*), the error strategy tends to apply first, unless it is interrupted by the child's attention processes. Because these strategies are intertwined, being elicited by some common situational cues, they constitute a misleading situation. Given this state of affairs, the conflict represented in **f3** cannot be solved favorably unless infants actually achieve the strategy symbolized in **f2**. This strategy demands, however (see formula **f2**), that mental capacity (*M*-operator) simultaneously boosts as many as four sensorimotor schemes (the schemes underlined in formula **f2**). Our developmental model predicts that this is not possible until a child reaches *M*-level of *Me*=4 at or after 12 months of age. This age of acquisition, first discovered by Piaget and often replicated (e.g., Diamond, 2001; Stiles, 2001), is usually not explained. Our model of growth in developmental *M*-capacity gives a plausible explanation. This is important, because as Diamond (2001, p. 436) asserts, "There is no behavioral task more firmly linked to DLPFC [dorsolateral prefrontal cortex] than the A-not-B/delayed response task." She summarizes the diverse and very compelling experimental evidence that selectively links this task's solution strategy with both activity in the dorsolateral prefrontal region (Brodmann area 46) and the dopaminergic (DA) pathway. We can thus say that application of the strategy symbolized in formula **f3** expresses a solution process that in the brain requires use of DLPFC.

Two Prefrontal Tasks: N-Back and Self-Ordered Pointing

Let us compare the task analyses of two related (but mental-symbolic, not just sensorimotor) tasks that mobilize the DLPFC, both involving working memory, but exhibiting clear neuroscientific differences. These are the *n-back task*, known to activate dorsolateral PFC and frontopolar cortex (Owen, McMillan, Laird, & Bullmore, 2005) versus Petrides' *self-ordered pointing task*, which mobilizes the dorsolateral but not frontopolar cortex (Petrides, Alivisatos, Evans, & Meyer, 1993).

In the n-back task, subjects are presented with a series of stimuli one at a time. They must PRESS a "yes" button when the current stimulus item is the same as the one presented n (e.g., 1, 2, or 3) items earlier (MATCH), or a "no" button if it does not match. Because the subject must compare the present stimulus item with an item previously presented (which keeps changing), this is a mental-symbolic task. To model the two-back version (i.e., match to two items earlier), we will call n the current item and call $n - 1$, $n - 2$, and so on, the preceding items; we call $n+1$ the item coming after n. Because the stream of stimuli keeps coming, subjects must continually mentally update (RESET) the items $n - 1$ and $n - 2$ to be remembered. They must do this each time a new target—t—stimulus is being considered for matching with the set of criterion—c—stimuli.

For this updating, subjects may experience interference from other item-schemes found in the brain's field of activation (the set of currently activated schemes—circuits, neuron assemblies). Under these conditions, salient item-schemes[5] may cause confusion and errors and may perhaps lead subjects to quit (we symbolize such possibility by a logical disjunction V connecting the alternatives). We denote such failing (false, ^) strategy as *MATCHsalient/QUIT* and symbolize it in formula **f4**. Here, salient c-item stands for a criterion item that is salient within the current situation (Sit) and is directly activated by Sit (subscripted) but is not appropriate; t-item is the target-item currently under consideration.

$$\text{PRESS}_{L1}\ (\underline{\text{MATCH}}^{L1}\ (\{\wedge\text{salient c-item}\}_{Sit},\ \underline{\text{t-item})}\ \text{V QUIT}) \rightarrow \wedge\text{RESPONSE} \qquad \textbf{(f4)}$$

The true succeeding ($) strategy *MATCH/2-back* that leads to correct performance we represent as follows:

$$\text{(1) PRESS}_{L2}\ (\underline{\text{IDEN\&MATCH}}^{L2}\ (\#\underline{\text{cn-1}}_1,\ \underline{\text{tn}}_1\text{: }\underline{\text{cn-2}}_1)) \rightarrow \$\text{RESPONSE}_1$$

$$\text{(2) }\underline{\text{IDEN\&RESET}}^{L1}((\{\text{cn-1}_2\text{: cn-2}_2\}_{L1} \leftarrow \underline{\text{tn}}_1\text{: }\underline{\text{cn-1}}_1)\ \underline{\text{tn}}_2)$$

$$\text{(3) PRESS}_{L2}(\underline{\text{MATCH}}^{L2}(\#\underline{\text{cn-1}}_2,\ \underline{\text{tn}}_2\text{: }\underline{\text{cn-2}}_2)) \rightarrow \$\text{RESPONSE}_2 \qquad \textbf{(f5)}$$

This formula **f5** has three steps listed in the task's order. The first step shows the beginning: if we number the stimuli so that the current target (t) stimulus is n (which we denote as tn), the participant will have already seen the temporally first-to-appear criterion (c) stimulus cn - 2_1, and then seen the following criterion stimulus cn-1_1 just before the appearance of tn (i.e., the first target stimulus, tn_1). The participant must identify and match (IDEN&MATCH) stimuli cn-2_1 and tn_1; to do so he or she uses cn–1_1 as a parameter (#) that points to the currently relevant criterion stimulus (i.e., cn-2_1) as the one that appeared preceding cn–1_1 in the sequence (sequence incompletely symbolized by :). This match leads to a correct (true, symbolized by $) $RESPONSE_1$, that is, pressing (PRESS) the appropriate button.

The second step in formula **f5** occurs when another target stimulus tn_2 appears. At this point the participant must RESET the criterion stimulus so that the just previous sequence (:) of stimuli tn1: cn–1_1 is converted (←) into the currently relevant sequence of criterion stimuli cn–1_2: cn–2_2. Once this resetting of stimuli is done, participants will attend and IDENtify tn2 and cn–2_2. This identification leads to step (3), where tn_2 and cn–2_2 are matched to determine if they are the same or different, which leads to the correct (true, $) $RESPONSE2. The superscripts and subscripts L_1 and L_2 indicate, as always, that schemes involved have been chunked into a single (*LCLM*) complex scheme via intratask learning, because every time a MATCH occurs the subject must PRESS the button, and so chunking occurs (thus boosting one of them with *M*-capacity should automatically activate both).

This is indeed a complex mental or internal processing task (Christoff & Gabrieli, 2000) that requires symbolic/executive (i.e., not just sensorimotor) mental attention. In it, the RESET and MATCH operations are intertwined and must be performed quickly, because the stream of stimuli keeps coming. This has led to interpreting this task as a dual-task paradigm (Watter, Geffen, & Geffen, 2001). We can estimate its mental (*M*-) demand by counting distinct schemes (schemas, chunks) that must be simultaneously boosted into activation by *M*; these schemes/schemas are underlined in formula **f5**. Each step requires four schemes to be boosted with *M*, causing an *M*-demand of $e+4$ (the *M*-capacity of 9-year-olds). However, there is more: because at the beginning of this recursive task-sequence, participants must always remember to coordinate RESET with MATCH operations, they may need to simultaneously boost with *M*-capacity scheme RESET or some related parameter (e.g., *#RESET before MATCH*). They may need to do so as a reminder (but only if their executive know-how is not good enough). This added parameter would cause an *M*-demand of $e+5$ (accessible to 11-year-olds); 9-year-olds could also learn the task but need more time to learn with added guidance and corrections. Also, at the beginning of the task, before the operative chunk L_2 has been formed, inexperienced subjects may need to separately boost with *M* the MATCH and

PRESS operations, which could even raise the *M*-demand to *e*+6—or make the learning of the task much longer. The relation between the two strategies discussed appears in formula **f6**.

^*F,LC* [(**f4**) *MATCHsalient/QUIT*] **V̲** $*E,M,I* [(**f5**) *MATCH/2-back*] **(f6)**

In this formula we express that strategy **f4**, which is false (^), is in conflict (**V̲**) with the correct ($) strategy **f5**. Strategy **f4** is boosted by hidden operators *F* and *LC*, whereas **f5** is boosted by the executive operator (*E*), which mobilizes mental-attentional capacity (*M*) and attentional interruption/inhibition (*I*). Because the operators of **f4** are faster to apply in perceptual situations, participants tend to fail the task unless they can inhibit **f4** as they synthesize strategy **f5**. Strategy **f5** carries initially a higher *M*-demand (*Md*=*e*+5 or *e*+6 maximum), but the basic mental demand, if proper training is given, is *Md*=*e*+4 (i.e., the mental-attentional capacity of normal children of at least 9 and 10 years of age).

Results congruent with our analysis were obtained by Johnson, Howard, and Pascual-Leone (2009; Howard, 2009) in a study that compared gifted versus mainstream children of 9 to 10 and 13 to 14 years of age. Gifted and mainstream children of both ages performed well on the easier one-back task (in which match is only between cn–1 and tn items and does not use cn–2, which reduces *M*-demand to *e*+3—see formula **f5**), although gifted children scored higher than mainstream. In the more difficult two-back task only gifted 13- and 14-year-olds performed above chance. Because gifted children have a superior executive repertoire (and may, but not always, have superior *M*-capacity), these data suggest that *M*-demand of the two-back task may indeed exceed *e*+4 if relevant prior experience or executive know-how are limited.

Normal adults tend to use in their everyday performances no more *M*-capacity than *e*+5, although their maximum capacity is *e*+7 (Arsalidou et al., 2013; Pascual-Leone, 1970; Pascual-Leone & Johnson, 2005, 2011). Because of the need to continually switch quickly between the MATCH and RESET steps, the two-back task has high executive demand (*E*-demand). For this reason our model should predict a high level of dorsolateral prefrontal and possibly frontopolar processing, in both left and right hemispheres. Christoff and Gabrieli's (2000) findings with adults are congruent with this model.

Notice that formulas **f3** and **f6** are similar. From a metasubjective perspective (from within participants' own operations), this is a mental-strategy formula of tasks that engage prefrontal (often dorsolateral) cortex; when *E*-demand and *M*-demand are high, the frontopolar cortex may also be engaged. Such functional-structure model of a novel task using dorsolateral PFC is shown by an fMRI experiment by Goel and Vartanian

(2005). They studied adults solving match problems (Guilford, 1967). This is a classic divergent-thinking task, a task of operative mobility (Pascual-Leone, 1989). Guilford considered such tasks to index a person's *adaptive flexibility,* because participants continuously must shift operative strategies looking (this is the rule) for novel solutions to the same problem. This task is scored for accuracy (proportion correct), and strategies already used turn into "bad" strategies when they are no longer novel. Pascual-Leone (1969, 1987) has shown, via exploratory factor analysis and task analysis, that adaptive flexibility tasks have a functional structure similar to those of formulas **f3** and **f6**.

Goel and Vartanian (2005) demonstrated in this sort of task that number of novel solutions per item, relative to the subjects' general average, was associated with activity in the right dorsolateral prefrontal cortex (BA 46) and cerebellum—activation increasing with number of novel solutions found. In contrast, when fMRI analysis focused on the contrast between finding one solution versus not finding any (i.e., participants were not penalized for failing to find more novel solutions), fMRI activity was found exclusively in the right ventrolateral PFC (BA 47). This is a test of our model, because in the latter case "bad" (i.e., not novel) strategies are no longer penalized. With such scoring the task's functional structure is no longer that of formula **f6**, because now this is a facilitating task—there is no longer a competing false strategy. Indeed, with this lenient score, the solution requires only perceptual attention (Pascual-Leone, 1989) with low *M*-demand and no attentional inhibition (*I*-demand).

Consider now the *self-ordered pointing task* (Petrides et al., 1993). Subjects are presented with a series of cards (or screens), one at a time. Every card has printed the same set of eight abstract designs, but their location in the cards varies randomly. Subjects must POINT to a different design on each card, avoiding repetitions. At the end of the eight-card series, a blank card appears to signal beginning of a new game series, with a different pointing function but the same designs. In this new game all designs must be considered novel and should be pointed once again. Neuropsychology of memory suggests two distinct but complementary strategies that may intervene in this task (e.g., Broadbent, Clark, Zola, & Squire, 2002). The simplest one (here called POINT-the-unfamiliar) involves pointing to designs that do not evoke a strong feeling of familiarity, because the subject has not recently pointed to them. This strategy is perceptual and uses the ventrolateral brain (ventrolateral prefrontal, paleocortical limbic, occipitotemporal pathways). The strategy is simple and is facilitated by perceptual content learning (*C*, *LC*) and motivation (*A*, *B*).

The other, more trustworthy, complex strategy we call CATEGORIZE & POINT. We describe it in formula **f7** by modeling several successive mental-processing steps of the task needed to produce a correct performance (symbolized by a $ prefixed):

(1) POINT (#NON-POINTEDdesign, design$_1$) ➔ \$RESPONSE-NP$_1$
(2) POINT (#NON-POINTEDdesign, CATEG[designN-P/designP], design$_2$) ➔ \$RESPONSE-NP$_2$
(3) POINT (#NON-POINTEDdesign, CATEG[designN-P/designP], design$_3$) ➔ \$RESPONSE-NP$_3$
....
(9) RESET^{L1} (#blankCard, {8=designN-P← designP}$_{L1?}$) ➔ Reset[designN-P=8]
(10) POINT (#NON-POINTEDdesign, design$_1$) ➔ \$RESPONSE-NP$_1$ **(f7)**

Participants following this strategy seek to POINT to designs they have not pointed to before in the game (#NON-POINTEDdesign). To do so, they would categorize (CATEG) them as either nonpointed (designN-P) or pointed (designP). When the blank card appears, they RESET their game and begin again to treat all designs as nonpointed. Notice that RESET marks the beginning of a new game cycle (a "new task"), after the POINT/CATEG operations. Thus *M*-demand of the RESET operation ($Md=e+3$, and with practice $e+2$) should not be added to that of the POINT/CATEG operation, which has a basic *M*-demand of $e+4$ (i.e., the schemes underlined in formula **f7**). This is a less complex strategy than found in the n-back task, both in terms of *M*-demand and *E*-demand. The task is easier at the beginning and a bit harder later on in every game (because then number of already pointed designs increases), and the alternative facilitating perceptual strategy (feeling the designs' familiarity/unfamiliarity) may wear out as the task advances. Nonetheless, with some intratask practice, the *M*-demand may decrease because this is essentially a perceptually facilitating task (no misleading or distracting aspects). Initially, the task has moderate *M*-demand and *E*-demand and the strategy should then require the use of prefrontal pathways—dorsolateral cortex (BA 46), but not frontopolar. Petrides et al.'s (1993) data are congruent with our model. The functional relationship between strategies in this task can be represented as follows:

\$ *C,LC,A* [*POINTunfamiliar*] **V** \$ *E,M,LM* [(**f7**) *CATEGORIZE & POINT*] **(f8)**

A comparison of formulas **f6** and **f8** shows one major difference. Strategies of **f8** are both adaptive (\$, i.e., true, valid) and are mutually compatible (**V**, i.e., one and/or the other strategy can contribute to the solution). This is a facilitating situation, so its perceptual strategy using habitual schemes may use predominantly the right hemisphere, although its second (mental-categorization) strategy uses also the left hemisphere (see data from Petrides et al., 1993). In contrast, formula **f6** exhibits a purely misleading situation, where the two strategies involved contradict one another.

Diamond (2001) investigated a child version of the same **f8**-type task, along with other tasks such as the A-not-B-task described in formulas **f1**, **f2**, and **f3**. Among the child samples of Diamond was one with children suffering from phenylketonuria (PKU). In this (biogenetic) medical condition, children lack an enzyme for converting the amino acid phenylalanine into the amino acid tyrosine, which is needed to produce the neurotransmitter dopamine (DA). Diamond has shown that, even when treated, these PKU children exhibit a deficiency of DA in their prefrontal lobes. This DA deficiency should lower their ability to perform in (misleading—we emphasize) tasks that require use of dorsolateral PFC. In psychological terms, these are tasks demanding mental attention (which some researchers equate with executive-driven working memory plus inhibition). Indeed, Diamond found that PKU children underperformed in tasks demanding full use of mental attention (mental activation and inhibition—often called *mental effort*), of which the A-not-B task is an example. However, to her surprise, PKU children did not underperform in the perceptually facilitated self-ordered pointing task (SOPT). According to our model, *M*-capacity is needed in SOPT, but attentional interruption (*I*-operator inhibition) is not, because SOPT is not a misleading task.

Yet Diamond (2001, p. 454) wrote, "We still do not understand, however, *why* that is the case." Comparison of the mental strategies in formulas **f3**, **f6**, and **f8** in the light of our cognitive model of neurotransmitters (see table 11.1 of chapter 11) yields a plausible explanation. The DA arousal stream is needed in cognitively misleading situations that require mental-executive (symbolic) attention, such as **f3** or **f6**, and so DA is necessary in the A-not-B task, which should cause underperformance in PKU children. However, because the acetylcholine arousal system is the main neurotransmitter within perceptual and cognitively facilitating situations, which require only perceptual attention but not inhibition, and because this neurotransmitter can also mobilize *M*-capacity via its arousal mechanism, PKU children should not exhibit underperformance in tasks such as self-ordered pointing, using strategies like **f8**.

After this summary of our general TCO model, as it relates to neuropsychology or neuroscience, the next chapter attempts to show how the TCO interprets known functional networks of neuroscience and how it interprets within the subjects' agency/praxis the semiotics of key brain areas and neurotransmitters.

11 The Working Mind Inside a Working Brain: Functional Dimensions in Brain Semiotics

We use the constructivist framework of the book to interpret semantic dimensions and functional networks of the brain. The focus is on seven semantic systems or networks: (1) operative/procedural versus figurative/declarative processing and learning (anterior vs. posterior lobes); (2) dorsal-attention versus ventral-attention networks (frontal and parietal dorsal vs. ventral); (3) symbolic or effortful mode of problem solving/learning (left hemisphere) versus signalic or automatic (right hemisphere); (4) levels of reductive abstraction in the brain; (5) functional hierarchies of operative symbolic streams; (6) mental-attentional activation and inhibition brain mechanisms; and (7) representational streams. We also briefly discuss cortical arousal, neurotransmitters, and the limbic-system's role in cognition.

Reality is best construed as a universe of resistances to the subject's goal-directed activity (praxis)—both positive (Gibson's affordances) and negative (encumbrances or obstacles to praxis). Construction of experienced reality and novelty in performance are explained by the persons' capacity to synthesize, represent, and re-represent to themselves patterns of coactivation (and activation sequences) of their own schematic processes (which tend to express, directly or indirectly, Reality resistances). Reflective abstraction is Piaget's name for this important way of synthesizing flexible hierarchical levels of abstraction (see chapters 1, 5, and 10). To facilitate learning processes, the brain comes prepared by evolution with distinct architectural locations or sites, often specialized for distinct modes of processing. To determine performance, these modes of processing tend to suppress (i.e., inhibit) one another when in competition (e.g., Walsh & Pascual-Leone, 2003).

Before proceeding further, we should comment on the difference between resting-state network "maps," which express cofunctional and often coactivated (usually well-learned) functional connections existing in the cortex (schemes/schemas linking these sites), and the semantic brain "mapping" of our constructivist constructs. Resting-state network configurations may be found because their linking processes are innate (hard wired, causing joint oscillations of connected regions), or they are well learned/

overlearned, or they have been recently used and remain coactivated. The cortical functional relations we investigate in this book stem from a developmental-constructivist interpretation of brain growth and functioning. This developmental-constructivist functionality may lead to patterns of coactivation, learning, and overlearning, eventually expressed in resting state brain patterns. However, unless it is innate, well learned, or overlearned, developmental constructivist functionality may not be expressed in resting-state networks. We now discuss functional modes (and sites) particularly relevant for our theory (Pascual-Leone, 1995).

Operative/Procedural versus Figurative/Declarative Learning

We believe that schemes stored in the frontal lobe (i.e., anterior to the central or Rolando's fissure/sulcus) are operative or transformational schemes. Schemes stored posterior to this central sulcus (in the parietal, temporal, or occipital lobes) are, we believe, figurative-state processes. The latter often are representational declarative schemes standing for physical or mental objects, and simple or complex signalic/symbolic patterns, which include numbers and their interrelations. The *operative mode of processing* (referred to by neurologists as motor) includes mentation directed to external or internal acts of local transformation, or procedures of any sort. For instance, executive schemes (according to our definition) are operative, so they appear in relation with prefrontal activity. From this site, they coordinate activation or inhibition power of mental/endogenous attention, applying it to activate (or inhibit) task-relevant (or misleading) action schemes elsewhere in the cortex. The *figurative mode* consists of schemes standing for signalic or symbolic states (i.e., objects of any sort) or adjunct operative information (schemes—parameters—prescribing how an operative scheme should be applied). To clarify the functional significance, we use "operative versus figurative" in lieu of neurology's terms "motor" versus "sensorial" processes. We do so because "most of the brain's neurons are, strictly speaking, neither sensory nor motor" (Nauta & Feirtag, 1986, p. 38), but intermediate/associative components of the cortex.

To explicitly represent change, the human brain (mathematical calculus illustrates that well) proceeds to model dynamic flow of change by using sequences of idealized state-descriptions (i.e., figurative schemes or momentary-state models), sequencing representation moments of this flow. Each state-description is reached by applying idealized transformational/procedural units (i.e., operative schemes, operators, or action models) to the previous figurative state. These sequences can then be abstracted or mapped into sequence-representing schemes, which we call *fluents,* because they embody the flow of temporally structured experience (Johnson, Fabian, & Pascual-Leone, 1989; Pascual-Leone, 1984). Fluent schemes can function as either figurative (perhaps task parameters) or operative processes, depending upon the context, and they may be stored in higher complexity-abstraction areas (tertiary or quaternary, i.e., high tertiary). As Kant and other classic rationalists (Hegel, Bergson, Cassirer) emphasized, temporal structures (our fluents) are essential for coordinating or constructing complex mental schemas.

An intuitive illustration of this function of temporal-process structuring is offered by construction of complex spatial structures. Spatial schemas use as data the step-by-step products of visual scanning or mental representation; products that are effortlessly (automatically) coordinated into not-yet-learned intuitive spatial schemas, via temporal-structuring processes (possibly in medial temporal brain regions, our *T*-operator). Only after these sequentially acquired components have been brought together by the working mind can mental attention (Matt) and Schemes' Overdetermination of Performance (*SOP)* synthesize here-and-now spatial schemas/structures encompassing the functional totality (which may generate, for instance, a spatial mental map of our city). Neuroscientists recognize this constructive function of temporal structuring. For instance, Alvaro Pascual-Leone, distinguished behavioral neurologist from Harvard University, believes that these temporal factors (coded by patterns of brain oscillations) must be considered together with spatial structures in various levels of processing (primary, secondary, tertiary, quaternary). For him (personal communication, August 2019) temporal constructive factors of complex spatial-temporal organizations serve as "fingerprints," which highlight activity (agency/praxis) needed to construct complex schemas.

Cortical control of operative/procedural versus figurative/declarative processes can be seen in language, in the contrasted cortical-control location of verbs versus nouns. Verbs clearly are linguistic expressions of operative schemes (fluent schemes in the mind of trained speakers), but nouns express figurative schemes. Shapiro, Pascual-Leone, Mottaghy, Gangitand, and Caramazza (2001) conducted "virtual-lesion" experiments with rTMS (repetitive transcranial magnetic stimulation), which produces temporary changes that result in transitory interruption of local cortical processing. These experiments showed that participants voiced their grammatical answers rapidly for nouns (e.g., singular vs. plural form: song, songs) when rTMS was applied to the prefrontal cortex (midfrontal gyrus anterior and superior to Broca's area, Brodmann areas [BA] 44 and 45). They were significantly slowed, however, when asked instead to give the third-person singular or plural of a verb (e.g., sings, sing). Related findings about verb versus noun processing have been reported using functional magnetic resonance imaging (fMRI) and lesion research (e.g., Damasio & Damasio, 1994).

Dorsal-Attention versus Ventral-Attention Networks in the Cortex

Piaget described two self-propelling modes of functioning in schemes: the assimilation and accommodation tendencies that are important to explain adaptive cognitive and developmental change. Assimilation is the tendency of schemes (i.e., neuronal assemblies or brain circuits) to apply and produce performance, whether mental or behavioral. It induces performance to follow the schemes' constraints for response and expectations. Accommodation is a tendency of schemes to adapt by incorporating to them new aspects/features of the situation or ongoing dominant process, when relevant for affective goals of the organism. These functional modes may be innately prepared, since they appear semiotically (semantic-pragmatically) expressed by two distinct brain networks also found in monkeys and apes. These are, respectively, the dorsal-attention versus ventral-attention networks—both basically frontal and parietal (Austin, 2009, 2010; Corbetta, Patel, & Shulman, 2008; Corbetta & Shulman, 2002; Sara & Bouret, 2012; Stöttinger et al., 2015). Functional complementarity of these two networks is well studied.

The dorsal-attention network carries out top-down two sorts of activity: it pursues endogenous cognitive goals for the intended praxis and explores unexpected and salient, not-task-relevant, exogenous aspects of situations. The ventral-attention network notices, bottom-up, task relevant (or affectively relevant) aspects of the situation (part of this network overlaps with the "salience network," Seeley et al., 2007); it appraises mismatches with the intended praxis. Corbetta and Shulman (2002, p. 208) described this contrast as follows: "Neurophysiological studies indicate that the dorsal frontoparietal network, which is recruited for top-down selection, is also modulated by the bottom-up distinctiveness of objects in a visual scene." They subsequently added (p. 210), "[T]he ventral frontoparietal network is modulated by detection of unattended or low-frequency events, independent of their location, sensory modality of presentation or response demands. These stimuli reorient attention, but not necessarily spatially." Stöttinger et al. (2015, p. 119) clarified further the function of this ventral-attention pathway: "a more plausible assumption is that the observed [ventral-attention] network is involved in processing alternative options after a mismatch has been detected. In order to update a mental representation … one has to flexibly decide either to stay with the current model [representation] or to explore new, alternative options." Waters and Tucker (2013) pointed out that the functioning in these dorsal versus ventral areas may relate, respectively, to Piaget's assimilation versus accommodation.

Without reviewing data, we note that the ventral-attention pathway tends to strongly use the right hemisphere, whereas the dorsal-attention pathway predominately uses the left hemisphere, where executive network and mental attention dominate (Austin,

2010; Corbetta et al., 2008; Corbetta & Shulman, 2002). Thus, one could say that the mode of assimilation, in Piaget's sense, is best expressed in the brain by the dorsal-attention pathway of the left hemisphere, whereas accommodation is expressed by the ventral-attention pathway and the right hemisphere (accommodation must adapt previously automatized or well-learned schemes—neuroplasticity). These are two functional systems of brain representation in continuous dialectical interaction that interrupt one another as needed (Corbetta et al., 2008). These approaches and their interaction are necessary in science and life (e.g., "top-down" versus "bottom-up" thinking, "field-independent" versus "field-dependent" attitudes).

In two meta-analyses Yuan and Brown (2015) contrasted cortical activity (activation likelihood estimation) during adults' drawing and writing. This study is important for us, because (as Piaget emphasized) drawing largely involves accommodation, but writing involves assimilation. Consistent with this interpretation, Yuan and Brown found that drawing showed a tendency to activate the ventral parietal region (inferior parietal lobe, BA 40), whereas writing tended to activate the dorsal parietal region (BA 7), although they share many sensorimotor networks. Because the brain systems are innately prepared, so should be Piaget's assimilation versus accommodation modes. The complementarity and intertwining of assimilation and accommodation is a key developmental-learning characteristic of neuropsychological schemes.

LM (Symbolic or Effortful, Left Hemisphere) versus *LC* (Signalic or Automatic, Right Hemisphere) Learning

LM and *LC* are two distinct but nonexclusive modes of processing (see chapter 5, Pascual-Leone, 1995, 1996b, 2000b; Pascual-Leone & Johnson, 2005, 2011). *LM* is logical-structural learning caused by effortful use of mental attention; it is relational-symbolic. *LC* learning is relational-associative, forming logical-structural content schemes due to repetition. These two modes are at times conflated by researchers with the modes mentioned above (figurative or declarative/representational vs. operative or procedural learning). Such conflation may explain why researchers often claim that declarative memory is always explicit (which *LM* mode tends to be), whereas procedural and other nondeclarative memory processes are implicit (e.g., Fuster, 1995; Nelson & Webb, 2003), as *LC* learning often is. We think this correlation is correct only as a tendency (i.e., not categorically), because there are operative (procedural) processes that turn explicit after being implicit, and there are figurative processes that begin implicit (*LC*) but eventually become explicit (*LM*). For example, Alvaro Pascual-Leone, Grafman, and Hallett (1994) studied development of implicit and then explicit knowledge in a visuomotor learning task (organized as a serial reaction time task). Participants were regularly probed verbally about their location knowledge of a repeated (twelve-item) sequence of stars that appeared on a computer screen, to which they responded by pressing suitable buttons. Implicit knowledge was reflected in decreasing response times. Explicit knowledge, when it appeared, was shown in participants' expressed anticipation of the screen position of the star next to come. Transcranial magnetic stimulation (TMS) was performed repeatedly during the experiment, showing that the cortical motor-output map of muscles involved in the task became progressively larger during implicit learning but returned to usual topography and size when knowledge became explicit, although improvement in performance continued.

It is well established that linguistic and conceptual tasks tend to involve the left hemisphere, whereas visuospatial tasks often involve the right hemisphere (Kolb & Whishaw, 2003; Trevarthen, 1990). This difference may be due to lateralization of distinct modes of processing likely to intervene in the domains in question. Processing done in the left hemisphere (LH) is usually described as analytic, logical, categorical, abstract/conceptual, digital, sequential, and serial. In contrast, the right hemisphere (RH) is seen as specialized in holistic, global, coordinating, parallel, analog processing (Gazzaniga, Ivry, & Mangun, 1998; Goel, Shuren, Sheesley, & Grafman, 2004; Kosslyn, 1994; Pascual-Leone, 1976b; Stephan et al., 2003; Vartanian & Goel, 2005). In a similar vein, it has been suggested that the LH is better in processing spatial high-frequency information and thus finds it easier to notice parts and details in a pattern (or notice fast changing ones), whereas the RH is better with low-frequency information, noticing overall patterns and perhaps missing details in visuospatial figures (Kitterle, Christman, & Hellige, 1990; Piazza & Silver, 2017; Wolford, Miller, & Gazzaniga, 2000).

From our constructivist developmental viewpoint (Arsalidou, Pascual-Leone, Johnson, Morris, & Taylor, 2013; Arsalidou, Pawliw-Levac, Sadeghi, & Pascual-Leone, 2018; Pascual-Leone, 1995), propensities fostered by LH may be characterized as problem-solving analytical (using endogenous mental-attention and executive-driven), often within misleading novel situations. This formulation agrees with the "abstract/conceptual" (LH) versus "perceptual" or overlearned (RH) distinction offered by Stevens, Kahn, Wig, and Schacter (2012); the characterization "explanation/inferential" (LH) versus "conflict detection" and "context monitoring" (RH) given by Marinsek, Turner, Gazzaniga, and Miller (2014); and novel relational reasoning versus more global and perhaps overlearned concrete reasoning, as offered by Vendetti, Johnson, Lemos, and Bunge (2015). It also is congruent with Goel's (2019) model of LH as a top-down interpreter or problem solver versus an RH as a bottom-up analyst in (often facilitating) situations that are perceptually clear and well learned or very unclear or indeterminate as life can be. Such lateralization of modes of processing may result from prewired differences in microneuroanatomy (e.g., Gazzaniga et al., 1998). A case in point is the anatomical finding that RH (when compared with LH) has greater functional connectivity with the brain as a whole (e.g., Gotts et al., 2013) and can therefore more easily abstract holistic general patterns, in facilitating contexts, and automatize them.

Sherman, Seth, and Kanai (2016) have shown, using fMRI and a perceptual processing experiment, that top-down expectancies can be evaluated against sensorial experience to reach decisions or perceptual confidence. Top-down perceptual expectations (expressed in orbito-frontal cortex activity) and bottom-up sensory inputs (perhaps expressed by activity in BA 17) are then contrasted (match/mismatch processing), yielding subjective evaluation of confidence. Prefrontal lobe of the RH may play a decision-making role in this appraisal.

Our model says that effortful *LM* learning processes (which involve mental-attentional control, or *M*-operations) usually take place in the LH (Arsalidou et al., 2013, 2018; Pascual-Leone 1989, 1995). *LM* processes can produce symbols and can explicitly engage in symbolic processing. In contrast RH seems more able to store signalic *LC* processes (logical learning of content-schemes, a form of chunking), although it may also store automatized *LM*-schemes (which we call *LCLM* learning; see chapter 5). *M*-operations and *LM* learning may explain the higher aptitude of LH to interpret sequences of events that predict outcomes (Wolford et al., 2000).

When mental demand of the task at hand is greater than mental capacity of the person (*M*-power smaller than *M*-demand), left-hemisphere *LM*-schemes may not be able to cope with the task. Right-hemisphere *LC*-schemes may then be used, tacitly seeking help from automatized processes. This is consistent with Goel (2015, 2019; Goel & Grafman, 2000), who found that mentally demanding, ill-defined or indeterminate cognitive problems often engaged the RH in adults. For a similar reason, early language learning, prior to 17 months of age, takes place in the RH (Mills, Coffey-Corina, & Neville, 1993; Molfese, Wetzel, & Gill, 1993; Thal et al., 1991). The LH becomes dominant in language only after the age of 20 to 35 months, when growth of *M*-capacity permits language-learning to form *LM*-schemes, driven by mental attention. Analogously, amateurs can process music mainly in the RH, but professional musicians process it predominantly in the LH (Bever & Chiarello, 1974; Ohnishi et al., 2001). Note that the examples of language in infants and music in professionals argue against a common idea (e.g., Goel, 2019): that language localization in LH may cause the top-down interpretive problem-solving powers in LH.

The functional distinction RH/LH has a long historical record: for Hughlings Jackson (1915), a founder of modern clinical neurology, the LH was involved in volitional/conscious aspects of language, but both hemispheres contributed to automatic language. Other recent theoreticians of brain lateralization have expressed related views: Eccles (1980) thought that the "conscious self" is in LH, and Gazzaniga and others (Marinsek et al., 2014; Goel, 2015, 2019) believe that the LH contains a creative causal "interpreter" of felt experiences, whereas the RH instead appraises issues (Goel's, 2015, 2019, "indeterminacy") to notice contradicting interpretations. The two hemispheres interact and share task processing. To summarize (see table 10.1), the LH is dominant in misleading problem-solving situations that it can handle, to produce analytical and often conscious processes. The RH (globally experiential) predominates in either of two circumstances: (1) when the intended performance is more or less automatized, or (2) when the mental-attentional demand of the task is too high for the LH to cope with it alone, so help from the RH's automatized schemes could be useful.

Levels of Reductive Abstraction: Experiential versus Conceptual and Complex Episodic Processes

The modes of processing (functional dimensions of variation) we have described as reflective abstraction (i.e., levels of abstraction-complexity, see Table 5.2) can be examined from a semantic meaning-reduction perspective: some schemes are more concretely particular, whereas others are abstract and more generic (Goldstein & Sheerer, 1941). The amount of meaning reduction that takes place to ensure effective complexity of schemes varies with personal style and circumstances. For this reason, *reductive abstraction* is a distinct dimension of variation that is subject to individual differences. This dimension expresses how much meaning is reduced (or eliminated) when reflective abstraction changes from concrete, local, global schemes, to more generic and analytical schemes—schemes that often have greater scope and semiotic complexity but are more abstract. This meaning-reduction perspective marks the dimension of variation called experiential versus conceptual. Conceptual learning, broadly defined (e.g., Pascual-Leone, 1995; Pascual-Leone & Irwin, 1998), is a constructive (comprehensive) or reductive (actively selective) abstraction of aspects of any sort (e.g., perceptual, semantic, functional), found in represented objects or individuals during experience in situations. As Cassirer, quoting Burkamp, put it, "If I understand a new mechanical contrivance, it is a concept for me, even if I do not give it a name. The functional relation, transferable to an indetermined manifold [as functional invariant], is a concept" (Cassirer, 1929/1957, p. 328). The word *manifold* in this quote refers to Kant's sense, which means the collection of distinct sorts of particular relations to/from which the concept is applicable or is inferred, inducted, or abducted. Cassirer referred to the concept as a functional relation. It is a relation, because the concept does not preserve any particulars, but it abstracts generically some functional/descriptive relation(s) critical to the activity of the person who introduced or uses the concept.

Particulars also can be extracted from experience (i.e., constructed, construed) contingent to an implicit decision of relevance, but they are not concepts. Particulars are abstracted from much more narrowly defined types of situations and emerge as packages of functional invariances from here-and-now perception or mentation; that is, they are always situated. Piaget called *infralogical* the most concrete manner of extracting particulars from the here-and-now of experience. We use the term infralogical in this sense: the subrepertoire of schemes that embody and stand for particulars: that is, the here-and-now purely concrete things or actions of experience during agency/praxis. In contrast, concepts are generic, constructive/reductive abstractions from coordinated infralogical schemes (extracted across types of situations).

Starting from infralogical processing (which is virtually the sole mode of processing until about 8 months of age) conceptual processing, at first implicit and practical, emerges progressively following the stages of Piaget and Case summarized in figure 7.2. The greater the number of essential or necessary and distinct types of situations from which a given concept is abstracted, the higher will be the concept's level. With reflective abstraction, two very different, but complementary, modes of processing develop, although the first mode takes precedence during the sensorimotor period. These are the *mereological/intuitive* (more or less generic but not verbal) versus the *logological/linguistic* (verbal particular or generic and later propositional) modes of knowledge modeling. These two modes are prepared in the brain by phylogenetic evolution, and they occur in neurologically distinct brain locations. The three key modes of reductive abstraction (i.e., infralogical, mereological, logological[1], see table 5.2) are dialectically intertwined and collaborate in developing knowledge. The infralogical mode is the *final referential domain* (the ultimate source of semiotic referents) for the mereological and logological modes (with language included as part of the latter).

The mereological mode refers to part-whole relations found in objects or collections of objects (e.g., the body has "arms" and "a head"). Its modeling is based on intuitive spatial-temporal, at times causal, representations, action procedures, and so forth that reflect intuitive part-whole relations found in infralogical particulars (i.e., low-level schemes), within experience and across situations. Episodic memory is a product of mereological complex schemes (schemas) that capture details of life experiences with their sequencing and experiential flow. As also happens with language-based logological schemes, mereological schemes could express complex high-level abstractions (schemes of schemes of schemes of greater scope—schemas), while tending to preserve generic details from the actual here-and-now experiences (with less richness than infralogical schemes).

We call logological a mode that Piaget, among others, called "logical" ("logic" here stands for analytically essential functional structures). We say "logo" because this constructive/reductive abstraction produces, and is aided by, language. Properly speaking, all meaningful schemes, even infralogical ones, express the semiotics (semantics, pragmatics) of experiencing and knowing situations.

Language is common but not essential for complex conceptual abstraction, as suggested by Cassirer's quote mentioned earlier. Abstraction of concepts can also be done mereologically, by using intuitive/imaginal (e.g., visual, auditory, motor) quasi-representations and feelings, which may constitute mental/internal models (Johnson-Laird, 2010; Kosslyn, 1994). The latter in turn can be expressed in language. For instance, one develops language-based concepts of complex distal objects (abstracted entities) of the world-as-represented, such as supermarkets, universities, airports, and organizations, including their sociocultural aspects. These are mereological entities when first abstracted: the city has, among other aspects, houses, streets, hydro power, and underground water mains. When both figurative and operative processes are coordinated into complex mental/internal models, they become tools for intuitive causal abstraction and effective agency/praxis.

Brain Areas Serving Modes of Reductive Abstraction

The brain has distinct processing levels, from information about particulars to progressively more complex concepts. These developmentally emerging heterarchies of levels are stored in functionally nested brain areas (simplifying: primary, secondary, tertiary, or high tertiary/quaternary—maximally polymodal) that may incorporate temporally structured (fluent) schemas. The mereological mode involves, among others, interactions among prefrontal, frontal, lateral parietal, and temporal lobes. The logological mode involves (among others) interactions between the prefrontal, frontal, and temporal lobes. Left hemisphere prefrontal, frontal, or temporal areas, such as Broca (BA 44, 45) and Wernicke (BA 22, 39, 40), often intervene in logological or language-based abstraction. In contrast, middle, inferior, and medial parietal areas, plus other temporal and occipital areas, are involved in intuitive mereological abstractions (often within the right hemisphere, if they are suitably automatized or too complex). The spiral of levels of abstraction is represented within heterarchies of nested primary, secondary, and polymodal (i.e., tertiary or quaternary) areas. These levels are often attained, we speculate, by using the mental-attentional system of operators (i.e., *E, M, I, F*) located in the prefrontal lobes (see chapters 5 and 7).

Neuroscience and cognitive science tend to collapse the various intertwined modes of processing into two polar extremes: *semantic* (conceptual, logological) processes supported by language and often using reductive abstraction versus *episodic* (experiential and temporally structured, mereological, or infralogical) processes, which often retain some of the here-and-now fluent concreteness of experience. Either of these processes could vary from high-level cognition to low-level cognition.

Because experiential processes are also semantic, and episodic processes could reach an effective complexity as high as logological-linguistic processes (produced by complex mereological schemas—explicit internal mental models), we can reduce these dimensions to two developmental mode categories (i.e., experiential vs. conceptual processes) and two corresponding modes of learning. Experiential learning (infralogical or low mereological) tends to be descriptive of life experience and often occurs in facilitating situations. Under these conditions it can be achieved by means of *LC* learning (which may occur predominantly in the RH), because learning in this case is data driven. Complex or misleading conceptual (logological or high mereological) learning is a very different case. Reductive abstraction often causes schemes not to be descriptive of concrete situations, making situations misleading or harder to grasp; this may occur due to situation-elicited unwanted schemes. Thus, people may mobilize executive-driven mental attention (i.e., Matt, with inhibition/interruption) to achieve logological or high-mereological learning. In such case, *LM* learning (LH) should be involved. Teachers, coaches, and trainers could minimize tasks' mental-attentional demand by changing these learning situations to render them more facilitating. After practice, logological and high-mereological learning could mostly use global strategies that are automatized or well-learned (*LC* learning) and stored in the RH.

Complexity Levels in Operative Symbolic or Executive Streams

The vast literature on the prefrontal brain shows that prefrontal lobes are decisive links between limbic system's affective priorities or implicit values (i.e., affective goals) and the person's actual performance—whether in internal mental processing (i.e., mentation) or external/overt responses of the organism (perceptual, motor; i.e., within the brain-interpreted environment and its causal texture, Tucker, 2001). Adaptation of performance to serve affective goals in situations is accomplished by the prefrontal lobe using the currently dominant set of activated and compatible executive schemes with their executive function (which we call *E*-operator), as well as other hidden operators like those of mental attention (see chapters 4, 5, and 7). Current literature often subsumes these constructs under executive processes, organized in flexible hierarchies in term of their effective complexity (Badre, 2008; Christoff & Gabrieli, 2000; Christoff, Keramatian, Gordon, Smith, & Mädler, 2009; Miller & Cohen, 2001). Executive schemes (*E*) control "central" activatory or inhibitory functions, helping to coordinate pathways. Researchers distinguish at least five prefrontal cortex (PFC) regions, where mental operative activities take place: *ventrolateral* (BA 44, 45, 47, 47/12), *dorsolateral* (BA 9, 46, 9/46), *frontopolar* (BA 10), *orbital or orbito-frontal or anterior insula* (BA 11, 12, 13, 14), and *medial* (BA 25, 32, which are part of the anterior cingulate region). If we consider the frontomedial region as distinct, we must add other structures closely interconnected with the prefrontal lobe: the *anterior and posterior cingulate gyri* (BA 24, 32, 33, and BA 23, 31). Anterior cingulate is a major constituent of the limbic system that is also part of the neocortex (e.g., Allman, Hakeem, Erwin, Nimchinsky, & Hop, 2001), and it could be seen as transforming affective goals (limbic and frontomedial processes) into cognitive goals (prefrontal/frontal processes), as we have proposed (Arsalidou & Pascual-Leone, 2016; Arsalidou et al., 2018; Pascual-Leone, Pascual-Leone, & Arsalidou, 2015).

Christoff and Gabrieli (2000), elaborating on Petrides' theory (Owen, Evans, & Petrides, 1996; Petrides, 1994), proposed a flexible nested hierarchy of processing levels in the prefrontal lobes, constituted by (1) ventrolateral PFC, (2) dorsolateral PFC, and (3) frontopolar PFC. They saw this cumulative hierarchy of levels as being caused by, and indexing, task complexity (which increases from areas 1 to 3). Our neuropsychological model (supported by task analyses of neuroscience research) suggests that this hierarchy depends on both the tasks' mental demand (the need to mobilize and allocate mental attention) and the executive demand. This possibility may not have been recognized, or not fully embraced, by Christoff and Gabrieli (2000). As readers of this book know, by mental attention (**Matt**) we mean the system <*E, M, I, F*>. Let us restate this model: (**Matt.1**) Matt has a currently activated set *E* of executive schemes (carriers of executive function), possibly generated or stored in ventrolateral, dorsolateral, or frontopolar areas of PFC, which controls allocation of attentional "energy" (mental activation capacity) and attentional interruption (inhibition). (**Matt.2**) It has a mental-attentional "energy" or capacity (*M*-operator) driven at the prefrontal lobe to boost activation of task-relevant schemes elsewhere in the cortex, and this serves perceptual/automatic attention or symbolic attention-demands of tasks. (**Matt.3**) It has mental-attentional inhibition or interruption, an attentional inhibition utility that we call *I*-operator. (**Matt.4**) Its dynamic syntheses are helped by local lateral-inhibition processes (found in layer four of cortical sites) that we identify with an internal "field factor" or *F*-operator. This *F*-factor, in coordination with the Schemes' Overdetermination of Performance (*SOP* principle, a winner-takes-all principle of brain functioning) causes dynamic syntheses of nonautomatized novel performances.

All cognitive goals are prompted by implicit or explicit affective goals (Pascual-Leone & Johnson, 2004; Pascual-Leone et al., 2015). For as long as performance is novel (in problem solving) and is not automatized, operative mental-attentional streams should be needed to cause performance. Monitored by the anterior cingulate gyrus (which brings about a conversion of affective goals to cognitive goals) these mental-attentional pathways serve to synthesize (with help from the *F-SOP* operator, i.e., lateral inhibition) intended novel performances.

To emphasize the relations of both affective/limbic system and mental-attentional demand with the cognitive-complexity hierarchical model of Christoff and Gabrieli (2000; Christoff et al., 2009), we propose to expand their model by distinguishing four different complexity levels—related, we think, with Sanides' (1970) three evolutionary differentiation-trend systems. Following Goldberg (1985), Tucker (2001), and others, we call these PFC evolutionary systems paleocortical (level one), archicortical (levels two and three), and neocortical (level four).

Complexity Level One

Level one of complexity processing is constituted by the *ventrolateral PFC and comes from the paleocortical system* (evolutionarily taking origin in the insular cortex, and related to the piriform—olfactory—cortex). It often is used in facilitating situations relative to internal needs (which the insula possibly synthesizes—Arsalidou et al., 2018; Damasio, 2012; Sescousse, Caldú, Segura, & Dreher, 2013) whether instinctive, affective, or cognitive. The ventrolateral PFC system mediates emotion-arousal, perhaps in interpersonal situations (Rolls, 1999). It also appears to participate in control of response inhibition of cognitive tasks that are already practiced (*LC* learned) and/or are relatively easy (Aron, Robbins, & Poldrack, 2004; Levy & Wagner, 2011).

Complexity Level Two

Level two of complexity processing is constituted by *medial PFC*. It belongs to the *archicortical* system (evolutionarily taking origin in the cingulate cortex, and related to the hippocampus). This level we add to Christoff and Gabrieli's three cognitive complexity levels to highlight a task-complexity level when anterior cingulate and inferior lateral but not dorsolateral PFC are activated—tasks with some interference but not complex or too misleading. It is used, in addition to level one, when inhibitory control processes are needed—in intersubjective, partly automatic but somewhat conflicting, situations—often during instinctive, feeling/emotive, interpersonal, or reward/

punishment exchanges (e.g., Bechara, Tranel, & Damasio, 2000; Stuss, Gallup, & Alexander, 2001). These initial two levels can serve more or less holistic or automatized (instinctual, feeling/emotive, empathic, cognitive) goals, by engaging in perceptual attention (external processing) and figurative (object-based) motivation in the context of not-too-complex tasks (Goel & Vartanian, 2005; Tucker, 2001).

These first two complexity levels often are driven by the lateral limbic system, the amygdala being a key source for motivation (the amygdala may set up positive or negative criteria of relevance, or priority, for the motivation in current situations; see Pascual-Leone et al., 2015; Sander, Grafman, & Zalla, 2003). The orbitofrontal cortex (OFC) is the site where lateral-limbic affective goals are appraised in terms of external circumstances relative to organism values, in interaction with amygdala, ventrolateral, and medial PFC. The inhibitory control of the orbital cortex is well recognized (Drevets, 1999; Rolls, 1999). When goals are to be turned into motivated action, the link of OFC with the ventral striatum (nucleus accumbens) may help to regulate choice of performance relative to affective goals (e.g., the amygdala's criterion of relevance for events/experiences, and possibly actions).

Complexity Level Three

Level three of processing is associated with the *dorsolateral PFC*, which evolutionarily belongs to the *archicortical system*; it is activated in situations that are relatively complex and often misleading. It can also be activated, however, in less complex situations (that promote the inferior lateral PFC) if some misleadingness occurs. There are neurons in the dorsolateral prefrontal cortex (e.g., Fuster & Alexander, 1971) that remain active during the interval between presentation of a task-relevant piece of information (e.g., a cue, a location, or identity of a stimulus) and the later use of this information in performance (e.g., execution of a contingent response). This is an expected property if the task's executive constraints (*E*-schemes) are to be sustained for as long as needed. The unique property of these PFC neurons is their capacity to actively maintain a task-relevant representation despite intervening distractions (Bunge, Hazeltine, Scanlon, Rosen, & Gabrieli, 2002; E. Miller & Cohen, 2001). In contrast, sustained activity of neurons in the inferotemporal or posterior parietal cortex is easily disrupted by distractors (Petrides, 2000).

Complexity levels three and four are driven by the medial limbic system. In this system the anterior cingulated cortex (ACC) is a site where affective goals are converted into cognitive goals, and performance errors relative to these goals are detected (Bush, Luu, & Posner, 2000; Luu, Tucker, Derryberry, Reed, & Poulsen, 2003). This ACC cognitive-goal control often occurs within complex misleading or conflict situations that demand mental-executive, not just perceptual/automatic attention (Albright, Jessell, Kandel, & Posner, 2001; Kerns et al., 2004).

Complexity Level Four

This level of processing is constituted by the *frontopolar* (*rostrolateral prefrontal*) *cortex*, BA 10. Christoff and Gabrieli (2000) contrasted their level-two region (complexity level three for us) with the frontopolar region (level three for them, complexity level four for us), in the following way. The dorsolateral cortex tends to be engaged in tasks strongly anchored in externally generated information, such as word recognition tasks, or cued-recall of paired associates, or Petrides' self-ordered pointing task (Petrides, Alivisatos, Evans, & Meyer, 1993), or matching visible properties of objects (Christoff et al., 2009). In contrast, the frontopolar cortex (supported by dorsolateral cortex) is engaged in more complex (but not necessarily more difficult) mentally demanding tasks anchored in internally generated information. Examples are an updating n-back task (Christoff & Gabrieli, 2000) or judging whether the inferred dimension of change from one object to another corresponds to the dimension of change between two other objects (Christoff et al., 2009).

The evolutionary origin of the frontopolar cortex (neocortical) is different and more recent than the paleocortical and archicortical levels of complexity processing just described (Goldberg, 1985; Sanides, 1970). This suggests that frontopolar cortex has a more complex or refined function than dorsolateral cortex. Consistent with the qualitative interpretation of Christoff and Gabrieli (i.e., "internally generated information") there is evidence suggesting that frontopolar cortex, greater in humans than in other apes, is the highest level of self-reflective consciousness (e.g., Stuss & Levine, 2002). Evolutionary qualitative differences of this sort may, however, be side effects of deeper complexity-processing differences in mental-attentional capacity and its executive-processing capacity—which limits tasks' *M*-demand and *E*-demand (Case, 1998; Morra, 2000; Pascual-Leone, 2000a, 2000b; Pascual-Leone & Johnson, 2004, 2017). To examine this possibility more closely, at the end of chapter 10 we contrasted task analyses of two functionally related tasks: one that induces activity restricted to our level-three (i.e., dorsolateral PFC; Petrides' self-ordered pointing task) and another that is related to activity of our level-four frontopolar PFC (the n-back updating task, e.g., Christoff & Gabrieli, 2000; Owen, McMillan, Laird, & Bullmore, 2005). We refer readers to chapter 10.

Two remarks about this heterarchical organization (i.e., paleocortical level one, archicortical level two, archicortical level three, and neocortical level four) are necessary. First, some PFC structures are functionally close to the limbic system and can be seen as belonging to the paralimbic cortex (Mesulam, 2003); examples are the orbitofrontal or cingulate gyrus (but also temporal regions like temporopolar cortex, insula, BA 13, parahippocampus complex). They are very densely interconnected among each other, including extensive connections with tertiary and quaternary areas (as Edelman [1987] would say, they are polymodal). In contrast, association (isocortex) areas such as the ventrolateral, dorsolateral and frontopolar PFC, and more so the primary cortical areas, are progressively more isolated and specialized in modality-specific processing of figurative/representation or operative/motor performance (Mesulam, 2003; Tucker, 2001). This architecture functionally unifies the four PFC levels for the control of cortical processes during fluent praxis and mentation.

Second, all four main specific arousal systems (indexed by their distinct neurotransmitters, discussed below) are in dynamic interaction with the PFC complexity levels just described; each neurotransmitter contributes its own qualitatively different sort of processing. Indeed, acetylcholine and dopamine are involved predominantly in cognitive-act processing, that is, processes geared, mediately or immediately, to operative praxis (goal-directed intentional activities, often anticipating near or distant future events or outcomes), with the participation of striatum and cerebellum when action is to be implemented or when complex operative schemes should be coordinated (Cotterill, 2001; Goel & Vartanian, 2005). In dialectical complementarity with those neurotransmitters, norepinephrine and serotonin (also dopamine) are involved in affective processes, that is, the vital (organismic, life-oriented) evaluation of situations, which assigns here-and-now motivation value to objects and figurative representations (we call *figurative motivation* these situation-induced evaluations, very often driven, we think, by the amygdala).

Operative praxis and figurative motivation are important in the PFC functioning. Although PFC complexity levels one and two contribute more to the latter, and complexity levels three and four to the former (Tucker, 2001), operative praxis and figurative motivation are intertwined: all neurotransmitter pathways and all prefrontal systems cooperate. Such cooperation is nowhere clearer than in the case of language, in conceptual development and mental life, or even in the production of meaningful movement sequences. It is now well recognized (Arbib, 2005; Leonard, 2003) that language shares the ventrolateral PFC sites BA 45 and 44 (Broca area) with symbolic meaning-carrying motor sequences (e.g., reaching for and grasping a given object, or imitation).[2] These sites contain special neurons, the "mirror neurons," that become activated whenever meaning-carrying sequences of actions are either observed in others or are performed by the subject him- or herself (such equivalence of others with self explains the "mirror" name, but see Hickok, 2014 for "the myth" of mirror neurons). This unique mirroring function suggests to us that what is being expressed here are functionally invariant (probabilistically recurrent) and relevant (motivated and meaningful) temporally structured sequences of actions, irrespective of who makes them or with what.

For example, in the case of grasping, "the trajectory of the relation of parts of the hand to the object rather than the visual appearance of the hand" is expressed (Arbib, 2005). The same actions performed meaninglessly (e.g., reaching without grasping an object) would not activate these special neurons. Rizzolatti, Arbib, and others (see Arbib, 2005) claim that these mirror neurons in prefrontal (BA 45, 44), inferior parietal (BA 40, 39), and superior temporal sulcus (all quaternary areas) are involved in intentional grasping, complex imitation, symbolic performance (e.g., pantomime), and the emergence of language. Notice that there is also an operative (PFC) fluent/temporal stream for relating, interpreting, and producing (rather than perceiving or representing) sequences of operative schemes. This stream is located in BA 44, 9, 8, 6 (e.g., Leonard, 2003). Fuster (1995, 2004) and Luria (1973, 1962/1980) are earlier researchers who recognized the importance of prefrontal lobes for fluent/temporal structuring of operative plans and procedures. To serve these and other cognitive functions, the temporal and parietal lobe pathways, are, respectively, intertwined with the paleocortical and the archicortical PFC, as we discuss next.

How Mental-Attentional Activation (*M*-Operator) and Inhibition (*I*-Operator) Are Expressed in the Brain

In this section we discuss how our complex system of constructs for mental endogenous attention (Matt = <*E, M, I, F*>), and in particular the *M*- and *I*-operators, may be expressed in the brain. This dialectical-constructivist model of mental attention explicates common cognitive psychology constructs such as executive function, working memory, and cognitive inhibition processes. These constructs often are subsumed under the label *executive function* and currently are used as central explanatory constructs, with claims at times that one or another construct is not necessary to explain the data (Buss & Spencer, 2014). In the literature these three complex information-processing constructs are often unclear, however, and their systemic coordination must be clarified. The key issues seem to be three: (1) Can the *M*-capacity component of mental attention (often wrongly equated with working memory) explain the apparent cognitive inhibition performances of children and adults, or do we also need separate constructs for attentional inhibition? (2) If an attentional-inhibition construct is necessary, will one suffice, or do we need to distinguish effortful inhibition versus automatic inhibition? (3) Can the attentional inhibition construct(s) function independently, or are they functionally intertwined within a mental-attention (Matt) complex system endowed of activation and inhibition operators?

Howard, Johnson, and Pascual-Leone (2014) reviewed the pertinent psychological literature (which we omit here) and tested these and other hypotheses with a set of relevant differential tasks, using confirmatory factor analysis. Their analyses support our Theory of Constructive Operators (TCO) by confirming that mental attention is necessary, including distinct psychological activation (*M*) and inhibition (*I*) processes. The factor-analytical model with only inhibitory processes was inferior to one with intertwined activation and inhibition processes. The modeling also supported two distinct inhibitory mechanisms: effortful interruption/inhibition (I_e) and automatic interruption/inhibition (I_{au}). However, there was support as well for a single inhibitory factor, which agrees with our claim that I_e and I_{au} are two distinct-but-complementary functions of a unified complex inhibitory/interruption construct *I* (Johnson, Im-Bolter, & Pascual-Leone, 2003; Pascual-Leone, 1984; Pascual-Leone & Johnson, 2005, 2011). They differ in that I_e is slower and attentionally effortful and can be very selective in its application, whereas I_{au} is fast, effortless, and its application is global and not discriminating (excluding all that is not being attended to).

Here, we interpret the *M* and *I* constructs within the brain, an exercise in scientifically informed speculation. *M* and *I* emerge developmentally in concurrent interaction (complementary functions, a dialectical pair). They serve to control and regulate one another, both using executive schemes as controls. Constituents of mental attention, <*E, M, I, F*>, can be mapped in the working brain (as Luria, 1973, may have intuitively felt). We first address mechanisms of general (affective and/or cognitive) arousal.

fMRI shows mental attention as the activation/inhibition of neural energy (Arsalidou et al., 2013). This mental energy is conjointly produced, in our model, by three distinct and complementary sources of activation regulated in part by the PFC. (**Source 1**) First is the *brainstem arousal system*, with nonspecific and specific sources (each specific source related to one of the four main slow neurotransmitters). A nonspecific fast-neurotransmitter source is now regarded as most important to cause cortical arousal; this is the glutamatergic activation input from the basal forebrain, which has as inhibitory source the fast neurotransmitter GABA (Fuller, Sherman, Pedersen, Saper, & Lu, 2011; Saper & Fuller, 2017). The specific networks and their corresponding slow neurotransmitters (dopamine, norepinephrine, serotonin, acetylcholine) are now regarded as modulators of arousal within specific sorts of situations. Activation is brought to the cortex via glutamatergic circuits, which reformulates the up-to-now standard model of a reticular ascending system (RAS) of activation. (**Source 2**) Second there are *three pathways* that relate PFC, basal ganglia (BG), and thalamus: the direct pathway (this is, we think, the BG-link that enables production of mental power/energy—*Mp*); the BG indirect pathway (i.e., the mediator of effortful attentional interruption, I_e); and the BG hyperdirect pathway (i.e., automatic, unspecific attentional interruption, I_{au}). (**Source 3**) Third is the *corticocortical connections from PFC* to the relevant site for the task in question, as well as other cortical-loop connections with the thalamus, cerebellum, and other parts of the brain. Our model assumes that this cumulative series of three activation/inhibition sources occurs in the order described (1, 2, 3), and the amount of initiating energy needed by each source decreases in the order of mention, although the specificity of this effect (specific vs. global) increases with the order mentioned.

We will discuss specific arousal (RAS) network processes (**source 1**) below, when we examine our model for the four key neurotransmitters. We now discuss in more detail the BG-source of attentional activation/inhibition (**source 2**). We hypothesize that *M* and *I* are expressed in the brain (within the second, BG, source) by a complex network of connections that relate anterior cingulate gyrus (ACC), prefrontal lobes (PFL), and pre-supplementary motor areas (preSMA), to the basal ganglia (BG) and/or subthalamic nucleus (STN), which are linked to the thalamus. Thalamic nuclei regulate strength and variety of information (the information-bearing schemes or circuits) that at any time can reach the cortex. According to reviews by Cockburn and Frank (2011), Greenhouse, Swann, and Aron (2011), and Holroyd and Yeung (2012), there are three distinct pathways, which we relate to our constructs: BG's *direct pathway* (our *M*), *indirect pathway* (our I_e), and *hyperdirect pathway* (our I_{au}).

These pathways are schematized in figure 11.1, which elaborates on a figure from Cockburn and Frank (2011). The figure represents functional anatomy of the BG circuit. The *direct pathway* (*M*), originates in dopaminergic-D1 (GO-D1) receptor units in the striatum (BG) as they receive excitatory activation from the cortex (PFC and presupplementary motor area [preSMA]); it then gives a neural inhibitory input to the internal globus pallidus (GPi—part of BG). GPi has a permanent tonic inhibitory link with the thalamus (Thal) that suppresses/reduces the messages from Thal to cortex (Cx). Thus, inhibition of GPi weakens the inhibitory GPi-Thal link and consequently boosts communication between Thal and Cx, which facilitates afferent cortical reception of thalamic messages—a psychological-processing activation boost. By negating a negation, the direct pathway has a positive effect, boosting selectively Thal-Cx links as required by the prefrontal lobes and preSMA. Notice however that "disinhibiting thalamic units only permits those units to become active if those same units also receive top-down excitation from cortex" (Cockburn & Frank, 2011, p. 317). This is the neurological effect that may underpin the executive-driven psychological boosting that *M*-operator achieves. Notice further that (under this interpretation) psychological inhibition and neurological

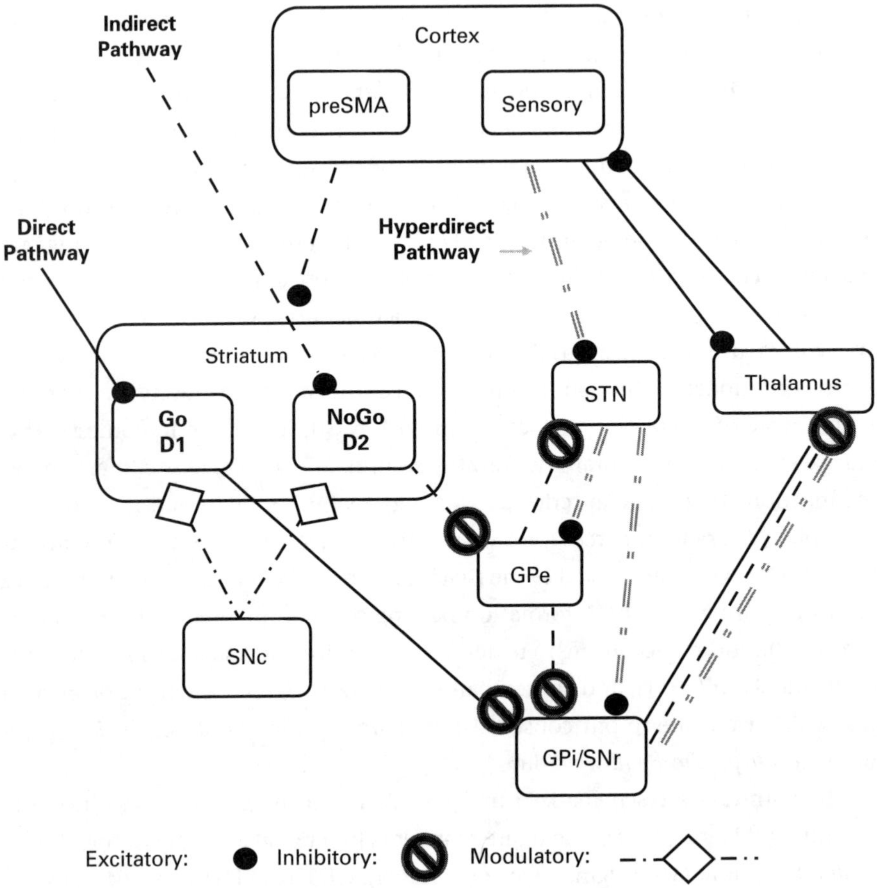

Figure 11.1
Functional anatomy of the basal ganglia (BG) circuit. (Adapted from Cockburn, J., & Frank, M.J. [2011]. Reinforcement learning, conflict monitoring, and cognitive control: An integrative model of cingulate-striatal interactions and the ERN. In R. Mars, J. Sallet, M. Rushworth, & N. Yeung [Eds.], *Neural basis of motivational and cognitive control* [p. 316]. MIT Press. Copyright 2011 by MIT Press. Adapted with permission.)

inhibition are distinct and different functions, which should not be confused. BG modulates only processes that originate in the cortex. Other competing figurative or operative schemes are then immediately suppressed in the cortex by lateral inhibition. This lateral inhibition is the process of *F-SOP* resolution that we explained in chapter 10 using Edelman's model. The currently active executive cortex (*E*-operator in our theory) may select its own action/scheme options in misleading situations with the help of this BG direct pathway (this would be a neuroscience aspect of *M*-operator's power capacity).

The indirect pathway (which psychologically expresses I_e—effortful attentional interruption) goes from PFC and preSMA, via dopaminergic-D2 receptor units in the striatum (BG), which give direct and selectively specific inhibitory input to external globus pallidus (GPe—part of BG). Thus, GPe weakens (in a selective, specific manner) its usual inhibitory input of GPi, weakening selectively the link of GPi on Thal. Here the negation of a negation (disinhibition) of GPi produces positive enhancement of GPi's inhibitory power over the thalamus, causing stronger (neurological and psychological) selective inhibition, which reduces the flow of thalamocortical activity. Dynamic balance between the direct pathway (our *M*-operator) and this indirect pathway (our *I*-operator) "determines the probability that the action [or figurative process] in question is gated" (Cockburn & Frank, 2011, p. 317). This neural connectivity aspect is congruent with our hypothesis that *M* and *I* constitute a functionally intertwined dialectical pair (Howard et al., 2014).

The hyperdirect pathway (which psychologically expresses automatic attentional interruption I_{au}) originates in the cortex and goes through the subthalamic nucleus (STN), bypassing the striatum (BG). STN receives excitatory input from the cortex and forwards excitatory input to the GPi, which inhibits further the thalamocortical activity. The projections of STN on GPi are diffuse, and so STN provides a "global no-go" signal to the thalamus, providing nonspecific inhibition to the thalamocortical pathway (Cockburn & Frank, 2011).

The hyperdirect pathway, the indirect pathway, and the direct pathway work in coordination and also in a proactive manner: if there are several alternative specific responses (schemes) in competition, and one has been chosen by the direct pathway (*E* and *M* processes), the STN hyperdirect pathway (I_{au}) will dynamically acquire greater strength to inhibit globally the unwanted schemes. The latter is aided by the indirect pathway (I_e) that modulates and contributes specificity to the inhibitory action (congruently with the neurological model of Cockburn & Frank, 2011; Greenhouse et al., 2011). If we translate these neurofunctional patterns into dialectical constructivist psychology, this means (as we theoretically implied above and in chapters 5 through 7) that the brain mechanisms for hidden operators *E*, *M*, I_e, and I_{au} are functionally intertwined, constituting a dialectical quartet in continuous coordination. The dorsolateral prefrontal cortex seems to do this coordination (Greenhouse et al., 2011), particularly when the task is complex or misleading.

Consider now the *corticocortical connections* (**source** 3), as well as other corticothalamocortical connections and loops via the cerebellum. We cannot review this considerable literature and will not appraise our hypothesis here. We posit that there are corticocortical connections from PFC to cortical sites relevant or irrelevant for the task at hand, which will add to excitatory/inhibitory activation that comes from the two sources of activation/inhibition mentioned above (i.e., **1** and **2**). This effect may be reinforced by corticothalamocortical connections and (to ensure coordinating operative balance) loops with the cerebellum. When **source 3** so defined applies, task-relevant sites may (or may not) have already been activated by sources of activation **1** and **2**, but with the three sources cumulating, the effect of this selective **3** corticocortical excitation can finally complete the cumulative process of activation/inhibition in the specific directions intended by the prefrontal processes (the executive *E*-operator).

This full effect of mental attention's activation (*Mp*) and effortful inhibition (I_e) may have the complex neuroscientific foundation just described, but there is another important aspect to *M*-capacity: the conjoint coactivation of schemes that could co-exist in the internal focus of mental centration without interfering with one another. This aspect, closer to the intuition of a working memory, also requires (we think) that the coactivated scheme processes in question, which carry the information (in an electroencephalogram [EEG] this may appear as fast waves, e.g., gamma waves), be protected from interfering with one another. Such protection often is achieved (as we explain in more detail later on) by the coupling of these gamma waves (the coactivated schemes) with theta waves. This coupling of gamma with theta waves would prevent interference among coactivated but distinct gamma waves and allow their effective concurrent coactivation with *M*-capacity. This coupling of EEG waves is further discussed in the next section. We will comment on the mental-attention **source 1** (i.e., the brainstem arousal system with nonspecific excitatory processes, together with specific modulator processes each with one of the four main neurotransmitters) in the section Arousal, Neurotransmitters, and the Limbic System's Role in Cognition.

Representational (Figurative) Streams

The posterior cortex is involved in processing figurative schemes/schemas (e.g., representations of objects, events, situations), as well as semantic relations holding among them (logological/linguistic or mereological/experiential relations). The causal texture of the environment is psychologically made by these internal/mental models and their cognitive/truth relations (e.g., Johnson-Laird, 2010). Mental models are constructively and reductively abstracted from probabilistic invariants internalized from experience. Relations holding among objects, and objects themselves, cannot therefore be absolute but relative to goals, agency, or praxis—consciously intended agency (Gonseth, 1936/1974; Johnson-Laird, 2010; Pascual-Leone, Escobar, & Johnson, 2012). In this respect, Leibniz was right in the eighteenth century, maintaining against Newton/Clarke (Alexander, 1976, p. xxv) that space is "the order of coexistent phenomena." In our terms, space is a constructed referential framework formed by relations of coexistence (caused by our innately prepared hidden *S*-operator) among agency-related schemes/schemas stored within the organism. Likewise, Leibniz took the view that time is "the order of successive phenomena" (Alexander, 1976, p. xxv). That is, time is a constructed referential framework formed by the flow of sequential relations—in our model caused by a hidden Time (*T*-) operator among schemes and schemas. Kant thought that space and time thus defined were purely transcendental[3] categories of intuition. This is what we today would describe as *natural* dimensions of experiential variation (prepared in the brain connectivity, partly innate, partly constructed). Against these natural dimensions we organize our, initially mereological, synthesis of experience that internalizes Reality resistances. This is important, because sensorial experience reaches us only through a vast collection of perceptual snapshots that must be synthesized into cognitive schemes during our reflective abstraction process.

Neuroscience has confirmed a modern version of the idea that Leibniz and Kant initially introduced: it has been shown that there are two innate streams of processing usually called representational, dorsal (occipito-parietal) stream and ventral (occipito-temporal) stream. They constitute, respectively, neuroscience's current version of the Leibniz-Kant space and time (a historical connection often not recognized). An alert human organism can recognize, without much effort and at times automatically, the flow of sequences of processes induced by experience. The various sensorial channels are mapped topographically and serially onto the brain, preserving dynamically basic relations of coexistence and sequencing among sensorial traces. Spatial and temporal relations can be meaningfully constructed with little effort, relating actual resistances that express these particular dimensions among/within packages of external resistances. Relations of coexistence may be constructed from meaningful pragmatic (in the sense of goal-directed, praxis-driven) sequences and interrelations among experienced resistances, relative to situations (context sensitive). They are pragmatic, because they express relations among resistances that emerge in agency or praxis.

The representational ventral stream is, in our view, the *figurative-representation temporal stream*.[4] It is often thought (perhaps too restrictively) as a perceptual or object-identification stream, because it answers questions about the "what" or "how" of experience (i.e., What is this? How to do it?; Atkinson & Braddick, 2003; Milner & Goodale, 1995).[5] Its location is occipito-temporal. In our theory, this is a site of fluent-schemes created by the *T*-operator, an automatic flow- or sequence-reader and learning factor. In addition to visual centers (e.g., for shape, color, or orientation), it involves the inferotemporal region, temporal lobe, and superior temporal sulcus. These centers send information to the perirhinal cortex (BA 35, 36), which adds its own processing (perhaps sequential patterning, although the nature of this processing may not be clear). Then, information is transferred to the entorhinal cortex (BA 28)—medial temporal cortex. The entorhinal cortex is a "major zone of convergence, both for object information arriving from the perirhinal cortex and for spatial information arriving from the parahippocampal cortex" (Suzuki, 2002, p. 294). Suzuki points out that entorhinal neurons signal memory for objects and spatial information in similar ways: "they respond selectively to particular objects or cued spatial locations (i.e., stimulus-selective response), and memory is signaled by enhancing or suppressing these stimulus-selective responses" (Suzuki, 2002, p. 294). In turn, the entorhinal cortex is in close bilateral connection with the hippocampus, which abstracts a higher-level, spatiotemporal, representation (Suzuki, 2002). This is a representation of a here-and-now environmental "context" intertwined with internal action processes (Jiang, Brashier, & Egner, 2015; Pribram, 1991).

The role of hippocampus in the explicit sequential structuring of navigational learning has been shown by Ruediger, Spirig, Donato, and Caroni (2012) using individual mice learning to navigate a water maze. The mice learned sequentially the steps of the track, and these steps were explicitly and sequentially represented in distinct (progressively engaged, as learning proceeded) sites in the hippocampus (H): from ventral-H (in humans anterior-H), to intermediate-H, to dorsal-H. Ruediger et al. (2012, p. 1570) put it in these words: "We provide evidence that maze navigation involves a stereotyped sequence of subtasks that involve learning at distinct hippocampal subdivisions." This discovery in mice is important, because it shows how the brain could register temporal-sequence scheme learning (the constitution of fluent schemes) that we hypothesized when positing the existence of a *T*-operator.

There is also indirect evidence that this sort of mechanism may be found in humans. For instance, according to Olsen et al. (2015, p. 5342) "The hippocampus may be necessary when the item elements need to be bound within and across episodes, to form a lasting representation that can be used flexibly." They investigated this idea by studying a young woman with developmental amnesia and bilateral hippocampus damage, as well as some control participants, with a series of cards reproducing a face taken from different changing perspectives versus a series of faces from a fixed perspective. Despite repeated presentations, the patient could not learn to recognize the face presented with changing perspectives, but she could learn the fixed perspective face. This is consistent with our model: whereas the former must be learned by retaining the face features and coordinating them (sequentially and spatially as one distal object) via the flow of temporally situated scanpaths (sequences), the latter could be learned as a complex pattern or template (a spatial proximal object). Thus *T*-operator would be needed only in the former. As the quote shows, these authors are coming very close to our idea: the hippocampus is necessary for reductively abstracting (as a probabilistic invariant) the common face structure across the face-episode sequence, when the changing perspectives prevent a simpler pattern-learning.

A problem remains, however: how can this spatial-and-temporal structure, this analytical sequential totality, be communicated (preserving the distinctiveness of sequential parts) across the brain, from one brain center to the other? Buzsaki (2006) and Lisman and Jensen (2013) suggest that the brain could do this by using a neural code that intertwines theta (or alpha—*slow EEG wave frequencies*) with *fast frequency waves*, which carry information—for example, gamma rhythms of activity (theta or alpha waves would serve to frame and mark the sequencing of gamma or beta waves). These interlocked ("multiplexed") rhythms embody the temporal sequencing of segregated gamma units, because this fast (gamma) rhythm is sequentially ordered and temporally framed by the slow theta or alpha waves. This framing with slow waves is found in the hippocampus and transferable elsewhere. Such multiplexed rhythms may be an essential substratum for our *T*-operator.

As Buzsaki (2006) suggests, slow wave framing of fast waves (which, we add, carry experienced information—schemes) may also be essential to explain in part the magical number 7; that is, the maximum capacity of mental attention (i.e., capacity of our *M*-operator), which Buzsaki calls working memory. As Buzsaki (2006, p. 352) says, "The number of items that can be stored by the multiplexed gamma-theta model is identical with the magical number 7 (+/–2)." A theoretical micro-analysis of mental-attentional knowing (in both perception and mentation) suggests this intertwining of *T*-operator and *M*-operator: most knowledge acquisition (using scanning, focal attention) occurs sequentially. Then the schemes/events experienced during this sequential processing are dynamically synthesized into a single mental centration compound (i.e., simultaneous attention points within the focus of working memory, whose number, when appropriate tasks are used, is our measure of *M*-capacity). According to Buzsaki (2006) the theta period can encompass seven to nine items multiplexed on successive gamma cycles, "However, in the case of the faster alpha oscillator [alpha rhythm], the span is shorter, with a capacity limit of four or five nested cell assemblies in a single alpha wave" (p. 353).[6] Thus, the maximum or reserve mental-attentional (*M*-) capacity of adults (which generally is seven) may be fixed by the encompassing capacity of these theta waves, but the well-known commonly exhibited, functional *M*-capacity of normal adults (which is four or five) may be fixed by the encompassing capacity of their alpha wave cycles.

Hippocampus, however, may not be essential for all sorts of processing in the ventral pathway (Atkinson & Braddick, 2003). This ventral stream is closely connected with the ventrolateral areas (paleocortical system), predominantly BA 45 and 12, but also BA 47, 11/12, 13, 8. We think of this pathway as a figurative-temporal (sequence-patterning) stream that synthesizes distal objects of praxis out of interrelated proximal objects of perception. In addition to synthesizing meaning for objects, the ventral pathway may participate in the learning of sequences of appropriate responses and rewards or outcomes, which would lead to the formation of expectancies (fluent schemes). These are perhaps good reasons to claim that these fluent schemas are critical for neural representation, serving as spatiotemporal "fingerprints" (Alvaro Pascual-Leone, personal communication, August 2019).

The temporal lobe has other important streams, which are neurologically and semantically different, but also involve sequences, all of which could perhaps be subsumed under the concept of a figurative-temporal (*T*-) operator. This hidden *T*-operator is temporal, because it can (effortlessly) capture the flow of events, by synthesizing dynamic sequence-structuring schemes that can be learned as fluents.[7] This *T*-operator works cognitively in coordination with space (*S*-operator) by the mediation of the hippocampus. From this perspective one may perhaps say that receptive language is precisely "located" on the left superior-and-posterior temporal lobe (Wernicke area, posterior BA 22), because language processing shares with the ventral pathway the need to abstract and retain meaningful sequences—sequences of words in language. Arbib, Rizzolatti, and others (see Arbib, 2005) proposed that language derives evolutionarily from imitation, gesture, communication, and pantomime (all fluent spatiotemporal activities), and this processing is facilitated by special neurons that sustain their activation to mediate the processing of sequences and fluent representations. These are, for instance, mirror and canonical neurons, located in the prefrontal lobe (BA 45, 44) but also in temporal-parietal regions such as superior temporal sulcus, entorhinal cortex, or intraparietal sulcus (Arbib, 2005; Hickok, 2014; Suzuki, 2002). Similar arguments about abstraction of sequences can be made with reference to sound patterns and music. Music processing seems to occur (besides the auditory centers) in the superior and anterior temporal lobe—anterior BA 22 (Zatorre & Peretz, 2001).

In dialectical complementarity with this ventral (figurative-temporal) stream, and richly interconnected with it (e.g., Stiles, 2001), is the *representational dorsal stream*. This is a space-structuring pathway, which constructively abstracts relations of coexistence (our hidden *S*-operator) at the service of operative processing (e.g., goal-driven activity). It (along with the prefrontal lobe when mental attention is needed) represents the location of objects in space from an egocentric perspective, that is, the perspective of the persons' goal-directed action or praxis, suggesting "how" to proceed (Milner & Goodale, 1995). From studies in monkeys (Suzuki, 2002) we know that, along with numerous visual-processing sites, this stream includes several temporal/spatial sites: middle temporal area (MT—also called V5), the dorsal bank of superior temporal sulcus (STS), posterior superior temporal gyrus, inferior or posterior parietal areas (BA 40, 39, 7), and lateral intraparietal region (LIP). The STS, BA 40, 39, and LIP are polymodal (high tertiary or quaternary) areas that may contain the special mirror and canonical neurons mentioned above (Arbib, 2005; Hickok, 2014). These dorsal centers converge toward the parahippocampal cortex along with input from the insula and the cingulate gyrus. Superordinate to both perirhinal cortex (ventral stream) and parahippocampal cortex (dorsal stream), the entorhinal cortex synthesizes the complex information prepared by these two pathways (Damasio & Damasio, 1994; Suzuki, 2002) and brings it to the hippocampus, which is the apex of this heterarchy. The hippocampus may be the site where sequences of events (schemes) are organized into a *causal context of the environment*, a "cognitive map" of relations of coexistence (in space and/or time) among the various identified objects. This well-known context-generating function of the brain expresses our space/time intertwining processes, also necessary for figurative (declarative memory) and relational learning (Shimamura, 2002).

As mentioned above, the hippocampus is a major structure of the medial (archicortical) limbic system, maintaining bilateral connections with prefrontal neocortical structures such as the dorsolateral prefrontal area (BA 46). Whenever a visuospatial task is complex or is misleading enough to demand mental attention, this prefrontal cortex is activated.

Arousal, Neurotransmitters, and the Limbic System's Role in Cognition[8]

The arousal mechanism helps to modulate readiness of neuronal synapses to fire; this mechanism relates to our lateralization neuropsychological model of activation/inhibition mentioned above and in chapter 10. The brainstem's mesencephalic reticular formation (MRF—also called the reticular ascending/activating system, or RAS) functions largely as a modulator for cortical and subcortical processes driven also by glutamatergic circuits, as mentioned before. It regulates arousal level by means of (direct and indirect) reciprocal connections that are activatory or inhibitory. For instance, connections that MRF sends to the reticular nucleus (NR) of the thalamus are inhibitory; the NR acts as a gate that filters sensory input arriving at the thalamus on the way to the cortex. This is done by inhibiting the thalamic relay nuclei that transmit sensory input to the cortex. Thus, the inhibitory connection from MRF to NR opens (or regulates) the gate by inhibiting (or releasing) this inhibition. The prefrontal lobe can regulate activity of MRF by directly activating it also via the mostly activatory prefrontal connection with NR nucleus (Mesulam, 2003). J. Skinner and Yingling (1977), who pioneered discussion of the NR-gating mechanism, interpreted their data as meaning that the MRF-NR link participates in causing tonic arousal, and the prefrontal-NR link in modulating attention. Modulation here means regulating effects of other sources of activation/inhibition so that they together codetermine outcomes (this is part of the functional infrastructure for *SOP* organismic dynamics, i.e., the "winner takes all" brain principle already mentioned, e.g., chapter 10).

This unspecific function of attention is further modulated by four specific pathways that together with the MRF system constitute an ascending reticular activating system (ARAS—Mesulam, 2003; also called RAS). These pathways have neurotransmitters that mediate synaptic transmissions within them (Kolb & Whishaw, 2003; Robbins & Everitt, 1995; Stahl, 2000). We will mention four neurotransmitters: acetylcholine (cholinergic or ACh system), norepinephrine (noradrenergic or NE system—also called NA system), serotonin (serotoninergic or 5-HT system), and dopamine (dopaminergic or DA system). Each has a specific function and is compatible with the others: "These reticular systems are simultaneously active in varying degrees, to optimize processing capacity and facilitate response output" (Robbins & Everitt, 1995, p. 715). The four of them intervene in attention by serving as modulators (neuromodulators) of cortical activity. Even though more research is needed to clarify their psychological significance (Richards, 2003; Stahl, 2000; Ullsperger, 2011), we believe that they can be understood in their interactions as part of dynamic meta-function controls (different from hidden operators) that can reset the brain for specialized ways of functioning. Ullsperger (2011) seemed to be asking for this sort of theoretical effort: "This clearly calls for models that try to integrate more than one neuromodulator" (p. 43). Our process-analytical constructivist model for these neurotransmitters (sketched in table 11.1) highlights their functional significance as *slow* (from many milliseconds to seconds) *and often sustained neuromodulators of arousal* (Stahl, 2000). Lack of space prevents discussion of the vast, relevant and (when interpreted) supportive, empirical evidence.

Table 11.1
A physiological interpretive model of the extrinsic arousal neurotransmitters

	Facilitating Situations	Misleading Situations
Perceptual/Automatic Attention	ACETYLCHOLINE: Habitual processes; mediates the body's parasympathetic system. Pribram's arousal and Posner's orienting.	NOREPINEPHRINE: Novel affective orienting reactions; mediates the body's sympathetic system. Pribram's activation and Posner's alerting.
Mental/Effortful Attention	SEROTONIN: Related to well-being and active interruption (inhibition) of irrelevant processes and negative affects. Regulates norepinephrine and dopamine systems.	DOPAMINE: Related to agency, value assignment (reinforcement), and praxis (reward anticipation, etc.). Pribram's effort and Posner's executive attention.

Note:

Cognitive Processes: Basal ganglia, etc. relate to operative praxis	**Affective Processes**: Amygdala, etc. relate to figurative motivation

We describe each neurotransmitter in terms of three dimensions of variation, or categories of processing, which characterize temporary effects on neurons that each neurotransmitter may promote (see table 11.1). The three dimensions are (1) perceptual/automatic attention versus mental/effortful attention (see chapters 6 and 7 and below); (2) facilitating situations versus misleading situations (a distinction explained repeatedly in this book); and (3) cognitive processes (important in operative agency/praxis) versus affective processes (important in figurative motivation; see chapter 5). The first dimension concerns the type of attention, perceptual versus mental. Perceptual/automatic attention involves cognitive processes applied to information entering through sensorial receptors. It is the activity Posner (Petersen & Posner, 2012; Rothbart & Posner, 2001) assigned to his orienting and alerting networks. This may involve activity in the posterior parietal lobe, the frontal eye area (BA 6, 8), the right frontal lobe (superior BA 6), right parietal lobe, and the brainstem locus coeruleus (a main source of the NE system). Perceptual attention is closely related to the classic orienting reaction, which leads the animal to attend automatically whenever some novel event happens in its perceptual field (Kahneman, 1973; Pribram, 1971, 1991). We speculate that the ACh system (acetylcholine) and the NE system (norepinephrine) intervene as a pair of complementary and dialectically interacting arousal processes in perceptual/automatic attention. ACh is released with respect to cognitively facilitating aspects of situations. In this context mental "cool" arousal (which we interpret as *M*-activation) is potentiated by ACh. Also, during sleep, dreams (REM sleep) are enabled by the cholinergic network (Platt & Riedel, 2011). NE, in contrast, is released in response to affectively misleading (disturbing, surprising, confusing, emotion arousing) aspects of situations and causes its own different form of ("hot") arousal, which possibly also activates *M*-capacity; NE activity interrupts sleep, including REM sleep.

Mental attention involves activation and control of arousal to enable thinking, planning, goal-directed activity, conflict resolution, and other activity. These are internally mediated (i.e., mental) processes that the brain maintains in relative autonomy from the perceptual input. Posner and others call this sort of attention "executive attention" or executive-control network (Heitz, Unsworth, & Engle, 2005; Petersen & Posner, 2012; Rothbart & Posner, 2001). As mentioned in chapters 3 and 7, we prefer the term *mental* instead of *executive*, because mental-attentional processes already exist in children of 2 to 3 months of age, when "executives" have not yet appeared. The cingulate gyrus and midline frontal areas (e.g., anterior cingulate gyrus, supplementary motor area in BA 6), prefrontal cortex, and basal ganglia often are involved in mental attention (as contrasted with perceptual/automatic attention, e.g., Rothbart & Posner, 2001). We speculate that in "cool" (i.e., not affectively loaded) mentation and complex cognition, mental-attentional processes are specially (not exclusively) supported by the DA system (dopamine). The DA system is in dialectical complementarity (mutual regulation) with the serotonin system. In "hot" (affectively loaded) cognition the norepinephrine system may join in or replace the DA system, although serotonin continues to be the complementary or regulating system.

Notice that perceptual/automatic attention is involved predominantly in figurative representation of the input and its affective/emotional appraisal. This emotional appraisal gives affect-values (vital organismic values) to the representation of inputs, values that we call *figurative motivation* and others have called "salience." Mental (in contrast to perceptual) attention is often involved in intentional/volitional activity (praxis), or in mentation that is preparatory for subsequent or future action/praxis. As Piaget, Vygotsky, and many others have claimed, covert mental processes (mentation) can be interpreted as internalized action, or preparation for action; this is operative activity, whether strictly motor or not. Particularly when applying on novel and suitably complex situations, mental-attentional processes tend to engage left frontal or prefrontal activities. In contrast, perceptual attention (which is more automatized, *LC* schemas) engages predominantly parietal (and temporal) processes, often with right frontal processes. Bilateral processing occurs when tasks are only partly automatized, or too demanding, for the subject's own mental (*M*) capacity. The three intertwined dimensions presented in table 11.1 are crossed: each neurotransmitter prepares the organism for specified positions on each of the three dimensions, all at the same time.

Acetylcholine (ACh)

Cholinergic (acetylcholine) mediation occurs in not-too-complex perceptual or cognitive tasks (Störmer, Passow, & Biesenack, 2012) to mobilize arousal and mental attention (Matt, *M*-capacity). It is mobilized in the context of perceptual attention to relevant aspects of agency/praxis, often within facilitating situations. As Störmer et al. (2012, p. 885) said, "Precisely, most studies using tasks that require orienting processes, such as spatial cueing tasks, reported effects of the cholinergic system. Tasks involving the maintenance or updating of information, on the contrary, have mostly been found to depend on dopaminergic neurotransmission and in some cases on the interaction of the cholinergic and the dopaminergic system."

Robbins and Everitt (1995) described an experiment with rats that illustrates the difference among the four neurotransmitters we are considering. Using a special apparatus, rats were trained to detect brief visual stimuli (brief light flashes) presented unpredictably at one of five locations. Five seconds after the rat obtained its reward, the next stimulus occurred; a session contained one hundred such trials. Rats that had a chemically induced 70% reduction of activity in the cholinergic cortical pathways showed important deficits, relative to normal rats, in detecting the location of randomly presented light flashes. Similar results with a similar task were obtained with patients with probable Alzheimer dementia (who usually exhibit a cortical cholinergic reduction). As happened with the experimental rats, problems of these patients in the task could be reversed if an anticholinesterase (a chemical that retards decomposition of ACh in the tissues) was administered.

Because acetylcholine is the main neurotransmitter in neuronal pathways of the autonomous nervous system, for which the vagus is the key cranial nerve, vagus activity increases whenever the cholinergic system is high in the orbitofrontal cortex (Richards, 2002), causing the heart rate to decelerate. Consequently, a slow heart rate is reliable indicator of sustained perceptual attention and openness to environmental cues (Richards, 2002). Robbins and Everitt suggested that the ACh system "aids the processing of stimuli at the cortical level by enhancing the impact of salient information" (1995, p. 715). This increment in processing efficiency may be caused by the ACh system's enhancement of the cortical lateral-inhibition processes discussed in previous chapters. Cortical lateral-inhibition processes correspond to our *F*-operator. If this hypothesis were correct, the ACh system would be one way in which our model's *F*-operator potency is increased, regulated upward.

Also, the ACh system may be able to mobilize mental attention (dopaminergic system) within facilitating situations (Störmer et al., 2012). Intriguing experiments point in this direction. Castro-Alamancos and Gulati (2014) have shown in rats that the ACh system increases cortical activity while increasing activity in the thalamus, which we should expect if the cholinergic system increases openness and sensitivity to facilitating perceptual situations. In contrast, these authors show that the norepinephrine (NE) system increases cortical activity but decreases activity in the thalamus. This should diminish perceptual openness to external stimulation in general—as expected if NE prepares the organism for selective perceptual processing in misleading situations (where misleading perceptual factors should be inhibited).

Norepinephrine (NE System)

Norepinephrine, also called noradrenaline, is a key neurotransmitter for the sympathetic division of the autonomous nervous system. Epinephrine (adrenaline), a close variant, is released into the bloodstream by suprarenal glands in alarm or peak emotional reactions (Richards, 2002, 2003). The NE-system mediates in the brain's orienting reactions (Pribram, 1971, 1991; Sara & Bouret, 2012), that is, reactions to unexpected novelty and good or bad surprising events, and participates in emotions initiated by the amygdala in the limbic organization (Pribram, 1971, 1991; Robbins & Everitt, 1995; Sara & Bouret, 2012). Activation of the NE-system may be accompanied by such reactions as a fast heart rate, a skin conductance response (SCR or galvanic skin response), or pupil dilation. Norepinephrine-sensitive (NE) neurons are best activated by "new, unexpected, nonpainful stimuli in the animal's environment" and "are least active when the animals are not vigilant, just sitting around quietly" (Bear, Connors, & Paradiso, 2001, p. 515). NE can make neurons of the cortex more responsive to salient sensorial stimuli, making the sensory and motor systems more efficient, "improving the 'signal' by decreasing the 'noise'; related to tonic activity of the LC [Locus Coeruleus—key source of the NE system] nucleus" (Howells, Stein, & Russell, 2010, p. 14). Howells et al. further suggest that feeling of mental effort, common in misleading situations, is produced by tonic activity of the norepinephrine system.

In the Robbins and Everitt (1995) experiment described above, rats with a damaged NE-system performed well under normal conditions, but as soon as a distractor or stressor was added (e.g., white noise) their performance collapsed. The problem with NE-damaged rats was distraction, whereas that of ACh-depleted rats was accuracy (Robbins & Everitt, 1995). NE and ACh complement one another: ACh occurs in easy cognitive processes and facilitating (nonconflictive) situations, whereas NE occurs in affective processes and misleading, conflictive situations. These (and the other two) key neurotransmitters complement each other by adapting the brain function to different sorts of situations and contingencies.

Prefrontal lobes (operative/executive processes) are needed to apply mental attention to misleading or complex—affective or cognitive—situations (as Luria, 1973, early emphasized). The NE system is an important modulator of prefrontal activity, and its activation is possibly counterbalanced by the serotoninergic system. This form of arousal corresponds to Posner's alerting (Rothbart & Posner, 2001) and to Pribram's activation. The NE system is involved in highly active affective processes (our *A*-operator) and possibly intervenes in learning driven by affect (*LA*-learning). Organismic affective/cognitive reactions to experience may give (via this NE system) a feeling of mental or affective effort, which, if extreme, could lead to stress with its distinct neurobiological characteristics (Hermans et al., 2011). Under neurobiological stress, or under salient/relevant unexpected circumstances, the NE system can cause a reorienting reaction (Corbetta et al., 2008; Sara & Bouret, 2012) as it changes cortical control of attention from dorsal-attention (frontoparietal) network orientation (involved in focal selective attention and in task-directed activity—analytical alertness with *phasic firing* of norepinephrine neurons) to the ventral-attention network. The latter is characterized by global alertness, *tonic firing* of norepinephrine-sensitive neurons, affect-driven perception, interoception (heightened organismic feelings—resulting from the fronto-insular cortex), and activity in the dorsal anterior cingulate cortex, medial prefrontal cortex, amygdala, and other areas of the brain (Corbetta et al., 2008; Hermans et al., 2011; Sara & Bouret, 2012).

Dopamine (DA System)

Dopamine is implicated in processing rewards and motivated behavior, hunger, thirst, sexual arousal, endogenous (positive or negative) emotions, muscular activity, and so on (Walton, Gan, & Phillips, 2011). In cognition it mediates arousal in misleading situations, which demands mental attention (Matt, working memory). A well-studied domain for these situations is financial appraisal, or economic behavior, in cost/benefit decision making (Treadway et al., 2012). The DA system mediates connections between the basal ganglia (caudate nucleus, putamen, globus pallidus, nucleus accumbens), the thalamus, the prefrontal lobe, and the limbic organization. Via basal ganglia connection with the limbic system, the DA system is involved in "reward" evaluations that reinforce or assign organismic/vital value to (adaptive or nonadaptive) behaviors, energizing relevant circuits and focusing attention when needed (Bear et al., 2001;

H. Fischer et al., 2010; Rolls, 1999). The DA system helps to form *criteria of relevance* for encountered events or mentations with the nonexclusive participation of insula, orbital prefrontal cortex, nucleus accumbens, and amygdala (Bear et al., 2001; Pascual-Leone et al., 2015). Via orbitofrontal, anterior cingulate, and amygdala (etc.) connections, the DA system helps to focus attention and exert praxis on relevant events. DA processes enable mental/executive attention, which is diminished when the DA system decreases in potency either because of old age or because younger people have received some DA-D1 receptor antagonist (H. Fischer et al., 2010; Petersen & Posner, 2012; Pribram, 1991; Sagvolden, Johansen, Aase, & Russell, 2005). The DA system has a well-recognized age-developmental pattern of growth and of decrementation with aging (H. Fischer et al., 2010; S.-C. Li, 2012), possibly related to development of mental-attentional capacity in humans and of individual differences in DA system potency.

The emergence of motivation (both operative—internal, enactive— and figurative—external, contextual) is, as mentioned, related to these neurotransmitters. Motivation may involve a heterarchy of sites that include periaqueductal gray area, anterior insula, orbitofrontal cortex, amygdala (criterion of relevance), and anterior cingulate, which is where affective goals (implicit motives) are converted, we think, into cognitive goals (Arsalidou et al., 2018; Pascual-Leone et al., 2015).

Mental effort, related to mental (*M*) activation and/or mental interruption (*I*), and caused by the NE system's tonic activation, expresses the activity of prefrontal lobes (e.g., lateral, dorsolateral, frontal polar). This function of the prefrontal lobes was recognized by Luria (1973) and is now generally admitted. Such executive planning function of prefrontal lobes is mediated by dopaminergic, cholinergic, noradrenergic, and serotonergic arousal processes (perhaps in that priority order).

Serotonin (5-HT System)

Serotonin is much less understood in its effect on performance control and on thinking, but it appears to be involved in affective regulation, motivational processes, cognitive flexibility, and impulse control (Ullsperger, 2011). Often in interaction with dopamine, it seems to have some controlling effect on such issues as impulsive/aggressive behavior, suicide tendencies, and individuals with borderline personality disorders (Dalley & Roiser, 2012). In our model, we assume a soothing control effect (table 11.1) in more or less facilitating situations that require mental attention and affective processing. Another cognitive function may be the control (reduction) of task-irrelevant, perhaps not misleading, processes. For instance, in the rat's five-choice attentional task already mentioned, the effect of central serotonin loss was a significant increase in premature or impulsive responding without change in the percent correct (Robbins & Everitt, 1995).

The serotonin pathway plays a role in the regulation of mood—emotional behavior such as (in rats, primates, or humans) aggression or anxiety, and regulation of sleep (sleep/awake are controlled by ACh, NE, and serotonin pathways—Bear et al., 2001). Research in primates suggests the serotonin system may control, regulate, or reduce emotive expression of the NE system, simultaneously facilitating social-cognitive interactions with congeners (members of the same species) and improving the mood or social skills in monkeys (Pribram, 1991). In humans, deficiencies in brain serotonin may cause depression (also anorexia nervosa or bulimia). We interpret serotonin as expressing an interruption mechanism (an automatic *I*-operator), often used with affective processes, which is activated independently from, but coordinated with, the effortful mental-attentional operators primed by the DA system. It may be regulated by the prefrontal lobe (e.g., orbitofrontal, medial frontal cortex). Serotonin and dopamine systems seem to interact and complement one another (Benes, 2001).

Brain Expression of Hidden Operators: Summary

We briefly review our hypothesis about how the prefrontal lobe accomplishes the executive activation (or inhibition) of task-relevant (or -irrelevant) areas of the cortex. This explanation complements the one provided earlier in the chapter. Once the criteria of relevance (expressing dominant affective/cognitive goals) are clear in a task or situation (likely accomplished by anterior cingulate gyrus and its connections), the prefrontal lobe induces activation of the mesencephalic reticular formation (MRF), which causes RAS processes (reticular ascending/activating system activity, including specific pathways for the four discussed neurotransmitters). RAS activation reaches relevant regions of the cortex priming or modulating activation, either in a diffuse or a regionally specific manner (in accord with currently dominant neurotransmitters). Complex or misleading cognitive situations need use of mental attention (Pascual-Leone, 2000b, 2019; Pascual-Leone & Johnson, 2005, 2011, 2017). These cognitive situations are controlled by the prefrontal executive network via dopaminergic connections that relate basal ganglia with thalamus (Cockburn & Frank, 2011). This is done via three links: psychological-activatory direct pathway (related to our *M*-operator), psychological-inhibitory indirect pathway (related to our effortful *I*-operator), and psychological-inhibitory hyperdirect pathway (related to our automatic *I*-operator). These three specific sources of activation/inhibition ensure that specific (locally focused) task-relevant areas of the cortex will be further primed, whereas irrelevant areas are inhibited. Because of these subcortical sources of priming, the prefrontal lobe may achieve enough activation of specifically relevant cortical sites by using cortico-cortical connections (possibly mediated by neurotransmitters glutamate—activatory—and GABA—inhibitory), and other connections, to recruit into action the needed processes (Dunnett & Robbins, 1992; Edelman & Tononi, 2000; Shimamura, 2002).

Congruent with the perceptual/automatic attention versus mental attention distinction (two complementary modes), the limbic organization seems to have two evolutionarily distinct limbic systems. These have reciprocal connections with various parts of the frontal lobe: the lateral (basolateral) circuits versus the medial circuits (Bakchine, 2000; Goldberg, 1985; Pascual-Leone, 1995; Thatcher, 1994; Tucker, 2001). As summarized by Bakchine (2000) and Thatcher (1994), the evolutionarily older (paleocortical) lateral limbic circuit consists principally of the amygdala, dorsomedial thalamic nuclei, orbitofrontal, insular, and temporal-polar cortices, with premotor area connections as their motor link. It also has extensive connections with the dorsolateral prefrontal cortex and posterior parietal association cortex. It processes information concerning the external world, contributing implicit affective evaluation (global affective goals and their bodily manifestations, Damasio's [1999] somatic markers) to here-and-now situations, motivating acts and social-personal interactions. Pribram's (1991) work suggests that this system, centered on the amygdala, defines and emphasizes (somatic markers) the affective importance of subjects or objects within here-and-now experience, helping to constitute a criterion of relevance. Neurotransmitters acetylcholine and epinephrine, with serotonin and dopamine in complex situations, seem important here (see table 11.1).

The evolutionarily recent medial limbic circuit (archicortical cortex) includes the hypothalamus, anterior thalamic nuclei, hippocampus, cingulate gyrus, and related tracks, having as main motor link the supplementary motor areas, and drawing arousal from the brainstem reticular formation (Bakchine, 2000; Goldberg, 1985; Thatcher, 1994). It mediates important aspects of learning, memory, and mental attention. Via hippocampus, it intervenes in representations of space and time, and in relational (*LM* or *LC*) learning, although it cannot alone identify objects. The importance of this medial limbic system is apparent in episodic-memory research (Moscovitch, Cabeza, Winocur, & Nadel, 2016). For Pribram (1991) the hippocampus helps to demarcate the context, not the object, of experience, by construing situations as analytical wholes (a "context"). Fuster (1995) holds related views, interpreting the hippocampus as a convergence zone (Damasio & Damasio, 1994) "involved in the binding—in part through non-cortical neural stations—of widely dispersed representations" (Fuster, 1995, p. 103).

For us the hippocampus complex (including entorhinal, perirhinal, and parahippocampal cortex) can coordinate the replicable activation sequences of schemes (neocortical processes), with their relations of coexistence (e.g., spatial, temporal) and retain these relational patterns as local dynamic "cognitive maps" at the service of agency/praxis (Pribram, 1991). These spatial-temporal maps internalize the causal texture of the environment (Tolman & Brunswik, 1935). This function of relational-temporal fluent learning (organized by the hippocampus) expresses in our theory the intertwining of *T*-operator (temporal structuring) and *S*-operator (relations of coexistence).

The medial limbic circuit intervenes in representing space-time and is essential for planful, predictive, and appropriate action. In contrast to the here-and-now relevance of the lateral system, this medial system is involved in anticipatory or predictive actions (prospective operative praxis) guided by expectancies of how the world will be in the future (Goldberg, 1985; Thatcher, 1994). Dopamine is a major neurotransmitter in the medial limbic system and helps to define criteria of relevance in complex and misleading situations.

Conclusions

We began the two neuropsychology chapters by outlining a problem of current cognitive neuropsychology: the lack of an organismic (i.e., general and causal) theory or model for psychology—an analytical formulation of neuropsychological processes within high cognition. This organismic model should explain the macro (molar) levels of performance and have a clear (micro-level) neuroscientific interpretation. Further, this organismic model should include methods of process representation and task analysis, to yield causal descriptions of psychological processes interpretable as neurological activity. We suggested how the theory of constructive operators (TCO), its constructs and principles, can serve this purpose. We have discussed how brain research supports developmental neo-Piagetian constructs such as mental attention, which explicates the functional origin of working memory (Pascual-Leone, 2000b; Pascual-Leone & Baillargeon, 1994; Pascual-Leone & Johnson, 2005). The growth throughout infancy and childhood, until adolescence, of mental attentional capacity (our model <*E*, *M*, *I*, *F*>) explains emergence of developmental cognitive stages and the growth of learning potential, conceptual abstraction, and language.

In this chapter we have discussed seven major functional dimensions of variation in the cortex: (1) cortically anterior versus posterior lobes; (2) dorsal-attention (frontoparietal cognitive assimilation) versus ventral-attention (frontoparietal cognitive accommodation); (3) left versus right hemisphere; (4) experiential versus conceptual knowing (primary, secondary, tertiary, and quaternary or polymodal areas); (5) functional hierarchies of operative symbolic streams; (6) mental-attentional activation and inhibition brain mechanisms; and (7) representational-dorsal (spatial) versus representational-ventral (temporal) streams. Coordination of these dimensions clarifies how the brain encodes and expresses meaning (semiotics). Localization separates (to minimize interference) different functional modes of processing that are complementary and collaborate to overdetermine human performance. We have also discussed key neuromodulators (extrinsic neurotransmitters) to show how they contribute to adapt processing to situations and to internal states.

To clarify evolutionary grounding of some functional dimensions, we briefly discussed how neurotransmitters and processes relate to the paleocortical (lateral) and the archicortical (medial) limbic subsystem, perhaps clarifying interactions among cognitive/cortical, affective/limbic, and arousal/neuromodulator processes. We hope to have intimated how the brain can produce affectively "cool" versus "hot" cognitive processes and how it can cope with facilitating versus misleading situations and deal with simple versus complex situations and with figurative (representational) versus operative processes.

We have also intimated how abstract relations of coexistence among schemes (space, dorsal-representational pathways) coordinating with the flow of sequential relations (time, ventral-representations) can constitute the environmental "context" (Pribram, 1991) or "causal texture of the environment" (Tolman & Brunswik, 1935). Our conclusions presented in chapters 10 and 11 are based on task-analyses of neuroscience research findings—research that for brevity we have omitted. Findings discussed are those chosen to illustrate ideas and theoretical analyses of this scientific essay.

12 The Working Mind Model across Human Domains

The Theory of Constructive Operators (TCO) is a process-qualitative causal model and, in method, might be compared with Darwin's theory of evolution. We address how our theory can be useful and discuss potential contributions to cognitive psychology, neuroscience, individual differences and cognitive styles, psychotherapy, and current notions of consciousness, self, and Will. We suggest that TCO may be helpful to human science and education because of its causal-organismic approach and methods of task analysis.

> Those who have treated of the sciences have been either empiricists or dogmatists. Empiricists, like ants, simply accumulate and use; Rationalists, like spiders, spin webs from themselves; the way of the bee is in between: it takes material from the flowers of the garden and the field; but it has the ability to convert and digest them.
> —Bacon, 1620/2004, XCV, p. 79[1]

> In summary, empiricism begins with the record of plain facts, science denounces this evidence to discover hidden laws. *There is no science but of that which is hidden.*
> —Bachelard, 1949/1975, p. 38; translation by JPL

> Come back, Piaget: Much (but not all) is forgiven.
> —Adey & Shayer, 1993, p. 29

The aim of this final chapter is to provide a rationale for the book and to summarize some of its key innovative ideas. First we address the approach to knowledge (epistemology) underlying our theory and methods. Most explanatory theories in psychology are empiricist or meta-empiricist. *Empiricist* theories are grounded in perception and aim to describe/analyze from an observer's (outside) perspective the phenomena relevant for the problems to be understood. *Meta-empiricist* theories are sophisticated versions of empiricism; they are mindful that the organism (in psychology) or nature (in physics) contains processes that interact in hidden ways to produce observable phenomena. This is the dominant approach in science. There is, however, a complementary, equally important approach, empirical rationalist *constructivism*.

Unlike empiricism the constructivist approach does not aim to describe phenomena from an observer's (outside) perspective. It accepts phenomena described by empiricism and meta-empiricism (including its own empirical-rationalist observations) and proceeds to investigate empirically-inferable hidden processes (in nature or in the organism), whose interactions may serve as organismic causes of the phenomenal manifestations. This alternative complementary approach is *constructivism*. The constructivist approach attempts to model hidden processes from within; this is what we have called a *metasubjective* perspective. It is complementary with, but radically distinct from, the meta-empiricist perspective. Empiricist or meta-empiricist theories tend to be local and descriptive, whereas constructivist ones are general and causal-organismic.

The two types are incommensurable in method and theory. We cannot present the one by using presentational methods of the other. For instance, one does not make constructivist theories clearer by first critically reviewing empiricist research (from an empiricist perspective) and then judging whether constructivism, in terms of empiricist standards, meets the empirical/theoretical constraints. Their respective methods are different: the old empiricist presumption that theories emerge by contrasting and comparing empiricist research data across situations is not valid for causal-organismic (or causal-nature) inferences. Constructivist theories demand other epistemological foundations in biology and evolution, to guide organismic-causal modeling. As with the theory of evolution, when it was first presented, constructivist theories must be argued using metatheoretical biological arguments and by interpreting many concrete and diverse examples that use the same constructivist organismic-causal model—as we have done repeatedly by way of metasubjective task analysis.

Copernicus may have been the first physicist to adopt a perspective from within nature's own processes—a metasubjective perspective, as we discussed in chapter 1. In biology the first scientist to adopt an explicit constructivist approach may have been Darwin. His theory and methods are empirical rationalist. His purpose is to understand hidden causal-organismic processes that enabled species of animals to evolve over the millennia from simple to complex and become more versatile and adapted. The hidden operators and principles in Darwin's evolution were inferred abductively (empirical rationalism), adopting a from-within-the-organism (metasubjective) perspective. Paraphrasing Stephen Jay Gould (2002), we can say that Darwin's theory has as metatheory a causal-organismic framework of three complementary processes in dialectical interaction (a dialectical trinity), which cause natural selection in evolution.

These three interacting constituents of natural selection are first, *overproduction of offspring* and the *agency* in these *organisms* that struggle "for their own personal advantages as expressed in differential reproductive success" (Gould, 2002, pp. 14–20). Second is spontaneous biogenetic *variation* of individual qualities and characteristics, which achieves *efficacy* "as the positive mechanism of evolutionary novelty" and can "create the fit" as well as "eliminate the unfit" by slow accumulation of useful variations across generations. Third is *hereditability*, which can ensure transmission of gains across generations, with a *scope* that explains the full diversity of continuous evolutionary change "through geological time." According to Gould, these three causal-organismic dialectical factors produce, in their interaction, natural selection. Thus, Darwin' theory is not empiricist but constructivist, and his empirical research aims to exhibit across species the anatomical and functional phenomena recurrent across individuals or species that support, in a constructivist manner (often through Peirce's abductions), the causal-organismic functional framework just described.

In the Theory of Constructive Operators, a dialectical trinity of causal-organismic factors is found, somewhat analogous to what Gould (2002) described for Darwin's theory. These three intertwined dialectical factors are first, *overproduction of schemes/structures* that internalize *resistances* from *Reality*, are self-propelling, and spontaneously seek *agency*. This is Piaget's assimilation function. Because they actively adapt (Piaget's accommodation) as experience accumulates, more and more schemes/schemas are formed adapting to distinct circumstances. Second is *variation of schemes promoted by other causal-organismic factors*, or innate constraints (Edelman, 1989, p. 185). In our theory, these constraints are hidden operators and organismic principles, which appear in the brain with evolution to ensure that enough schemes adapt with *efficacy* across expectable environments or milieux. Third, synthesis of all these effects produces a *functional totality* that unitizes individuals' performance and follows a principle of Schemes' Overdetermination. These dialectical factors codetermine together individuals' functioning and their learning and development. These functional syntheses in the psychological organism are the counterpoint of natural selection in humans' performance, development, and learning. Edelman (1987, 1989), with his theory of neural Darwinism, was first to formulate in neuroscience a related analogy between Darwin's theory of evolution and the brain's adaptive growth.

Notice that from this constructivist/evolutionary perspective, the various constructs here formulated as a dialectical trinity correspond epistemologically with Darwin's core dialectical system and other dialectical systems of lower level. For instance, all schemes can be decomposed in self-propelling dialectical systems such as *<:rc: ec: fc:>*. That is, they are constituted by a releasing component, an effecting component, and a functional component, as described in chapter 5 of this book. However, every scheme can also be reduced to a more basic dialectical system, such as <:animal[needs]:milieu [resistances]:>. That is to say, the animal with its needs is in dialectical exchange with the surrounding milieu, which presents positive resistances that satisfy needs (by way of affordances), while the animal manages obstacles and encumbrances (i.e., negative resistances). It is in such a biologically basic system that functional emergence of new schemes appears.

The TCO model may be important for human science: it helps to explain developmental transitions from perception (proximal objects) to high cognition (complex distal objects); or from sensorimotor processing in infancy to symbolic processing in school-age children, adolescents, and adults. It also helps to integrate cognitive and affective/personal processes and integrate cognitive development with neuroscience. It provides a method of "from within" (metasubjective) process/task analysis of situated mentations. This method can produce *causal-organismic models* of mentation in tasks, which may be deeper than the qualitative methods of phenomenology, hermeneutics, descriptive process/task analysis, and so on. Our causal-organismic models yield quantitative (at times interval-scale) estimates of mental-processing task complexity and allow analysis of real-life mentation and experiences.

The TCO is a framework for causal modeling and for relating cognition, affect, interpersonal relations, individual differences, learned automatisms, mental-executive attention, problem solving, learning, intelligence, developmental growth, and so forth. It does so by deconstructing organismic processes into schemes, hidden operators, and organismic principles.

Because generic rules of organismic *schemes* are invariant across content domains and participants, the method can create process/task analytical models to compare across domains and across samples, as we demonstrated in infancy (chapter 3) and in research with school-aged children (chapter 7). Furthermore, our research may help clarify causal-organismic factors for psychometric intelligence versus developmental intelligence. Individual differences can also be modeled with our analyses.

Hidden operators are purely organismic specific regulations or resources that control development, learning, and application of schemes; *organismic principles* (such as simplicity, overdetermination, equilibration, and reflective abstraction) express rules (general regulations) for coordinating schemes and operators in specific situations.

Unlike many cognitive-science theories, the TCO is truly developmental. This aspect has real advantages: developmental constructivism could make cognitive theorizing better grounded when ontogenetic and cultural developments are incorporated and modeled. For any developmental-constructivist cognitive science, a complex cognitive (i.e., distal) object is a manifold of coordinated functional invariants (invariant across chosen situations and contexts) abstracted during meaningful goal-directed actions. Cognitive objects emerge as targets of our activities and so are called distal.

The TCO emphasizes that perception and cognition, although deeply intertwined, are different psychologically and neurologically. Consider a simple example common in both perception and art: line drawings provide essential methods for object representation. To explain and convey deeper meanings of good figurative-art drawings, we must assume that artists and observers (i.e., creative artists versus observing artists) have internalized functional models in their brains of some distal objects. Efficient generic syntheses performed by the brain (which causes representation of distal objects) are needed to explicate meaning of meaning, even in perception. Indeed, researchers may use line drawings as unambiguous stimuli. Thus, distal objects must exist as models in the brain abstracted from the Real. TCO can explain this reflective abstraction.

Scheme units adapt (accommodate and assimilate) to specific Reality constraints, the resistances. Schemes incorporate the good/positive resistances and attempt to accommodate (control) unsuitable ones. Accommodation to resistances (in order to assimilate them) helps schemes to internalize relevant aspects of Reality, allowing them to serve as mental tools for agency/praxis, as Vygotsky claimed (Pascual-Leone, 1996b). Without this postulate of Reality resistances, a general model could not explain how schemes may be useful in outer Reality.

TCO and Cognitive Psychology

The TCO can explain, process analytically, differences across age-bound developmental stages and between facilitating versus misleading processes (or situations). It also explains the necessary complementarity between automatic attention versus mental attention (Matt, the largely maturational constituent of working memory) and can deconstruct mental attention into its four key operators (Matt=<*E, M, I, F*>—see chapter 7). These operators, plus other necessary learning operators (*C, LC, LA, LM*—see chapter 5), can provide causal specific models (via process/task analysis) for the emergence of working memory, thus explicating this unclear but important construct. From the perspective of the TCO, working memory appears as a descriptive construct; it is just an estimate of mental attention after the accumulated effect of prior learning is added (schemes' repertoire, including executive schemes). In contrast, TCO gives an organismic account of how mental attention evolves, from infancy to old age, changing performance and producing, among others, the complex symbolic function (see chapters 3 and 4).

TCO also explicates differences and complementarity among various dialectical pairs of complex descriptive constructs: problem solving versus constructivist learning, or developmental/maturational intelligence versus adult psychometric intelligence, or levels of reflective abstraction versus maturational levels (epilevels)—the latter explicating proper stages of cognitive development. These constructs can in turn explain dialectical complementarity of perception versus learning versus cognition and of cognitive versus affective versus personal, psychosocial processes.

The TCO can help to integrate adult-experimental with developmental psychology, and cognitive with clinical/personality psychology. It can clarify motivation, organismic-measurement of developmental intelligence, and how babies psychologically evolve into adults. Referring to our neo-Piagetian work, Barrouillet and Gaillard (2011, p. 270) said, "Neo-Piagetian theories have the potential to account for most of the learning difficulties and developmental disorders by cumulating the strengths of the functionalist and the structuralist approaches." Some working-memory researchers now support maturational growth in mental-attention capacity (which they conflate with WM), as illustrated by Cowan's experiments in the last ten years (e.g., Cowan, 2016; Cowan, AuBuchon, Gilchris, Ricker, & Saults, 2011; Cowan, Ricker, Clark, Hinrichs, & Glass, 2015).

Limited mental-activation capacity is recognized in meta-empiricist psychology but often reduced to a maximum of four to five units. People may not always mobilize all their capacity, and so four to five may be mental-attentional units some people commonly use. However, these reduced capacity-estimates, claimed by researchers, may have a different reason: researchers do not recognize the total mental demand in a task, because they fail to count operative schemes or procedures needed to manipulate objects. Barrouillet and Gaillard (2011, p. 264) acknowledge our method: "This conception [the TCO's task analysis] consists in taking into account both the figurative and the operative aspects of the cognitive functions." Researchers on working memory have recognized this problem, which we first formulated (e.g., Pascual-Leone, 1978, 1984, 2000a; Pascual-Leone & Johnson, 2005, 2011; Pascual-Leone & Sparkman, 1980).

Reluctance in accepting fully our mental model and measures may be caused by the need to conduct task analyses and quantify mental attention. A task analysis "from within" requires change toward an empirical-rationalist (constructivist) perspective. Nonetheless, the broad impact of our task analytical method can be recognized (e.g., Arsalidou, Pascual-Leone, & Johnson, 2010; Arsalidou, Pawliw-Levac, Sadeghi, & Pascual-Leone, 2018; Burtis, 1982; Case & Okamoto, 1996; de Ribaupierre & Bailleux, 2000; Greenberg & Goldman, 2019; Morra, 2000; Niaz, 2006; Antonio Pascual-Leone, Greenberg, & Pascual-Leone, 2009, 2014; Tsaparlis, 1998).

TCO and Neuroscience

TCO constructs and methods provide a heuristic tool to study organismic-causal significance of brain networks. Much research on brain networks uses a variety of tasks to study elicited brain activity. These tasks are typically analyzed using only intuitive common sense, which often is inadequate. In contrast, our process/task analyses construct metasubjective (from within) models of subjects' semiotic activity in a task, and such models could semantic-pragmatically explicate task-related patterns of brain activity. Metasubjective models clarify meaning of activated circuits. These semiotic analyses are illustrated in chapter 11 and could advance research on the brain (e.g., Arsalidou & Pascual-Leone, 2016; Arsalidou et al., 2018; Guevara, Arsalidou, Pascual-Leone, & Stevens, 2019). TCO can help to interpret brain processes, differentiating those of cognition versus perception, versus affects or emotions, versus motivation (Arsalidou & Pascual-Leone, 2016; Arsalidou et al., 2018; Pascual-Leone, 1987, 1989, 1995; Pascual-Leone, Pascual-Leone, & Arsalidou, 2015).

A person's processing limitations in cognitive-complexity (within and across domains) can be investigated in neural networks. Also, because mental-endogenous attention (*M*-capacity in particular) is measurable across content domains (Pascual-Leone, 2019; Pascual-Leone & Johnson, 2011), units of *M*-measurement can be compared across brain areas and domains.

TCO and Individual Differences or Cognitive Styles

Adopting a causal-organismic constructivist perspective, TCO modeling pioneered a from-within process-analytical investigation of individual differences and cognitive styles. Such pioneering of a "new approach to individual differences" was recognized by Case and Edelstein (1993, p. 5), who wrote referring to our analyses of cognitive conflict and cognitive style (field-dependence-independence): "The nature of this cognitive conflict, Pascual-Leone asserted, is the same as the conflict elicited by Witkin's classic embedded figures test." Case and Edelstein (1993, pp. 5–6) concluded their comments as follows:

> Pascual-Leone was able to predict the pattern of individual differences across a remarkably broad variety of Piagetian and psychometric tasks (Pascual-Leone, 1969). In the general theoretical system that he evolved, what was seen as universal about development was its inexorable move toward greater complexity [of processing], as a result of biennial increases in mental power (*M* power). What was seen as individual about development was the particular situations to which different children were sensitive, and the particular styles or strategies that

> they evolved for dealing with these situations. Field independence was just one of the styles Pascual-Leone studied. Others were *adaptive flexibility*... and *reflectivity*.

These authors appreciate our theoretical approach regarding individual differences (ID), which expresses differentials (functional specific differences) in various categories of organismic processes, yielding specific ID-formulas of metasubjective processes not emphasized in this book. The approach has had an impact on the (organismic) construal of ID variables, as illustrated by field-dependence-independence (ID cognitive style described by Witkin, e.g., Witkin & Goodenough, 1981; Witkin, Dyk, Faterson, Goodenough, & Karp, 1962). Messick (1994) explicitly sided with our conflict-analytical interpretation of FDI (Pascual-Leone, 1969, 1989), emphasizing that ID-cognitive styles were not competences or skills but dispositions (person-based unconscious probabilistic tendencies). That is, they are tendencies to adopt one of two contradictory perspectives: either pursuing *reflective cognitive restructuring* (which demands executive-driven mental attention) or reacting with *spontaneous field sensitivity* to salient aspects of experience (content-schemes sensitivity and the neo-Gestaltist field factor). As Messick (1994, p. 127) put it, echoing our TCO model, "Field independence as restructuring skill versus sensitivity to field effects." In field-independent (FI) persons, restructuring (driven by executive and mental-attentional processes) tends to be stronger; in field-dependent (FD) persons, the field-sensitivity factor (mediated by *C*- and *F*-driven processes) tends to dominate, controlling/reducing the restructuring factor (Pascual-Leone, 1989).

However, Messick (as well as Witkin & Goodenough, 1981, who also accepted our FDI formulation) points out that experimental psychologists tend to misconstrue cognitive style as a skill or competence (e.g., spatial ability). This miscontrual contradicts findings suggesting that FI and FD people are functionally different in their favored brain processes: FI people exhibit spontaneous propensity to favor use of the left hemisphere, whereas FD people have the propensity toward the right hemisphere (e.g., J. Levy, 1980, 1983). Goodman (1971) in Pascual-Leone's lab (see Witkin & Goodenough, 1981, who summarize these data) found that FD adults were much worse at disambiguating meaning of ambiguous sentences than were FI adults, thus establishing that FDI is not just a spatial structuring skill. This important finding was replicated by Lefever and Ehri (1976). Mindful of the need for causal-organismic ID research in psychology, Dr. Michael Shayer (Kings College, University of London, UK; personal communication to JPL, 2014) writes, "Another aspect of your work *should* have found its way into the education literature: that is the relation between field-dependence and your *M*-measures."

The impact of our cognitive-style TCO research, done in connection with Pascual-Leone's lab (e.g., Goode, Goddard, & Pascual-Leone, 2002; Johnson, Prior, & Artuso,

2000; Johnson & Rosano, 1993; Pascual-Leone, 1969, 1989) has been felt mostly in Europe and Latin America.

TCO and Education

Our metasubjective model of children, sensitive to individual and style differences, could be useful in education. It can help teachers to design plausible educational interventions that are different for distinct sorts of children or adults. Michael Shayer raised this issue in a personal communication (2014): "For me, in the educational intervention field, the key contribution of your work would be showing that cognitive demand of a learning task can be estimated without the backing of empirical testing of the students' success on it: that is, using an independent theory (and MTA)." By MTA Shayer means our metasubjective task analysis.

Some of TCO's influence on education and educational interventions was mediated by mentees of Pascual-Leone, such as Robbie Case (1985, 1992, 1998; Case & Okamoto, 1996), Anik de Ribaupierre (de Ribaupierre, Fagot, & Lecerf, 2011; de Ribaupierre & Lecerf, 2006; 2017) and Olivier Houdé[2] (1995; Houdé et al., 2011). Case and Edelstein (1993, p. 5) said, "The first step in this direction [i.e., specific structural analysis of individual differences relevant to Education] came from the work of Pascual-Leone (1969). In addition to carrying out more local process analyses of individual tasks, Pascual-Leone began to group the results of his analyses into broader categories."

A new line of work in science education influenced by the TCO was initiated by Niaz (1995, 2006; Niaz & Robinson, 1992; Tsaparlis, Kousathana, & Niaz, 1998), inspired by our theory and task-analytical method. Much research in science education has been directed to show relationships between science achievement and students' mental-attentional capacity, including analysis of logical-structures and their *M*-demand in science problems (e.g., Niaz, 1992; Stamovlasis & Tsaparlis, 2012; Tsaparlis, 1998; Tsaparlis, Kousathana, & Niaz, 1998). Tsarparlis and Angelopoulus (2000) have shown that there is a line of British research in science education initiated by Johnstone (Johnstone & Al-Naeme, 1991; Johnstone & El-Banna, 1986, 1989) that uses one of our *M*-measures. These authors have used our figural intersections task and have adopted (although with too much simplification and lack of explicitness) constructs similar to participants' *M*-capacity versus tasks' *M*-demand. Research in our own lab has investigated the role of *M*-capacity in children's understanding of multiplication (Agostino, Johnson, & Pascual-Leone, 2010; Pascual-Leone, Johnson, & Agostino, 2010) and use of genres in writing (Balioussis, Johnson, & Pascual-Leone, 2012).

TCO, Psychotherapy, and Emotional Change

Our work on understanding psychological processes has had an impact on psychotherapy research. Theories of clinical practice try to explain human emotional change, in terms of processes (therapeutic insight, self-knowledge, behavior therapy). Contributions to psychotherapy research initiated by Rice (1974) and greatly expanded by Greenberg, founder of emotion-focused therapy, use a dialectical-constructivist approach and methods inspired by us (Greenberg & Goldman, 2019; Greenberg & Pascual-Leone, 1995, 2001; Greenberg, Rice, & Elliott, 1993). This therapy relates to the TCO in its interpretation of affects, emotions, cognition, and method of task analysis.

To illustrate how TCO may contribute to ground theory in psychotherapy, consider Greenberg's key principle in therapy: "Changing emotion with emotion." We all have different modes/styles of experiencing and of mentation (the working mind). One is purely cognitive, another is personal (cognitive and affective)—the mode where emotions emerge. To have therapeutic effect, real human encounters (both experiential and existential) must engage actual schemes of original personal concrete experiences or problematic issues (the schemes of relevant affects and emotions re-enacted). To achieve such personal encounter, client and therapist should direct their mental attention (Matt=$<E, M, I, F>$; see chapters 7 and 10) toward a mode/style of processing embedded in personal and empathic re-experiencing, very different from cool associative remembering. These therapy-relevant intuitive personal schemes may initially be weak and need boosting with Matt to re-experience and synthesize new insights about old disturbing issues. New personal schemes formed in the human-relational (client and therapist) encounter may lead to better emotion control (via self-monitored overdetermination and reflective abstraction), and the soothing personal presence and relational modeling by the therapist could aid clients to use their attentional *M*- and *E*-power to reconstruct empathically the original emotion situation. "Changing emotion with emotion" is to rework original problem situations with open mental attention and a clear personal mind (Greenberg & Goldman, 2019). Our theory's working-mind modeling could also be useful to other psychotherapies. Despite different theories of change, all clinicians see clients with varied but related problems; metasubjective modeling of the therapy process could be useful to many sorts of therapy (e.g., Antonio Pascual-Leone et al., 2009, 2014).

TCO, Consciousness, and the Self

The TCO can contribute to understanding of consciousness as we know it, primary consciousness and self-consciousness. Paraphrasing the language of Edelman (Edelman, 1989; Edelman & Tononi, 2000) consciousness is caused by groups of highly activated and closely interconnected neurons (functionally integrated by many reentrant corticocortical and corticothalamic dual links) that, in each act of consciousness, produce a momentarily integrated *dynamic core* of activated informational processes (i.e., schemes), causing the conscious act. These core schemes are heterarchically organized with relatively high-level cofunctional neurons, which carry much information complexity (Edelman & Tononi, 2000). Similar conceptions of consciousness are found in other current neuroscientists and psychologists. Damasio (2012) and Dehaene (2014) agree with this view of Edelman, as we and many others do (Pascual-Leone, 2000a). On a first approximation (Pascual-Leone, 2000a), consciousness is the aptitude of an organism to have a distinct representation of some constituents of its own mentation (perception or cognition), feeling (affects and emotions), or willing (deliberate intentions). Damasio (2012, p. 167) has defined consciousness as "*a state of mind in which there is knowledge of one's own existence and the existence of surroundings.*" Demetriou and Spanoudis (2018), using the term *cognizance*, have emphasized this intentional or causal (operative self-agency) aspect of consciousness.

Even in primary consciousness (where the self is only implicit) intentional and causal aspects can be found. In self-consciousness this is much more so. The self is always, implicitly or explicitly, at the core of consciousness. As James Mark Baldwin (1894/1968) first claimed and we all accept today, a coordinating core cluster of personal (affective and cognitive) self-schemes emerges, as self-referential functional invariants, in interaction with others and the world. When a dynamic core of consciousness is constituted by highly activated coordinating self-schemes, the self may appear in introspection—as William James emphasized; particularly the Me-self of James, that is, the person's figurative or representational self.

This Me-self appears early, in the preschool period (4 to 5 years). Our grandson, Jasper, in a surprising utterance to his father Antonio, illustrates this well. Antonio described the following exchange, when Jasper was close to 5 years old. In a dispute with his dad about nap time, Jasper suddenly retorted, "You're not the boss of me!! The inside ME is the boss of me! The inside of you thinks that you're the boss of me, but you're not!" Damasio (2012) localizes processes for this representational self in the posteromedial cortex (posterior cingulate, retrosplenial cortex, precuneus). However, there is also an operative self (James' I-self). We hypothesize this operative self to have a processing "centre" in the medial prefrontal cortex, possibly including the anterior cingulate gyrus.

Notice that these constructs of an operative self and a figurative self are not in contradiction with the Buddhist or the current psychoanalytic insistence (Safran, 2003) that the self tends to become a *no-self* and functionally disappears from phenomenological expression as persons mature (with or without meditation methods) and approach wisdom. The I-self (operative) and the Me-self (figurative) are not just phenomenological (i.e., subjectively descriptive) but are also causal-organismic—metasubjective factors in executive control. As psychological maturity and wisdom progress, these self-executive processes persist, albeit hidden, and they mature toward a low-key non-self-subjectivity/phenomenology, often providing easier-than-before empathic identification with others.

It is this causal-organismic I-self that common sense considers an agent of volitional, intentional activities (initiator of good or bad praxis). The *unconscious* (the manifold repertoires of schemes, together with hidden operators and principles) exists, but it is not a main cause of intentional behavior. Conscious intentionality, willful activity, is mostly caused by the operative self (I-self)—our repertoire of self-referring executives, action schemes, and schemas. This operative consciousness can be causally effective, because of its dominant, high level and encompassing, schemes (Edelman's dynamic core), which often are boosted by endogenous/maturational mental attention.

This is the model of intentional consciousness that our book implies, radically different from other current theories of consciousness. Edelman and Dehaene do not have an independent activation-boosting function for schemes and schemas as found in mental attention (i.e., *M*). They only seem to recognize language, and a limited unexplicated working memory, as organizer and energizer of consciousness. There is now some consensus that this limited aspect of working memory may be caused (e.g., Buzsaki, 2006) by limits in the temporal conjoint holding of schemes, which may be expressing limits in the coupling of slow (e.g., theta) with fast (e.g., gamma) brain waves (see chapter 11). This coupling would keep separate the gamma waves (which express schemes) and enable the schemes to remain both distinct and coactivated. However, to explain the causal power of intentional consciousness there is also need for an independent scheme-boosting function (*M*); otherwise, particularly in misleading situations, task-relevant schemes with low activation level could not be boosted enough to be part of the dynamic core of consciousness, the functionally dominant schemes included in the couplings. Damasio (e.g., 1999, 2012) seems to suggest that bodily affects, *primordial feelings*, may be the tacit scheme-boosters of consciousness. This is indeed important, but it would not suffice in misleading situations, because affects are often misleading in these situations; intentional consciousness must have a separate mental-attention booster if it is to attain control in difficult situations.

In chapter 3 we presented four cumulative developmental stages of emergent consciousness (see also Pascual-Leone, 1990a, 1990b, 2000a, 2000b; Pascual-Leone & Irwin, 1998). These developmental stages complement views of Damasio and other non-developmental modelers of self and consciousness. We also introduced, via task analysis, an explicit causal-organismic developmental perspective—predicated on maturational growth of mental attention.

Our model has four initial cumulative stages of consciousness, followed by other stages after self-consciousness has been achieved (refer to chapter 3 for details): *sentience* (from before birth), *protoself* (at about 8 months of age), *self1* (at about 18 months), and *self2.1* (at about 35 months). In sentience, infants can feel external stimulation without having a distinction between object and self-subject (Pascual-Leone, 2000a). In protoself, consciousness of the object (but not the self-subject) appears; in self1, there is experience of the object as well as the own self-subject, distinct and physically separate from the experienced objects. In self2.1 (at 3 years of age), the child is now distinctly aware of the object, the self-subject, and of a minimal reflective (operative) self that can intuitively contemplate self-subject and distal object co-existing in the mind. In self2.2 (at 5 years of age), the growth of mental attention allows self-consciousness to evolve into Damasio's *autobiographical self* (Damasio, 2012). The person can now reflect about her or his own life, consciously recollecting memories. Children can now recall three symbolic schemes/schemas concurrently: the object of experience, their self-subject (the one who knows), and their own reflective inner self (both I-self and Me-self) that can contemplate object and self-subject as distinct and coexistent.

Up to adolescence *M*-capacity continues growing and brings more activation energy and greater effective complexity to self-judgment and the autobiographic self, as chapter 3 explained. Damasio does not mention these developmental aspects. Later developments of consciousness are caused by growth of reflective abstraction via constructivist learning, which could increase experience and wisdom. Consciousness (including inner self) is regarded by common sense and many scientists (including William James) as a key causal determinant for planning and good reasoning.

The TCO explicates these constructs: consciousness results from strong activation of high-level and informationally complex schemes (*self-schemes*), which are often executives. These self-schemes carry multiple relevant meanings, are boosted by mental attention, and are served by progressively (developmentally) stronger knowledge schemas. Operative consciousness relates to frontopolar activity in some complex tasks, and to both frontomedial cortex (I-self) and posteromedial cortex (the Me-self—see chapter 9 of Damasio, 2012), it can serve as supervisory system (highest executive level of control) of ongoing mental processes. This characteristic has been interpreted as a mental "global workspace" or "blackboard" accessible to other cortical processes (e.g., Dehaene, 2014).

TCO and the Will

The Will exists when the self (that dynamic core of operative and figurative self-schemes) is able to judge, decide, and act intentionally on personal decisions. Freedom of choice exists if the person can judge, mindfully decide, and act congruently with the decision.

Judgment is essential in acts of Will (psychologically, judgments are dynamic syntheses that appraise alternatives, driven by conscious processes, executives, the self, and mental-attentional effort). Unconscious automatisms or overlearned habits (shadow schemes) do not override freedom when the Will (inner self) is strong enough—when people are project driven and responsible. Common sense agrees with this conclusion, but some psychologists (e.g., B. F. Skinner), neuroscientists, and philosophers think of humans as refined versions of animals of habit and associative learning. These researchers doubt the scientific reality of Will, self, and freedom. It is possible to justify Will and the inner self scientifically. We shall not review this literature, to which we made a small contribution (Pascual-Leone, 1990a, 1990b). Neuroscience is beginning to see where these hubs of scheme-coordination and structuring are found in the cortex (e.g., Damasio, 2012; Edelman & Tononi, 2000).

Jaspers, the distinguished existential philosopher and psychiatrist, maintained similar views when he wrote, "Volition is a *relation to oneself,* a self-awareness in which I am the object of my active conduct rather than my contemplative scrutiny.... as Kierkegaard put it: 'The more will the more self'" (Jaspers, 1970, p. 135). Thus, power of the Will stems from the person's *operative self,* related to the use of mental attention to boost task-relevant complex operative schemes/schemas driven by executive schemes. As William James (1892/1961) wrote, "*Effort of attention is thus the essential phenomenon of the will*" (his own italics, p. 317).

Arguments against Will and against freedom (B. F. Skinner, 1971) show that the key issue is not so much existence of Will but its capacity to overcome many overlearned bad habits, addictions, unconscious misleading dispositions, and automatic social conventions (i.e., shadow schemes and what has been called *outer self* or *persona*). These are tendencies in the person that may oppose/disrupt willful plans and cause failure. Unless we can explain how truly novel (creative) willful achievements can be reached against shadow schemes and other misleading factors (which established theories often fail to do) the existence of human freedom may be denied by science, contradicting key assumptions of democratic societies. The TCO explains Will and freedom by recognizing, in its theory and in brain processes, constructs and sites for executive processes, mental attention (Matt), the inner self (operative I-self and figurative Me-self), as well as the outer self or persona. TCO has constructs and methods of process/task analysis to identify relevant neuropsychological constructs by interpreting neuroscience.

The Matt system can boost task-relevant schemes and inhibit irrelevant ones, and after training by others and via self-reflection, it can lead the inner self to seek new social contexts and social environments (suitable new milieux) that could compel the person (even against bad habits and dispositions) to acquire needed missing schemes good for the willed project. Complex professional careers of all sorts proceed willfully in such intentional trial-and-error. Affects, emotions, life projects, the willful inner self explain (together) the seeking and choice of new milieux that appear project-suitable and could develop the person (outer and inner self) as was intended by the inner self—the milieu offering opportunities for learning, social networking, and acquisition of suitable skills.

The active working of a primitive but willful self can already be found in infancy. A good example was given in chapter 2, in the section The Meaning of Meaning. Piaget's (1948/1963, p. 249) Observation 133, about his 9-month-old daughter, Jacqueline, illustrates this intertwining of affect and incipient willful self-choice in young infants (when the Will is likely boosted by sensorimotor mental attention). Talking about her, Piaget said, "She likes the grape fruit in a glass but not the soup in a bowl.... When the spoon comes out of the glass she opens her mouth wide, whereas when it comes from the bowl, her mouth remains closed" (see discussion in chapter 2). Thus, Jacqueline acts intentionally, using circumstantial cues, to assert and bring about her "project"—her choice of food.

We can briefly illustrate more advanced self-heuristic strategies by describing how an adult (JPL) willfully changed his life from being a medical psychiatrist to becoming a research psychologist and co-author of this book. This was done with a succession of conscious decisions that used circumstantial cues from the milieu's situational conditions (as Jacqueline did) to progressively (with some good luck) transform his life and himself in the chosen/intended direction that he envisioned in his life project. JPL was a young psychiatrist in Spain during the 1950s, who (right or wrong) felt psychiatry often lacked analytic insights into the person's (inner self) feelings, thinking, and causal individuation or developmental processes. JPL's overambitious life project was to help correct this deficiency. He reckoned that working with Piaget (an author discussed in a significant psychiatry handbook he had studied) could help advance his project. He sought and obtained a Spanish fellowship to study with Piaget in Geneva, Switzerland. This four-year experience made him a Piagetian developmental research psychologist. Then, recognizing the importance of individual differences and styles, he discovered (in psychiatry books and with advice from Inhelder, Piaget's key collaborator) the work of Herman Witkin in New York. He wrote to Witkin, who obtained a fellowship for JPL to spend the following year (1964) in New York. Thus, in 1964 he worked in Witkin's New York lab and tested (with help from three lab assistants) New York children and adults for his doctoral thesis.

This experience made JPL knowledgeable in cognitive-style and individual-difference research, including psychometric intelligence, which he related to Piaget's developmental intelligence. He was now a true cognitive-developmental researcher with original ideas. When, after a year, he was forced to leave the United States because of his visitor's visa, he decided to take a job in Canada—to continue using English, because his New

York data had to be analyzed. Two jobs were offered to him: one as psychiatrist in a hospital in Quebec, the other as assistant professor in psychology at the University of British Columbia. He took the latter (with less pay) to advance his knowledge of North American psychology, postponing a return to psychiatry. JPL was naive; he never returned to psychiatry. Four years later he was associate professor of psychology at York University and was doing research on which this book—more than fifty years later—is based. As in Jacqueline, but with more complexity, acts of Will to change milieux, and the milieux reactions (and some amount of good luck) brought JPL to be the one that his inner self and Will had foreseen—via willful existential trial and error.

At every turn, Jacqueline and JPL had been confronted with decisions about what to do next; they consistently decided in a direction closer to their own preferences or intended goals. Thus, good advice, a bit of luck, and many acts of Will (Matt-operations, executive-driven, and motivated by affect-loaded long-range plans) can bring willful goal-seekers to new milieux that progressively help to change the person and approach his or her life choices.

Brief Conclusions

In this final chapter we have examined epistemological aspects of our theory, the TCO, and compared its constructivist (empirical rationalist) approach to that of Charles Darwin's theory of evolution, to contrast differences between constructivist (empirical rationalist) science and empiricist science. Then we briefly discussed contributions that TCO and its methods could make to relevant areas of human science: cognitive psychology, neuroscience, education, psychotherapy and emotional change, consciousness, the self, and the Will.

For the purpose of brevity, we have not summarized the book, because the abstracts of chapters already do so. Instead, we re-emphasized constructivism in the context of two very distinct sciences, biology and psychology, following up the reference to Darwin's theory (Edelman, 1987). We hope to have given pointers to how TCO could be used as tool for thinking and analysis across domains, including education, where causal-organismic theories are still rare and not often available. Kurt Lewin, a Gestalt psychologist and eminent researcher on human relations, thought that there is nothing as practical as a good theory. Jean Ullmo (a French constructivist epistemologist and physicist) reached the same idea in a more concrete manner: "An experiment is a theory in action; an instrument, a materialized theory; a scientific concept (the sun, the atom of carbon) an objectified theory" (Ullmo, 1967, p. 657, translated by JPL).

Appendix: On Metasubjective Task Analysis (MTA)

Throughout the book we present process/task analyses. These are operative models of thinking processes, done "from within"; that is, from the perspective of thinking processes themselves as the person copes with a task. A *metasubjective task analysis* (MTA) is an explicit model representing these analyses. Such models facilitate attention during analysis and lighten analysts' working memory load—as readers who use the notation will notice. Our logical notation for MTA is analogous in its qualitative function to what high-level programming languages do, or what notational calculus does for mathematical-physics analysis, although our notation is informal and its meaning is qualitative and concrete.

We represent sets and sequences of schemes of various sorts, which represent how task processes reach for a solution or attempt one. The task analyst in MTA must know concretely the sort of task analyzed to provide suitable representation and interpretations. A superordinate *operative* (OP) applies to *figuratives* or *operatives* placed inside the adjunct parentheses OP(...); parentheses demarcate scope of application for the superordinate operative they are attached to.

In any MTA formula, active concrete processing during the task begins with the *rightmost operative processes*, represented in the formula (proceeding from right to left in the formula's sequence). This is so because rightmost schemes in the formula are the most concrete, whereas leftmost schemes are more abstract and usually superordinate. This right-to-left order of applying operations during task-solution processes complements the convention that all *operative schemes in the formula apply to the schemes to their immediate right* (schemes within their scope of application being marked by the parentheses enclosure). Whereas the control function of schemes by other, more abstract, superordinate schemes moves from left to right in the formula sequencing, the abstraction level of schemes moves in the formula from the right to the left direction—schemes to the right being more concrete. When a rightmost process has been completed, the resulting subjective process (symbolized in the formula's complex) will tend to be retained by the subject and held in his or her mind (within mental attention—i.e., *M-centration*), as the next-left operative process is enacted during the task. This continues until, from right to left, the whole formula has been implemented to produce an action/representation result (good or bad).

We explain here key notational conventions used in the book. We define various optional symbols that can be used in MTA formulas. Analysts could design and use other symbols; these are the ones we have used. MTA formulas are best kept as simple as possible, and many symbols described below are used rarely. To give these rules more generality, no specific content schemes are used in this presentation. Notational rules are presented generically on types of schemes (OPERATIVE or OP, figurative or fig1, fig2...etc.). Unrepresented schemes are simply replaced by suspension points. In these definitions, *X* and *Y* stand for schemes or scheme-formulas of any kind. The sign (:=) means "equal by definition."

Task-Analytical Notation

OPERATIVE := Operative scheme; uppercase notation is used to name any operative.

OPERATIVE(...) := Operative schemes always apply on figurative, operative, or parameter schemes that are to their right and within the scope of the parentheses next to them (suspension points stand here for schemes). For operative schemes, parentheses demarcate scope of application.

figurative or ***figurative** or **figurative*** := figurative scheme. Lowercase notation is used to signify figurative schemes (of any sort and complexity). One can also pre-superscript or post-superscript figuratives with an asterisk * when wishing to emphasize the figurative type. These three notational forms are equivalent.

$^{fl\text{-}}$**OPERATIVE**, or ***OPERATIVE****, or* $^{fl\text{-}}$**figurative**, or ***figurative*** := fluent scheme. We use a prefixed $^{fl\text{-}}$ or, more commonly, use italics to represent temporally structured schemes (operative or figurative), which we call *fluents*. Fluents embody temporal change and produce *expectancies*—figurative or operative anticipatory expectations. Expectancies (fluent schemes) function as figuratives or operatives depending on the context.

#OPERATIVE, **#figurative**, **#*OPERATIVE***, **#*figurative*** := parameter. Parameters are schemes that indicate conditions under which the operative scheme to their left should be applied on its figurative(s) to achieve intended results. Fluents (temporally structured schemes) and other relational schemes can function as parameters. We symbolize parameters with a number sign (#) placed before the scheme.

^{E}OPERATIVE := Executive scheme. Anterior E-superscript indicates that the operative in question stands for an executive.

***figurative[...]** or **OPERATIVE[...]** := The content inside square brackets specifies semantic aspects of the scheme to which the brackets attach (i.e., schemes inside the brackets are related to, or part of, the scheme characteristics). We represent *adjunct information* relative to a figurative or operative by affixing to the scheme this information enclosed in square brackets. For instance, in the state of affairs described by the sentence, "Boy who dogs chase feeds cat," the relative clause (i.e., "who dogs chase") is adjunct information referring to the boy, which we can represent as *boy[CHASE(*dogs, *boy)].

$figurative or **$OPERATIVE** := The scheme is true, i.e., concretely real (examples: $dog, $cookie-grasped). This meaning of $ applies to any sort of scheme. It indicates that the scheme, here-and-now, is *perceptually and cognitively true* and concrete in its scheme application.

λfigurative or **λOPERATIVE** := The lambda prefix indicates that the figuratives or operatives are *language schemes*, expressed by words or phrases. They may or not have a concretely real meaning referent (then they could also be marked by prefix $).

figurative[K: ...] or OPERATIVE[K: ...] := The symbol **K** here stands for knowledge and indicates that the schemes that follow (i.e., ...) are activated in the mind, represented in consciousness as mentation, but are not applied to any actual situation or action. They are mental without perceptual or motor applications of schemes. For example: Sally[K: choco : Loc1] or Sally[NOT.K: choco : Loc2]. In English, these translate as: "Sally knows that chocolate is in Location1," or "Sally does not know that chocolate is in Location2."

X.Y := X is semantically related with Y in the current context (schemes of any sort).

X :: Y := The scheme X precedes and (in the present process) leads to scheme Y (i.e., X and Y are an ordered pair with a causal link).

X : Y : Z… := Schemes X, Y, Z temporally follow one another and may be coordinated. They can also be used to mark distinct dimensions of variation or aspects, with various parameter values, within a functional totality (i.e., a coordinated set of schemes in a hierarchical or complex application). Names of the dimensions/aspects are often omitted. For instance in: "PLACE(Anne : chocolate : Loc2)." This means "PLACE(*Who*Anne: *What*chocolate: *Where*Loc2)."

X:.Y := X controls Y. Thus when X is a figurative scheme, it relates semantically to Y; when X is an OPERATIVE, it applies on and controls Y.

X ← Y := Scheme Y leads to the local emergence of scheme X, or Y brings attention to X within the task process, perhaps because X is precondition of Y (or is otherwise related to it).

X… → Y := The thin arrow (→) indicates results of the locally preceding (X…) operation. With a similar meaning we can also use (:>). For instance, X… :> Y.

OP(#fig1, OP2(fig2), …) := The comma (,) separates schemes (parameters, operatives, or figuratives) that, within the scope of a given operative OP, are distinct and locally independent.

XX ➔ Y := The formula of schemes XX produces, by the principle of Schemes' Overdetermination of Performance (*SOP*), the response or outcome Y.

M [OP (#scheme-1, *scheme-2, *scheme-3, …)] ➔ scheme-n := Any underlined scheme within a formula (here *M*[…]) is being boosted in its activation by the *M*-operator, which applies on the schemes *enclosed by square brackets*. Thus operative scheme OP, parameter #, and figurative schemes *…have boosting activation by *M*. In most of these formulas, *M*-operator and its square brackets (*M* […]) are omitted for simplicity; but operatives, figuratives, and parameter(s) boosted by *M* are underlined to indicate *M*-boosting.

M [OP (#scheme-1, *scheme-2, *scheme-3, …)] ➔ **$RESPONSE** := The correct, true overt RESPONSE (i.e., the task-appropriate overt response) is symbolized with a prefix $ to mean *in truth a correct overt Response* (or perhaps a correct external referent). When the overt Response in question (or external referent) is inappropriate/*false* (i.e., there is mismatch between scheme's expectancies and real effect or referent) we would instead prefix the *false* negative sign ^. For instance, ^RESPONSE.

{.... *M* [OP (#scheme-1, *scheme-2, *scheme-3, ...)] ➔ scheme-n } := Braces {...} without a posterior subscript stand for the *field of activation*—that is, all activated schemes in the person's total repertoire (activated long-term memory). Here suspension points stand for activated schemes that are omitted. This representation of a task is useful when one wants to present several strategies that are either collaborating or competing to control performance. In this case, outside the M-centration (or *M*-space), that is, outside *M*[...] but within { YY...*M*[...]}, one would present other schemes YY that stand for alternative strategies. We can signal whether strategies are competing with the main strategy because they are false (prefixing *false* sign, e.g., ^YY), or are compatible with it (prefixing a *true* sign, e.g., $XX).

OPERATIVE^{L1} := An *L-superscripted* scheme (whether operative, figurative, or expectancy/fluent) is *L-structured* (overlearned, chunked) with another scheme that has subscripted the same *L1* or *L2,* and so on. In this case, the first (*L-superscripted*) scheme is boosting the activation of the second (*L-subscripted*) scheme. This boosting allows the second to be highly activated without using power of mental attention (*M*-boosting). Thus *L*-operator (not the *M*-operator) would be boosting the scheme. Presence of a question mark (*L*?) next to the second scheme (subscripted *L*?) indicates that this *L*-boosting is uncertain and may not exist.

{....}$_{L1,\,F}$:= Braces with subscript indicate that enclosed scheme(s) is(are) not being boosted by *M*, but boosted instead by operators posteriorly subscripted to the braces—in the example *L1* and *F*. When braces would contain only one scheme, they can be omitted and superscripts attached to the scheme, i.e., scheme$_{L1,\,F}$.

{....}$_{Sit,\,F}$:= Subscripts *Sit* and *F* indicate that schemes inside the braces are not boosted by *M* but by local factors in the current situation (*Sit*) such as sensorial/perceptual salience or automatized perception. *Sit,* as local cause or boosting factor, would include sensorial-field factors (F_s) but exclude *intellective* cognitive-field and mental factors represented as *F, M, LM, E,* and so on.

Figurative!, **Figurative**!, or ***Operative***! := An exclamation point (!) postfixed to a scheme, perhaps a fluent, indicates that the expectancy attached to this scheme has been violated. For example, the hand with a toy goes behind screen **b** and then reappears without the toy: {*toy-in-hand-TO-**b***}^{L1,Sit} ← {no-toy-at-hand-after-**b**!}$_{L1,Sit}$. In this example, two schemes (demarcated by braces {...}) are activated in an empirically reverse sequence (←), which exhibits violation of the initial scheme's expectancy. We symbolize this expectancy-violation by ! because subjects should experience surprise when toy-in-hand went to b but toy does not reappear behind b as expected.

X…<::> Y… := Strategies X…and Y…are in dialectical/contradictory relation and may be in conflict or compete, that is, be misleading to each other.

<: x: y: z :> := **x, y, z** constitute a dynamic dialectical system with three components, each dialectically coordinated (: :) with the other two.

^{**figurative[…]**$_I$ }$_{sit}$ or ^{**OPERATIVE(…)**$_I$ }$_{sit}$:= The negative sign ^ prefixed to schemes inside braces {…}, which have situational boosting factors (*sit*) subscripted to the second brace (i.e., ^{…}sit), means the schemes are not task relevant and are perhaps misleading, although strongly activated by *sit* or by hidden operators. These schemes should be attentionally interrupted (inhibited—our *I*-operator) to solve the task. We have indicated so by subscripting *I* after the second bracket. In this manner, one could highlight misleading characteristics of any situation appearing in a formula. A parameter (#) may also be needed, boosted by *M,* to remind the subject to do this inhibition (subscripted *I*).

<u>Scheme-2?</u> := Scheme-2 may or may not need to be boosted by *M*-capacity (the underline suggests *M*-boosting). This uncertainty is signaled by the question mark. The estimate of *M*-demand should then be a range of values that include or exclude it from *M*-marked schemes. Thus, in this example: *M* [<u>OP</u> (#<u>scheme-1</u>, *<u>scheme-2</u>?, *<u>scheme-3</u>, …)], the *M*-demand would be either 4 or 3, both being plausible.

[[…]] := The content of these double square brackets is an analyst's commentary, description, or statement. Although relevant to the task analysis, it is not part of it (i.e., it does not belong to, but is outside, the metasubjective model description).

Formulas for Mental Strategies

To represent pragmatic relations found among strategies in a given task (e.g., a conflict/misleading situation among schemes that constitute contradictory strategies within the same misleading situation), we place the strategies' name inside brackets (e.g., [(fa) YY], meaning "formula number a, of strategy YY"). We then connect the strategies by a logical connective to indicate the sort of interrelation. For example, <u>V</u> is a logical incompatibility symbol signifying that the strategies activated are in conflict; & is a conjunction indicating compatible co-activation; V is a disjunction indicating compatibility but often separate activation; <::> is a dialectical relation between strategies, and so forth.

Example: ^*F,LC* **[(fa) YY] V** $*E,M,I* **[(fb) XX]**

To the left-side bracket of each strategy we name the hidden operators (organismic resources) that boost activation to schemes of their strategy (in this example, the field factor *F* and the automatized-learning operator or structures *LC* boost strategy YY). To the left of the hidden operators we place a symbol to indicate whether the strategy in question is valid/true for the task (symbolized by $) or is invalid/false (symbolized by ^). The full generic formula in the example is the key analytical expression of a *misleading situation*. All misleading situations satisfy this generic formula (Pascual-Leone, 1989), which is also found in infancy: a false (^) strategy boosted by *F* and *LC* opposed to a true ($) strategy boosted by *E*, *M*, and *I*.

This sort of formula expresses different sorts of relation among strategies. For instance, in *facilitating situations*, where various activated strategies can apply together complementarily to produce the intended (valid, true = $) performance, the logical connective chosen to relate them would be an inclusive disjunction (V). In the example below, both YY and XX could contribute to a correct performance, although XX is more analytical, attentionally effortful, and perhaps more sophisticated. Often the less sophisticated strategies serve two functions in facilitating situations: they can speed up performance via overdetermination, and they can add a personal or psycho-cultural style to actions or representations occurring in the task, an individual-expression possibility that dynamic-analytical psychologies and artists often utilize.

Example: $*C,LC,A* **[(fa) YY] V $***E,M,LM* **[(fb)XX]**

Glossary

abstraction (reflective): This abstraction extracts and learns invariant patterns (types) and individually variable characteristics (tokens) from actual experience, within and across situations. Reflective abstraction often occurs in levels, the upper levels abstracting from lower ones. This abstraction process is reflective in the sense that upper levels mirror (reflect patterns or characteristics of) the lower levels.

afferent versus efferent processes in the brain: Afferent processes are bio-electric information signals that reach the brain from outside (usually from the sensorial receptors); efferent are processes that leave the brain (usually from motor operative effectors) to go to muscles and other parts of the body.

attention (automatic): Automatic attention, often perceptual or overlearned operative processes, is mobilized spontaneously and without effort.

attention (mental, executive, endogenous): Mental attention (Matt) is the brain capacity to enhance or boost activation of some scheme (information-bearing and/or affect) processes. This *mental or effortful attention* (also called *focal* or *executive attention*) is endogenously driven by brain activation processes. Mental attention is needed to boost task-relevant schemes and interrupt/suppress irrelevant schemes (particularly within misleading situations). It is constituted by a functional system of *four hidden operators* (brain resource processes): *E* (executive schemes), *M* (mental-attentional activation booster), *I* (mental-attentional inhibition or interruption process), *F* (internal field factor minimizing complexity of the schemes' application and synthesis/integration). That is: Matt=*<E, M, I, F>*.

centration/recentration/decentration: Centration is an act of mental attention. Recentration is a change in centration that changes content but not level of analysis (often called *executive updating*). Decentration is a change of centration that changes both content and level of analysis (often called executive *shifting* or *switching*).

constructivism (empirical rationalism): A form of epistemology (approach to knowledge) that can integrate *empiricism* (the assumption that all knowledge can be acquired via objective observation of outside things out there) *with rationalism* (assuming knowledge is conditioned by innate structures). Constructivism models processes from within the subject's own mind (called *metasubjective* modeling).

constructivist (or cognitive) learning: The theoretical view that knowledge is not directly innate nor directly acquired from experience, but dynamically constructed via reflective abstraction. This constructivist (dynamic dialectical) learning occurs spontaneously in facilitating situations but needs mental attention in misleading situations.

epistemic/epistemological: Referring to the act of knowing. Epistemology is the discipline investigating basic principles of knowing.

equilibration (regulations, disequilibration): In Piaget's theory *equilibration* is a dynamic-balance system of the psychological organism, which will spontaneously initiate dynamic processes to restore internal equilibrium, when its balance is disturbed by one or another event. *Regulations* are Piaget's name for the sources of equilibration. *Disequilibration* results from (external or internal) disturbances.

field of activation/*F*-factor: The internal *field of activation* is the set of currently activated schemes in the repertoire (long-term memory) of a subject. Coactivated schemes interact to synthesize a minimally complex dynamic resolution. The organismic factor (hidden operator) doing so is the neo-Gestaltist *F*-factor (*F*-operator, which is produced by neuronal lateral-inhibition processes).

figuratives, operatives, and parameters: These are different *sorts of schemes*. *Figuratives* represent objects or entities of any sort. *Operatives* represent transformations, operations that can be applied on figurative schemes. *Parameters* stipulate conditions for successfully applying operatives onto figuratives.

functional structures (e.g., schemes, schemas—affective, cognitive, and personal): Emerging, and later internalized, patterns of structured information or procedures found at a given level of abstraction; they are often reflectively abstracted from the person's own (external or mental) actions. Schemes/schemas are learned (*internalized*) functional structures. As we use the term, schemas are *complex schemes*. Functional structures emerging from performance are not in the repertoire of the person until internalized into schemes/schemas. Schemes/schemas come in four sorts: *cognitive-operative* (procedural), *cognitive-figurative* (which express the *truth/falsity* of experienced reality), *affective* (expressing organismic *vital values—feelings*), and *personal* (expressing both truth/falsity and vital values).

functional totality: The totality of distinct operations (operators) that can bring about change in a given domain of application, while preserving invariant a set of characteristics (the *functional invariant*).

hierarchy, heterarchy: *Functional invariants,* in repeatable situations of variation, *emerge within levels of analysis* (or abstraction) *in which their own structure is clear*. Distinct levels emerge in hierarchies, within the same (or across) content domains. The rank of a level (from low/concrete to high/abstract) is the position it holds in the hierarchy. Heterarchy (or *flexible hierarchy*) is one where ranking of levels can change with the domain of application.

intelligence: The aptitude for solving problems, particularly in novel and misleading situations. Aptitude to anticipate with accurate expectancies (elicited in the situation) that for such purpose or goal a given inference/behavior is good.

internalization/internalized: Term introduced by Vygotsky referring to constructive processes of reflective abstraction (constructivist learning) producing schemes that *epistemically reflect*, often in their own configuration and characteristics, important reality *resistances* and how to control them. Internalization places in the repertoire (long-term memory) new schemes, new *functional invariants* defined at some level of abstraction where their structure is clear.

invariants (functional): Functional invariants are invariants obtained in *repeatable situations* where some aspects are being varied or exchanged. They occur in variations such that at least one set or system of characteristics is preserved, invariantly, across repeatable exchanges. When all possible types of operations that would produce this variation are grouped into a *functional totality*, all these operations constitute a (mathematical) group of variation, whose functional invariant they preserve as a structure.

invariants (types): Invariants are probabilistic patterns of representation, meaning, or action, found to be stable across situations. Peirce and others call them *types* (signs constant across situations that often are symbols). They are complemented by *variants/tokens*—characteristics that change across situations or individuals.

learning (modes): Simple content learning (*C learning*) is substantive-associative and internalizes simple resistances of Reality. Relational-associative learning (*LC*) abstracts simple patterns as well as relational-structural (Logical) learning; with enough repetitions it creates overlearned associative links. Relational-symbolic Logical learning (*LM*) is effortful (uses mental attention, *M-boosting*), producing fast learning and complex schemes. Other sorts of *L learning* use affect (*A*), like *LA* learning. All forms of learning can be automatized: *LC, LCLM, etc.*

levels versus epilevel: A *level of knowing* (*epistemic level*) is a level of analysis that maximizes the clarity of relevant functional invariants or structures. These levels are relative to the knowing process. We call *epilevels* levels that are enforced not just by the act of knowing but also by Reality itself: these are *epistemo-ontological* levels. For instance, *proper developmental levels*, such as sensorimotor versus symbolic (i.e., preoperational, concrete operational, formal operational) *are in fact epilevels.*

logic/psycho-logic: *Logic* is the discipline that studies whether and how knowledge and linguistic statements convey truth, and how to draw true inferences from statements or state descriptions. Truth is the capability of knowledge representations, and of statements, to express information (or fail to express in the case of falsity) that is suitably accurate or appropriate for the object of knowledge. We distinguish two sorts of logic. The first, such as found in *formal logic or mathematical logic*, constitutes a collection of generic systems that represent and analyze thought in a formal (focused on form rather than content or local meaning), categorical, and systematic manner. These generic logical systems serve to formulate possible "objective" relations among meaning-bearing structures of inference in thought and language; to characterize them as true or false.

The second sense of logic is found in *dialectical and other approaches to philosophical/epistemological logic, or psycho-logic* (e.g., Piaget's developmental psychology modeling). This alternative intuitive sort of logic may study temporal (diachronic) emergence and evolution of logical systems in the minds of humans who mature and develop—a sort of constructivist process analysis that is qualitative but probabilistic.

meaning (of an object or situation): All anticipations or expectancies, that is, all schemes, that can be estimated to apply to an object or situation, and all those estimated not to apply.

mediation, mediational processes: Mediation is the aptitude of a process (or physical entity) to serve as functional link relating two other terms (processes, objects). It can mean *external mediation*, such as psychosocial relations that a person has with special others. Mediation can also be *internal*—for example, cognitive or perceptual processes that (inside a working mind) link the here-and-now situation (as represented) with thinking processes or actions for coping with the task. For Kant schemas are a mediation between the mind's abstract dimensions or categories (e.g., space, time, quantity, quality, substance, causality) and the concrete raw reality.

metasubjective analysis or perspective: In addition to *objective* (describing from outside) and *subjective* (describing from within the processes), we have the *metasubjective* analysis or perspective, which adopts a *from-within-process perspective* to model (in a rational reconstruction, not just a description) processes that the subject uses vis-à-vis a task.

misleading/facilitating situations: A situation is *facilitating* when its salient aspects (activated cue-schemes, habits, schemas) make it easy to attain intended goals. A situation is *misleading* when salient situational aspects interfere with, or strongly distract from, intended goals.

model (or mental model): A system of signs/symbols that stands for either complex entities of reality, or procedures (blueprints for action), or both. Mental models are functional structures, and often complex schemes (schemas).

motivation: Is the conversion of affective motives into cognitive goals. Motivation functionally intertwines affect/emotion, cognition, and Reality/reality, which leads advanced animals to *internalize* suitable scheme packages and their interdependencies, thus learning the environment's *causal texture*.

objects, proximal and distal: An object is a coordinated functional totality (a "package" or grouping) of characteristics or figurative schemes that together constitute a functional invariant in perception (*proximal object*) or in action and cognition (*distal object*).

operators (brain resources): An operator is a causal agent in the organism that brings about change. We distinguish two distinct sorts: *subjective operators* and *hidden operators*. Subjective operators are schemes, which can be recognized in consciousness because of their effects. Hidden operators are not like schemes; they are regulatory resources, often hidden from consciousness, that can change schemes and their manifestations. Examples of hidden operators are: *A* (affective factors), *B* (psychosocial affective schemas) *M* (mental-attentional capacity), *I* (attentional inhibition), *S* (spatial factor for structuring relations of coexistence), *T* (temporal factor for sequential/episodic structuring), *C, LC, LM*, and so on.

organismic (causal organismic): These are characteristics pertaining to the organism as a *functional totality* within psychology, behavioral neurology, or neuroscience. Organismic processes are causal (*causal-organismic*) if they can predict outcomes produced by the organism (or the brain's functional totality) by using *organismic factors* with their interactions.

overdetermination (*SOP*): A form of causation in which the outcome results from many factors (schemes, etc.) that could concurrently apply to produce it; for example, multiple organismic causes dynamically synthesizing the result. Organismic causation is always overdetermined. One such instance is the *Schemes' Overdetermination of Performance* (*SOP*).

psychological organism (or psychological brain): The functional organismic totality constituted by the brain and bodily processes.

Reality versus reality: The raw Resistances of experience within the Reality out there (or inside us) are presented in our theory with a capital R. We present with lowercase *r* the resistances of reality as already coded and interpreted (*represented*) by the psychological organism.

relations (of coexistence; their complexity): Relations are *functional structural patterns* that coordinate objects or aspects of experience. Their complexity depends on the number of objects that necessarily must be coordinated, via reflective abstraction, to internalize a given activity or state of affairs. *Relations of coexistence* are abstracted relations that express a set of objects or actions (i.e., schemes), which in given situations are often coactivated and work together to produce an intended result.

representation (levels of): This is the classic term referring to reflective abstraction of experientially encountered invariants and stable variable aspects.

Resistances: These are invariant (or variable) characteristic aspects of situations that impose necessary constraints on a person's actions or representations, relative to some goals in a type of situation. Resistances are often relative to each species, within a given type of situation. Written with capital *R*, Resistances are aspects of Reality; if written with lowercase *r*, resistances are schemes' coded characteristics. Facilitating Resistances/resistances are often called *affordances*; interfering or misleading ones are called *encumbrances* or *obstacles*.

schemes (of action, versus executive, versus ephemeral): Schemes are anticipations, that is, action patterns (and/or affective feelings) that can be transferred from one situation or object or action to another—in recognition of similarity relations. *Executive schemes* are those that carry executive functions of control or planning. *Action schemes* (*figurative, operative, affective,* or *personal*) are schemes that represent objects, conditions (parameters), procedures, or affective values. *Ephemeral schemes* are not-yet-learned (schematized) functional structures; that is, figurative, operative, executive, affective/personal schemes currently synthesized by the psychological organism, but not yet internalized into the repertoire of permanent schemes.

sign, signal, symbol: A *signal* is a sign (usually a figurative scheme) that has associative links with another scheme (*the referent*—object or event) so that it can evoke this other scheme before it arrives. A *symbol* is a sign functionally detached from its referent, which allows it to evoke *the possibility* (without the actuality) of its referent.

stages (development): An organismic developmental stage is a stable predictable level of complexity-processing across content domains. It is manifested in critical landmark performances across domains, which are predictable by an organismic-developmental theory or general process model.

structure: A system of relations holding among elements such that if the elements are exchanged with elements of a different suitable kind, the same system of relations is preserved as a functional invariant. A face is therefore a structure, because identification of a face can be done across all animals.

task analysis: A method of formulating the meaning and procedural process-structure for solving tasks by using the subject's own schemes. Metasubjective task analysis (MTA) is one such method for modeling "from within" the subject's task solution.

TCO: The Theory of Constructive Operators, which is the subject of this book. See preface and chapter 1.

Notes

Preface

1. We place *Reality* and *Resistances* in capitals to emphasize that we are talking about actual Reality out there imposing on us its Resistances. This is different from modeling. We write *reality*, in lowercase, to signify any interpreted modeling.

2. What follows is translated and modified from Pascual-Leone (2015).

Chapter 1

1. A situation is an immediate context, objective and subjective, where there are one or more persons, one or more objects, and physical or mental activity. When at least one person in the situation also has goals to pursue, then the situation contains a task.

2. Remarks on Piaget's work come with modifications from Pascual-Leone (2012a).

3. Not all Raven's items require consideration of all dimensions of variation. In simpler items just one or two distinct dimensions must be considered to triangulate the relational invariant features to be abstracted, which the missing lower-right section should also have. These simpler items are often (see table 3 of Unsworth & Engle, 2005) more strongly correlated with working memory tasks (Operation Span) because in Raven's they have less executive demand.

Chapter 2

1. Mediation as a technical term has two distinct senses in psychology. It can mean external mediation, that is, the social interactions and psychosocial relations that a person has with others that support thinking or action. Vygotsky used the term in this manner. Alternatively mediation can be internal, that is, the cognitive or perceptual processes that in a working mind link (mediate between) the here-and-now situation (as represented) and the response—that is, thinking processes or action needed to cope with the task at hand. This second sense, favored by Hebb and others, is our intended meaning.

2. Stages are distinct developmental levels of processing in which mental schemes or schemas are processed by other schemes/schemas and by the brain's general-purpose resources. This processing is progressively more complex as hierarchically nested stages advance.

3. Infrastructure is a term engineers and economists use for functionally essential aspects of complex objects, relative to the praxis and function served by these objects.

4. This is an important point usually not appreciated. Contemporary researchers often wrongly interpret classic theoretical statements as consistent with contemporary views. For instance, Case (1985, p. 33) and Barrouillet and Gaillard (2011, p. 1) have interpreted Baldwin's (1894/1968) concept of a (mental) attention that increases developmentally, as if it were the same as Pascual-Leone's (1970) construct of mental attention. They fail to notice that Baldwin, like his contemporaries, thought of growth in the power or breadth of attention as the operative/motor "excess function" that develops as a consequence of neuronal coordinations, due to associative learning. Baldwin thought of mental attention as being a consequence (not a maturational cause) of higher learning and cognitive growth. That mental attention is cause, and not just consequence of learning, was first proposed by us (Pascual-Leone, 1970; Pascual-Leone & Smith, 1969).

5. Driven by both situation-specific motivations and Baldwin's excess function—which is analogous to Piaget's operative-assimilation roots of all knowing.

6. G. Miller (1956) ends his paper with a theoretical disclaimer: "Perhaps there is something deep and profound behind all these sevens [his maximum estimate of mental attention span], something just calling out for us to discover. But I suspect that it is only a pernicious, Pythagorean coincidence" (p. 96).

7. A minority of these preoperational children may also claim the contrary, that there is more in the sausage than in the ball, because it is longer. Thus, preoperational children do not mentally attend to both length and width of the object simultaneously, and fail to coordinate both dimensions.

8. There are many different examples of these vicariance structures. A familiar example corresponds to the concept of number. If we take for instance the number 7, we could say that 7 is the name for the grouping of covariances constituted by all arithmetic operations that preserve as functional invariant the number 7, i.e., $3+4=2+5=9-2=\ldots$ Each of these operations can be interpreted as facets of the same arithmetic distal object 7.

9. Maturation, like the concept of development itself, has two different related meanings in psychology. One is *biogenetic*: maturation, in this biological sense (like development *in its narrow sense*, as Piaget emphasized) refers to changes largely caused by biogenetic unfolding. The other sense is *psychological*. It refers to mental or behavioral maturing/growth in individual ways of coping with situations; this is also the *broad sense* of development for Piaget. In the present text, unless stated otherwise, we mean the biogenetic sense of maturation.

10. These are action schemes boosted in their activation by affective schemes motivating them.

Chapter 3

1. Benson-Hamstra used means and mean differences because our *M*-operator is not a latent variable but a *hidden variable*, that is, an organismic variable that is *codetermining* performance with other process variables like several sorts of learning (e.g., *LC*- and *LM*-operators), the neo-Gestaltist Field factor (*F*-operator), and so forth. This hidden variable maintains some nonlinear organismic interactions with other codetermining variables as they generate performance. Thus theory-predicted age-bound values of *M* cannot be found directly in performance. They can only be found indirectly, when quantitative relations among various mean performances are contrasted in items in which subjects' *M*-power minus the items' *M*-demand are systematically varied and compared.

2. Interrater reliability was assessed by having a second rater transcribe complete videotapes for two children in every age group. There was 86.6% agreement in assignment of semantic relation codes to the video transcriptions, and 91.7% agreement on the total score of the tasks. Differences were resolved by discussion between raters.

Chapter 4

1. Note that *subjective* in the term *metasubjective* refers not just to psychological subjectivity but more broadly to adoption of a "from within" perspective to study the subject's own processes. There is no intended reference to psychological subjectivity, in the sense of private or relative to the subject, although this too can be studied metasubjectively. "Metasubjective" refers here to thinking and theorizing (metacognition) *from within* the perspective of a given subject-matter.

2. The terms *resemblance, likeness,* or *similarity* capture a key aspect of iconic codes for resistances. The old term *similarity (resemblance)* is ambiguous, as has been long noticed (from Aristotle to Watanabe—1969, see his "Theorem of the Ugly Duckling," p. 376). As Watanabe indicates, similarity refers to selective organismic biases in recognizing stipulated characteristics that are perhaps relevant for the individual's life (biases coming from evolution and/or experience). These characteristics emerge in praxis as recurrent resistances (recognized in the form of similarity relations) that constitute context-relative functional invariants.

3. *Semiotics*, which Peirce founded, is the logic of signs: the systematic analysis of patterns of interpretive behavior.

Chapter 5

1. Piaget gives two complementary definitions. In 1957 he said (JPL's translation), "The scheme of an action with regard to a class of actions equivalent from the perspective of a subject, is the common structure that characterizes this equivalence" (Apostel et al., 1957, p. 46. Df. 7). A year later, he wrote: "schemes [are] the common structure of actions equivalent from the perspective of the subject (according to equivalence classes more or less large or narrow)" (Jonckheere, Mandelbrot & Piaget, 1958; p. 56, JPL's translation).

2. By *praxis* we mean conscious goal-directed activity addressed to the environment, to satisfy relatively *central* and *intrinsic* needs (direct *affective goals*).

3. For instance, imagine that walking in the forest we suddenly hear a loud sound-of-a-tree-falling (a figurative expectancy or sign) telling us that an actual tree just fell, and we must be guarded. This is because of a package of interrelated schemes we had previously internalized. This particular package gives us the causal texture of the forest environment, raising the question of whether wind or axe, or some other force caused the tree to fall.

4. Notice that *epireflection/epireflect* is a more encompassing function than Peirce's *iconic* (or *iconicity*). In our definition, epireflection can be obtained by either of two mutually compatible processes (see chapter 4): Peirce's iconic function and his indexical (i.e., associative learning) function. Either can inform (inject form into) a scheme or sign so that it reflects epistemically some relevant aspect(s) of the Real object or process at hand.

5. Updating is the executive process of changing the current content of working memory (WM, produced by mental-attentional centration) so as to drop now-irrelevant schemes and replace them with currently relevant ones; doing so without (this condition is important) changing the level or "grain" of descriptive analysis (e.g., to replace the previously heard two digits with the newly last-heard two digits). Shifting/switching is the executive function in which the content of WM is changed so that the level or "grain" of descriptive analysis is changed as well (i.e., shifted). An example of the latter is shifting attention from overall patterning to local details (e.g., from the overall face to an eye or the lips).

6. This hypothetical mechanism is clearly related to Hebb's (1949/1961) rule, expressed by the phrase "neurons that fire together wire together" (i.e., become coordinated within the same scheme).

7. This backward-and-forward analysis we use in task analysis is reminiscent of the classic method of geometry (Hintikka & Remes, 1974), as we discuss in chapter 8. A key difference of our method is that it investigates semantic-pragmatic situational *resistances* imposed by the task performance (its constraints). Our analysis does so by representing schemes that express these constraints. Further, auxiliary constructions in our method are not new figures, as in geometry constructions, but they are new schemes of various sorts that we posit and use to model the task at hand.

8. Notice that "expectancies," as used originally by Tolman and by many researchers today (implicit functional hypotheses), do not involve consciousness or conceptualization. In a footnote Tolman (1938, p. 163) said, "Let me emphasize again that an 'expectation' does not require words nor consciousness—that it is just a 'set' for a certain environmental object-sequence."

9. This is so because (with repeated *C* learning) high activation of all *C*-schemes tends to become equalized across schemes within the activated set, since they reach asymptotic levels.

10. The possibility for mental attention to effortfully interrupt or inhibit (*I*) task-irrelevant schemes can facilitate and make sharper *LM* learning.

11. This simple four-way categorization of cortical areas is useful to grasp basic differences in processing and so was used by early neuroscientists (e.g., Eccles, 1980; Luria, 1973), but it is no longer popular in neuroscience. It has been replaced by more complex and concretely refined local brain levels. We use the old categorization here because it helps to capture general functional relations between brain processes, cognitive development, and semiotics that more recent descriptions do not show as well because of their greater specificity.

Chapter 6

1. A situation is a state of affairs in which there is an experiencing subject, object(s), and an external/internal context, in which some events and experiences (e.g., cognitions, feelings, emotions) occur to the subject.

2. An activated scheme is misleading when it is task-irrelevant and strongly competing with task-relevant schemes for application. In that case, the probability of subsequent application of task-relevant schemes is reduced, unless misleading schemes are actively suppressed (Pascual-Leone, 1969, 1989).

3. This formulation comes from Sherrington (1906) and was originally restricted to motor neurons (see also McFarland & Sibly, 1975).

Chapter 7

1. From our perspective the input sources A and B could be understood as carrying information about figurative (object) schemes, operative (procedural) schemes, and/or adjunct-information schemes (parameters or expectancies).

Chapter 8

1. Mereologic is the logical study of objects and their part-whole relations.

2. The task used for testing children in conservation of identity does not present a ball **A**, only a **B** ball that is transformed into a sausage in front of the child.

3. Notice that in figure 8.1 and formula (**f1**), the relation $TB=B'$ is seen by the child as meaning that $B=B'$.

4. Such judgment $A>B'$ occurs in children when the thickness of **A** appears more perceptually salient than the thinness of **B′**. When the thinness and length of **B′** are more salient, the child's judgment may instead be $A<B'$.

Chapter 10

1. Mereologic (its adjective mereological) is a branch of logic founded by Lesniewski that studies part-whole relations, that is, the constitution of objects and their interrelations.

2. "Epistemic" etymologically refers to knowing; "ontic" and "ontological" refer to being, that is, Reality prior to its knowledge representation. Thus "epistemo-ontological" refers to a knowing that is truly grounded into (or true) to Reality.

3. This equilibration principle resembles other principles of general adaptation that dynamic system theories in computational rationality and neuroscience currently express (e.g., Gershman et al., 2015; Friston, 2010; Schöner, 2014).

4. Notice that we are here adopting the definition of schema currently found in cognitive science and neuroscience, which we think usually corresponds to complex schemes. Piaget had defined a schema as different from a scheme, standing for a configural representation or simplified image. We disregard such definition of schema and instead talk of figurative schemes/schemas versus operative schemes/schemas.

5. The saliency is caused, for instance, by the field—F_s – perceptual-salience factor, or by primacy or recency effects (*C, LC* learning), affective factors (*A*), and so on.

Chapter 11

1. The logological mode generates a *linguistic* mode in the context of communication; as a communication tool at the service of logological processing. We omit separate reference to this important language mode and treat it as part of logological processes, although a refined study would have to separate them, because language has partly distinct but related brain expressions.

2. The learning and production of motor sequences involves a prefrontal circuit constituted (in monkeys) by F5; PreSMA; and SMA (supplementary motor area). These areas correspond, respectively, in humans to BA 44, 45; BA 9, 8; and BA 6 (Arbib, 2005; Leonard, 2003).

3. In Kant's sense, transcendental means essentially categories or intuitive dimensions that refer not to actual experience but to processes of knowing making any experiences possible (such as, we say, innate neuroscience determinations), thus establishing possibilities and limits to experience.

4. There is also an "operative temporal" stream for producing (rather than perceiving or representing) sequences, which is located, as we mentioned earlier, in BA 44, 9, 8, 6 (e.g., Leonard, 2003).

5. Notice that to cognitively construct objects via experience one has to store temporal sequences of acts and actions in which these objects in fact intervene, so as to cognize them. Thus the identity of objects may be a by-product of this time or *T*-operator's functioning.

6. This important book is available on ResearchGate: https://www.researchgate.net/profile/Gyorgy_Buzsaki/publication/223130267_Rhythms_of_The_Brain/links/00b4952bb0ae609ac9000000/Rhythms-of-The-Brain.pdf

7. Fluents are schemes that reflect the experiential flow (Johnson et al., 1989).

8. We are grateful to Marie Arsalidou and Anton Alexandru, who separately sought out references on neurotransmitters and wrote relevant reports. Their review work has informed our neurotransmitters' model.

Chapter 12

1. We received this quote about Bacon's "constructivist" bee from Dr. Muhammad Ali Khalidi, Department of Philosophy and Cognitive Science Program, York University.

2. Olivier Houdé was not a formal student of JPL; he attended seminars given by JPL in France and then adopted JPL's theory as his initial approach to research.

References

Adey, P., & Shayer, M. (1993). An exploration of long-term far-transfer effects following an extended intervention program in the high school science curriculum. *Cognition and Instruction, 11*(1), 1–29.

Adey, P., & Shayer, M. (1994). *Really raising standards: Cognitive intervention and academic achievement*. London: Routledge.

Agostino, A., Johnson, J., & Pascual-Leone, J. (2010). Executive functions underlying multiplicative reasoning: Problem type matters. *Journal of Experimental Child Psychology, 105*(4), 286–305.

Albright, T. D., Jessell, T. M., Kandel, E. R., & Posner, M. I. (2001). Progress in the neural sciences in the century after Cajal (and the mysteries that remain). In P. C. Marijuán (Ed.), *Cajal and consciousness: Scientific approaches on the centennial of Ramón y Cajal's Textura* (pp. 11–40). New York: New York Academy of Sciences.

Alexander, H. G. (Ed.). (1976). *The Leibniz-Clarke correspondence*. New York: Barnes & Noble.

Allman, J. M., Hakeem, A., Erwin, J. M., Nimchinsky, E., & Hop, P. (2001). The anterior cingulate cortex: The evolution of an interface between emotion and cognition. In A. R. Damasio, A. Harrington, J. Kagan, B. S. McEwen, H. Moss, & R. Shaikh (Eds.), *Unity of knowledge: The convergence of natural and human science* (pp. 107–117). New York: The New York Academy of Sciences.

Alp, I. E. (1988)."Mental capacity and working memory in 1 to 3 year olds" [Unpublished doctoral dissertation]. York University.

Alp, I. E. (1994). Measuring the size of working memory in very young children: The Imitation Sorting Task. *International Journal of Behavioral Development, 17*(1), 125–141.

Anderson, A. K., & Phelps, E. A. (2001). Lesions of the human amygdala impair enhanced perception of emotionally salient events. *Nature, 411*(6835), 305–309.

Anderson, N. H. (1981). *Foundations of information integration theory*. New York: Academic Press.

Andrews, G., & Halford, G. S. (2011). Recent advances in relational complexity theory and its application to cognitive development. In P. Barrouillet & V. Gaillard (Eds.), *Cognitive development and working memory: A dialogue between neo-Piagetian and cognitive approaches* (pp. 47–68). New York: Psychology Press.

Anglin, J. (1977). *Word, object, and conceptual development.* New York: Norton.

Apel, K-O. (1995). *Charles S. Peirce: From pragmatism to pragmaticism.* Atlantic Highlands, NJ: Humanities Press.

Apostel, L. (1967). Syntaxe, semantique et pragmatique [Syntax, semantics, and pragmatics]. In J. Piaget (Ed.), *Logic et connaissance scientifique* (pp. 290–311). Paris: Editions Gallimard.

Apostel, L., Mays, W., Morf, A., & Piaget, J. (1957). *Les liaisons analytiques et synthetiques dans les comportements du sujet* [Analytic and synthetic connections in the subject's behavior]. Paris: Presses Universitaires de France.

Arbib, M. A. (2005). From monkey-like action recognition to human language: An evolutionary framework for neurolinguistics. *Behavioral and Brain Sciences, 28*(2), 125–167.

Arbib, M. A., Érdi, P., Szentágothai, J. (1998). *Neural organization: Structure, function, and dynamics.* Cambridge, MA: MIT Press.

Aron, A. R., Robbins, T. W., & Poldrack, R. A. (2004). Inhibition and the right inferior frontal cortex. *Trends in Cognitive Sciences, 8*(4), 170–177.

Aronson, R. (1987). *Sartre's second critique.* Chicago: University of Chicago Press.

Arsalidou, M. (2013). Working memory capacity: The need for process task-analysis. *Frontiers in Psychology, 4*, 257. https://doi.org/10.3389/fpsyg.2013.00257

Arsalidou, M., & Im-Bolter, N. (2017). Why parametric measures are critical for understanding typical and atypical cognitive development. *Brain Imaging and Behavior, 11*(4), 1214–1224.

Arsalidou, M., & Pascual-Leone, J. (2016). Constructivist developmental theory is needed in developmental neuroscience. *npj Science of Learning, 1*,16016. https://www.nature.com/articles/npjscilearn201616

Arsalidou, M., Pascual-Leone, J., & Johnson, J. (2010). Misleading cues improve developmental assessment of working memory capacity: The color matching task. *Cognitive Development, 25*(3), 262–277.

Arsalidou, M., Pascual-Leone, J., Johnson, J., Morris, D., & Taylor, M. J. (2013). A balancing act of the brain: Activations and deactivations as a function of cognitive load. *Brain and Behavior, 3*(3), 273–285.

Arsalidou, M., Pawliw-Levac, M., Sadeghi, M., & Pascual-Leone, J. (2018). Brain areas associated with numbers and calculations in children: Meta-analyses of fMRI studies. *Developmental Cognitive Neuroscience, 30*, 239–250.

Atkinson, J., & Braddick, O. (2003). Neurobiological models of normal and abnormal visual development. In M. de Haan & M. H. Johnson (Eds.), *The cognitive neurosciences of development* (pp. 43–71). New York: Psychology Press.

Attneave, F. (1959). *Applications of information theory to psychology.* New York: Holt, Rinehart & Winston.

Audi, R. (Ed.). (1995). *The Cambridge dictionary of philosophy*. Cambridge, UK: Cambridge University Press.

Austin, J. H. (2009). *Selfless insight: Zen and the meditative transformations of consciousness*. Cambridge, MA: MIT Press.

Austin, J. H. (2010). The thalamic gateway: How the meditative training of attention evolves toward selfless transformations of consciousness. In B. Bruya (Ed.), *Effortless attention: A new perspective in the cognitive science of attention and action* (pp. 373–407). Cambridge, MA: MIT Press.

Bachelard, G. (1975). *Le rationalisme applique* [Applied rationalism]. Paris: Presses Universitaires de France. (Original work published 1949)

Bachelard, G. (1980). *Le matérialism rationnel* [Rational materialism]. Paris: Presses Universitaires de France.

Bachelard, G. (1981). *La philosophie du non* [The philosophy of no]. Paris: Presses Universitaires de France.

Bachelard, G. (1987). *Le nouvel esprit scientifique* [The new scientific spirit]. Paris: Presses Universitaires de France. (Original work published 1934)

Bacon, F. (2004). *The new organon* (L. Jardin & M. Silvertthorne, Eds.). Cambridge, UK: Cambridge University Press. (Original work published 1620)

Badre, D. (2008). Cognitive control, hierarchy, and the rostro-caudal organization of the frontal lobes. *Trends in Cognitive Sciences, 12*(5), 193–200.

Baillargeon, R. [Raymond], Pascual-Leone, J., & Roncadin, C. (1998). Mental-attentional capacity: Does cognitive style make a difference? *Journal of Experimental Child Psychology, 70*(3), 143–166.

Baillargeon, R. [Renée]. (1987). Object permanence in 3½- and 4½-month-o1d infants. *Developmental Psychology, 23*(5), 655–664.

Baillargeon, R. [Renée]. (2008). Innate ideas revisited: For a principle of persistence in infants' physical reasoning. *Perspectives on Psychological Science, 3*(1), 2–13.

Baillargeon, R. [Renée], Scott, R. M., & Bian, L. (2016). Psychological reasoning in infancy. *Annual Review of Psychology, 67*, 159–186.

Baillargeon, R. [Renée], Spelke, E. S., & Wasserman, S. (1985). Object permanence in five-month-old infants. *Cognition, 20*(3), 191–208.

Bakchine, S. (2000). Temporal lobe behavioral syndromes. In J. Bogousslavsky & J. L. Cummings (Eds.), *Behavior and mood disorders in focal brain lesions* (pp. 369–398). New York: Cambridge University Press.

Baldwin, J. M. (1968). *Mental development in the child and the race: Method and processes*. New York: Augustus M. Kelley. (Original work published 1894)

Balioussis, C., Johnson, J., & Pascual-Leone, J. (2012). Fluency and complexity in children's writing: The role of mental attention and executive function. *Rivista di Psicolinguistica Applicata / Journal of Applied Psycholinguistics, 12*(3), 33–45.

Baltes, P. B., Lindenberger, U., & Staudinger, U. M. (2006). Life span theory in developmental psychology. In W. Damon, & R. M. Lerner (Eds.), *Handbook of child development: Theoretical models of human development* (pp. 569–664). New York: Wiley.

Barrouillet, P., & Gaillard, V. (Eds.). (2011). *Cognitive development and working memory: A dialogue between neo-Piagetian and cognitive approaches.* New York: Psychology Press.

Barsalou, L. W. (2008). Grounded cognition. *Annual Review of Psychology, 59,* 617–645.

Barth, J., & Call, J. (2006). Tracking the displacement of objects: A series of tasks with great apes (Pan troglodytes, Pan paniscus, Gorilla gorilla, and Pongo pygmaeus) and young children (Homo sapiens). *Journal of Experimental Psychology: Animal Behavior Processes, 32*(3), 239–252.

Bear, M. F., Connors, B. W., & Paradiso, M. A. (2001). *Neuroscience: Exploring the brain* (2nd ed.). Baltimore: Lippincott Williams & Wilkins.

Bechara, A., Tranel, D., & Damasio, H. (2000). Characterization of the decision-making deficit of patients with ventromedial prefrontal cortex lesions. *Brain, 123*(11), 2189–2202.

Beiser, F. C. (1993). Introduction: Hegel and the problem of metaphysics. In. F. C. Beiser (Ed.), *The Cambridge companion to Hegel* (pp. 1–24). Cambridge, UK: Cambridge University Press.

Benes, F. M. (1994). Development of the corticolimbic system. In G. Dawson & K. W. Fischer (Eds.), *Human behavior and the developing brain* (pp. 176–206). New York: Guilford.

Benes, F. M. (2001). The development of prefrontal cortex: The maturation of neurotransmitter systems and their interactions. In C. A. Nelson & M. Luciana (Eds.), *Handbook of developmental cognitive neuroscience* (pp. 79–92). Cambridge, MA: MIT Press.

Bennett, M. R., & Hacker, P. M. S. (2003). *Philosophical foundations of neuroscience.* Malden, MA: Blackwell.

Benson, N. (1989). "Mental capacity constraints on early symbolic processing: The origin of language from a cognitive perspective" [Unpublished doctoral dissertation]. York University.

Benson, N., & Pascual-Leone, J. (1997). "Mental capacity constraints on early symbolic processing" [Unpublished manuscript]. Department of Psychology, York University, Toronto, Canada.

Bereiter, C. (1985). Toward a solution of the learning paradox. *Review of Educational Research, 55*(2), 201–226.

Berthoz, A. (2003). *La decisión* [The decision]. Paris: Odile Jacob.

Berthoz, A. (2012). *Simplexity: Simplifying principles for a complex world.* New Haven, CT: Yale University Press.

Beth, E. W., & Piaget, J. (1961). *Epistémologie mathématique et psychologie* [Mathematical epistemology and psychology]. Paris: Presses Universitaires de France.

Bever, T., & Chiarello, R. J. (1974). Cerebral dominance in musicians and non-musicians. *Science, 185*(4150), 537–539.

Bickhard, M. H. (2015a). Toward a model of functional brain processes I: Central nervous system functional micro-architecture. *Axiomathes, 25*, 217–238.

Bickhard, M. H. (2015b). What could cognition be, if not computation ... or connectionism, or dynamic systems? *Journal of Theoretical and Philosophical Psychology, 35*(1), 53–66.

Bloom, L. (1973). *One word at a time: The use of single word utterances before syntax*. The Hague: Mouton.

Bohm, D. (1980). *Wholeness and the implicate order*. New York: Routledge.

Borsboom, D., Mellenbergh, G. J., & van Heerden, J. (2004). The concept of validity. *Psychological Review, 111*(4), 1061–1071.

Bower, T. G. R. (1974). *Development in infancy*. San Francisco: Freeman.

Broadbent, N. J., Clark, R. E., Zola, S., & Squire, L. R. (2002). The medial temporal lobe and memory. In L. R. Squire & D. L. Schacter (Eds.), *Neuropsychology of memory* (3rd ed.) (pp. 3–23). New York: Guilford.

Bruner, J. (1964). The course of cognitive growth. *American Psychologist, 19*(1), 1–15.

Bruya, B. (Ed.). (2010). *Effortless attention*. Cambridge, MA: MIT Press.

Buchler, J. (1955). *Philosophical writings of Peirce*. New York: Dover.

Bunge, S. A., Hazeltine, E., Scanlon, M. D., Rosen, A. C., & Gabrieli, D. E. (2002). Dissociable contributions of prefrontal and parietal cortices to response selection. *NeuroImage, 17*(3), 1562–1571.

Burtis, P. J. (1982). Capacity increase and chunking in the development of short term memory. *Journal of Experimental Child Psychology, 34*(3), 387–413.

Bush, G., Luu, P., & Posner, M. (2000). Cognitive and emotional influences in anterior cingulate cortex. *Trends in Cognitive Neurosciences, 4*(6), 215–222.

Buss, A. T., & Spencer, J. P. (2014). The emergent executive: A dynamic field theory of the development of executive function. *Monographs of the Society for Research in Child Development, 79* (Serial No. 313).

Buzsaki, G. (2006). *Rhythms of the brain*. New York: Oxford University Press.

Cabeza, R., Anderson, N. D., Locantore, J. K., & McIntosh, A. R. (2002). Aging gracefully: Compensatory brain activity in high-performing older adults. *NeuroImage, 17*(3), 1394–1402.

Callari, A., & Ruccio, D. F. (Eds.). (1996). *Postmodern materialism and the future of Marxist theory*. Hanover, NH: University Press of New England.

Calvo, A. (2004). "Detection of latent giftedness by means of mental capacity testing" [Unpublished master's thesis]. York University.

Camba, R. (2014). La conservazione del numero secondo la teoria della catastrofe [Number conservation according to catastrophe theory]. *Giornale Italiano di Psicologia, XLI*(3), 555–571.

Campbell, J. (Ed.). (1971). *The portable Jung*. New York: Viking.

Carey, S., Zaitchik, D., & Bascandziev, I. (2015). Theories of development: In dialog with Jean Piaget. *Developmental Review, 38*, 36–54.

Case, R. (1975). Gearing the demands of instruction to the developmental capacities of the learner. *Review of Educational Research, 45*(1), 59–87.

Case, R. (1985). *Intellectual development: Birth to adulthood*. New York: Academic Press.

Case, R. (1992). *The mind's staircase*. Hillsdale, NJ: Erlbaum.

Case, R. (1998). The development of conceptual structures. In W. Damon (Series Ed.) & D. Kuhn & R. S. Siegler (Vol. Eds.), *Handbook of child psychology: Vol. 2. Cognition, perception, and language* (5th ed., pp. 745–800). New York: Wiley.

Case, R., Demetriou, A., Platsidou, M., & Kazi, S. (2001). Integrating concepts and tests of intelligence from the differential and developmental traditions. *Intelligence, 29(4)*, 307–336.

Case, R. & Edelstein, W. (1993). Introduction: Structural approaches to individual differences. In R. Case & W. Edelstein (Eds.), *The new structuralism in cognitive development: Theory and research on individual pathways* (pp. 1–10). Basel, CH: Karger.

Case, R., & Okamoto, Y. (1996). The role of conceptual structures in the development of children's thought. *Monographs of the Society for Research in Child Development, 61* (Serial No. 246).

Case, R., Okamoto, Y., Henderson, B., & McKeough, A. (1993). Individual variability and consistency in cognitive development: New evidence for the existence of central conceptual structures. In R. Case & W. Edelstein (Eds.), *The new structuralism in cognitive development: Theory and research on individual pathways* (pp. 71–100). Basel, CH: Karger.

Casasanto, D. (2003). Hemispheric specialization in prefrontal cortex: Effects of verbalizability, imageability and meaning. *Journal of Neurolinguistics, 16*(4–5), 361–382.

Cassirer, E. (1938). Le concept de groupe et la theorie de la perception [The group concept and the theory of perception]. *Journal de Psychologie Normale et Pathologique, XXXV*, 368–383.

Cassirer, E. (1953). *Substance and function* (W. C. Swabey & M. C. Swabey, Trans.). New York: Dover. (Original work published 1923)

Cassirer, E. (1957). *The philosophy of symbolic forms, Vol. 3: The phenomenology of knowledge*. New Haven, CT: Yale University Press. (Original work published 1929)

Cassirer, E. (1966). *An essay on man*. New Haven, CT: Yale University Press. (Original work published 1944)

Cassirer, E. (1996). *The philosophy of symbolic forms, Vol. 4: The metaphysics of symbolic forms*. New Haven, CT: Yale University Press. (Original work published 1940)

Castro-Alamancos, M. A., & Gulati, T. (2014). Neuromodulators produce distinct activated states in neocortex. *The Journal of Neuroscience, 34*(37), 12353–12367.

Catania, A. C., & Harnad, S. (Eds.). (1988). *The selection of behavior: The operant behaviorism of B. F. Skinner*. Cambridge, UK: Cambridge University Press.

Christoff, K., & Gabrieli, J. D. E. (2000). The frontopolar cortex and human cognition: Evidence for a rostrocaudal hierarchical organization within the human prefrontal cortex. *Psychobiology, 28*(2), 168–186.

Christoff, K., Keramatian, K., Gordon, A. M., Smith, R., Mädler, B. (2009). Prefrontal organization of cognitive control according to levels of abstraction. *Brain Research,* 1286, 94–105.

Claparede, E. (1946). *Psychologie de l'enfant et pédagogie expérimentale: I. Le developpement mental* [Child psychology and experimental pedagogy: Mental development]. Neuchâtel, CH: Delachaux et Niestlé. (Original work published 1915)

Cockburn, J., & Frank, M. (2011). Reinforcement learning, conflict monitoring, and cognitive control: An integrative model of cingulate-striatal interactions and the ERN. In R. B. Mars, J. Sallet, M. F. S. Rushworth, & N. Yeung (Eds.), *Neural basis of motivational and cognitive control* (pp. 311–331). Cambridge, MA: MIT Press.

Collette, F., Germain, S., Hogge, M., & van der Linden, M. (2009). Inhibitory control of memory in normal ageing: Dissociation between impaired intentional and preserved unintentional processes. *Memory, 17*(1), 104–122.

Conway, A. R. A., Kane, M. J., Bunting, M. F., Hambrick, D. Z., Wilhelm, O., & Engle, R. W. (2005). Working memory span tasks: A review and a user's guide. *Psychonomic Bulletin and Review, 12*(5), 769–786.

Corbetta, M., Patel, G., & Shulman, G. L. (2008). The reorienting system of the human brain: From environment to theory of mind. *Neuron, 58*(3), 306–324.

Corbetta, M., & Shulman, G. L. (2002). Control of goal-directed and stimulus-driven attention in the brain. *Nature Reviews Neuroscience, 3*(3), 201–215.

Cotterill, R. M. J. (2001). Cooperation of the basal ganglia, cerebellum, sensory cerebrum and hippocampus: Possible implications for cognition, consciousness, intelligence and creativity. *Progress in Neurobiology, 64*(1), 1–33.

Cowan, N. (2005). *Working memory capacity*. New York: Psychology Press.

Cowan, N. (2016). Working memory maturation: Can we get at the essence of cognitive growth? *Perspectives on Psychological Science, 11*(2), 239–264.

Cowan, N., AuBuchon, A. M., Gilchrist, A. L., Ricker, T. J., & Saults, J. S. (2011). Age differences in visual working memory capacity: Not based on encoding limitations. *Developmental Science, 14*(5), 1066–1074.

Cowan, N., Ricker, T. J., Clark, K. M., Hinrichs, G. A., & Glass, B. A. (2015). Knowledge cannot explain the developmental growth of working memory capacity. *Developmental Science, 18*(1), 132–145.

Csikszentmihalyi, M., & Nakamura, J. (2010). Effortless attention in everyday life: A systematic phenomenology. In B. Bruya (Ed.), *Effortless attention: A new perspective in the cognitive science of attention and action* (pp. 179–189). Cambridge, MA: MIT Press.

Cullenberg, S. (1996). Althusser and the decentering of the Marxist totality. In A. Callari & D. F. Ruccio (Eds.), *Postmodern materialism and the future of Marxist theory* (pp. 120–149). Hanover, NH: University Press of New England.

Cunning, S. (2003). "The direction-following task: Assessing mental capacity in the linguistic domain" [Unpublished doctoral dissertation]. York University.

Dalley, J. W., & Roiser, J. P. (2012). Dopamine, serotonin and impulsivity. *Neuroscience, 215*, 42–58.

Damasio, A. R. (1999). *The feeling of what happens*. San Diego, CA: Harcourt.

Damasio, A. R. (2012). *Self comes to mind: Constructing the conscious mind*. New York: Vintage.

Damasio, A. R., & Damasio, H. (1994). Cortical systems for retrieval of concrete knowledge: The convergence zone framework. In C. Koch & J. L. Davis (Eds.), *Large-scale neuronal theories of the brain* (pp. 61–74). Cambridge, MA: MIT Press.

Davis, S. W., Dennis, N. A., Daselaar, S. M., Fleck, M. S., & Cabeza, R. (2008). Que PASA? The posterior-anterior shift in aging. *Cerebral Cortex, 18*(5), 1201–1209.

da Vinci, L. (1959). *Selections from notebooks of Leonardo da Vinci* (I. A. Richter, Ed.). London: Oxford University Press.

Deary, I. J. (2000). *Looking down on human intelligence: From psychometrics to the brain*. New York: Oxford University Press.

Dehaene, S. (2014). *Consciousness and the brain: Deciphering how the brain codes our thoughts*. New York: Viking.

de Hevia, M. D., Izard, V., Coubart, A., Spelke, E. S., & Streri, A. (2014). Representations of space, time, and number in neonates. *Proceedings of the National Academy of Sciences of the USA, 111*(13), 4809–4813.

De Houwer, J., Barnes-Holmes, D., & Moors, A. (2013). What is learning? On the nature and merits of a functional definition of learning. *Psychonomic Bulletin & Review, 20*(4), 631–642.

DeLoache, J. S. (1989). Young children's understanding of the correspondence between a scale model and a larger space. *Cognitive Development, 4*(2), 121–139.

Demetriou, A., Makris, N., Spanoudis, G., Kazi, S., Shayer, M., & Kazali, E. (2018). Mapping the dimensions of general intelligence: An integrated differential-developmental theory. *Human Development, 61*(1), 4–42.

Demetriou, A., & Spanoudis, G. (2018). *Growing minds: A developmental theory of intelligence, brain, and education*. New York: Routledge.

Demetriou, A., Spanoudis, G., & Shayer, M. (2014). Inference, reconceptualization, insight, and efficiency along intellectual growth: A general theory. *Enfance, 3*(3), 365–396.

de Ribaupierre, A., & Bailleux, C. (2000). The development of working memory: Further note on the comparability of two models of working memory. *Journal of Experimental Child Psychology, 77*(2), 110–127.

de Ribaupierre, A., Fagot, D., & Lecerf, T. (2011). Working memory capacity and its role in cognitive development. In P. Barrouillet & V. Gaillard (Eds.), *Cognitive development and working memory: A dialogue between neo-Piagetian and cognitive approaches* (pp. 105–133). New York: Psychology Press.

de Ribaupierre, A., & Lecerf, T. (2006). Relationships between working memory and intelligence from a developmental perspective: Convergent evidence from a neo-Piagetian and a psychometric approach. *European Journal of Cognitive Psychology, 18*(1), 109–137.

de Ribaupierre, A., & Lecerf, T. (2017). Intelligence and cognitive development: Three sides of the same coin. *Journal of Intelligence, 5*(2), 14. doi:10.3390/jintelligence5020014

Diamond, A. (2001). A model system for studying the role of dopamine in the prefrontal cortex during early development in humans: Early and continuously treated phenylketonuria. In C. A. Nelson & M. Luciana (Eds.), *Handbook of developmental cognitive neuroscience* (pp. 433–472). Cambridge, MA: MIT Press.

Dormashev, Y. (2010). Flow experience explained on the grounds of an active approach to attention. In B. Bruya (Ed.), *Effortless attention* (pp. 287–333). Cambridge, MA: MIT Press.

Drescher, G. L. (2003). *Made-up minds. A constructivist approach to artificial intelligence*. Cambridge, MA: MIT Press.

Drevets, W. C. (1999). Prefrontal cortical-amygdalar metabolism in major depression. *Annals of the New York Academy of Sciences, 877*, 614–637.

Dunnett, S. B., & Robbins, T. W. (1992). The functional role of the mesotelencephalic dopamine systems. *Biological Reviews of the Cambridge Philosophical Society, 67*(4), 491–518.

Eccles, J. C. (1980). *The human psyche*. Berlin: Springer.

Edelman, G. M. (1987). *Neural Darwinism*. New York: Basic Books.

Edelman, G. M. (1989). *Remembered present: A biological theory of consciousness*. New York: Basic Books.

Edelman, G. M., & Mountcastle, V. B. (1978). *The mindful brain: Cortical organization and the group-selective theory of higher brain function*. Cambridge, MA: MIT Press.

Edelman, G. M., & Tononi, G. (2000). *A universe of consciousness*. New York: Basic Books.

Ekman, P., & Davidson, R. J. (Eds.). (1994). *The nature of emotion*. New York: Oxford University Press.

Elman, J. L. (1993). Learning and development in neural networks: The importance of starting small. *Cognition, 48*(1), 71–99.

Elman, J. L., Bates, E. A., Johnson, M. H., Karmiloff-Smith, A., Parisi, D., & Plunkett, K. (1996). *Rethinking innateness: A connectionist perspective on development.* Cambridge, MA: MIT Press.

Engle, R. W. (2002). Working memory capacity as executive attention. *Current Directions in Psychological Science, 11*(1), 19–23.

Engle, R. W. (2018). Working memory and executive attention: A revisit. *Perspectives on Psychological Science, 13*(2), 190–193.

Engle, R. W., Tuholski, S. W., Laughlin, J. E., & Conway, A. R. A. (1999). Working memory, short-term memory and general fluid intelligence: A latent variable approach. *Journal of Experimental Psychology: General, 128*(3), 309–331.

Espy, K. A. (Ed.). (2016). The changing nature of executive control in preschool. *Monographs of the Society for Research in Child Development, 81* (Serial No. 323).

Evans, B. T., & Stanovich, K. E. (2013). Dual-process theories of higher cognition: Advancing the debate. *Perspectives on Psychological Science, 8*(3), 223–242.

Feyerabend, P. (1978). *Science in a free society.* London: New Left Books.

Fillmore, C. J. (1968). The case for case. In E. Bach & R. T. Harms (Eds.), *Universals in linguistic theory* (pp. 1–90). New York: Holt, Rinehart, & Winston.

Fischer, H., Nyberg, L., Karlsson, S., Karlsson, P., Brehmer, Y., Rieckmann, A., MacDonald, S. W. S., Farde, L., & Bäckman, L. (2010). Simulating neurocognitive aging: Effects of a dopaminergic antagonist on brain activity during working memory. *Biological Psychiatry, 67*(6), 575–580.

Fischer, K. W. (1980). A theory of cognitive development: The control and construction of hierarchies of skills. *Psychological Review, 87*(6), 477–531.

Flavell, J. H. (1963). *The developmental psychology of Jean Piaget.* New York: Van Nostrand Reinhold.

Fodor, J. A. (1981). The present status of the innateness controversy. In J. Fodor (Ed.), *RePresentations: Philosophical essays on the foundations of cognitive science* (pp. 257–316). Cambridge, MA: MIT Press.

Ford, R. M., Keating, S., & Patel, R. (2004). Retrieval-induced forgetting: A developmental study. *British Journal of Developmental Psychology, 22*(4), 585–603.

Fraisse, P. (1963). Les emotions [Emotions]. In J. Nuttin, P. Fraisse, R. Meili, & P. Roubertoux, *Traite de psychologie experimentale. V. Motivation, emotion et personnalite* (pp. V-83–V-153). Paris: Presses Universitaires de France.

Fredrickson, B. L. (2001). The role of positive emotions in positive psychology. *American Psychologist, 56*(3), 218–226.

Frege, G. (1980). *The foundations of arithmetic: A logico-mathematical enquiry into the concept of number* (J. L. Austin, Trans.). Evanston, IL: Northwestern University Press.

Frijda, N. H. (1987). Emotions, cognitive structure, and action tendency. *Cognition and Emotion, 1*(2), 115–144.

Friston, K. (2010). The free-energy principle: A unified brain theory? *Nature Reviews Neuroscience, 11*(2), 127–138.

Fuller, P. M., Sherman, D., Pedersen, N. P., Saper, C. B., & Lu, J. (2011). Reassessment of the structural basis of the ascending arousal system. *The Journal of Comparative Neurology, 519*(5), 933–956.

Fuster, J. M. (1995). *Memory in the cerebral cortex*. Cambridge, MA: MIT Press.

Fuster, J. M. (2004). Upper processing stages of the perception-action cycle. *Trends in Cognitive Sciences, 8*(4), 143–145.

Fuster, J. M., & Alexander, G. E. (1971). Neuron activity related to short-term memory. *Science, 173*(3997), 652–654.

Gadamer, H.-G. (1971). *Hegel's dialectic*. New Haven, CT: Yale University Press.

Gallistel, C. R., & Matzel, L. D. (2013). The neuroscience of learning: Beyond Hebbian synapse. *Annual Review of Psychology, 64*, 169–200.

Gallup, G. G. (1982). Self-awareness and the emergence of mind in primates. *American Journal of Primatology, 2*(3), 237–248.

Garcia, R. (1980). Postface: Dialectique, psychogenèse et histoire des sciences [Postscript: Dialectics, psychogenesis and the history of science]. In J. Piaget, *Les formes élémentaires de la dialectique* (pp. 229–246). Paris: Gallimard.

Gazzaniga, M. S., Ivry, R. B., & Mangun, G. R. (1998). *Cognitive neuroscience: The biology of the mind*. New York: Norton.

Gell-Mann, M. (1994). *The quark and the jaguar: Adventures in the simple and the complex*. New York: Freeman.

Gershman, S. J., Horvitz, E. J., & Tenenbaum, J. B. (2015). Computational rationality: A converging paradigm for intelligence in brains, minds, and machines. *Science, 349*(6247), 273–270.

Gibson, J. J. (1979). *The ecological approach to visual perception*. Boston: Houghton Mifflin.

Giuliano, M., Johnson, J., & Pascual-Leone, J. (2015, June 4–6). *Does Spanish word order enhance performance on a direction following task?* [Poster presentation]. Jean Piaget Society Annual Conference, Toronto, ON, Canada.

Goel, V. (2015). Indeterminacy tolerance as a basis of hemispheric asymmetry within prefrontal cortex. *Frontiers in Human Neuroscience, 9*, 326. doi: https://dx.doi.org/10.3389%2Ffnhum.2015.00326

Goel, V. (2019). Hemispheric asymmetry in prefrontal cortex for complex cognition. In M. D'Esposito & J. H. (Eds.), *Handbook of Clinical Neurology: Vol. 163. The frontal lobes* (pp. 179–196). Amsterdam: Elsevier.

Goel, V., & Grafman, J. (2000). The role of the right prefrontal cortex in ill-defined problem solving. *Cognitive Neuropsychology, 17*(5), 415-436.

Goel, V., Shuren, J., Sheesley, L., & Grafman, J. (2004). Asymmetrical involvement of frontal lobes in social reasoning. *Brain, 127*, 783–790.

Goel, V., & Vartanian, O. (2005). Dissociating the roles of right ventral lateral and dorsal lateral prefrontal cortex in generation and maintenance of hypotheses in set-shift problems. *Cerebral Cortex 15*(8), 1170–1177.

Gogtay, N., Giedd, J. N., Lusk, L., Hayashi, K. M., Greenstein, D., Vaituzis, A. C., Nugent, T. F., Herman, D. H., Clasen, L. S., Toga, A. W., Rapoport, K., & Thompson, P. M. (2004). Dynamic mapping of human cortical development during childhood through early adulthood. *Proceedings of the National Academy of Sciences of the USA, 101*(21), 8174–8179.

Goldberg, G. (1985). Supplementary motor area structure and function: Review and hypotheses. *Behavioral and Brain Sciences, 8*(4), 567–616.

Goldstein, K. (2000). *The organism.* New York: Zone Books. (Original work published 1934)

Goldstein, K., & Sheerer, M. (1941). Abstract and concrete behavior: An experimental study with special tests. *Psychological Monographs, 53*(2, Whole No. 239).

Gonseth, F. (1974). *Les mathematiques et la réalité* [Mathematics and reality]. Paris: Albert Blanchard. (Original work published 1936)

Goode, P. E., Goddard, P. H., & Pascual-Leone, J. (2002). Event-related potentials index cognitive style differences during a serial-order recall task. *International Journal of Psychophysiology, 43*(2), 123–140.

Goodman, D. R. (1971). "Cognitive style factors in linguistic performance with ambiguous sentences" [Unpublished master's thesis]. York University.

Gopnik, A., & Meltzoff, A. N. (1992). Categorization and naming: Basic-level sorting in eighteen-month-olds and its relation to language. *Child Development, 63*(5), 1091–1103.

Gordon, A. C. L., & Olson, D. R. (1998). The relation between acquisition of a theory of mind and the capacity to hold in mind. *Journal of Experimental Child Psychology, 68*(1), 70–83.

Gottfredson, L. S. (2016). A g theorist on why Kovacs and Conway's process-overlap theory amplifies, not opposes, g theory. *Psychological Inquiry, 27*(3), 210–217.

Gotts, S. J., Jo, H. J., Wallace, G. L., Saad, Z. S., Cox, R. W., & Martin, A. (2013). Two distinct forms of functional lateralization in the human brain. *Proceedings of the National Academy of Sciences of the USA, 110*(36), E3435–E3444.

Gould, S. J. (2002). *The structure of evolutionary theory.* Cambridge, MA: Harvard University Press.

Granott, N., & Parziale, J. (Eds.). (2002). *Microdevelopment: Transition processes in development and learning.* New York: Cambridge University Press.

Greenberg, L. S., & Goldman, R. N. (2019). *Clinical handbook of emotion focused therapy*. Washington, DC: American Psychological Association.

Greenberg, L. S., & Pascual-Leone, J. (1995). A dialectical constructivist approach to experiential change. In R. A. Neimeyer & M. J. Mahoney (Eds.), *Constructivism in psychotherapy* (pp. 169–191). Washington, DC: American Psychological Association.

Greenberg, L. S., & Pascual-Leone, J. (2001). A dialectical constructivist view of the creation of personal meaning. *Journal of Constructivist Psychology, 14*(3), 165–186.

Greenberg, L. S., Rice, L. N., & Elliott, R. (1993). *Facilitating emotional change: The moment-by-moment process*. New York: Guilford.

Greenhouse, I., Swann, N., & Aron, A. R. (2011). Fronto-basal ganglia circuits for stopping action. In R. B. Mars, J. Sallet, M. F. S. Rushworth, & N. Yeung (Eds.), *Neural basis of motivational and cognitive control* (pp. 189–207). Cambridge, MA: MIT Press.

Guevara, M. A., Arsalidou, M., Pascual-Leone, J., & Stevens, W. D. (2019, March 23–26). *Neural underpinnings of mental attentional capacity* [Poster presentation]. Cognitive Neuroscience Society Annual Meeting, San Francisco, CA, United States.

Guilford, J. P. (1967). *The nature of intelligence*. New York: McGraw-Hill.

Halford, G. S., Cowan, N., & Andrews, G. (2007). Separating cognitive capacity from knowledge: A new hypothesis. *Trends in Cognitive Sciences, 11*(6), 236–241.

Halford, G. S., Wilson, W. H., Andrews, G., & Phillips, S. (2014). *Categorizing cognition: Toward conceptual coherence in the foundations of psychology*. Cambridge, MA: MIT Press.

Hansell, N. K., Halford, G. S., Andrews, G., Shum, D. H. K., Harris, S. E., Davies, G., Franic, S., Christoforou, A., Zietsch, B., Painter, J., Medland, S. E., Ehli, E. A., Davies, G. E., Steen, V. M., Lundervold, A. J., Reinvang, I., Montgomery, G. W., Espeseth, T., Pol, H. E. H., … Wright, M. J. (2015). Genetic basis of a cognitive complexity metric. *PLoS ONE, 10*(4), e0123886. https://doi.org/10.1371/journal.pone.0123886

Harlow, H. F. (1959). Learning set and error factor theory. In S. Koch (Ed.), *Psychology: A study of a science* (Vol. 2, pp. 492–537). New York: McGraw-Hill.

Hebb, D. O. (1961). *Organization of behavior*. New York: Wiley. (Original work published 1949)

Hederich-Martínez, C., & Camargo-Uribe, A. (2016). Cognitive style and educational performance. The case of public schools in Bogotá, Colombia. *Educational Psychology, 36*(4), 719–737.

Heitz, R. P., Unsworth, N., & Engle, R. W. (2005). Working memory capacity, attention control, and fluid intelligence. In O. Wilhelm & R. W. Engle (Eds.), *Handbook of understanding and measuring intelligence* (pp. 61–77). Thousand Oaks, CA: Sage.

Hermans, E. J., van Marle, H. J. F., Ossewaarde, L., Henckens, M. J. A. G., Qin, S., van Kesteren, M. T. R., Schoots, V. C., Cousjin, H., Rijpkema, M., Oostenveld, R., & Fernández, G. (2011). Stress-related noradrenergic activity prompts large-scale neural network reconfiguration. *Science, 334*(6059), 1151–1153.

Heraclitus. (2001). *Fragments: The collected wisdom of Heraclitus* (B. Haxton, Trans.). New York: Viking.

Hickok, G. (2014). *The myth of mirror neurons: The real neuroscience of communication and cognition.* New York: Norton.

Hildebrand, D. K., Laing, J. D., & Rosenthal, H. (1977). *Analysis of ordinal data.* Newbury Park, CA: Sage.

Hintikka, J., & Remes, U. (1974). *The method of analysis.* Dordrech, NL: D. Reidel.

Hochberg, J. E. (1964). *Perception.* Englewood Cliffs, NJ: Prentice-Hall.

Holroyd, C. B., & Yeung, N. (2012). Motivation of extended behaviors by anterior cingulate cortex. *Trends in Cognitive Sciences, 16*(2), 122–128.

Holyoak, K., & Cheng, P. W. (2011). Causal learning and inference as rational process: The new synthesis. *Annual Review of Psychology, 62,* 135–163.

Hommel, B., Colzato, L. S., & van den Wildenberg, W. P. M. (2009). How social are task representations? *Psychological Science, 20*(7), 794–798.

Horn, J. L. (1998). A basis for research on age differences in cognitive capabilities. In J. J. McArdle & R. W. Woodcock (Eds.), *Human cognitive abilities in theory and practice* (pp. 57–91). Mahwah, NJ: Erlbaum.

Horn, J. L., & Hofer, S. M. (1992). Major abilities and development in the adult period. In R. J. Sternberg & C. A. Berg (Eds.), *Intellectual development* (pp. 44–99). New York: Cambridge University Press.

Horn, J. L., & Noll, J. (1994). A system for understanding cognitive capabilities: A theory and the evidence on which it is based. In D. K. Detterman (Ed.), *Current topics in human intelligence: Vol. 4. Theories of intelligence* (pp. 151–203). Norwood, NJ: Ablex.

Houdé, O. (1995). *Rationalité, développement et inhibition: Un nouveau cadre d'analyse* [Rationality, development, and inhibition: A new frame of analysis]. Paris: Presses Universitaires de France.

Houdé, O., Pineau, A., Leroux, G., Poirel, N., Perchey, G., Lanoë, C., Lubin, A., Turbelin, M.-R., Rossi, S., Simon, G., Delcroix, N., Lamberton, F., Vigneau, M., Wisniewski, G., Vicet, J.-R., & Mazoyer, B. (2011). Functional MRI study of Piaget's conservation-of-number task in preschool and school-age children: A neo-Piagetian approach. *Journal of Experimental Child Psychology, 110*(3), 332–346.

Howard, S. J. (2009). "Factors underlying cognitive giftedness: Mental vs. perceptual attention" [Unpublished master's thesis]. York University.

Howard S. J., Johnson J., & Pascual-Leone J. (2014). Clarifying inhibitory control: Diversity and development of attentional inhibition. *Cognitive Development, 31,* 1–21.

Howells, F. M., Stein, D. J., & Russell, V. A. (2010). Perceived mental effort correlates with changes in tonic arousal during attentional tasks. *Behavioral and Brain Functions, 6,* Article 39. https://doi.org/10.1186/1744-9081-6-39

Huddy, V. C., Aron, A. R., Harrison, M., Barnes, T. R. E., Robbins, T. W., & Joyce, E. M. (2009). Impaired conscious and preserved unconscious inhibitory processing in recent onset schizophrenia. *Psychological Medicine: A Journal of Research in Psychiatry and the Allied Sciences, 39*(6), 907–916.

Husserl, E. (1970). *The crisis of European science and transcendental phenomenology*. Evanston, IL: Northwestern University Press.

Husserl, E. (1973). *Experience and judgment: Investigations in a genealogy of logic*. Evanston, IL: Northwestern University Press.

Im-Bolter, N., Johnson, J., Ling, D., & Pascual-Leone, J. (2015). Inhibition: Mental control process or mental resource? *Journal of Cognition and Development, 16*(4), 666–681.

Im-Bolter, N., Johnson, J., & Pascual-Leone, J. (2006). Processing limitation in children with specific language impairment: The role of executive function. *Child Development, 77*(6), 1822–1841.

Inhelder, B., Garcia, R., & Voneche, J. (Eds.). (1977). *Epistémologie génétique et équilibration* [Genetic epistemology and equilibration]. Neuchatel, CH: Delachaux et Niestlé.

Jackson, J. H. (1915). On the nature of the duality of the brain. *Brain, 38*, 80–103.

James, W. (1961). *Psychology: The briefer course* (G. Allport, Ed.). New York: Harper Torchbook. (Original work published 1892)

Janet, P. (1889). *L'Automatisme psychologique* [Psychological automatism]. Paris: Felix Alcan.

Jaspers, K. (1959). *Truth and symbol*. New Haven, CT: College and University Press.

Jaspers, K. (1962). Kant. In *The great philosophers: Vol 1. The foundations*. New York: Harcourt.

Jaspers, K. (1970). *Philosophy* (Vol. 2). Chicago: University of Chicago Press.

Jennings, F. G. (1967). Education in America, Jean Piaget: Notes on learning. *The Saturday Review*, May 20, 81–83.

Jedrzkiewicz, J. A. (1983). "Adult development and mental effort: A neo-Piagetian developmental analysis" [Unpublished master's thesis]. York University.

Jiang, J., Brashier, N. M., & Egner, T. (2015). Memory meets control in hippocampal and striatal binding of stimuli, responses, and attentional control states. *The Journal of Neuroscience, 35*(44), 14885–14895.

Johansen, J. (1993). *Dialogic semiosis: An essay on signs and meaning*. Bloomington: Indiana University Press.

Johnson, J., Fabian, V., & Pascual-Leone, J. (1989). Quantitative hardware-stages that constrain language development. *Human Development, 32*(5), 245–271.

Johnson, J., Howard, S., & Pascual-Leone, J. (2009, April 2-4). *Factors underlying cognitive giftedness: Mental vs. perceptual attention* [Poster presentation]. Society for Research in Child Development Biennial Meeting, Denver, CO, United States.

Johnson, J., Im-Bolter, N., & Pascual-Leone, J. (2003). Development of mental attention in gifted and mainstream children: The role of mental capacity, inhibition, and speed of processing. *Child Development, 74*(6), 1594–1614.

Johnson, J., & Pascual-Leone, J. (1989). Reply to A. Karmiloff-Smith. *Human Development, 32*(5), 276–278.

Johnson, J., Pascual-Leone, J., Im-Bolter, N., & Verrilli, E. (2004, July 11–15). *Executive functions and mental attention in cognitively gifted children* [Poster presentation]. International Society for the Study of Behavioral Development Biennial Meeting, Ghent, Belgium.

Johnson, J., Prior, S., & Artuso, M. (2000). Field dependence as a factor in second language communicative production. *Language Learning, 50*(3), 529–567.

Johnson, J., & Rosano, T. (1993). Relation of cognitive style to metaphor interpretation and second language proficiency. *Applied Psycholinguistics, 14*(2), 159–175.

Johnson, J., Tsvetkov Kristen, I., & Pascual-Leone, J. (2014, May 29-31). *Language factors in measurement of working memory capacity* [Poster presentation]. Jean Piaget Society Annual Meeting, San Francisco, CA, United States.

Johnson-Laird, P. N. (2010). Mental models and human reasoning. *Proceedings of the National Academy of Sciences of the USA, 107*(43), 18243–18250.

Johnstone, A. H., & Al-Naeme, F. F. (1991). Room for scientific thought? *International Journal of Science Education, 13*(2), 187–192.

Johnstone, A. H., & El-Banna, H. (1986). Capacities, demands and processes—a predictive model for science education. *Education in Chemistry, 23*(3), 80–84.

Johnstone, A. H., & El-Banna, H. (1989). Understanding learning difficulties—a predictive research model. *Studies in Higher Education, 14*(2), 159–168.

Jonckheere, A., Mandelbrot, B., & Piaget, J. (1958). *La lecture de l'expérience* [The reading of experience]. Paris: Presses Universitaires de France.

Jorna, R. J. (1995). The semiotic (sign-oriented) aspects of connectionism. In I. Lubek, R. van Hezewijk, G. Pheterson, & C. W. Tolman (Eds.), *Trends and issues in theoretical psychology* (pp. 233–238). New York: Springer.

Juckes, T. J. (1991). Equilibration and the learning paradox. *Human Development, 34*(5), 261–272.

Kagan, J. (2002). *Surprise, uncertainty, and mental structures*. Cambridge, MA: Harvard University Press.

Kagan, J. (2008). In defense of qualitative changes in development. *Child Development, 79*(6), 1606–1624.

Kahneman, D. (1973). *Attention and effort*. Englewood Cliffs, NJ: Prentice-Hall.

Kahneman, D. (2011). *Thinking, fast and slow*. New York: Farrar, Straus and Giroux.

Kanizsa, G. (1955). Margini quasi-percettivi in campi con stimolazione omogenea [Quasi-perceptual margins in homogenously stimulated fields]. *Rivista di Psicologia, 49*, 7–30.

Kant, I. (1965). *Critique of pure reason* (N. K. Smith, Trans.). New York: St. Martin's Press. (Original work published 1929)

Karmiloff-Smith, A. (1992). *Beyond modularity: A developmental perspective on cognitive science*. Cambridge, MA: MIT Press.

Kelso, J. A. S. (2016). On the self-organizing origins of agency. *Trends in Cognitive Sciences, 20*(7), 490–499.

Kerns, J. G., Cohen, J. D., MacDonald, A. W., Cho, R. Y., Stenger, V. A., & Carter, C. S. (2004). Anterior cingulate conflict monitoring and adjustments in control. *Science, 303*(5660), 1023–1026.

Kim, T. H., Pascual-Leone, J., Johnson, J., & Tamim, H. (2016). The mental-attention Tai Chi effect with older adults. *BMC Psychology, 4*, 29. doi: 10.1186/s40359-016-0137-0

Kitterle, F. L., Christman, S., & Hellige, J. B. (1990). Hemispheric differences are found in the identification, but not the detection, of low versus high spatial frequencies. *Perception and Psychophysics, 48*(4), 297–306.

Klein, S. B. (1987). *Learning*. New York: McGraw-Hill.

Koch, C., & Crick, F. (1994). Some further ideas regarding the neuronal basis of awareness. In C. Koch & J. L. Davis (Eds.), *Large-scale neuronal theories of the brain* (pp. 93–110). Cambridge, MA: MIT Press.

Koffka, K. (1963). *Principles of Gestalt psychology*. New York: Harcourt. (Original work published 1935)

Kolb, B., & Whishaw, I. Q. (2003). *Fundamentals of human neuropsychology* (5th ed.). New York: Worth.

Kosslyn, S. M. (1994). *Image and brain: The resolution of the imagery debate*. Cambridge, MA: MIT Press.

Kozulin, A. (1990). *Vygotsky's psychology*. Cambridge, MA: Harvard University Press.

Krantz, D. H., Luce, R. D., Suppes, P., & Tversky, A. (1971). *Foundations of measurement: Vol. I. Additive and polynomial representations*. New York: Academic Press.

Lechuga, M. T., Moreno, V., Pelegrina, S., Gomez-Ariza, C. J., & Bajo, M. T. (2006). Age differences in memory control: Evidence from updating and retrieval-practice tasks. *Acta Psychologica, 123*(3), 279–298.

Lefever, M. M., & Ehri, L. C. (1976). The relationship between field independence and sentence disambiguation ability. *Journal of Psycholinguistic Research, 5*, 99–106.

Legerstee, M. (1991). The role of person and object in eliciting early imitation. *Journal of Experimental Child Psychology, 51*(3), 423–433.

Legerstee, M. (1998). Mental and bodily awareness in infancy: Consciousness of self-existence. *Journal of Consciousness Studies, 5*(5–6), 627–644.

Leibniz, G. W. (1966). *Nouveaux essais sur l'entendement humain* [New essays on human understanding]. Paris: Garnier-Flammarion.

Leonard, C. M. (2003). Neural substrate of speech and language development. In M. de Haan & M. H. Johnson (Eds.), *The cognitive neurosciences of development* (pp.127–155). New York: Psychology Press.

Levy, B. J., & Wagner, A. D. (2011). Cognitive control and right ventrolateral prefrontal cortex: Reflexive reorienting, motor inhibition, and action updating. *Annals of the New York Academy of Science, 1224*, 40–62.

Levy, J. (1980). Cerebral asymmetry and the psychology of man. In M. C. Wittrock (Ed.), *The brain and psychology* (pp. 245–321). New York: Academic Press.

Levy, J. (1983). Individual differences in cerebral hemisphere asymmetry: Theoretical and experimental considerations. In J. Hellige (Ed.), *Cerebral hemisphere asymmetry: Method, theory, and application* (pp. 465–497). New York: Praeger.

Lewis, C. I. (1960). *A survey of symbolic logic*. New York: Dover.

Li, S.-C. (2012). Neuromodulation of behavioral and cognitive development across the life span. *Developmental Psychology, 48*(3), 810–814.

Li, Y., Wu, S., Zhu, J. & O'Boyle, M. W. (2014). Sex and ability differences in neural activation for disembedding figures: An EEG Investigation. *Learning and Individual Differences, 35*, 142–146.

Liljeholm, M., & O'Doherty, J. P. (2011). Subcortical contributions to the motivational and cognitive control of instrumental performance by Pavlovian and discriminative stimuli. In R. B. Mars, J. Sallet, M. F. S. Rushworth, & N. Yeung (Eds.), *Neural basis of motivational and cognitive control* (pp. 149–162). Cambridge, MA: MIT Press.

Lisman, J. E., & Jensen, O. (2013). The theta-gamma neural code. *Neuron, 77*(6), 1002–1016.

Llinás, R., & Ribary, U. (2001). Consciousness and the brain: The thalamocortical dialogue in health and disease. In P. C. Marijuán (Ed.), *Cajal and consciousness: Scientific approaches on the centennial of Ramón y Cajal's Textura* (pp. 166–175). New York: New York Academy of Sciences.

Lövdén, M., & Lindenberger, U. (2005). Development of intellectual abilities in old age: From age gradients to individuals. In In O. Wilhelm & R.W. Engle (Eds.), *Handbook of understanding and measuring intelligence* (pp. 203–221). Thousand Oaks, CA: Sage.

Lukacs, G. (1978). *The ontology of social beings: 1. Hegel*. London: Merlin Press.

Luria, A. R. (1973). *The working brain*. New York: Penguin.

Luria, A. R. (1979). *The making of the mind*. Cambridge, MA: Harvard University Press.

Luria, A. R. (1980). *Higher cortical functions in man* (2nd ed.). New York: Basic Books. (Original work published 1962)

Luu, P., Tucker, D. M., Derryberry, D., Reed, M., & Poulsen, C. (2003). Electrophysiological responses to errors and feedback in the process of action regulation. *Psychological Science, 14*(1), 47–53.

Marinsek, N., Turner, B. O., Gazzaniga, M., & Miller, M. B. (2014). Divergent hemispheric reasoning strategies: Reducing uncertainty versus resolving inconsistency. *Frontiers in Human Neuroscience, 8,* 839. doi: https://dx.doi.org/10.3389%2Ffnhum.2014.00839

Mars, R. B., Sallet, J., Rushworth, M. F. S., & Yeung, N. (Eds.). (2011). *Neural basis of motivational and cognitive control.* Cambridge, MA: MIT Press.

Martí, E. (2012). Thinking with signs: From symbolic actions to external systems of representation. In E. Martí & C. Rodriguez (Eds.), *After Piaget* (pp. 151–169). New Brunswick, NJ: Transaction Publishers.

Marx, K., Engels, F., & Lenin, V. (1977). *On dialectical materialism.* Moscow: Progress Publishers.

Matsuzawa, T. & Yamakoshi, G. (1996). Comparison of chimpanzee material culture between Bossou and Nimba, West Africa. In A. E. Russon, K. A. Bard, & S. Taylor Parker (Eds.), *Reaching into thought: The mind of the great apes* (pp. 211–232). Cambridge, UK: Cambridge University Press.

McClure, E. R., Chentsova-Dutton, Y. E., Holochwost, S. J., Parrott, W. G., & Barr, R. (2018). Look at that! Video chat and joint visual attention development among babies and toddlers. *Child Development, 89*(1), 27–36.

McFarland, D. J., & Sibly, R. M. (1975). The behavioural final common path. *Philosophical Transactions of the Royal Society of London, Biology, 270*(907), 265–293.

Meltzoff, A. N. (1990). Foundations for developing a concept of self: The role of imitation in relating self to other and the value of social mirroring, social modeling, and self practice in infancy. In D. Cicchetti & M. Beeghly (Eds.), *The John D. and Catherine T. MacArthur foundation series on mental health and development. The self in transition: Infancy to childhood* (pp. 139–164). Chicago: University of Chicago Press.

Meltzoff, A. N., & Moore, M. K. (1999). Persons and representation: Why infant imitation is important for theories of human development. In J. B. G. Nadel (Ed.), *Imitation in infancy: Cambridge studies in cognitive perceptual development* (pp. 9–35). New York: Cambridge University Press.

Mendonça Filho, E. J. (2017). "Evidências de validade relacionadas à estrutura interna da escala cognitiva do Inventário Dimensional de Avaliação do Desenvolvimento Infantil" [Validity evidence based on internal structure of the cognitive scale of the Dimensional Inventory for Child Development Assessment] [Unpublished master's thesis]. Universidade Federal do Rio Grande do Sul.

Merleau-Ponty, M. (1968). *The visible and the invisible.* Evanston, IL: Northwestern University Press.

Mesulam, M.-M. (2003). Some anatomical principles related to behavioral neurology and neuropsychology. In T. E. Feinberg & M. J. Farah (Eds.), *Behavioral neurology and neuropsychology* (pp. 45–56). New York: McGraw-Hill.

Messick, S. (1994). The matter of style: Manifestations of personality in cognition, learning, and teaching. *Educational Psychologist, 29*(3), 121–136.

Mieville, D., & Vernant, D. (Eds.). (1996). *Stanislaw Lesniewski aujourd'hui: No. 16. Recherches sur la Philosophie et le Langage* [Stanislaw Lesniewski today: Research on philosophy and language]. Paris: Librairie Philosophique Vrin.

Mihalits, D. S., & Valsiner, J. (2020). Dialectics of influence: How agency works. *Human Arenas.* doi: https://doi.org/10.1007/s42087-020-00126-6

Miller, E. K., & Cohen, J. D. (2001). An integrative theory of prefrontal cortex function. *Annual Review of Neuroscience, 24*, 167–202.

Miller, G. A. (1956). The magical number seven, plus or minus two: Some limits on our capacity for processing information. *Psychological Review, 63*(2), 81–97.

Miller, G. A., Galanter, E., & Pribram, K. H. (1960). *Plans and the structure of behavior.* New York: Henry Holt.

Miller, J. F., Chapman, R. S., Branston, M. B., & Reichle, J. (1980). Language comprehension in sensorimotor stages V and VI. *Journal of Speech and Hearing Research, 23,* 284–311.

Miller, R. (2011). *Vygotsky in perspective.* Cambridge, UK: Cambridge University Press.

Milner, A. D., & Goodale, M. A. (1995). *The visual brain in action.* Oxford, UK: Oxford University Press.

Mills, D. L., Coffey-Corina, S. A., & Neville, H. J. (1993). Language acquisition and cerebral specialization in 20-month-old infants. *Journal of Cognitive Neuroscience, 5*(3), 317–334.

Miyake, A., Friedman, N. P., Emerson, M. J., Witzki, A. H., & Howerter, A. (2000). The unity and diversity of executive functions and their contributions to complex "frontal lobe" tasks: A latent variable analysis. *Cognitive Psychology, 41*(1), 49–100.

Molenaar, P. C., & van der Maas, H. L. J. (1994). Commentary. *Human Development, 37*(3), 177–189.

Molfese, D. L., Wetzel, W. F., & Gill, L. A. (1993). Known versus unknown word discriminations in 12-month-old human infants: Electrophysiological correlates. *Developmental Neuropsychology, 9*(3-4), 241–258.

Morra, S. (2000). A new model of verbal short-term memory. *Journal of Experimental Child Psychology, 75*(3), 191–227.

Morra, S. (2008). A test of a neo-Piagetian model of the water-level task. *European Journal of Developmental Psychology, 5*(3), 369–400.

Morra, S. & Borella, E. (2015). Working memory training: From metaphors to models. *Frontiers in Psychology, 6,* 1097. doi: 10.3389/fpsyg.2015.01097

Morra, S., Camba, R., Calvini, G., & Bracco, F. (2013). Italians do it better? M-capacity measurement and cross-linguistic differences in the direction following task (DFT). *Rivista Di Psicolinguistica Applicata/Journal of Applied Psycholinguistics, 13*(1), 9–24.

Morra, S., Gobbo, C., Marini, Z., & Sheese, R. (2008). *Cognitive development: Neo-Piagetian perspectives*. New York: Erlbaum.

Morra, S., & Panesi, S. (2017). From scribbling to drawing: The role of working memory. *Cognitive Development, 43*, 142–158.

Morris, C. W. (1946). *Signs, language, and behavior*. New York: G. Braziller.

Moscovitch, M. (1992). Memory and working-with memory: A component process model based on modules and central systems. *Journal of Cognitive Neuroscience, 4*(3), 257–267.

Moscovitch M., Cabeza, R., Winocur, G., & Nadel, L. (2016). Episodic memory and beyond: The hippocampus and neocortex in transformation. *Annual Review of Psychology, 67*, 105–134.

Munakata, Y., Herd, S. A., Chatham, C. H., Depue, B. E., Banich, M. T., & O'Reilly, R. C. (2011). A unified framework for inhibitory control. *Trends in Cognitive Sciences, 15*(10), 453–459.

Munakata, Y., & Stedron, J. M. (2001). Neural network models of cognitive development. In C. A. Nelson & M. Luciana (Eds.), *Handbook of developmental cognitive neuroscience* (pp. 159–171). Cambridge, MA: MIT Press.

Nadel, J., Guerini, C., Peze, A., & Rivet, C. (1999). The evolving nature of imitation as a format for communication. In J. Nadel & G. Butterworth, G. (Eds.). *Imitation in infancy* (pp. 209–234). Cambridge, UK: Cambridge University Press.

Naglieri, J. (1985). *Matrix Analogies Test (Short Form): Examiner's manual*. San Antonio, TX: Psychological Corporation.

Nauta, W. J. H., & Feirtag, M. (1986). *Fundamental neuroanatomy*. New York: W.H. Freeman.

Nelson, C. A., & Webb, S. J. (2003). A cognitive neuroscience perspective on early memory development. In M. de Haan & M. H. Johnson (Eds.), *The cognitive neurosciences of development* (pp. 99–125). New York: Psychology Press.

Nelson, K. (1974). Concept, word, and sentence: Interrelations in acquisition and development. *Psychological Review, 81*(4), 267–285.

Newton, I. (1974). *Newton's philosophy of nature. Selections from his writings* (H. S. Thayer, Ed.). New York: Hafner Press.

Niaz, M. (1992). From Piaget's epistemic subject to Pascual-Leone's metasubject: Epistemic transition in the constructivist-rationalist theory of cognitive development. *International Journal of Psychology, 27*(6), 443–457.

Niaz, M. (1995). Cognitive conflict as a teaching strategy in solving chemistry problems: A dialectic-constructivist perspective. *Journal of Research in Science Teaching, 32*(9), 959–970.

Niaz, M. (2006). Can the study of thermochemistry facilitate students' differentiation between heat energy and temperature? *Journal of Science Education and Technology, 15*(3), 269–276.

Niaz, M., & Robinson, W. R. (1992). From "algorithmic mode" to "conceptual Gestalt" in understanding the behavior of gases: An epistemological perspective. *Research in Science and Technological Education, 10*(1), 53–64.

Noll, J. G., & Horn, J. L. (1998). Age differences in processes of fluid and crystallized intelligence. In J. J. McArdle & R. W. Woodcock (Eds.), *Human cognitive abilities in theory and practice* (pp. 263–281). Mahwah, NJ: Erlbaum.

Nöth, W. (1990). *Handbook of semiotics*. Bloomington: Indiana University Press.

Nozick, R. (2001). *Invariances: The structure of the objective world.* Cambridge, MA: Belknap.

Ochsner, K. N., Bunge, S. A., Gross, J. J., & Gabrieli, J. D. E. (2002). Rethinking feeling: An fMRI study of the cognitive regulation of emotion. *Journal of Cognitive Neuroscience, 14*(8), 1215-1229.

Oerter, R. (2000). Activity and motivation: A plea for a human frame motivation. *Advances in Psychology, 131,* 57–78.

Ohnishi, T., Matsuda, H., Asada, T., Aruga, M., Hirakata, M., Nishikawa, M., Katoh, A., & Imabyashi, E. (2001). Functional anatomy of musical perception in musicians. *Cerebral Cortex, 11*(8), 754–760.

Olsen, R. K., Lee, Y., Kube, J., Rosenbaum, R. S., Grady, C. L., Moscovitch, M., & Ryan, J. D. (2015). The role of relational binding in item memory: Evidence from face recognition in a case of developmental amnesia. *Journal of Neuroscience, 35*(13), 5342–5350.

Ortega, J. (1980). *El hombre y la gente* [Man and people]. Madrid, ES: Revista de Occidente en Alianza Editorial. (Original work published 1935)

Owen, A. M., Evans, A. C., & Petrides, M. (1996). Evidence for a two-stage model of spatial working memory processing within the lateral frontal cortex: A positron emission tomography study. *Cerebral Cortex, 6*(1), 31–38.

Owen, A. M., McMillan, K. M., Laird, A. R., & Bullmore, E. (2005). N-back working memory paradigm: A meta-analysis of normative functional neuroimaging studies. *Human Brain Mapping, 25*(1), 46–59.

Panksepp, J., & Biven, L. (2012). *The Norton series on interpersonal neurobiology. The archaeology of mind: Neuroevolutionary origins of human emotion.* New York: Norton.

Pascual-Leone, A. [Alvaro], Grafman, J., & Hallett, M. (1994). Modulation of cortical motor output maps during the development of implicit and explicit knowledge. *Science, 263*(5151), 1287–1289.

Pascual-Leone, A. [Alvaro], & Hamilton, R. (2001). The metamodal organization of the brain. *Progress in Brain Research, 134,* 427–445.

Pascual-Leone, A. [Alvaro], Freitas, C., Oberman, L., Horvath, J. C., Halko, M., Eldaief, M., Bashir, S., Vernet, M., Shafi, M., Westover, B., Vahabzadeh-Hagh, A. M., & Rotenberg, A. (2011). Characterizing brain cortical plasticity and network dynamics across the age-span in health and disease with TMS-EEG and TMS-fMRI. *Brain Topography, 24*(3-4), 302–315.

Pascual-Leone, A. [Antonio], Greenberg, L. S., & Pascual-Leone, J. (2009). Developments in task analysis: New methods to study change. *Psychotherapy Research, 19*(4-5), 527–542.

Pascual-Leone, A. [Antonio], Greenberg, L. S., & Pascual-Leone, J. (2014). Task analysis: New developments for programmatic research on the process of change. In W. Lutz & S. Knox (Eds.), *Quantitative and qualitative methods in psychotherapy research* (pp. 249–273). New York: Routledge.

Pascual-Leone, J. (1969). "Cognitive development and cognitive style: A general psychological integration" [Unpublished doctoral dissertation]. University of Geneva.

Pascual-Leone, J. (1970). A mathematical model for the transition rule in Piaget's developmental stages. *Acta Psychological, 32*(4), 301–345.

Pascual-Leone, J. (1976a). A view of cognition from a formalist's perspective. In K. F. Riegel & J. Meacham (Eds.), *The developing individual in a changing world* (pp. 89–100). The Hague: Mouton.

Pascual-Leone, J. (1976b). Metasubjective problems of constructive cognition: Forms of knowing and their psychological mechanism. *Canadian Psychological Review, 17*(2), 110–125.

Pascual-Leone, J. (1978). Compounds, confounds and models in developmental information processing: A reply to Trabasso and Foellinger. *Journal of Experimental Child Psychology, 26*(1), 18–40.

Pascual-Leone, J. (1983). Growing into human maturity: Toward a metasubjective theory of adulthood stages. In P. B. Baltes & O. G. Brim (Eds.), *Life-span development and behavior* (Vol. 5, pp. 117–156). New York: Academic Press.

Pascual-Leone, J. (1984). Attention, dialectic, and mental effort: Toward an organismic theory of life stages. In M. Commons, E. Richards, & C. Armon (Eds.), *Beyond formal operations: Late adolescent and adult cognitive development* (pp. 182–215). New York: Praeger.

Pascual-Leone, J. (1987). Organismic processes for neo-Piagetian theories: A dialectical causal account of cognitive development. *International Journal of Psychology, 22*(5–6), 531–569.

Pascual-Leone, J. (1988). Affirmations and negations, disturbances and contractions in understanding Piaget: Is his later theory causal? *Contemporary Psychology, 33*(5), 420–421.

Pascual-Leone, J. (1989). An organismic process model of Witkin's Field-Dependence-Independence. In T. Globerson & T. Zelniker (Eds.), *Cognitive development and cognitive style* (pp. 36–85). Norwood, NJ: Ablex.

Pascual-Leone, J. (1990a). An essay on wisdom: Toward organismic processes that make it possible. In R. J. Sternberg (Ed.), *Wisdom: Its nature, origins, and development* (pp. 244–278). New York: Cambridge University Press.

Pascual-Leone, J. (1990b). Reflections on life-span intelligence, consciousness and ego development. In C. Alexander & E. Langer (Eds.), *Higher stages of human development: Perspectives on adult growth* (pp. 258–285). New York: Oxford University Press.

Pascual-Leone, J. (1991a). A commentary on Boom and Juckes' on the learning paradox. *Human Development, 34*(5), 288–293.

Pascual-Leone, J. (1991b). Emotions, development and psychotherapy: A dialectical constructivist perspective. In J. Safran & L. Greenberg (Eds.), *Emotion, psychotherapy and change* (pp. 302–335). New York: Guilford.

Pascual-Leone, J. (1995). Learning and development as dialectical factors in cognitive growth. *Human Development, 38*(6), 338–348.

Pascual-Leone, J. (1996a). Piaget, Vygotski, y la función del símbolo [Piaget, Vygotsky and the symbolic function]. *Substratum, III (8–9),* 63–87.

Pascual-Leone, J. (1996b). Vygotsky, Piaget, and the problems of Plato. *Swiss Journal of Psychology, 55*(2-3), 84–92.

Pascual-Leone, J. (1998). To appraise developmental difficulty, or mental demand, relational complexity is not enough. *Behavioral and Brain Sciences, 21*(6), 843–844.

Pascual-Leone, J. (2000a). Mental attention, consciousness, and the progressive emergence of wisdom. *Journal of Adult Development, 7*(4), 241–254.

Pascual-Leone, J. (2000b). Reflections on working memory: Are the two models complementary? *Journal of Experimental Child Psychology, 77*(2), 138–154.

Pascual-Leone, J. (2006). Mental attention, not language, may explain evolutionary growth of human intelligence and brain size. *Behavioral and Brain Sciences, 29*(1), 19.

Pascual-Leone, J. (2012a). Piaget as a pioneer of dialectical constructivism: Seeking dynamic processes for human science. In E. Martí & C. Rodriguez (Eds.), *After Piaget* (pp. 15–41). New Brunswick, NJ: Transaction Publishers.

Pascual-Leone, J. (2012b). What is the *M*-capacity demand of car driving? A small essay on constructivist process/task analysis. In J. A. García Madruga, R. Kohen, C. del Barrio, I. Enesco, & J. Linaza, J. (Eds.), *Construyendo mentes. Ensayos en homenaje a Juan Delval* (pp. 69–84). Madrid, ES: UNED.

Pascual-Leone, J. (2013). Can we model organismic causes of working memory, efficiency and fluid intelligence? A meta-subjective perspective. *Intelligence, 41*(5), 738–743.

Pascual-Leone, J. (2014). Dialectics. In T. Teo (Ed.), *Encyclopedia of critical psychology* (pp. 421–428). New York: SpringerReference.

Pascual-Leone, J. (2015). Prólogo: Principios de constructivismo psicogenético [Prologue: Principles of psychogenetic constructivism]. In L. M. Rodriguez-Salazar & Z. Monroy Nasr (Eds.),

Psicología para epistemólogos, epistemología para psicólogos (pp. xi–xv). Mexico City: Corinter y Editorial Gedisa Mexicana.

Pascual-Leone, J. (2019). Growing minds have a maturing mental attention: A review of Demetriou and Spanoudis (2018). *Intelligence, 72,* 59–65.

Pascual-Leone, J., & Baillargeon, R. (1994). Developmental measurement of mental attention. *International Journal of Behavioral Development, 17*(1), 161–200.

Pascual-Leone, J., Escobar, E. M. R., & Johnson, J. (2012). Logic: Development of logical operations. In W. Hirstein (Ed.), *Encyclopedia of human behavior* (2nd ed., pp. 538–549). New York: Elsevier.

Pascual-Leone, J., & Goodman, D. (1979). Intelligence and experience: A neo-Piagetian approach. *Instructional Science, 8*(4), 301–367.

Pascual-Leone, J., Goodman, D. R., Ammon, P., & Subelman, I. (1978). Piagetian theory and neoPiagetian analysis as psychological guides in education. In J. M. Gallagher & J. Easley (Eds.), *Knowledge and development (Vol. 2): Piaget and education* (pp. 243–289). NY: Plenum.

Pascual-Leone, J., & Irwin, R. R. (1998). Abstraction, the will, the self, and modes of learning in adulthood. In M. C. Smith & T. Pourchot (Eds.), *Adult learning and development: Perspectives from educational psychology* (pp. 35–66). Hillsdale, NJ: Erlbaum.

Pascual-Leone, J. & Johnson, J. (1991). The psychological unit and its role in task analysis. A reinterpretation of object permanence. In M. Chandler & M. Chapman (Eds.), *Criteria for competence: Controversies in the assessment of children's abilities* (pp. 153–187). Hillsdale, NJ: Erlbaum.

Pascual-Leone, J., & Johnson, J. (1999). A dialectical constructivist view of representation: Role of mental attention, executives, and symbols. In I. E. Sigel (Ed.), *Development of mental representation: Theories and applications* (pp. 169–200). Mahwah, NJ: Erlbaum.

Pascual-Leone J., & Johnson, J. (2004). Affect, self-motivation, and cognitive development: A dialectical constructivist view. In D. Y. Dai & R. J. Sternberg (Eds.), *Motivation, emotion, and cognition* (pp. 197–235). Mahwah, NJ: Erlbaum.

Pascual-Leone, J., & Johnson, J. (2005). A dialectical constructivist view of developmental intelligence. In O. Wilhelm & R. W. Engle (Eds.), *Handbook of understanding and measuring intelligence* (pp. 177–201). Thousand Oaks, CA: Sage.

Pascual-Leone, J. & Johnson, J. (2011). A developmental theory of mental attention: Its application to measurement and task analysis. In P. Barrouillet & V. Gaillard (Eds.), *Cognitive development and working memory: A dialogue between neo-Piagetian and cognitive approaches* (pp. 13–46). New York: Psychology Press.

Pascual-Leone, J. & Johnson, J. (2017). Organismic causal models "from within" clarify developmental change and stages. In N. Budwig, E. Turiel, & P. Zelazo (Eds.), *New perspectives on human development* (pp. 67–87). Cambridge, UK: Cambridge University Press.

Pascual-Leone, J., Johnson, J., & Agostino, A. (2010). Mental attention, multiplicative structures, and the causal problems of cognitive development. In M. Ferrari & L. Vuletic (Eds.), *The developmental relations among mind, brain and education* (pp. 49–82). New York: Springer.

Pascual-Leone, J., Johnson, J., Baskind, S., Dworsky, S., & Severtston, E. (2000). Culture-fair assessment and the processes of mental attention. In A. Kozulin & Y. Rand (Eds.), *Experience of mediated learning: An Impact of Feuerstein's theory in education and psychology* (pp. 191–214). New York: Pergamon.

Pascual-Leone, J., Johnson, J., & Calvo, A. (2004, June 3–5). *Can mental attentional capacity predict the Canadian Cognitive Abilities score of school children?* [Poster presentation]. Jean Piaget Society Annual Conference, Toronto, ON, Canada.

Pascual-Leone, J., Johnson, J., Calvo, A., & Verrilli, E. (2005, August 3–7). *Can latent giftedness be indexed by mental-attentional capacity?* [Poster presentation]. 16th Biennial World Conference of the World Council for Gifted and Talented Children, New Orleans, LA, United States.

Pascual-Leone, J., & Morra, S. (1991). Horizontality of water level: A neo-Piagetian developmental review. In H. W. Reese (Ed.), *Advances in child development and behavior* (pp. 231–276). New York: Academic Press.

Pascual-Leone, J., Pascual-Leone, A., & Arsalidou, M. (2015). Neuropsychology still needs to model organismic processes "from within." *Behavioral and Brain Sciences, 38*, e83.

Pascual-Leone, J., & Smith, J. (1969). The encoding and decoding of symbols by children: A new experimental paradigm and a neo-Piagetian model. *Journal of Experimental Child Psychology, 8*(2), 328–355.

Pascual-Leone, J., & Sparkman, E. (1980). The dialectics of empiricism and rationalism: A last methodological reply to Trabasso. *Journal of Experimental Child Psychology, 29*(1), 88–101.

Pepper, S. C. (1942). *World hypotheses: A study of evidence*. Berkeley: University of California Press.

Pessoa, L. (2013). *The cognitive-emotional brain: From interactions to integration*. Cambridge, MA: MIT Press.

Petersen, S. E., & Posner, M. I. (2012). The attention system of the human brain: 20 years after. *Annual Review Neuroscience, 35*, 73–89.

Petrides, M. (1994). Frontal lobes and behavior. *Current Opinion in Neurobiology, 4*(2), 207–211.

Petrides, M. (2000). Dissociable roles of mid-dorsolateral prefrontal and anterior inferotemporal cortex in visual working memory. *The Journal of Neuroscience, 20*(19), 7496–7503.

Petrides, M., Alivisatos, B., Evans, A. C., & Meyer, E. (1993). Dissociation of human mid-dorsolateral from posterior dorsolateral frontal cortex in memory processing. *Proceedings of the National Academy of Sciences of the USA, 90*(3), 873–877.

Piaget, J. (1954). *The construction of reality in the child* (M. Cook, Trans.). New York: Basic Books.

Piaget, J. (1956). Discussion entre les rapporteurs [Discussion among participants]. In P. Osterrieth, J. Piaget, R. de Saussure, J. M. Tanner, H. Walon, R. Zazzo, B. Inhelder, & A. Rey, *Le problem des stades en psychologie de l'enfant* (p. 57). Paris: Presses Universitaires de France.

Piaget, J. (1958). Assimilation et connaissance [Assimilation and knowledge]. In A. Jonckheere, B. Mandelbrot, & J. Piaget, *La lecture de l'expérience* (pp. 49–108). Paris: Presses Universitaires de France.

Piaget, J. (1959). Apprentissage et connaissance [Learning and knowledge]. In P. Greco & J. Piaget, *Apprentissage et connaissance* (pp. 21–67). Paris: Presses Universitaires de France.

Piaget, J. (1963). *Origins of intelligence in children* (M. Cook, Trans.). New York: Norton. (Original work published 1948)

Piaget, J. (Ed.). (1967). *Logique et connaissance scientifique* [Logic and scientific knowledge]. Paris: Gallimard.

Piaget, J. (1970). *Structuralism* (C. Mashler, Trans.). New York: Basic Books.

Piaget, J. (Ed.). (1974). *Recherches sur la contradiction. 2: Les relations entre affirmations et négations* [Research on contradiction: Relations among affirmations and negations]. Paris: Presses Universitaires de France.

Piaget, J. (1980). *Les formes élémentaires de la dialectique* [Elementary forms of dialectics]. Paris: Gallimard.

Piaget, J. (1985). *The equilibration of cognitive structures: The central problem of intellectual development* (T. Brown, Trans.). Chicago: University of Chicago Press. (Original work published 1975)

Piaget, J., & Garcia, R. (1983). *Psychogenèse et histoire des science* [Psychogenesis and the history of science]. Paris: Flammarion.

Piaget, J., & Garcia, R. (1987). *Vers une logique des significations* [Toward a logic of meaning]. Geneva, CH: Murionde, Science Nouvelle.

Piaget, J., & Inhelder, B. (1969). *The psychology of the child* (H. Weaver, Trans.). New York: Basic Books.

Piaget, J., & Morf, A. (1958). Les isomorphismes partiels entre les structures logiques et les structures perceptives [The partial isomorphisms between logical and perceptual structures]. In J. S. Bruner, F. Bresson, A. Morf, & J. Piaget (Eds.), *Logique et perception* (pp. 49–116). Paris: Presses Universitaires de France.

Piazza, E. A. & Silver, M. A. (2017). Relative spatial frequency processing drives hemispheric asymmetry in conscious awareness. *Frontiers in Psychology, 8*, 559. https://doi.org/10.3389/fpsyg.2017.00559

Platt, B., & Riedel, G. (2011). The cholinergic system, EEG and sleep. *Behavioural Brain Research, 221*(2), 499–504.

Polya, G. (1973). *How to solve it: A new aspect of mathematical method* (2nd ed.). Princeton, NJ: Princeton University Press.

Pribram, K. H. (1971). *Languages of the brain*. Englewood Cliffs, NJ: Prentice-Hall.

Pribram, K. H. (1991). *Brain and perception*. Hillsdale, NJ: Erlbaum.

Prigogine, I., Allen, P. M., & Herman, R. (1977). The evolution of complexity and the laws of nature. In E. Laszlo & J. Bierman (Eds.), *Goals in a global community. A report to the Club of Rome: Volume I. Studies on the conceptual foundations* (pp. 1–63). New York: Pergamon.

Proctor, R. W., & Reeve, T. G. (1990). *Stimulus-response compatibility*. Amsterdam: North-Holland.

Rapaport, D. (1960). *The structure of psychoanalytical theory: A systematizing attempt*. New York: International Universities Press.

Rebotier, T. P., & Elman, J. L. (1996). Explorations with the dynamic wave model. In D. Touretzky, M. Mozer, & M. Haselmo (Eds.), *Advances in neural information processing systems 8* (pp. 549–556). Cambridge, MA: MIT Press.

Reinecke, M. & Fogel, A. (1994). The development of referential offering in the first year. *Early Development and Parenting, 3*(3), 181–186.

Rescorla, R. A., & Solomon, R. L. (1967). Two-process learning theory: Relationships between Pavlovian conditioning and instrumental learning. *Psychological Review, 74*(3), 151–182.

Reuchlin, M. (1962). *Les méthodes quantitatives en psychologie* [Quantitative methods in psychology]. Paris: Presses Universitaires de France.

Reuchlin, M. (1964). L'intelligence: Conception génétique opératoire et conception factorielle [Intelligence: Operational genetic and factorial conceptions]. *Revue Swiss de Psychologie, 23*(2), 113–134.

Rice, L. N. (1974). The evocative function of the therapist. In D. A. Wexler & L. N. Rice *Innovations in client centered therapy* (pp. 289–318). New York: Wiley.

Richards, J. E. (2002). Attention in young infants: A developmental psychophysiological perspective. In C. A. Nelson & M. Luciana (Eds.), *Handbook of developmental cognitive neuroscience* (pp. 321–338). Cambridge, MA: MIT Press.

Richards, J. E. (2003). The development of visual attention and the brain. In M. de Haan & M. H. Johnson (Eds.), *The cognitive neurosciences of development* (pp. 73–98). New York: Psychology Press.

Robbins, T. W., & Everitt, B. J. (1995). Arousal systems and attention. In M. S. Gazzaniga (Ed.), *The cognitive neurosciences* (pp. 703–720). Cambridge, MA: MIT Press.

Rock, I. (1983). *The logic of perception*. Cambridge, MA: MIT Press.

Rolls, E. T. (1999). The functions of the orbitofrontal cortex. *Neurocase, 5*(4), 301–312.

Romero Escobar, E. M. (2006). "A model of free recall using the mental-attention memory task" [Unpublished master's thesis]. York University.

Rothbart, M. K., & Posner, M. I. (2001). Mechanisms and variation in the development of attention. In C. A. Nelson & M. Luciana (Eds.), *Handbook of developmental cognitive neuroscience* (pp. 353–363). Cambridge, MA: MIT Press.

Rouse Ball, W. W. (1905). *Mathematical recreations and essays* (4th ed). New York: Macmillan.

Ruchkin, D. S., Grafman, J., Cameron, K., & Berndt, R. S. (2003). Working memory retention systems: A state of activated long-term memory. *Behavioral and Brain Sciences, 26*(6), 709–728.

Ruediger, S., Spirig D., Donato F., & Caroni, P. (2012). Goal-oriented searching mediated by ventral hippocampus early in trial-and-error learning. *Nature Neuroscience, 15*(11), 1563–1571.

Russell, J., & Hanna, R. (2012). A minimalist approach to the development of episodic memory. *Mind & Language, 27*(1), 29–54.

Russon, A. E., Bard, K. A., & Taylor Parker, S. (1996). *Reaching into thought: The minds of great apes.* Cambridge, UK: Cambridge University Press.

Saby, J. N., Bouquet, C. A., & Marshall, P. J. (2014). Young children co-represent a partner's task: Evidence for a joint Simon effect in five-year-olds. *Cognitive Development, 32,* 38–45.

Safran, J. (2003). *Psychoanalysis and Buddhism.* Somerville, MA: Wisdom Publications.

Sagvolden, T., Johansen, E. B., Aase, H., Russell, V. A. (2005). A dynamic developmental theory of attention-deficit/hyperactivity disorder (ADHD) predominantly hyperactive/impulsive and combined subtypes. *Behavioral and Brain Sciences, 28*(3), 397–419.

Sander, D., Grafman, J., & Zalla, T. (2003). The human amygdala: An evolved system for relevance detection. *Reviews in the Neurosciences, 14*(4), 303–16.

Sanides, F. (1970). Functional architecture of motor and sensory cortices in primates in the light of a new concept of neocortex evolution. In C. Noback & W. Montagna (Eds.), *The primate brain: Advances in primatology* (pp. 137–208). New York: Appleton-Century-Crofts.

Saper, C. B., & Fuller, P. M. (2017). Wake-sleep circuitry: An overview. *Current Opinion in Neurobiology, 44,* 186–192.

Sara, S. J., & Bouret, S. (2012). Orienting and reorienting: The locus coeruleus mediates cognition through arousal. *Neuron, 76*(1), 130–140.

Schaefer, S. M., Jackson, D. C., Davidson, R. J., Aguirre, G. K., Kimberg, D. Y., & Thompson-Schill, S. L. (2002). Modulation of amigdalar activity by the conscious regulation of negative emotion. *Journal of Cognitive Neuroscience, 14*(6), 913–921.

Scheler, M. (1961). *Man's place in nature* (H. Meyerhoff, Trans.). Boston: Beacon Press.

Schöner, G. (2014). Dynamical systems thinking. In P. C. M. Molenaar, R. N. Lerner, & K. N. Newell (Eds.), *Handbook of developmental systems theory and methodology* (pp. 188–217). New York: Guilford.

Searle, J. R. (1983). *Intentionality: An essay in the philosophy of mind.* New York: Cambridge University Press.

Seeley, W. W., Menon, V., Schatzberg, A. F., Keller, J., Glover, G. H., Kenna, H., Reiss, A. L., & Greicius, M. D. (2007). Dissociable intrinsic connectivity networks for salience processing and executive control. *The Journal of Neuroscience, 27*(9), 2349–2356.

Sescousse, G., Caldú, X., Segura, B., & Dreher, J.-C. (2013). Processing of primary and secondary rewards: A quantitative meta-analysis and review of human functional neuroimaging studies. *Neuroscience and Biobehavioral Reviews, 37*(4), 681–96.

Shanks, D. (2010). Learning: From association to cognition. *Annual Review of Psychology, 61*, 273–301.

Shapiro, K. A., Pascual-Leone. A., Mottaghy, F. M., Gangitand, M., & Caramazza, A. (2001). Grammatical distinctions in the left-frontal cortex. *Journal of Cognitive Neuroscience, 13*(6), 713–720.

Shayer, M., Demetriou, A., & Pervez, M. (1988). The structure and scaling of concrete operational thought: Three studies in four countries. *Genetic, Social, and General Psychology Monographs, 114*(3), 307–375.

Sherman, M. T., Seth, A. K., & Kanai, R. (2016). Predictions shape confidence in right inferior frontal gyrus. *Journal of Neuroscience, 36*(40), 10323–10366.

Sherrington, C. S. (1906). *The integrative action of the nervous system.* New York: Scribner's.

Sherrington, C. S. (1940). *Man on his nature.* London: Cambridge University Press.

Shimamura, A. P. (2002). Relational binding theory and the role of consolidation in memory retrieval. In L. R. Squire & D. L. Schacter (Eds.), *Neuropsychology of memory* (3rd ed., pp. 61–72). New York: Guilford.

Shipstead, Z., Lindsey, D. R. B., Marshall, R. L., & Engle, R. W. (2014). The mechanisms of working memory capacity: Primary memory, secondary memory, and attention control. *Journal of Memory and Language, 72*(1), 116–141.

Shirer, W. R., Ryali, S., Rykhlevskaia, E., Menon, W., & Greicius, M. D. (2012). Decoding subject-driven cognitive states with whole-brain connectivity patterns. *Cerebral Cortex, 22*(1), 158–165.

Shrager, J., & Johnson, M. H. (1996). Dynamic plasticity influences the emergence of function in a simple cortical array. *Neural Networks, 9*(7), 1119–1129.

Sigel, I. E. (1993). The centrality of a distancing model for the development of representational competence. In R. R. Cocking & K. A. Renninger (Eds.), *The development and meaning of psychological distance* (pp. 141–156). Hillsdale, NJ: Erlbaum.

Silva, M. A., Mendonça Filho, E. J., & Bandeira, D. R. (2019). Development of the Dimensional Inventory of Child Development Assessment (IDADI). *Psico-USF, 24*(1), 11–26. http://www.scielo.br/scielo.php?script=sci_arttext&pid=S1413-82712019000100011&tlng=en

Silva, M. A., de Mendonça Filho, E. J. & Bandeira, D. R. (in press). *Inventário Dimensional de Avaliação do Desenvolvimento Infantil* [Dimensional Inventory for Child Development Assessment]. São Paulo, BR: Vetor Editora.

Simon, T. J. (1997). Reconceptualizing the origins of number knowledge: A "non-numerical" account. *Cognitive Development, 12*(3), 349–372.

Singer, W. (2001). Consciousness and the binding process. In P. Marijuan (Ed.), *Cajal and consciousness* (pp. 123–146). New York: New York Academy of Sciences.

Skinner, B. F. (1971). *Beyond freedom and dignity*. New York: Bantam Books.

Skinner, J. E., & Yingling, C. D. (1977). Central gating mechanisms that regulate event-related potentials and behavior. In J. E. Desmedt (Ed.), *Progress in clinical neurophysiology: Attention, voluntary contraction and event-related cerebral potentials* (Vol. 1, pp. 30–69). Basel, CH: Karger.

Skuy, M., Gewer, A., Osrin, Y., Knunou, D., Fridjhon, P., & Rushton, J. P. (2002). Effects of mediated learning on Raven's matrices scores of African and non-African university students in South Africa. *Intelligence, 30*(3), 221–232.

Spanoudis, G., Demetriou, A., Kazi, S., Giorgala, K., & Zenonos, V. (2015). Embedding cognizance in intellectual development. *Journal of Experimental Child Psychology, 132,* 32–50.

Spearman, C. (1927). *The abilities of man*. Oxford, UK: Macmillan.

Spelke, E. (1994). Initial knowledge: Six suggestions. *Cognition, 50*(1–3), 431–445.

Spencer, J. P., Perone, S., & Buss, A. (2011). Twenty years and going strong: A dynamic systems revolution in motor and cognitive development. *Child Development Perspectives, 5*(4), 200–266).

Stahl, S. M. (2000). *Essential psychopharmacology*. Cambridge, UK: Cambridge University Press.

Stamovlasis, D. & Tsaparlis, G. (2012). Applying catastrophe theory to an information-processing model of problem solving in science education. *Science Education, 96*(3), 392–410.

Stankov, L. (2002). g: A diminutive general. In R. J. Sternberg & E. L. Grigorenko (Eds.), *The general factor of intelligence: How general is it?* (pp. 19–37). Mahwah, NJ: Erlbaum.

Stephan, K. E., Marshall, J. C., Friston, K. J., Rowe, J. B., Ritzl, A., Zilles, K., & Fink, G. R. (2003). Lateralized cognitive processes and lateralized task control in the human brain. *Science, 301*(5631), 384–386.

Steriade, M. (1996). Arousal: Revisiting the reticular activating system. *Science, 272*(5259), 225–226.

Stern, W. (1977). The psychological methods of testing intelligence (G. M. Whipple, Trans.). In D. N. Robinson (Ed.), *Significant contributions to the history of psychology 1750–1920* (Series B, *4*[4]). Washington, DC: University Publications of America. (Original work published 1914)

Stevens, W. D., Kahn, I., Wig, G. S., & Schacter, D. L. (2012). Hemispheric asymmetry of visual scene processing in the human brain: Evidence from repetition priming and intrinsic activity. *Cerebral Cortex, 22*(8), 1935–1949.

Stevens, W. D., Kravitz, D. J., Peng, C. S., Tessler, M. H., & Martin, A. (2017). Privileged functional connectivity between the visual word form area and the language system. *Journal of Neuroscience, 37*, 5288–5297.

Stewart, L., & Pascual-Leone, J. (1992). Mental capacity constraints and the development of moral reasoning. *Journal of Experimental Child Psychology, 54*(3), 251–287.

Strauss, S. (Ed., 1982) *U-Shaped behavioral growth*. New York: Academic Press.

Striedter, G. F. (2005). *Principles of brain evolution*. Sunderland, MA: Sinauer.

Stiles, J. (2001). Spatial cognitive development. In C. A. Nelson & M. Luciana (Eds.), *Handbook of developmental cognitive neuroscience* (pp. 399–414). Cambridge, MA: MIT Press.

Störmer, V. S., Passow, S., & Biesenack, J. (2012). Dopaminergic and cholinergic modulations of visual-spatial attention and working memory: Insights from molecular genetic research and implications for adult cognitive development. *Developmental Psychology, 48*(3), 875–889.

Stöttinger, E., Filipowicz, A., Valadao, D., Culham, J. C., Goodale, M. A., Anderson, B., & Danckert, J. (2015). A cortical network that marks the moment when conscious representations are updated. *Neuropsychologia, 79*(Pt A), 113–122.

Stuss, D. T., Gallup, G. G., & Alexander, M. P. (2001). The frontal lobes are necessary for "theory of mind." *Brain, 124*(Pt 2), 279–286.

Stuss, D. T., & Levine, B. (2002). Adult clinical neuropsychology: Lessons from studies of the frontal lobes. *Annual Review of Psychology, 53*, 401–433.

Suzuki, W. A. (2002). Cortical memory systems in the nonhuman primate: An anatomical and physiological perspective. In L. R. Squire & D. L. Schacter (Eds.), *Neuropsychology of memory* (3rd ed., pp. 289–300). New York: Guilford Press.

Swanson, L. W. (2003). *Brain architecture: Understanding the basic plan*. New York: Oxford University Press.

Talmi, D., Seymour, B., Dayan, P., & Dolan, R. J. (2008). Human Pavlovian-instrumental transfer. *The Journal of Neuroscience, 28*(2), 360–368.

Terwee, S. J. S. (1995). Deconstructing social constructionism. In I. Lubek, R. van Hezewijk, G. Pheterson, & C. W. Tolman (Eds.), *Trends and issues in theoretical psychology* (pp. 189–194). New York: Springer.

Thal, D. J., Marchman, V., Stiles, Y., Aram, D., Trauner, D., Nass, D., & Bates, E. (1991). Early lexical development in children with focal brain injury. *Brain and Language, 40*(4), 491–527.

Thatcher, R. W. (1994). Cyclical cortical reorganization: Origins of human cognitive development. In G. Dawson & K. W. Fischer (Eds.), *Human behavior and the developing brain* (pp. 1232–266). New York: Guilford.

Thatcher, R. W. (1997). Human frontal lobe development: A theory of cyclical cortical reorganization. In N. A. Krasnegor, G. R. Lyon, & P. S. Goldman-Rakic (Eds.), *Development of prefrontal cortex* (pp. 85–113). Baltimore: Paul H. Brooks.

Thatcher, R. W. (2010). Higher-order network reworking—New findings. In M. Ferrari & L. Vuletic (Eds.), *The developmental relations among mind, brain and education* (pp. 83–104). New York: Springer.

Thatcher, R. W., North, D. M., & Biver, C. J. (2008). Intelligence and EEG phase reset: A two compartmental model of phase shift and lock. *NeuroImage, 42*(4), 1639–1653.

Thelen, E., & Smith, L. B. (1994). *A dynamic systems approach to the development of cognition and action*. Cambridge, MA: MIT Press.

Thelen, E., & Ulrich, B. D. (1991). Hidden skills: A dynamic systems analysis of treadmill stepping during the first year. *Monographs of the Society for Research in Child Development, 56* (Serial No. 223).

Todor, J. (1977). Cognitive development, cognitive style, and motor activity. In B. Kerr (Ed.), Human performance and behavior. *Proceedings of the 9th Canadian Psycho-Motor Learning and Sports Psychology Symposium*, Banff, AB, Canada.

Todor, J. (1979). Developmental differences in motor task integration: A test of Pascual-Leone's theory of constructive operators. *Journal of Experimental Child Psychology, 28*(2), 314–322.

Tolman, E. C. (1938). A reply to Professor Guthrie. *Psychological Review, 45*(2), 163–164.

Tolman, E. C. (1945). A stimulus-expectancy need-cathexis psychology. *Science, 101*(2616), 160–166.

Tolman, E. C. (1959). Principles of purposive behavior. In S. Koch (Ed.), *Psychology: A study of a science* (Vol. 2, pp. 92–157). New York: McGraw-Hill.

Tolman, E. C. (1961). *Behavior and psychological man*. Los Angeles: University of California Press.

Tolman, E. C., & Brunswik, E. (1935). The organism and the causal texture of the environment. *Psychological Review, 42*(1), 43–77.

Tomasello, M. (2008). *Origins of human communication*. Cambridge, MA: MIT Press.

Tomasello, M. (2018). Great apes and human development: A personal history. *Child Development Perspectives, 12*(3), 189–193.

Tomasello, M., & Call, J. (1997). *Primate cognition*. New York: Oxford University Press.

Treadway, M. T., Buckholtz, J. W., Cowan, R. L., Woodward, N. D., Li, R., Ansari, M. S., Baldwin, R. M., Schwartzman, A. N., Kessler, R. M., & Zald, D. H. (2012). Dopaminergic mechanisms of individual differences in human effort-based decision-making. *Journal of Neuroscience, 32*(18), 6170–6176.

Trevarthen, C. B. (Ed.). (1990). *Brain circuits and functions of the mind: Essays in honor of Roger W. Sperry*. Cambridge, UK: Cambridge University Press.

Trevarthen, C., Kokkinaki, T., & Flamenghi, G. A. (1999). What infant imitations communicate: With mothers, with fathers and with peers. In J. Nadel & G. Butterworth (Eds.), *Imitation in children* (pp. 127–185). Cambridge, UK: Cambridge University Press.

Troadec, B., & Martinot, C. (2003). *Le développement cognitif. Théories actuelles de la pensée en contextes* [Cognitive development: Current theories of thinking in context]. Paris: Belin.

Tsaparlis, G. (1998). Dimensional analysis and predictive models in problem solving. *International Journal of Science Education, 20*(3), 335–350.

Tsaparlis, G., & Angelopoulos, V. (2000). A model of problem solving: Its operation, validity, and usefulness in the case of organic-synthesis problems. *Science Education, 84*(2), 131–153.

Tsaparlis, G., Kousathana, M., & Niaz, M. (1998). Molecular-equilibrium problems: Manipulation of logical structure and of M-demand, and their effect in student performance. *Science Education, 82*(4), 437–454.

Tucker, D. M. (2001). Motivated anatomy: A core-and-shell model of corticolimbic architecture. In G. Gainotti (Ed.), *Handbook of neuropsychology: Vol. 5. Emotional behavior and its disorders* (2nd ed.). Amsterdam: Elsevier.

Tucker-Drob, E. M. (2009). Differentiation of cognitive abilities across the lifespan. *Developmental Psychology, 45*(4), 1091–1118.

Uhlhaas, P., Roux, F., Singer, W., Haenschel, C., Sireteanu, R., & Rodriquez, E. (2009). The development of neural synchrony reflects late maturation and restructuring of functional networks in humans. *Proceedings of the National Academy of Sciences of the USA, 106*(24), 9866–9871.

Ullmo, J. (1958). *La pensée scientifique modern* [Modern scientific thought]. Paris: Flammarion.

Ullmo, J. (1967). Les concepts physiques [Physical concepts]. In J. Piaget (Ed.), *Logique et connaissance scientifique* (pp. 623–705). Paris: Gallimard.

Ullsperger, M. (2011). Neurochemistry of performance monitoring. In R. B. Mars, J. Sallet, M. F. S. Rushworth, & N. Yeung (Eds.), *Neural basis of motivation and cognitive control* (pp. 37–49). Cambridge, MA: MIT Press.

Unsworth, N., & Engle, R. W. (2005). Working memory capacity and fluid abilities: Examining the correlation between operation span and Raven. *Intelligence, 33*(1), 67–81.

Uzgiris, I. C., & Hunt, J. McV. (1975). *Assessment in infancy: Ordinal scales of psychological development.* Champaign: University of Illinois Press.

Valsiner, J. (2006). Developmental epistemology and implications for methodology. In W. Damon & R. M. Lerner (Series Eds.) & R. M. Lerner (Vol. Ed.), *Handbook of child psychology: Vol. 1. Theoretical models of human development* (6th ed., pp. 166–209). New York: Wiley.

van der Maas, H. (1993). *Catastrophe analysis of stagewise cognitive development.* Amsterdam: University of Amsterdam.

Van der Veer, R., & Valsiner, J. (Eds.). (1994). *The Vygotsky reader.* Oxford, UK: Blackwell.

Vartanian, O., & Goel, V. (2005). Task constraints modulate activation in right ventral lateral prefrontal cortex. *NeuroImage, 27*(4), 927–933.

Vendetti, M. S., Johnson, E. L., Lemos, C. J., & Bunge, S. A. (2015). Hemispheric differences in relational reasoning: Novel insights based on an old technique. *Frontiers in Human Neuroscience, 9,* 55. https://doi.org/10.3389/fnhum.2015.00055

von Stein, A., & Sarnthein, J. (2000). Different frequencies for different scales of cortical integration: From gamma to long range alpha/theta synchronization. *International Journal of Psychophysiology, 38*(3), 301–313.

von Uexküll, J. (1957). A stroll through the worlds of animals and men: A picture book of invisible worlds. In C. H. Schiller (Ed. & Trans.), *Instinctive behavior* (pp. 5–80). New York: International Universities Press. (Original work published 1934)

von Uexküll, J. (1982). The theory of meaning. *Semiotica, 42*(1), 25–87.

von Uexküll, T. (1982). Introduction: Meaning and science in Jakob von Uexküll's concept of biology. *Semiotica, 42*(1), 1–24.

Vygotsky, L. (1978). *Mind in society: The development of higher mental processes.* Cambridge, MA: Harvard University Press.

Vygotsky, L., & Luria, A. R. (1994). Tool and symbol in child development. In R. van der Veer & J. Valsiner (Eds.), *The Vygotsky reader* (pp. 99–174). Oxford, UK: Blackwell.

Walsh, V., & Pascual-Leone, A. (2003). *Transcranial magnetic stimulation: A neurochronometrics of mind.* Cambridge, MA: MIT Press.

Walton, M. E., Gan, J. O., & Phillips, P. F. M. (2011). The influence of dopamine in generating action from motivation. In R. B. Mars, J. Sallet, M. F. S. Rushworth, & N. Yeung (Eds.), *Neural basis of motivation and cognitive control* (pp. 163–187). Cambridge, MA: MIT Press.

Wang, X. F., & Chen, G. (2003). Complex networks: Small-world, scale-free and beyond. *IEEE Circuits and Systems Magazine, 3*(1), 6–20.

Wapner, S., & Demick, J. (Eds.). (1991). *Field dependence-independence.* Hillsdale, NJ: Erlbaum.

Watanabe, S. (1969). *Knowing and guessing: A quantitative study of inference and information.* New York: Wiley.

Waters, A., & Tucker, D. (2013). Self-regulation of neural development. In B. Sokol, F. Grouzet, & U. Müller (Eds.), *Self-regulation and autonomy: Social and developmental dimensions of human conduct* (pp. 279–296). Cambridge, UK: Cambridge University Press.

Watter, S., Geffen, G. M., & Geffen, L. B. (2001). The n-back as a dual task: P300 morphology under divided attention. *Psychophysiology, 38*(6), 998–1003.

Watts, D. J., & Strogatz, S. H. (1998). Collective dynamics of "small-world" networks. *Nature,* 393(6684), 440–442.

Wellek, S. (2003). *Testing statistical hypotheses of equivalence.* Boca Raton, FL: Chapman and Hall/CRC.

Werner, H., & Kaplan, B. (1984). *Symbol formation.* Hillsdale, NJ: Erlbaum.

Wilhelm, O. & Engle, R. W. (Eds.). (2005). *Handbook of understanding and measuring intelligence.* Thousand Oaks, CA: Sage.

Wimmer, H., & Perner, J. (1983). Beliefs about beliefs: Representation and constraining function of wrong beliefs in young children's understanding of deception. *Cognition, 13*(1), 103–28.

Witkin, H. A., Dyk, R. B., Faterson, H. F., Goodenough, D. R., & Karp, S. A. (1962). *Psychological differentiation: Studies of development.* New York: Wiley.

Witkin, H. A., & Goodenough, D. R. (1981). *Cognitive styles, essence and origin: Field dependence and field independence.* New York: International Universities Press.

Wolford, G., Miller, M., & Gazzaniga, M. (2000). The left hemisphere's role in hypothesis formation. *Journal of Neuroscience, 20*(6), RC64.

Wynn, K. (1992). Addition and subtraction by human infants. *Nature, 358*(6389), 749–750.

Yamakoshi, G. (2004). Evolution of complex feeding techniques in primates: Is this the origin of great ape intelligence? In A. E. Russon & D. R. Begun (Eds.), *The evolution of thought: Evolutionary origins of great ape intelligence* (pp. 140–171). Cambridge, UK: Cambridge University Press.

Yaple, Z. A., Stevens, W. D., Arsalidou, M. (2019). Meta-analyses of the n-back working memory task: fMRI evidence of age-related changes in prefrontal cortex involvement across the adult lifespan. *NeuroImage, 196*, 16–31.

Yaremko, R. M., Harari, H., Harrison, R. C., Lynn, E. (1982). *Reference handbook of research and statistical methods in psychology.* New York: Harper & Row.

Yuan, Y., & Brown, S. (2015). Drawing and writing: An ALE meta-analysis of sensorimotor activations. *Brain and Cognition, 98*,15–26.

Zajonc, R. B. (1960). The concepts of balance, congruity, and dissonance. *Public Opinion Quarterly, 24*(2), 280–296.

Zatorre, R. J., & Peretz, I. (Eds.). (2001). The biological foundations of music. *Annals of the New York Academy of Sciences, 930*(1).

Zinchenko, V. P. (1985). Vygotsky's ideas about units for the analysis of mind. In J. V. Wertsch (Ed.), *Culture communication and cognition* (pp. 94–118). London: Cambridge University Press.

Zubiri, X. (1999). *Sentient intelligence* (T. B. Fowler, Trans). Washington, DC: The Xavier Zubiri Foundation of North America. https://www.scribd.com/document/57580653/Zubiri-Sentient-Intelligence

Zubiri, X. (2001). *Sobre la realidad* [About reality]. Madrid, ES: Alianza Editorial. (Original work published 1966)

Index

Note: Figures and tables are indicated by "f" and "t" respectively, following page numbers.

Abduction, 398
ascendent/abductive analysis, 283–285, 287, 289, 297
conservation and, 46–47, 264, 273
geometrical analysis and, 283–285, 287
in metasubjective task analysis, 283–285, 287, 289, 297
overdetermination and, 46–47
Peirce on, 14, 34, 46–47, 160, 264, 273, 284, 399
Abstraction, causal, 177t, 180, 322
in brain, 324–326
fluents used by, 325–326
Abstraction, constructive, 177t, 322
in brain, 179, 179t, 180, 324–325, 431n11
levels of, 178–179, 179t, 323
Abstraction, reductive, 177, 177t, 178, 322–324
in brain, 359, 368–369
levels of, 366–369
Abstraction, reflective, 124
in brain, 179, 179t, 180, 322, 324–326, 359, 368–369, 431n11
conceptual, 366–368
in constructivist learning, 320, 322, 422
epistemo-ontological levels and, 319, 322–323
equilibration and, 319, 322
infralogical, 177t, 178, 324–325
internalization in, 423
intratask learning and, 179–180
levels of, 178–179, 179t, 323, 366–369, 421
logological, 177, 177t, 178, 324–325, 367–368
mereological, 324–325, 367–368
in metasubjective task analysis, 295
modes of, 176–177, 177t, 178–179, 179t, 180, 324–325, 367–368
Piaget and, 37–38, 50, 123, 176–179, 179t, 180, 222, 319–320, 323, 349, 359, 367–368
relational-associative learning and, 177t, 178
relational-symbolic learning and, 177, 177t, 178–180
relations coordinated by, 24, 425
schemes and, 37–38, 176–177, 177t, 178–179, 179t, 180, 325–326
Accommodation, 13, 401
Baldwin on, 39–40
in brain, 362–363
content learning and, 168–169
encumbrances and, 25–26
Piaget on, 25, 40, 67–68, 168, 362
primary, 25–26, 67
Resistances/resistances and, 25–26
secondary, 25, 67–68

Acetylcholine (ACh), 386, 387t, 393–394
 arousal and, 387, 390
 norepinephrine and, 387–388, 390–391
Action schemes, 152–153, 428n10
 executive schemes and, 21, 152, 156, 425
 in motor rho task, 311, 312t, 313–314
Activation, 359, 375
 boosting of schemes', 117, 134, 209–212, 215
 in brain, 376–377, 378f, 379, 386, 392–393
 field of, 218, 219f, 252, 266, 333, 353
 in learning, 171–173, 430n9
 reticular ascending system of, 376–377
Activity, 201–202
Adaptive flexibility, 356
Adey, P., 228, 229f, 397
Ads (adjunct information), 143, 146–149, 153
Adults
 direction-following task scores of, 241, 241f, 242, 242f
 figural intersections task scores of, 241, 241f
 mental-attentional capacity in, 240–241, 241f, 242–243, 242f
 mental attention memory task scores of, 241, 241f
 working memory in, 240–241, 243
Affect
 facilitating/misleading task criterion and, 85–89
 motivation and, 22t, 55, 154, 199, 424
 A-operator of, 22t, 213, 214t, 215, 298–299, 391
Affective goals, 152, 154, 161, 430n2
 automatic attention and, 197–199
 in brain, 330, 369–372
 cognitive goals and, 62, 144, 197–199, 330, 370–372
 executive schemes and, 156–157
Affective learning, 22t, 171, 423
Affective-learning psychosocial processes (*B*-schemes), 62, 213, 214t, 216
Affective schemes (*A*-schemes), 152, 422, 425, 428n10
 attachment and, 298
 in brain, 147, 328, 330
 cognitive schemes and, 54–55, 62, 152–153, 252, 330–331
 feelings yielded by, 152–153
 processing complexity in, 166–167
Afferent processes, 333, 339, 377, 421
Affirmations, 11–13, 31
Affordances, 25–26, 326, 359
Agency/praxis, 374, 430n2
 activity and, 202
 attention and, 183
 James on, 24
 operative, 374
 Resistances/resistances and, 143–145, 326, 399, 401
 schemes and, 149–151, 167
 semiotic processes and, 123, 143–144
Agostino, A., 307
Alp, I. E., 97, 103–104
Analytical schemes, 147, 328
Animals
 meaning in, 119–120
 mental/effortful attention in, 207
 symbolic function in, 136–137
A-not-B-error task, 349–352
Antisaccade, 185–186
A-operator. *See* Affect
Arousal
 acetylcholine and, 387, 389
 mesencephalic reticular formation regulating, 385–386
 neurotransmitters of, 386–387, 387t, 388–393
Arsalidou, M., 244–245, 348–349
Ascendent/abductive analysis, 283–285, 287, 289, 297
Assimilation, 401
 in brain, 362–363
 content learning and, 168–170
 logical learning and, 170–171

overdetermination by, 13–14, 16
Piaget on, 16, 25, 61, 68, 141, 165, 169–170, 199, 295, 327, 333, 362–363, 399, 428n5
Schemes' Overdetermination of Performance and, 165, 190, 333
strength of, 151, 171
Associative learning, 163–164, 209–211
Attachment, 298–299
Attention
in brain, 203–204, 341–346, 346t, 347–349, 359, 362–363, 365, 370, 375–377, 394, 403
defining, 181–182
habit and, 182–184
Husserl on, 205–206
James on, 182–184
neuroscience on, 184–185, 209, 348
praxis and, 183
Attention, automatic. *See* Automatic attention
Attention, mental/effortful. *See* Mental/effortful attention
Austin, J. H., 203
Autobiographical self, 74, 77–78, 79t, 253, 409
Automatic attention, 213, 214t, 421
affective factors influencing, 197–199
in antisaccade, 185–186
content and relational-associative learning in, 196
in facilitating situation, 186, 189–190, 217, 336
habit and, 183–184
inhibition and, 197–198, 206, 375
intratask learning and, 174–175, 180, 196, 345
mental/effortful attention aiding, 201–204, 217
mental/effortful attention compared with, 181–183
motivation and, 199
neuroscience on, 184–185
neurotransmitters and, 387, 387t, 388–389
schemes and, 185–187, 196–199, 204
simple and complex, 186–188, 203–204
Automatization, 174–175, 180, 345
Bachelard, G., 3, 5, 397
Bacon, F., 397
Baillargeon, Renée
drawbridge screen task of, 90–91, 91f, 92
in neonativism, 89–91, 91f, 92–93
Bakchine, S., 394
Baldwin, J. M., 39–40, 49, 407, 428nn4–5
Bandeira, D. R., 61, 96
Barrouillet, P., 402, 428n4
Barth, J., 300, 303
Basal ganglia (BG), 376–377, 378f
Bennett, M. R., 318
Benson-Hamstra, Nancy, 61
sensorimotor mental-attentional capacity and research of, 94, 97–100, 100t, 102t, 103–104, 104f, 105, 105f, 106–107, 107f, 108–109, 109f, 110, 111t, 112, 429n1
Berthoz, A., 264, 317, 331
Bickhard, M. H., 252
Binding problem, 319, 334
Body schema, 63, 65t
Bohm, D., 221
B-operator (social Being), 62, 213, 216, 276, 298–299
Brain, 318. *See also* Neuropsychology/Neuroscience
abstraction levels in, 338–340
accommodation in, 362–363
activation in, 375–377, 378f, 379–380, 386, 392–393
affective goals in, 330, 369–372
affective-relevance calculation in, 198
affective schemes in, 147, 328, 330
assimilation in, 362–363
attention in, 203–204, 341–346, 346t, 347–349, 359, 362–363, 365, 370, 375–377, 387t, 394, 403
automatization in, 345
Berthoz on, 264, 317, 331
binding problem in, 319, 334
Brodmann areas of, 321f
causal abstraction in, 177t, 324–326
cognitive goals in, 330, 362, 370–372

Brain (cont.)
- constructive abstraction in, 177t, 179, 179t, 180, 324–325, 431n11
- cortex microanatomy of, 321f
- cortical areas in, 179, 179t, 180, 189, 264, 320f, 322f, 339, 352–353, 355–356, 358–359, 362–363, 370, 372–374, 431n11
- corticocortical connections in, 380, 393–394
- development and, 340–341
- dorsal-attention versus ventral-attention networks in, 359, 362–363
- dorsolateral prefrontal cortex, 352–353, 355–356, 358, 370, 372–374
- Edelman on, 190, 264, 319, 331–334, 339, 379, 399, 407–408
- executive schemes in, 147, 369–370
- facilitating situation and, 189–190, 371, 387, 387t, 389–390, 392, 396
- *F*-factor and, 264, 389–390
- figurative schemes in, 147, 328, 360–361, 380
- fluents in, 360–361
- frontopolar cortex, 353, 355, 373–374
- group competition in, 331–333
- group confinement in, 331–332
- group selection in, 190, 331–333
- heterarchy in, 322, 339, 368, 373–374, 392
- hippocampus, 382–385, 394–395
- infralogical domain of, 177t, 324–325
- inhibition in, 375–377, 378f, 379–380, 386, 392–393
- lateral inhibition in, 191–192, 264, 332, 334, 338, 370–371, 379, 389
- lateralization issues of, 177t, 320f, 321f, 324–325, 341–342, 343f, 344–346, 346t, 364–366, 372, 395
- limbic, subcortical centers in, 322f
- limbic organization, 390–391, 394–396
- logological, 177t, 324–325, 367–368
- medial right hemisphere, 320f, 321f, 322f
- meditation in, 203–204
- mental demand in, 354–355, 357, 365
- mental/effortful attention in, 341–346, 346t, 347–349, 354–355, 357–358, 365, 370, 375–377, 378f, 379–380, 383–384, 387t, 393–394, 403
- mereological domain, modes in, 177t, 324–325, 367–368
- metamodal processes of, 329, 334
- in misleading situation, 189–190, 390–391, 387t, 393, 395–396
- neuronal population processes in, 331–333
- norepinephrine in, 391
- operative schemes in, 147, 360–361, 374–375
- pathways in, 376–377, 378f, 379–380, 384–386, 389, 393
- prefrontal cortex, 352, 369–377, 379–380, 391, 393–394, 432n2
- prefrontal tasks for, 353–358
- quaternary areas in, 339–340
- Reality/reality and, 326–327
- reductive abstraction in, 177t, 359, 368–369
- reflective abstraction in, 177t, 179, 179t, 180, 322, 324–326, 359, 368–369, 431n11
- relational-associative (*LC*) learning in, 22t, 214t, 345, 363–365, 369
- relational-symbolic (*LM*) learning in, 22t, 214t, 345, 363–365, 369
- Resistances/resistances and, 326–327
- schemes in, 147, 328–330, 334–335, 341, 360–361, 369–370, 374–375, 380, 408
- semantic mapping of, 359–360
- space in, 22t, 200–201, 214t, 385, 394–395
- time in, 22t, 200–201, 214t, 381–385, 394–396
- topography of, 320f, 321f, 322f
- waves of, 342–346, 346t, 347–349

Brain resources. *See* Operators

Brown, S., 363

Brunswik, E., 25, 326–327

B-schemes (affective-learning psychosocial processes), 62, 213, 214t, 216

Buss, A. T., 206

Buzsaki, G., 383–384

Call, J., 82, 300
Calvo, A., 225
Canadian Cognitive Abilities Test (CCAT), 225–226, 226f, 227, 249, 250t
Casasanto, D., 324–325
Case, Robbie, 97, 209, 274
  Pascual-Leone, J., and, 20, 403–405
  staircase model and, 222, 222f, 223–225
  Thatcher and, 345–346, 348
Cassirer, E., 7, 119
  on concept, 366–368
  on invariants, 257
  on relational complexity, 20
  on Resistances/resistances, 24–25, 31
  on symbol, signal and, 125–126, 127f, 159
  on symbolic forms, 150
Causal abstraction. *See* Abstraction, causal
Causality
  causal texture of environment, 25, 145, 224, 326–327, 349, 396, 430n3
  multisided, 33, 40, 47
  Piaget on, 24
Causal-organismic processes. *See* Organismic processes
Causal theory/model, 27, 294, 318, 398, 400
  descriptive theory/model and, 5–6, 34, 36, 63, 175, 257–258
  Piaget's, 36–38, 40
Centration, 206
  decentration, 221–222, 421
  M-centration, 173, 218, 220, 292, 414, 417
  Piaget on, 40, 183, 221–222
  protracted *M*-centration, 173
  recentration, 287, 421
Cerebral cortex. *See* Brain
Chimpanzees, 75, 119, 136–137, 216
Cholinergic system. *See* Acetylcholine
Christoff, K., 355, 370–371, 373
Chunks, 172–173
Claparede, E., 183, 187
C learning. *See* Learning, content
Cockburn, J., 155–156, 377
Coexistence, relations of, 425
  Leibniz and Kant on, 200–201, 381
  S-operator of, 22t, 200–201, 213, 214t, 279, 326, 381, 385, 395
Cognition, 120
  emotions and, 54–56
  hidden operators and, 337–338
  high and low, 33, 47–49, 58, 123, 180, 329
Cognitive goals
  affective goals and, 62, 144, 197–199, 330, 370–372
  in brain, 330, 370–372
Cognitive learning. *See* Constructivist learning
Cognitive psychology, 401–402
Cognitive schemes, 56–57, 63, 329
  affective schemes and, 54–55, 62, 152, 252, 330–331
  cognitive-figurative, 153, 422, 425
  cognitive-operative, 422, 425
Cognitive styles, 403–405, 411
Cognizance, 217, 253, 407
Color-matching task (CMT)
  F-factor in, 243–244
  functional magnetic resonance imaging studies and, 245, 348–349
  logical learning in, 291–293
  mental demand in, 243–245, 290, 293–294
  mental-process reconstruction in, 291–294
  sequential and executive metasubjective task analysis in, 289–294
Communication paradoxes, 33, 50–51
Complexity. *See also* Simplexity
  crude, 258–259
  processing, 166–167
  relational, 20, 245, 268–270, 425
Complexity, effective, 258
  in metasubjective task analysis, 259–260, 265–266
  of neural networks, 340–341
Complex learning, 30–31
Complex schemes, 57, 328, 432n4
Computation, 328
Conceptual processes, 366–369, 395

Consciousness
- Damasio on, 73–74, 215, 251, 253, 407–408
- early development of, 72–78, 408–409
- Edelman on, 406–408
- functional totality and, 73–74
- mirror test and core, 215
- representation and, 251–253
- in sensorimotor period, 72–73
- stream of, 205–206

Conservation
- abduction and, 46–47, 264, 273
- distal object and, 45, 272
- as judgment task, 279
- learning paradoxes and, 42–46, 46f, 47, 428n7
- Piaget on, 42–45, 260–262, 262f, 263–264, 270–271, 271t, 272–274, 276, 284–285, 287, 431n2
- vicariances and, 44–45, 270–271, 271t, 272–274, 428n8

Conservation of substance task, 431n2
- *F*-factor in, 264, 270, 273
- metasubjective task analysis and, 260–262, 262f, 263–264, 270–271, 271t, 272–274, 276, 284–285, 287
- Piaget's model of, 270–271, 271t, 272–274structural model of, 262f

Constructivism (empirical rationalism), 3
- categorical, 8
- as causal-organismic, 397–398, 408
- Copernican Revolution and, 4, 398
- of Darwin, 398–399, 412
- developmental, neo-Piagetian, 18–21, 36, 400–401
- dialectical, 5–13, 49, 399–400, 406
- as empirical rationalism, 8, 398, 402, 412, 421–422
- functional invariants in, 6
- individual-difference intelligence and, 248
- of Piaget, 8–13, 30, 268–269, 319–320
- semantic brain mapping of, 359–360
- structural constructivist modeling, 268–269

Constructivist (cognitive) learning, 163
- Piaget on, 18–19, 36–38, 49, 62, 270–271, 276, 319–320
- reflective abstraction in, 320, 322, 422
- schemes and, 167–170
- theories of, 3, 19, 36–38, 49, 62

Constructivist maturational-attention theories, 3, 36

Content, 167

Content learning. *See* Learning, content

Context
- mental demand sensitive to, 243–247, 247f, 248
- semantic operative scheme and, 73, 80, 124, 125f, 127f, 128–130, 130f

Copernican Revolution, 4, 61, 398

Copy-the-stripe task (CST), 192–193, 193f, 194

Corbetta, M., 362

Core schemes, 407

Core self, 74, 76, 79t

Cortical areas. *See* Brain

Corticocortical connections, 380, 393–394

Counting, 86–88

Cowan, N., 209, 215, 269, 337, 402

Critical realism, 21, 23–24

Csikszentmihalyi, M., 187, 201

Cue, 39

Cullenberg, S., 33, 47

Damasio, A. R., 394
- on autobiographical self, 74, 77–78, 409
- on consciousness, 73–74, 215, 251, 253, 407–408
- on core self, 74, 76
- on primordial feelings, 74, 408
- on self, stages of, 74, 76–78, 253, 409

Darwin, Charles, 397–399, 412

Darwinism, neural, 319, 331, 339, 399

Da Vinci, Leonardo, 9, 11

Decalage, 43

Decentration, 221–222

Deduction. *See* Descendent/deductive analysis

Dehaene, S., 73, 251, 407–408

De Houwer, J., 163–164
DeLoache, J. S., 77, 79t, 82
Demetriou, A., 113, 167, 215, 230
  on cognizance, 217, 253, 407
  constructivist learning theory of, 19
  developmental *g* scores of, 233–234, 234f
Descendent/deductive analysis, 283–285, 287, 297
Descriptive theory/model
  causal theory/model and, 5–6, 34, 36, 63, 175, 257–258
Developmental growth, U-shaped curves of, 58
Dialectical system (DS), 9–10, 399–400
Dialectics
  constructivism and, 5–13, 49, 399–400, 406
  developmental spiral of, 345
  dialectical-constructivist epistemology, 5
  in metasubjective task analysis, 268–270
  of Piaget, 8–13
  types of, 14
Diamond, A., 350, 352, 357–358
Differentiation, 26, 168–169
Dimensional Inventory for Childhood Development Assessment (IDADI), 96–97, 98t, 99f, 232, 233f
Direction-following task (DFT), 237–238
  adult scores of, 241, 241f, 242, 242f
  F factor in, 246
  linguistic effect in, 245–246
  M-measurement invariance of, 238, 238f, 239, 239f, 240, 240f, 241, 241f, 242, 242f
  standardized ability tests correlating with, 249–250, 250t
Disequilibration, 3–4, 7, 422
Dopamine (DA), 386, 387t, 388, 392
  in limbic system, 391, 394–395
  serotonin and, 393
Dormashev, Y., 181–182, 196, 201–202
Dorsal-attention networks, 359, 362–363, 395
Dorsal stream, representational, 385, 395–396
Dorsolateral prefrontal cortex (DLPFC), 352, 370
  complexity and, 372–374
  in N-back task, 353, 355–356
  in self-ordered pointing task, 353, 358, 373
Drawbridge screen task, 90–91, 91f, 92
Dynamic synthesis, 6, 15, 151, 335
  binding problem and, 319, 334
  hidden operators in, 16–17
  regulations in, 14

Eccles, J. C., 339, 366
Edelman, G. M.
  consciousness and, 406–408
  on group selection, 190, 331–333
  on lateral inhibition, 264, 332, 334, 379
  on neural Darwinism, 319, 331, 339, 399
  on neuronal population processes, 331–333
Edelstein, W., 403–405
Education, 405, 412
Effecting component (*ec*), 147–148, 162, 276, 399–400
  in metasubjective task analysis, 158, 162, 260–262, 269
Efferent processes, 339, 421
Effort, 182–183
Electroencephalography (EEG), 380
  coherence in, 61, 94–96, 342–347, 347f, 348
  phase difference in, 342–344, 348
Elman, J. L., 209–211, 348
Emotions, 199
  emotion-focused therapy, 406
  personal schemes of, 54–57, 154, 216, 331
Empirical rationalism. *See* Constructivism
Empiricism, 397–398
Encumbrances, 25–26
Engels, F., 10
E-operator. *See* Executive schemes
Ephemeral schemes, 185, 425
Epinephrine. *See* Norepinephrine
Episodic processes, 367, 369
Episodic task. *See* Temporal-meaning operative integration task

Epistemic reflection (epireflection), 35, 38, 145, 430n4
Epistemology, 5, 422
Epistemo-ontological levels (epilevels), 423, 432n2
 in brain, 338–340
 levels and, 338–340, 423
 in neuropsychology, 319, 322–323
 reflective abstraction and, 177t, 319, 322–323
Equilibration, 151, 432n3
 affirmations and negations in, 12
 causality and, 37
 disequilibration, 3–4, 7, 422
 optimizing/progressive, 13–14, 17–18
 Piaget on, 3–5, 7, 9, 12–13, 19, 31, 37, 123, 264, 319, 422
 reflective abstraction and, 319, 322
 regulations in, 14, 422
Equilibration, principle of (*EQU*), 266–267, 276
Equivalence conservation. *See* Conservation of substance task
Everitt, B. J., 389–390
Executive function (EF), 206–207, 375
Executive mediation, 85–88
Executive schemes, 23f, 82, 147, 154–156, 325, 328, 415
 action schemes and, 21, 152–153, 156, 425
 affective goals and, 156–157
 in brain, 369–370
 in color-matching task, 289–294
 E-operator of, 22t, 207, 214t, 216, 338
 mental/effortful attention and, 217, 274, 376
 in metasubjective task analysis, 287, 289–294, 297, 304–311, 311f, 312, 312t, 313–315, 315f, 316
 in motor rho task, 311, 312t
 plans and controls in, 155
 relational-associative learning and, 154–155
 relational-symbolic learning and, 176
 relevance, criterion of, in, 155–156
 Resistances/resistances and, 142f, 143
 signalic and symbolic, 82, 124, 156–158, 158t, 159
 in subjective-2 analysis, 275
Expectancies, 365, 431n1
 affective, 152
 figurative, 53, 153, 164–165
 operative, 153, 160, 164–165
 Piaget on, 7, 62
 schemes and, 36–37, 414, 417–418
 Tolman on, 7, 141, 164, 430n8
 violation of, 417–418
Experiential learning, 369, 395
Explicit knowledge, 363–364
Expression, 188

Facilitating situation, 14–15, 39, 336, 419, 424
 automatic attention in, 186, 189–190, 217
 brain and, 189–190, 371, 390
 facilitating/misleading task criterion for, 85–90
 mental process in, 84–91, 91f, 92
 sensorimotor mental-attentional capacity in, 84–90, 92
 subjective-2 analysis and, 275
False belief theory-of-mind (ToM) task, 78, 82–83, 83t
Feelings, 55
 affective schemes yielding, 152–153
 impression and expression, 187–188
 primordial, 74, 408
Feyerabend, P., 4
F-factor (Field-factor), 201, 207, 338, 422
 binding problem and, 319
 brain and, 264, 389
 in color-matching task, 243–244
 in conservation of substance, 264, 270, 273
 in direction-following task, 246
 Gestalt, neo-Gestaltist psychology, and, 188, 191, 206, 243, 264
 inhibition and, 206
 in metasubjective task analysis, 277–279
 mini-max rule of, 191–192, 194–195, 213, 277–278, 295

Schemes' Overdetermination of Performance and, 17–18, 22t, 188, 190–192, 192f, 193, 193f, 194–196, 204, 213, 214t, 319, 333–334, 341, 370–371, 379
in Simon task, 246–247, 247f, 248
Field-dependence-independence (FDI), 193, 279, 337, 404
Field-factor (*F*-factor), neo-Gestaltist, 188, 191, 206, 264
Field of activation, 218, 219f, 252, 266, 333, 353
Figural intersections task (FIT), 225–226, 226f, 227, 246
adult scores of, 241, 241f
developmental intelligence in, 231, 231f, 234, 234f
individual-difference intelligence and, 231, 231f, 234, 234f, 249–250, 250t
mental demand in, 231f, 232
M-measurement invariance of, and, 237–238, 238f, 239, 239f, 240, 240f
standardized ability tests correlating with, 249–250, 250t
Figurative/declarative processing and learning, 359–361
Figurative expectancies, 53, 153, 162, 164–165
Figurative motivation, 387, 387t, 388
Figurative-representation temporal stream, 381–382
Figurative schemes, 21, 23f, 153,162, 431n1
ads, *obs*, and *pros*, 145–147, 153
in brain, 147, 328, 360–361, 380
distal object in, 21, 153, 157
in metasubjective task analysis, 413–418
as *model obs*, 145–146
in motor rho task, 311, 312t, 313
operative schemes and, 23f, 53–54, 153, 325, 329–330, 359–361, 422
Resistances/resistances and, 142f, 143, 145–147
in subjective-2 analysis, 274–275
Figurative self, 407–408
Figurative streams. *See* Representation
Figurative-temporal operator. *See* Time
Fillmore, C. J., 101
Flow, 187, 201, 217
Fluents, 384, 432n7
causal abstraction using, 325–326
in metasubjective task analysis, 414–415, 417
as sequence-representing schemes, 360–361
Fodor, J. A., 41–42
Fogel, A., 130, 130f
F-operator. *See* *F*-factor
Fraction word problems, 305–307
Frank, M., 155–156, 377
Frege, G., 124, 271, 271t
Freud, S., 15–16, 187
Frontopolar cortex, 353, 355, 373–374
Functional component (*fc*) of schemes, 147–148, 399–400
in metasubjective task analysis, 157, 162, 260–262, 269
Functional invariants. *See* Invariants, functional
Functional magnetic resonance imaging (fMRI) studies, 361, 365, 376
color-matching task and, 245, 348–349
Goel and Vartanian, 355–356
Functional-structuralist modeling, 268–269
Functional structures, 6–7, 147, 170, 224, 334, 422
Functional totality, 4, 73–74, 422
binding problem and, 319, 334
functional invariants in, 422–424
organismic processes of, 425
paradox of, 59
Schemes' Overdetermination of Performance in, 399
Fundamental measurement, 235–236

Gabrieli, J. D. E., 355, 370–371, 373
Gadamer, H.-G., 3
Gaillard, V., 402, 428n4
Galanter, E., 120

Gallup, G. G., 75–76, 215
Gell-Mann, M., 258–259
General theory/model, 121, 219, 265, 297, 317, 319
  local theory/model and, 5–6, 259
Geometry, analysis of figures in
  ascendent/abductive, 283–285, 287
  descendent/deductive, 283–285, 287
  metasubjective task analysis and, 282–286, 286f, 287, 430n7
Gestalt psychology, 6, 17, 35
  F-factor and, 188, 191, 206, 243, 264
  lateral inhibition and, 332
  neo-Gestaltist, 188, 191, 206, 243, 264
  Schemes' Overdetermination of Performance and, 191
G-factor, 225, 228, 230
  mental/effortful attention corresponding with, 233–234, 234f, 235, 250–251
Gibson, J. J., 25–26, 67, 168, 326, 359
Goel, V., 355–356, 364–365
Gonseth, Ferdinand, 281
Goodman, D., 151, 167
Gordon, A. C. L., 78
Gottfredson, L. S., 257
Gould, Stephen Jay, 398–399
Grafman, J., 363
Greenberg, L. S., 406
Group competition, in brain, 331–333
Group confinement, in brain, 331–332
Grouping of Vicariances. *See* Vicariances
Group selection, in brain, 190, 331–333

Habit
  attention and, 182–184
  James on, 39, 183
Hacker, P. M. S., 318
Halford, G. S., 20, 268–269
Hallett, M., 363
Hamilton, R., 329, 334
Hebb, D. O., 327, 340
Hegel, G. W. F., 3, 11, 121–123, 125
Heraclitus, 31, 317
Heterarchy (flexible hierarchy), 422
  in brain, 322, 339, 368, 373–374, 392
  of learning and scheme types, 175
  of schemes, 37–38, 175, 328–329, 335
Hidden operators. *See* Operators, hidden
Hierarchical reinforcement learning, 162–165
Hierarchy
  functional invariants in, 422
  nested, 37–38, 370
  of operative symbolic streams, functional, 359, 395
  in prefrontal cortex, 370
Hierarchy, flexible. *See* Heterarchy
Hintikka, J., 283
Hippocampus, 382–385, 394–395
Holistic schemes, 147, 328
Homunculus, 59, 151
Horn, J. L., 205, 251
Houdé, O., 93
Howard, S. J., 355
Hughlings Jackson, J., 366
Husserl, E., 197, 207
  on attention, 205–206
  on representation, presentation and, 252–253

Iconic function, 145
  impression and expression using, 188
  indexical function and, 126, 133, 134t, 135
  Peirce on, 35, 38, 122–123, 126, 128, 135, 188, 320, 430n4
  sign in, 122–123, 429n2
  symbol and, 126, 128, 135
Imitation sorting task (IST), 98, 103–106
  children passing, frequency of, 106–107, 107f
  metasubjective task analysis of, 115–118
  sensorimotor mental-attentional capacity in, 97, 105f, 106–107, 109f, 110, 111t, 112, 115–118

Implicit knowledge, 363–364
Impression, 187–188
Indexical function
iconic function and, 126, 133, 134t, 135
Peirce on, 122–123, 126, 127f, 135, 430n4
signals and symbols in, 135
Indexing processes, of *ads* and *obs*, 145–146
Individual-difference intelligence (ID), 205, 404, 411
constructivist developmental theory integrating with, 248–249
developmental intelligence and, 227–228, 229f, 230–231, 231f, 232, 233f, 234, 234f, 248–249
direction-following task correlating with, 249–250, 250t
figural intersections task and, 231, 231f, 234, 234f, 249–250, 250t
Individual differences, cognitive styles and, 403–405, 411
Infant development. *See* Sensorimotor period
Informational correspondence, 122–123
Infralogical processes
abstraction, 177t, 178, 324–325
in brain, domain of, 324–325
metadomains of, 280–281
Piaget on, 280, 324, 367
Infrastructure, 35, 428n3
Inhelder, B., 61, 192, 411
Inhibition, 359, 371–372
automatic, 197–198, 206, 375
in brain, activation and, 375–377, 378f, 379–380, 386, 392–393
effortful, 197–198, 206, 375
F-factor and, 206
I-operator of, 18, 22t, 197, 206–207, 214t, 215, 338, 375–377, 378f, 379–380, 393
Schemes Inhibition and Decay principle, 266–267
Inhibition, lateral. *See* Brain
Instrumental conditioning, 163–165
Intelligence, 205, 248, 423. *See also* Individual-difference intelligence
developmental, mental-attentional capacity measuring, 225–226, 226f, 227–228, 229f, 230–231, 231f, 232–233, 233f, 234, 234f, 235
individual-difference intelligence and developmental, 227–228, 229f, 230–231, 231f, 232, 233f, 248–249
mental/effortful attention and, 207–208
mental/effortful attention and tests of, 248–250, 250t, 251
Piaget and, 227–228, 229f, 230
Intelligence, fluid (Gf), 27, 225, 228, 230, 250–251. *See also* *G*-factor
Intentionality, 52–54, 408
Internalization, 207–208, 422
ads, *obs*, and *pros*, 80, 143, 145–149, 153
content learning and, 169
of functional invariants, 423
of *model obs*, 145–146
motivation leading to, 424
in reflective abstraction, 423
of Resistances/resistances, 24, 145, 430n3
Vygotsky on, 48, 349, 423
Internalized object representations. *See Obs*
Interruption. *See* Inhibition
Intratask learning. *See* Learning, intratask
Intuition, 196, 200, 282, 381
Invariants, 259, 400
Cassirer on, 257
types, 423
Invariants, functional, 320, 322
constructivist emphasis on, 6
in dialectical system, 10
in functional totality, 422–423
in structure, 426
I-self, 216, 407–409

James, William, 24, 409
on attention, 182–184
on habit, 39, 183

James, William (cont.)
  on I-self and Me-self, 216, 407
  on Will, 182, 216, 410
Janet, Pierre, 182–184, 196
Jaspers, Karl, 50, 410
Jealousy, 289, 307–309
Johnson, J., 225, 307, 355
Johnson, M. H., 211
Joint Simon task, 246–247, 247f, 248
Judgment task, 83, 83t, 279–280
Jung, C. G., 131, 136, 184, 187, 253

Kagan, J., 145–146
Kanai, R., 365
Kanizsa triangles, 143, 192, 192f
Kant, I., 141, 361, 366–367
  categorical constructivism of, 8
  on intuition, 200, 282, 381
  on knowing, forms of, 121
  on mediation, 34, 424
  on schema, 34, 36, 144, 424
  on space and time, 200, 381
  transcendental categories of, 381, 432n3
Kaplan, B., 123
Knowing, forms of, 121–122
Koffka, K., 153

LA learning. *See* Learning, affective
Language paradoxes, 33, 49–50
Language schemes, 70t, 114, 415
Lateralization. *See* Brain
LC learning. *See* Learning, relational-associative
LCLM learning. *See* Learning, automatization
Learning, 401. *See also B*-schemes
  associative, 163–164, 209–211
  associative versus cognitive, 163
  complex, 30–31
  conceptual, 366–367, 395
  development and, 162–163
  experiential, 369, 395
  figurative/declarative versus operative/procedural, 359–361
  functional-descriptive theory of, 164
  heterarchy of, 175
  hierarchical reinforcement, 162–165
  L-operator of, 171, 204, 338, 417
  outcome, 164
  of signal, 127, 134
  signalic and symbolic, 127, 134, 359
  space and time in, 200–201
Learning, affective (*LA* learning), 22t, 171, 423
Learning, automatization (*LCLM* learning), 216, 108, 109f, 117–118, 161, 174–175, 180, 196, 345
  reflective abstraction and, 179–180
Learning, constructivist. *See* Constructivist learning
Learning, content (*C* learning), 31, 127, 168–170, 423
  accommodation and, 168–169
  assimilation and, 168–170
  in automatic attention, 196
  conditions for, 169–170
  C-operator of, 22t, 201, 213, 214t, 338
  internalization and, 169
Learning, logical (*L* learning), 170–176
  assimilation and, 170–171
  in color-matching task, 291–292
  conditions for, 171
  L-operator of, 171, 338in metasubjective task analysis, schemes and, 296
Learning, relational-associative (*LC* learning), 23f, 31, 127, 161, 171–172, 423
  activation of, 172, 430n9
  in automatic attention, 196
  in brain, 363, 365, 369
  brain lateralization of, 345, 364
  characteristics of, 172
  executive schemes and, 154–155
  LC-operator of, 22t, 201, 213, 214t
  L-operator in, 171
  in object permanence task, 350–352
  reflective abstraction and, 177t, 178
  relational-symbolic learning versus, 363–366
  Tolman on, 171–172

Learning, relational-symbolic (*LM*), 171–176, 201, 430n10
automatization (*LCLM* learning) and, 174–175in brain, 345, 363–365, 369
brain lateralization and, 22t, 214t, 345
executive schemes and, 176
M-centration in, 173
mental-attentional capacity and, 22t, 31, 161, 172–176, 205, 214t, 216, 423
in motor rho task, 342, 343f
processing rules for, 173–176
reflective abstraction and, 177, 177t, 178–180
relational-associative learning versus, 363–366
symbol and, 134
Learning paradoxes
conservation and, 42–46, 46f, 47, 428n7
evolutionary, 42–45
Piaget and, 43–45, 49
strong, 42–46, 46f, 47, 428n7
Vygotsky and, 49
weak, 41–42
Leibniz, G. W., 200–201, 381
Lenin, V., 10
Lesniewski, S., 280, 431n1
Lewin, Kurt, 412
Life-world, subjective (Umwelt), 141–142
Limbic system, 322f, 390–391, 394–396
L learning. *See* Learning, logical
LM learning. *See* Learning, relational-symbolic
Local theory/model
general theory/model and, 5–6, 259
Logic, 423–424
Logical learning. *See* Learning, logical
Logological processes
abstraction, 177, 177t, 178, 324–325, 367–368
in brain, 324–325, 367–368
knowledge modeling, 367
linguistic mode and, 432n1
metadomains, 280–281
psycho-logic and, 281
L-operator. *See* Learning
Luria, A. R., 14, 317, 318, 339

Martí, E., 61
Maturational-attention theories, 19–20, 36
M-centration. *See* Centration
Meaning
in animals, 119–120
of object, 120
schemes and, 36–38, 52, 54, 424
signalic and symbolic, 119, 127f, 296
in subjective life-world, 141
von Uexküll on, 36, 119–120, 141
Means-end-readiness, 62, 68, 120, 164
Medial right hemisphere, 320f, 321f, 322f
Mediational processes
developmental growth in, 33–34
executive, 85–88
external and internal, 424, 427n1
schemes as, 34–36, 424
Meditation/contemplation techniques (MED), 202–204
Mendonça Filho, E.J., 61
research on developmental landmarks, 93–94, 96–97, 98t, 99f, 232, 233f
Mental-attentional capacity (*M*-capacity), symbolic (*Mk*), 21, 337–338
in adults, upper bound of, 21, 240–241, 241f, 242–243, 242f
in brain, 245, 341, 348–349, 354–355, 365, 403
brain waves and, 342–346, 346t, 347–349
developmental intelligence measured by, 225–226, 226f, 227–228, 229f, 230–231, 231f, 232–233, 233f, 234, 234f, 235
in education, 405
invariance of measure, 209, 237–238, 238f, 239, 239f, 240, 240f
measuring, 71, 205, 220, 232, 235–237
with mental demand, trade-off of, 58, 110, 112,173, 221, 245, 345, 365

Mental-attentional capacity (*M*-capacity), symbolic (*Mk*) (cont.)
mental demand and, 51, 58, 221, 223–224, 231f, 232, 243–246, 345, 346t, 354–355, 365
in misleading situation, 212, 217
in motor rho task, 315, 315f, 316, 342, 343f
Pascual-Leone, J., formulating, 21, 230
power of, 31, 211–212, 221
reflective abstraction and, 179–180
relational-symbolic learning and, 22t, 31, 161, 172–176, 214t, 216, 423
scale of (*Mk*), and sensorimotor scale (*Me*), 71, 84–85, 88, 94–97, 220, 223–224, 346
sensorimotor mental-attentional capacity in, 84, 220
stages of development in, 31, 200, 200t, 222, 222f
staircase model and, 221–222, 222f, 223–225
standardized ability score and, 225–226, 226f, 227, 249–250, 250t
working memory and, 215, 217, 235–237, 243, 337, 402
Mental-attentional capacity, sensorimotor (*Me*). *See also* Sensorimotor period
in facilitating situation, 84–93
measuring, 97–100, 102t, 103–108, 109f, 110
with mental demand, trade-off of, 97–98, 110, 111t, 112, 429n1
Piagetian substages and, 64, 65t
stages of development in, 61–62, 64–65, 65t, 66–72, 215–216
staircase model and, 221–222, 222f, 223–225
symbolic function and, 78–79, 79t, 80–83, 83t, 84
symbolic mental attention and, 71, 84–85, 94–97, 99f, 223–224, 346–347
Thatcher research and, 94–96
Mental-attentional demand. *See* Mental demand
Mental attention memory (MAM) task, 237
Adult scores of, 241, 241f
M-measurement invariance of, 238, 238f, 239, 239f, 240, 240f, 241, 241f
Mental demand (*M*-demand) (*Md*), 31, 207–208. *See also* Metasubjective task analysis
in color-matching task, 243–245, 290, 293–294
dynamic context-sensitive character of, 243–247, 247f, 248
in education, 405
in figural intersections task, 231f, 232
in fraction word problems, 305–307
jealousy and, 309
M-complexity score (*M*-score) of, 100, 108, 109f, 112, 225, 236–239
measuring, 235–236
with mental-attentional capacity, trade-off of, 58, 110, 111t, 112,173, 221, 245, 345, 365
mental-attentional capacity and, 51, 58, 221, 223–224, 231f, 232, 243–246, 345, 346t, 354–355, 365
in metasubjective task analysis, 235, 244, 296, 300–307
in motor rho task, 313–314
in prefrontal tasks, 354–355, 357
Simon task and, 246–247, 248, 304–305
Mental/effortful attention, 3, 197, 205–206, 217. *See also* Mental-attentional capacity
activation of, 211–212, 215, 375–377
in antisaccade task, 185–186
automatic attention aided by, 201–204, 217
automatic attention compared with, 181–183
in brain, 341–346, 346t, 347–349, 354–355, 357–358, 365, 370, 375–377, 378f, 379–380, 383–384, 387t, 393–394, 403
centration in, 40, 173, 183, 206, 218, 220–222, 287, 292, 414, 421
consciousness and, 251–253
in constructive abstraction, 323
evolutionary emergence of, 212–213, 214t, 215–217
executive function theories and, 206–207, 375
hidden operators and, 207, 212–213, 214t, 215–218, 337–338, 421

intelligence and, 207–208, 248–249
in misleading situation, 39, 186, 189, 196, 208, 212, 217–218, 336–337, 391, 393, 408
M-operator of, 18, 22t, 31, 207, 214t, 215–218, 235–236, 337–338, 370, 375–377, 378f, 379–380, 393, 401, 416, 418, 421, 429n1
model of, 217–219, 219f
maturational growth of, 18–20, 31, 36, 40, 209, 215, 223, 227–228, 230
neuroscience on, 184–185, 209, 348–349
neurotransmitters and, 387, 387t, 388–391
symbol and, 134
Will and, 182, 410–411
working memory and, 19–20, 27, 206, 209–212, 217, 235–237, 240–241, 243, 246, 402
Mental model. *See* Model
Mental-process reconstruction (*M*-construction)
in color-matching task, 291–294
in metasubjective task analysis, 275, 287, 289–294, 297, 306–309
in motor rho task, 310, 312, 314
Mental strategies, formulas for, 250–251, 352, 355–358, 418–419
Mereologic, 431n1
in abstraction, mode of, 324–325, 367–368
in brain, 324–325, 367–368
fallacy of, 318
metadomain of, 280–281
models of, 260, 367
processes of, 177, 177t, 178
Merleau-Ponty, M., 119, 205–207, 257
Me-self, 407–409
Mesencephalic reticular formation (MRF), 385–386, 393
Messick, S., 404
Metadomains, 280–281
Meta-empiricist theories, 4, 120, 257–258, 397–398
Metamodal processes, brain, 329, 334
Metastrategies, 280
Metasubject, 216–217, 219, 294–295
Metasubjective modeling, 4–5, 259, 398, 403, 422, 424, 429n1
causal-organismic theories and, 257–258, 400
Pavlovian-instrumental transfer and, 165–166
units of, 120
Metasubjective task analysis (MTA), 257, 275–279, 403, 422, 424, 426. *See also* Task analysis
action/operative task in, 279
ascendent/abductive, 283–285, 287, 289, 297
of attachment, sensorimotor period, 298–299
in color-matching task, sequential and executive, 289–294
conservation of substance and, 260–262, 262f, 263–264, 270–274, 276–279, 284–285
constraints in, sources of, 266–267
decalage in, 43
descendent/deductive, 283–285, 287, 297
dialectical steps in, 268–279
of drawbridge screen task, 90–92
effective complexity in, 259–260, 265–266
equilibration principle in, 266–267, 276
executive function theories compared with, 206
executive *M*-dimensional analysis, 287, 289–294, 297, 304–311, 311f, 312, 312t, 313–315, 315f, 316
of false belief theory-of-mind task, 82–83, 83t
F-operator in, 277–279
of fraction word problems, 305–307
fundamental measurement and, 235–236
geometry and, 282–286, 286f, 287
heuristics of, 269–270, 294–297
hidden operators in, 235–236, 273, 276–279, 286, 295, 297, 416, 418–419
of imitation sorting task (IST), 115–118
of jealousy, 307–309
judgment task in, 83t, 279–280
learning in, 173, 295–296
mental demand in, 235, 244, 296, 300–307

Metasubjective task analysis (MTA) (cont.)
mental strategies in, formulas for, 250–251, 278, 297, 352, 355–358, 418–419
metadomains in, 280–281
metasubject in, 294–295
methodological synthesis of, 307–312, 312t, 313–314
mini-max principle in, 266, 277–278
misleading situation in, 83–84, 132, 278–279, 418–419
mother leaving, anticipation of, in, 299–300
of motor rho task, 310–311, 311f, 312, 312t, 313–315, 315f, 316
neuroscience/neuropsychology, as tool in, 349–358
notation of, 69–70, 70t, 281–282, 413–419
object permanence task in, 300–304, 349–352
Pavlovian-instrumental transfer and model of, 165–166
Polya inspiring, 264–265
relational complexity and, 268–270
schemes, demarcation of, in, 260–263, 269–270, 294–297
Schemes Inhibition and Decay principle in, 266–267, 276
Schemes' Overdetermination of Performance in, 266, 276–278, 282, 286
of semantic complexity task (SCT), 113–115
sensorimotor mental-attentional capacity in, 69–70, 71–72, 84–93, 113–118
sequential, step *M*-construction in, 275, 287, 289–294, 297, 306–309, 310, 312, 312t, 313–314
shadow schemes in, 132–133, 270, 276, 278
Shayer on, 405
of Simon task, 247, 304–305
of small-number estimation task, 92–93
strategy in, 294–297, 352, 418–419
of symbol, 130, 130f, 131–132
symbolic mental attention in, 84–85
Miller, G., 40, 120, 220, 428n6
Miller, R., 170
Mini-max rule, 340
F-factor, 191–192, 194–195, 213, 277–278, 295
in metasubjective task analysis, 266, 277–278
Schemes' Overdetermination of Performance and, 191
Mirror neurons, 374–375
Mirror self-recognition, 75–76, 215
Misleading situation, 14–15, 424, 431n2
A-not-B-error task and, 350, 352
brain in, 189–190, 390–391, 393, 395–396
competing strategies in, 336–337
facilitating/misleading task criterion for, 85–90
mental/effortful attention in, 39, 186, 189, 196, 208, 212, 217–218, 336–337, 391, 393, 408
in metasubjective task analysis, 83–84, 132, 278–279, 418–419
neurotransmitters and, 387, 387t, 388, 390–391, 393, 395
in Simon task, 304–305
in staircase model, 223
subjective-2 analysis and, 275
symbolic mental attention in, 84–85, 88
Modal signs/symbols, and amodal, 120–121
Model, 136, 424
M-operator. *See* Mental/effortful attention
Morra, Sergio, 242, 245–246
Morris, C. W., 120–121
Moscovitch, M., 26–27, 29, 31
Mother leaving, anticipation of, 299–300
Motivation
affect and, 17, 22t, 25, 55, 57, 62, 145, 154, 199, 372, 424
automatic attention and, 198–199
figurative, 374, 387, 387t, 388
in metasubjective task analysis, 261, 267, 275
neurotransmitters and, 387, 387t, 388, 391–392

Motor rho task, 310–316, 342
latency data in, 314–315, 315f, 316, 342, 343f
mental demand in, 313–314
mental-process reconstruction in, 310, 314
metasubjective task analysis of, 310–311, 311f, 312, 312t, 313–315, 315f, 316
Moving object, tracking of, 85–86

Natural selection, 398–399
N-back task, 353–356, 373
Negations, 9, 11–13, 31
Neo-Gestaltist psychology, 17, 188, 191, 206, 264
Neonativism, 42, 45, 113
conceptual, 63–64
drawbridge screen task of, 90–91, 91f, 92
tasks of, 64, 89–91, 91f, 92–93
Neo-Piagetian theory, 5, 35, 112, 170, 179t, 395, 402
conceptual neonativism resisted by, 63–64
constructivist developmental, 18–21, 36, 400–401
developmental stages and, 40, 227–228, 348
functional-structuralist modeling in, 268–269
maturational-attention group of, 19–20, 36
Pascual-Leone, J., in, 20, 40
stages of, intelligence and, 227–228
staircase model of, 221–222, 222f, 223–225, 345, 367
structural constructivist modeling and, 268–269
Nested hierarchy, 37–38, 370
Neural networks, 211, 328–329, 331, 340–341
Neuronal population processes, 331–333
Neuropsychology/neuroscience. *See also* Brain
on automatic attention, 184–185
binding problem in, 319, 334
compared, 318
epistemo-ontological levels in, 177t, 319, 322–325, 338–340, 366–369
hidden operators in, 319, 337–338
on mental/effortful attention, 184–185, 211, 348–349
mereological fallacy and, 318
metasubjective task analysis and, 349–358
object permanence task in, 349–352
organismic model for, 318, 395, 403
Piaget and, 345
prefrontal tasks in, 353–358
schemes in, 319, 327–330
Schemes' Overdetermination of Performance in, 333–334, 338
Neurotransmitters, 376, 396
acetylcholine, 185, 374, 386–387, 387t, 388–391, 393–394
automatic attention and, 387, 387t, 388–389
dopamine, 185, 212, 352, 358, 374, 386, 387t, 388, 391–395
mental/effortful attention and, 377, 378f, 379, 387, 387t, 388–391
misleading situation and, 358, 387, 387t, 388, 390–391, 393, 395
motivation and, 387, 387t, 388, 391–392
norepinephrine, 185, 374, 386–387, 387t, 388, 390–391, 393–394
serotonin, 374, 386, 387t, 392–394
Neurotrophic dynamic factor, 211, 338, 348
Newton, I., 282–283
Niaz, M., 405
Noll, J. G., 205
Norepinephrine (NE), 185, 386, 387t, 390–391, 394
acetylcholine and, 387–388, 390–391
serotonin and, 374, 393
Novel performances, 6, 58–59

Object
Hegel on subject and, 122–123, 125
meaning of, 120
in mirror self-recognition, subject and, 75–76
Piaget defining, 67, 120

Object, distal, 25, 200, 207, 401
in figurative schemes, 153, 157
in metasubjective task analysis, 277
object permanence task and, 67–68, 300
proximal object contrasted with, 44, 67, 131, 424
as psycho-logical entity, 21
symbolic function and, 130f, 131
in tertiary circular reactions, 67–68
vicariances and, 44–45, 272, 428n8
Object, proximal, 44, 424
in secondary circular reactions, 67
symbolic function and, 131
Objective analysis, 268, 276, 281
Object permanence task
distal object and, 67–68, 300
in metasubjective task analysis, 300–304, 349–352
multiple invisible displacement, 303–304
multiple visible displacement, 302–303
in neuropsychology, 349–352
relational-associative learning in, 350–352
single visible displacement, 301–302
Obs (internalized object representations), 80, 143, 145–149, 153
in figurative schemes, 145–146, 153
indexing processes of, 145–146
in internalization, 143, 145–147
model, 145–146
in organismic schemes, 148–149
as protosymbol, 80
Olsen, R. K., 382–383
Olson, D. R., 78
Operative expectancies, 73, 153, 160, 164–165
Operative modes of processing. *See* Metastrategies
Operative/procedural learning, 359–361
Operative schemes, 21, 23f, 57, 147, 328, 422, 431n1
in brain, 360–361, 374–375
diagnosis of, 160–162
figurative schemes and, 23f, 53–54, 153, 325, 329–330, 359–361, 422
in metasubjective task analysis, notation of, 282, 413–418
as *model pros*, 146
in motor rho task, 311, 312t, 313–314
Resistances/resistances and, 142f, 143, 146
in subjective-2 analysis, 274–275
Operative self
figurative self and, 407–408
Will and, 216, 253, 410
Operative symbolic streams, 359, 395
Operative temporal stream, 432n4
Operators, hidden, 13, 21, 22t, 23f, 27, 64, 151, 212–213, 214t, 215–216, 337–338, 399–400. *See also* *F*-factor
A, 22t, 166, 213, 214t, 215, 298–299, 391
action schemes and, 21, 23f
B, 22t, 62, 213, 214t, 216, 276, 298–299
in brain, 22t, 214t, 319, 337–338, 375–377, 378f, 379–385, 393
C, 22t, 201, 213, 214t, 277, 338, 401
cognition and, 337–338, 401
in dynamic synthesis, 16–17
E, 22t, 206–207, 214t, 216, 279–280, 338
evolutionary emergence of, 22t, 212–213, 214t, 215–217
I, 18, 22t, 197, 206–207, 215, 324, 338, 370, 375–377, 378f, 379–380, 393, 401
L, 171, 204, 338, 401, 417
LC, 22t, 201, 213, 214t, 401
M, 18, 22t, 207, 214t, 215–218, 235–236, 337–338, 370, 375–377, 378f, 379–380, 393, 401, 416, 421, 429n1
in metasubjective task analysis, 235–236, 273, 276–279, 286, 295, 297, 416, 418–419
in neuropsychology, 319, 337–338
S, 22t, 200–201, 213, 214t, 279, 326, 381, 385, 395
schemes contrasted with, 14, 23f, 27, 150, 424
Schemes' Overdetermination of Performance and, 15–16, 21, 23f, 112
sensorimotor mental-attentional capacity and, 112–113

T, 22t, 200–201, 213, 214t, 326, 381–384, 395, 432n5
Operators, subjective, 16, 329, 424. *See also* Schemes
Optimizing/progressive equilibration (OE), 13–14, 17–18
Organismic processes, 4–5, 26–31, 64, 425
causal, 5–6, 26–31, 33, 47, 249, 257–258, 397–400, 403, 408, 425
causal mechanisms for, 13–18
causal-organismic, evolution and, 398–399
causal-organismic modeling, 257–258, 400
neuropsychology and, 318, 395
overdetermination of, 5–6, 13–14, 425
Organismic schemes. *See* Schemes
Ortega, J., 52
Orthogenetic principle, 123
Overdetermination, 40
abduction and, 46–47
as assimilation function, 13–14, 16
causal-organismic, 5–6, 13–14, 40, 425

Pappus, 283
Paradoxes
communication, 33, 50–51
epistemological/cognitive, 34
of functional totality, without totalizer, 59
language, 33, 49–50
learning, 41–46, 46f, 47, 49, 428n7
Piaget and, 33, 43–45, 49
of truly novel performances, 58–59
U-shaped curves, developmental growth, 58
of Vygotsky, 33, 48–49
Parameters, 21, 23f, 54, 69, 275, 422, 431n1
in A-not-B-error task, 350–351
in metasubjective task analysis, 282, 414–415
as *model ads*, 146
in motor rho task, 311, 312t, 313–314
Resistances/resistances and, 142f, 143, 145–146
Pascual-Leone, Alvaro
on explicit and implicit knowledge, 363–364
on metamodal brain processes, 329, 334
on temporal-structuring processes, 361
Pascual-Leone, J., 97, 151, 315, 355
Baldwin and, 428n4
brain-lateralization hypothesis of, 345
Case and Edelstein on, 403–405
copy-the-stripe task of, 192–193, 193f, 194
mental-attentional capacity formulated by, 21, 230
on motor rho task, 310–311, 311f, 312, 312t, 313–314, 342, 343f
in neo-Piagetian theory, 20
on Piagetian developmental stages, substages, 222–223
Pavlovian conditioning, 163–165
Pavlovian-instrumental transfer (PIT), 164–166
Peirce, C. S., 24, 423
on abduction, 14, 34, 46–47, 160, 264, 273, 284, 399
on iconic function, 35, 38, 122–123, 126, 128, 135, 320, 430n4
on indexical function, 122–123, 126, 127f, 135, 430n4
semiotics of, 35, 38, 120, 122–126, 127f, 128–129, 133, 135, 188, 320, 423, 429n3, 430n4
on sign, 124–126, 133
on symbol, 126, 127f, 128–129
Personal schemes, 154, 407, 422. *See also* *B*-operator
as action schemes, 152, 425
B-operator and, 62
emotions and, 54–57, 152, 216, 331
psychotherapy and, 406
Petersen, S. E., 184
Petrides, M., 353, 357, 370, 373
Phenylketonuria (PKU), 358
Piaget, J., 17, 26, 253, 397, 428n9
on accommodation, 25, 40, 67–68, 168, 327, 362
on affirmations and negations, 11–13, 31

Piaget, J. (cont.)
on assimilation, 16, 25, 61, 68, 141, 165, 169–170, 199, 295, 327, 333, 362–363, 399, 428n5
on causality, 24, 33, 36–38, 40–41
on centration, 40, 183
on conservation, 42–45
conservation of substance, equivalence model of, 270–271, 271t, 272–274, 276
conservation of substance task, 260–262, 262f, 263–264, 284–285, 287, 431n2
constructivism of, 8–13, 30, 268–269
on constructivist learning, 18–19, 36–38, 49, 62, 270–271, 276, 319–320
critical realism of, 21, 23–24
on decalage, 43
on decentration, 221–222
dialectical constructivism of, 8–13
on disequilibration, 3–4, 7, 422
on equilibration, 3–5, 7, 9, 12–13, 19, 31, 37, 123, 264, 319, 422
on expectancies, 7, 62
functional-structuralist modeling and, 268–269
on indicator and signal, 127f
on infant development, 61–62, 66–68, 112–113
on infralogical, 178, 280, 324, 367
intelligence and, 227–228, 229f, 230, 411
on intentionality, 52–54
on mediation, 34–35
on mother leaving, 299
neuropsychology and, 345
object defined by, 67, 120
object permanence task of, 67–68, 300, 303, 349–352
paradoxes and, 33, 43–45, 49
psycho-logic of, 8–9, 43, 268–270, 281, 424
reflective abstraction and, 37–38, 50, 123, 176–179, 179t, 180, 222, 319–320, 323, 349, 359, 367–368
on regulations, 14, 49, 422
on schemes, 7, 21, 23, 33–38, 52–55, 62, 66–67, 141, 144, 146, 150, 259–260, 268–269, 272, 328, 362, 429n1, 432n4
on sign, 127f, 133
stages, substages of, 31, 33, 39–41, 64, 65t, 71, 79t, 176, 178–179, 179t, 209, 220, 220t, 222–224, 227–228, 344
stage-transition problems of, 39–41
Thatcher and, 345
on vicariances, 44–45, 270–271, 271t, 272–274
water-level task of, 192–193, 193f
Plato, 47–48, 121–123, 182
Pointing, 80, 353, 356–358, 373
Polya, G., 267
geometrical analysis and, 283–286, 286f
heuristic methods of, 264–265, 283–284
Posner, M. I., 184–185, 387, 387t, 388, 391
Pragmatics, 120–121
Praxis. *See* Agency/praxis
Prediction logic model, 110, 112
Prefrontal cortex (PFC), 379, 432n2
activation regulated by, 376–377
complexity levels in, 369–375
corticocortical connections of, 380, 393–394
dorsolateral prefrontal cortex, 352–353, 355–356, 358, 370, 372–374
executive schemes in, 369–370
norepinephrine in, 391
regions, hierarchies of, 370
tasks involving, 353–358
Presentation, representation and, 252–253
Pribram, K. H., 120, 394
Primary circular reactions (PCR), 65, 65t, 66
Primordial feelings, 74, 408
Pros (procedures), 143, 146–149, 153
Protoself, 74, 79t, 253, 409
Protosymbol, 80, 128
Psychogenetic reconstructions (Rec), 261
Psycho-logic, 281, 424
functional-structuralist modeling and, 268–269
models of, 8–9, 268

object in, 21
Piaget on, 8–9, 43, 268–270, 281, 424
Psychological organism, 425
Psychometric intelligence. *See* Individual-difference intelligence
Psychotherapy, 406

Quaternary areas, 339–340

Radical communication, 50–51
Rasch analysis, 96–97, 99f, 232
Raven Matrices Test, 3, 27–28, 28f, 31, 249, 427n3
operative model of, 29–30, 30t
Reality/reality
capital and lowercase r, 142, 425, 427n1
functional packages of, 326–327
Resistances/resistances and, 6–7, 24–25, 142, 142f, 143–145, 168, 326–327, 359, 399, 401, 425, 427n1
Recentration, 287, 421
Referent, 72
semantic operative scheme and, 73, 80–82, 125, 125f, 127f, 128–129, 130f
signal and, 126, 127f, 128
sign functionally detached from, 133, 134t
symbol and, 127f, 129, 130f, 131
Regulations, 14, 27, 49, 59, 62, 64, 168, 221–222, 266, 319, 400, 422
Reinecke, M., 130, 130f
Relational-associative learning. *See* Learning, relational-associative
Relational complexity, 20, 245, 268–270, 425
Relational-symbolic learning. *See* Learning, relational-symbolic
Relations, reflective abstraction coordinating, 24, 425
Releasing component (*rc*) of schemes, 23, 126, 147–148, 399
in metasubjective task analysis, 157, 162, 260–262, 269
Relevance, criterion of, 155–156, 197, 335, 392
Remes, U., 283
Repetitive transcranial magnetic stimulation (rTMS), 361
Representation, 425. *See also Obs*
consciousness and, 251–253
interactive, 252
streams of, in the brain, 380–385, 395–396
Representational dorsal stream, 385, 395–396
Resistances/resistances
accommodation and, 25–26
agency/praxis and, 143–145, 326, 359, 399, 401
brain and, 326–327
capital and lowercase r, 142, 425, 427n1
Cassirer on, 24–25, 31
executive schemes and, 142f, 143
figurative schemes and, 142f, 143
internalization of, 145, 430n3
in metasubjective theory, 5
model ads of, 146
packages of, 326–327
Piaget on, 7
Reality/reality and, 6–7, 24–25, 142, 142f, 143–145, 168, 326–327, 359, 399, 401, 425, 427n1
schemes and, 142f, 143–144, 401
sign and, 143
Resource operators. *See* Operators, hidden
Resting-state networks, 359–360
Reticular ascending system (RAS), 376–377, 385–386, 393. *See also* Mesencephalic reticular formation
Reticular nucleus (NR), 386
Reuchlin, M., 248
Robbins, T. W., 389–390
Rouse Ball, W. W., 194
Ruediger, S., 382

Salience, 353, 362, 388, 432n5
Sartre, J.-P., 59
Schelling, F. W. J., 121–122
Schema, body, 63, 65t

Schemes, 141, 146–151, 425. *See also specific scheme types*
activation booster of, 209–212, 215
ads, *obs*, and *pros* in, 148–149
agency/praxis and, 149–151, 167
in A-not-B-error task, 350–351
assimilatory strength of, 151, 171
automatic attention and, 185–187, 196–199, 204
in brain, 147, 328–330, 334–335, 341, 360–361, 369–370, 374–375, 380, 408
causality and, 33, 36–37
in chimpanzees, 137
cluster of compatible, 37, 165, 190–191
components of, 146–148, 162, 260–262, 269, 276, 399–400
computation and, 328
constructivist learning and, 167–176
content (*C*) learning and, 168–170
in developmental stages, 35–36, 63, 428n2
diagnosis of, 159–162
dominant cluster of, 17, 165, 197, 333
effecting component of, 147–148, 162, 260–262, 269, 276, 399–400
expectancies and, 36–37, 414, 417–418
in facilitating/misleading task criterion, 85–89
functional component of, 147–148, 260–262, 269, 399–400
as functional structures, 6–7, 147, 170, 224, 334, 422
heterarchy of, 37–38, 175, 328–329, 335
hidden operators contrasted with, 14, 23f, 27, 150, 424
in infant development, 62, 64, 66–70, 71, 112–113
intentionality and, 52–54
internalization of, 144–147, 168–169, 207–208, 422–423
Kant and, 34, 36, 144, 424
in life-world, 141–142
logical (*L*) learning and,170–176
meaning and, 36–38, 52, 54, 424
mediation by, 34–36, 85, 424
in metasubjective task analysis, 69–70, 70t, 260–262, 262f, 263–270, 276–283, 294–297, 413–419
modalities, modes of, 329–330
in motor rho task, 310–312, 312t, 313–314
neuropsychological, 319, 327–330
Piaget on, 7, 21, 23, 33–38, 52–55, 62, 66–67, 141, 144, 146, 150, 259–260, 268–269, 272, 328, 362, 429n1, 432n4
as psychological units, 62, 120, 317, 319
psychosocial, 62–63
reflective abstraction and, 37–38, 176–177, 177t, 178–179, 180, 325–326
releasing component of, 148, 260–262, 269, 399
relevance, criterion of, in, 155–156, 197, 335
in representation, 252
Resistances/resistances and, 24, 142f, 143–144, 401
in Schemes' Overdetermination of Performance, synthesis of, 15–16, 20–21,190–191, 252, 333
signalic function and, 119, 126–127, 127f, 128–129, 156–157, 158t, 159, 296
in subjective-1 analysis, 273–274
in subjective-2 analysis, 274–275
as subjective operators, 16, 149, 329, 424
symbolic function and, 119, 126–127, 127f, 128–129, 156–157, 158t, 159, 296, 341, 409
Schemes Inhibition and Decay (*SID*), 266–267
Schemes' Overdetermination of Performance (*SOP*), 15–16, 190–191, 266, 425. *See also* Overdetermination
assimilation and, 16, 165, 190, 333
binding problem and, 319, 331–335
F-factor and, 17–18, 22t, 188, 190–192, 192f, 193, 193f, 194–196, 204, 213, 214t, 319, 333–334, 341, 370–371, 379
Freud and, 15, 190
hidden operators and, 15–16, 21, 23f, 112

in metasubjective task analysis, 266, 276–278, 282, 296
in neuropsychology, 16, 333, 338, 371, 379
Pavlovian-instrumental transfer and, 165–166
rules of, 190–191
schemes synthesized in, 15–16, 20–21
in water-level task, 192–193, 193f
in wine and water problem, 194–196, 208–209
Secondary circular reactions (SCR), 65t, 66–68
Self, 406
autobiographical, 74, 77–78, 79t, 253, 409
core, 74, 76, 79t
figurative, 407–408
I-self, 216, 407–409
Me-self, 407–409
operative, 216, 253, 407–408, 410
protoself, 74, 79t, 253, 409
stages of, 74, 253
Will and operative, 216, 253, 410
willful inner, 411
Self-ordered pointing task (SOPT), 353, 356–358, 373
Semantic brain mapping, 359–360
Semantic complexity task (SCT), 100–101, 102t, 105f, 108
children passing, frequency of, 103, 104f,
Metasubjective task analysis of, 113–115
sensorimotor mental-attentional capacity in, 97, 101, 103, 109f, 110, 111t, 112–115
Semantic operative scheme (SEM)
context and, 73, 80, 124, 125f, 127f, 128–130, 130f
referent and, 73, 80–82, 125, 125f, 127f, 128–129, 130f
signal and, 79t, 127–128, 127f
sign and, 73, 78–81, 124–125, 125f, 127f, 128, 130f
symbol and, 73, 78, 79t, 80–82, 127f, 129–130, 130f
token sign and, 73, 128–130, 130f
Semantic processes, episodic and, 369
Semiotics
function of, 122–125, 125f, 126, 133–134, 134t
of Peirce, 35, 38, 120, 122–126, 127f, 128–129, 133, 135, 188, 320, 423, 429n3, 430n4
schemes and, 73, 119, 162
Sensorimotor mental-attentional capacity. *See* Mental-attentional capacity, sensorimotor
Sensorimotor period, 112–113, 341
accommodation in, 67–68
attachment in, metasubjective task analysis of, 298–299
conceptual neonativism on, 63–64
consciousness, development of, in 73–78, 409
mental-attentional capacity growth in, 62, 64–65, 65t, 66–72, 108, 109f, 110–111, 111t, 112, 215–216, 346–348, 352
metasubjective tasks analysis in, 84–93, 113–118, 298–304, 350–352
Piaget on, 61–62, 66–68
Piagetian stages of, 64, 65t
schemes in, 62, 64, 66–70, 71, 112–113
secondary circular reactions in, 65t, 66–68
symbolic function in, 72–73, 78–79, 79t, 80–83, 83t, 84, 125, 125f, 126, 130, 130f, 131–133, 296
tertiary circular reactions in, 65t, 67–69
Sensorimotor (sm) processes
sensorimotor mental-attentional capacity in, 61–62, 64–65, 65t, 66–72
symbolic processes and, 62, 65, 71
Sequence-representing schemes. *See* Fluents
Sequential metasubjective task analysis. *See* Metasubjective task analysis
Serotonin (5-HT), 386, 387t, 392–394
Seth, A. K., 365
Shadow schemes, 131–133, 184
in metasubjective task analysis, 270, 276, 278
in subjective-1 analysis, 270
in subjective-2 analysis, 275

Shayer, Michael, 397
  developmental, individual-difference intelligence and, 228, 229f, 230–231
  on education, 404–405
  on metasubjective task analysis, 405
Sherman, M. T., 365
Sherrington, C. S., 16–17, 334
Shrager, J., 211
Shulman, G. L., 362
Sign
  constituents of, 124–125, 125f, 126
  functional detachment of, 133, 134t
  iconic, 122–123, 429n2
  indexical, 122–123
  in life-world, 141
  mental-process task analysis of, 119
  modal and amodal, 120–121
  Peirce on, 124–126, 133
  Piaget on, 127f, 133
  Resistances/resistances and, 143
  semantic operative scheme and, 73, 78–81, 124–125, 125f, 127f, 128, 130f
  signalic versus symbolic, 126, 127f, 425
  token, 72–73, 124, 128–130, 130f
Signal, 72
  configural and arbitrary, 134–135, 134t
  learning of, 127, 134
  ontogenetic development of, 119, 125–126
  referent and, 126, 127f, 128
  schemes and, 119, 126–127, 127f, 128–129, 156–157, 158t, 159, 296
  semantic operative scheme and, 79t, 127–128, 127f
  symbol and, 119, 125–127, 127f, 128–129, 133, 134t, 138, 156–157, 158t, 159, 296, 359, 365, 425
Signalic problem solving/learning, and symbolic, 359, 363, 365
Silva, M. A., 61, 96
Simon, T. J., 93
Simon task, 246–247, 247f, 248
  F-factor in, 246–247, 247f, 248
  Joint Simon task, 246–247, 247f, 248
  mental demand and, 246–247, 247f, 248, 305
  metasubjective task analysis of, 247, 304–305
  misleading situation in, 247, 304–305
Simplexity, 17, 264, 331
Situation, facilitating. *See* Facilitating situation
Situation, misleading. *See* Misleading situation
Situational boosting factors (*sit*), 70, 79, 83, 88, 91, 303, 351, 417–418
Small-number estimation task, 90, 92–93
Small-world networks, 341
Social Being. *See B*-operator
S-operator. *See* Coexistence, relations of
Space
  in brain, 200–201, 385, 394–395
  Kant on, 200, 381
  in learning, time and, 200–201
  Leibniz on, 200–201, 381
  schemas of, 361
Spanoudis, G., 167, 230, 233–234, 234f, 407
Spatial schemas, 361
Spelke, E. S., 90, 91f
Spencer, J. P., 206
Stages, developmental, 63, 74–84, 408–409, 426
  of Case, 20, 209, 222, 222f, 223–225, 345, 367
  facilitating situations and, 15, 41, 89–90
  misleading situations and, 15, 41, 212
  neo-Piagetian theory and, 19–20, 40, 227–228, 348
  of Piaget, substages and, 31, 33, 39–41, 64, 65t, 71, 79t, 176, 178–179, 179t, 209, 220, 220f, 222–224, 227–228, 344
  schemes in, 35–36, 63, 428n2
  sensorimotor mental-attentional capacity in, 61–62, 64–65, 65t, 66–70, 71–72, 74, 395
  symbolic mental-attentional capacity in, 31, 217, 220, 220t, 221, 222f, 223, 315, 349, 395
  transition problems of, 39–41

Staircase model
Case and, 222, 222f, 223–225
mental-attentional capacity and, 221–222, 222f, 223–225
Thatcher and, 345
Standardized ability score, 225–226, 226f, 227, 249–250, 250t
Stankov, L., 248
Stern, William, 248
Störmer, V. S., 389
Stöttinger, E., 362
Strauss, S., 58
Striedter, G. F., 189
Structuralist analysis, 268–270, 273
Structure, 426
Subjective-1 analysis, 268–270, 273–274, 276
Subjective-2 analysis, 274–276
Suzuki, W. A., 382
Symbol, 72
characteristics of, 128–129
conventional and universal, 133, 134t, 135–136
iconic function and, 126, 128, 134t, 135
indexical, 134t, 135
misleading factor/strategy in, 132
modal and amodal, 120–121
ontogenetic development of, 119, 125–126
Peirce on, 126, 127f, 128–129
protosymbol, 128
referent and, 127f, 129, 130f, 131
relational-symbolic learning and, 134
semantic operative scheme and, 73, 78, 79t, 80–82, 127f, 129–130, 130f
signal and, 119, 125–127, 127f, 128–129, 133, 134t, 138, 156–157, 158t, 159, 296, 359, 365, 425
task analysis of, 130, 130f, 131–132
Symbolic function, 123
in chimpanzees, 136–137
consciousness and, 72–73
development of, 64, 65t, 72–73, 78–79, 79t, 80–83, 83t, 84, 125, 125f, 126, 130, 130f, 131–133, 341
distal, proximal objects and, 130f, 131
executive, 82, 124, 156–158, 158t, 159
functional detachment of, 72–73, 129, 133, 134t, 136
mental-attentional capacity and, 78–79, 79t, 80–83, 83t, 83–84, 205
schemes and, 119, 126–127, 127f, 128–129, 156–157, 158t, 159, 296, 341, 409
sensorimotor mental-attentional capacity and, 78–79, 79t, 80–83, 83t, 84
in sensorimotor period, 72–73, 296
Symbolic problem solving/learning, 359
Symbolic (sym) processes
sensorimotor processes and, 62, 65, 71
symbolic mental attention in, 71, 84–85, 88, 94–97, 98t, 99f, 220–221, 224, 346

Task analysis, 426. *See also* Metasubjective task analysis (MTA)
objective, 31, 268, 276, 281
subjective-1, 268–270, 273–274, 276
subjective-2, 274–276
Temporal-meaning operative integration (episodic) task, 279–280
Tertiary circular reactions (TCR), 65t, 67–69
Test-Operate-Test-Exit (TOTE), 120
Thalamus, 376–377, 378f, 379, 386, 390
Thatcher, R. W., 61, 394
on brain lateralization, 344–345
on electroencephalography coherence, phase difference, 94–96, 342–347, 347f, 348
Theory of Constructive Operators (TCO), 20–21, 22t, 23f. *See also specific topics*
Theory-of-mind (ToM) task, 78, 82–83, 83t
Time, 279–280
in brain, 200–201, 381–385, 394–396
Kant on, 200, 381
in learning, space and, 200–201
Leibniz on, 200–201, 381
temporal-structuring processes, 361
T-operator of, 22t, 200–201, 204, 213, 214t, 326, 381–384, 432n5
Todor, John, 310, 313–315, 342

Token sign. *See* Sign
Tolman, E. C., 163, 333–334
  on causal texture, of environment, 25, 326–327
  on expectancies, 7, 141, 164, 430n8
  Piaget compared with, 62
  on relational-associative learning, 171–172
Tomasello, M., 80–81
T-operator. *See* Time
Topography, brain, 320f, 321f, 322f
Truly novel performances. *See* Novel performances
Truth-value appraisals, 152–153, 198–199
Types. *See* Invariants

Ullmo, Jean, 412
U-shaped curves, developmental growth, 58

Valsiner, J., 35
Vartanian, O., 355–356
Ventral-attention networks, 359, 362–363, 395
Vicariances
  conservation and, 44–45, 270–271, 271t, 272–274, 428n8
  distal object and, 44–45, 428n8
Vital-value appraisals, 152–153, 198–199
Von Uexküll, J.
  on meaning, 36, 119–120, 141
  on Umwelt, 141
Vygotsky, L., 14, 62, 401
  on high and low cognitive functions, 47–49, 180
  on internalization, 349, 423
  on mediation, 427n1
  paradox of, 33, 48–49

Wasserman, S., 90, 91f
Watanabe, S., 429n2
Water-level task (WLT), 192–193, 193f
Weight, 151
Werner, Heinz, 123
Will, 183
  James and, 182, 216, 410
  mental/effortful attention and, 182, 410–411
  operative self and, 216, 253, 410
  Theory of Constructive Operators and, 409–412
Wine and water problem, 194–196, 208–209, 217
Witkin, Herman, 411
Working memory (WM)
  in adults, upper bound of, 240–241, 243
  mental/effortful attention and, 206, 209–212, 215, 217–218, 219f, 235–237, 240–241, 243, 246, 402
  organismic processes of, 26–31
Wynn, K., 90

Yuan, Y., 363

Zen Buddhism, 203
Zubiri, X., 181, 188, 207